**Asok Kumar Mallik
Amitabha Ghosh
Günter Dittrich**

KINEMATIC ANALYSIS and SYNTHESIS of MECHANISMS

CRC Press
Boca Raton   Ann Arbor   London   Tokyo

Library of Congress Cataloging-in-Publication Data

Mallik, A.K. (Asok Kumar), 1947-
    Kinematic analysis and synthesis of mechanisms / Asok Kumar
Mallik, Amitabha Ghosh, Günter Dittrich.
        p.   cm.
    Includes bibliographical references and index.
    ISBN 0-8493-9121-0
    1. Machinery, Kinematics of.   I. Amitabha, G. (Gosh), 1941-
  II. Dittrich, Günter.   III. Title.
TJ175.M23   1994
621.8'11—dc20   93-37108
                                                                                        CIP

This book contains information obtained from authentic and highly regarded sources. Reprinted material is quoted with permission, and sources are indicated. A wide variety of references are listed. Reasonable efforts have been made to publish reliable data and information, but the author and the publisher cannot assume responsibility for the validity of all materials or for the consequences of their use.

Neither this book nor any part may be reproduced or transmitted in any form or by any means, electronic or mechanical, including photocopying, microfilming, and recording, or by any information storage or retrieval system, without prior permission in writing from the publisher.

CRC Press, Inc.'s consent does not extend to copying for general distribution, for promotion, for creating new works, or for resale. Specific permission must be obtained in writing from CRC Press for such copying.

Direct all inquiries to CRC Press, Inc., 2000 Corporate Blvd., N.W., Boca Raton, Florida 33431.

© 1994 by CRC Press, Inc.

No claim to original U.S. Government works
International Standard Book Number 0-8493-9121-0
Library of Congress Card Number 93-37108
Printed in the United States of America   2 3 4 5 6 7 8 9 0
Printed on acid-free paper

Dedicated to
The Alexander von Humboldt
Foundation

PREFACE

Design of mechanisms constitutes an integral part of the profession of mechanical engineering and has long since remained in the curriculum. In recent years, with increasing level of automation and mechanization, the importance and scope of this subject has increased tremendously. With the present day demand on accuracy and speed of operation of almost all machines, a systematic approach towards their design has become essential. Kinematic analysis and synthesis play a fundamental role in the design of mechanisms, and a thorough understanding of the subject is important towards achieving an optimal solution.

This book aims at providing the basic theory and goes on to show how the theory can be put to practice. A large number of solved examples and exercise problems is included. This makes the book suitable for self study and, therefore, can be used by both students and professional designers in the industry. For students, the material included in the text can be covered in two courses - one in the undergraduate level as a compulsory course and the other as an elective or a graduate course. The book presupposes a preliminary knowledge of mathematics and dynamics. The instructor can choose topics from each chapter according to his liking and the level of the students.

We felt that, even in the modern computer age, graphical methods still remain important for better understanding and appreciation of the subtlety of mechanism design. Further, very often a quick, approximate solution is obtained by graphical methods. Such a solution by itself may be acceptable or may provide the starting point to be improved upon by analytical tools. Consequently, the book attempts to provide equal emphasis on graphical and analytical methods. A balanced treatment of both analysis and synthesis has been provided. Cams and gears, besides linkages, are frequently used in mechanisms. Two chapters are devoted to cover the fundamentals of kinematics of cam-follower and geared systems.

The richness of the German literature in the field of kinematics of mechanisms is well known and we believe that this book provides some of it for the first time in the English language. The authors sincerely thank the Alexander von Humboldt Foundation for bringing them together, which made the writing of this book possible.

For a book on a traditional subject like this, one has to depend on the ideas and materials presented in a large number of books, too many to acknowledge individually. All such books are listed in the Bibliography. However, special mention must be made of the following books: *Getriebetechnik in Beispielen* by G. Dittrich and R. Braune; *Theory of Mechanisms and Machines* by A. Ghosh and A. K. Mallik; *Kinematic Synthesis of Linkages* by R. S. Hartenberg and J. Denavit, and *Getriebetechnik (Lehrbuch)* by J. Volmer.

We sincerely thank Mr. Peter Markert for his meticulous care and patience while preparing the final diagrams and Ms. Sandhya Mishra and Ms. Debjani Deb for putting the manuscript in the final form. Finally, we welcome comments and suggestions from the readers for further improvement and detection of any errors which have escaped our scrutiny.

A. K. Mallik
A. Ghosh
G. Dittrich

Contents

1 INTRODUCTION   1
    1.1 Mechanisms and Machines 1
    1.2 Kinematic Pairs 5
    1.3 Kinematic Chains and Linkages 11
    1.4 Structure and Kinematic Diagrams 12
    1.5 Planar and Spatial Mechanisms 16
    1.6 Limit and Disguise of Revolute Pairs 19
    1.7 Kinematic Inversion 21
    1.8 Four-Link Planar Mechanisms 22
    1.9 Equivalent Linkages 26
    1.10 A Note on Subsequent Chapters 28
    1.11 Exercise Problems 31

2 PLANAR KINEMATICS OF RIGID BODIES   37
    2.1 Introduction 37
    2.2 Plane Motion of a Particle 38
    2.3 Plane Motion of a Rigid Body 41
    2.4 Velocity and Acceleration Images 45
    2.5 Instantaneously Coincident Points 46
    2.6 Instantaneous Center of Velocity 49
    2.7 Aronhold-Kennedy Theorem 53
    2.8 Fixed and Moving Centrodes 58
    2.9 IC Velocity 67
    2.10 Theory of Path Curvature 70
       2.10.1 The Inflection Circle 70
       2.10.2 The Euler-Savary Equation 73
       2.10.3 The Cubic of Stationary Curvature 82

		2.10.4 Ball's Point	87
	2.11	Finite Movements	89
		2.11.1 Poles	90
		2.11.2 Pole Triangles	90
		2.11.3 Image Poles	91
		2.11.4 Opposite-pole Quadrilateral	93
		2.11.5 Circle-Point Curve and Center-Point Curve	95
		2.11.6 Burmester Points	103
	2.12	Exercise Problems	103

3 MOBILITY AND RANGE OF MOVEMENT   111

3.1	Kutzbach Equation and Grübler Criterion	111
3.2	Applications of Grübler Criterion	115
	3.2.1 Minimum Number of Binary Links	115
	3.2.2 Highest Order Link in an n-Link Mechanism	116
	3.2.3 Number Synthesis	117
3.3	Failure of Mobility Criteria	119
3.4	Grashof Criterion	122
	3.4.1 Inversions of a Grashof Chain	125
	3.4.2 Inversions of a Non-Grashof Chain	126
	3.4.3 Inversions of Chains with Uncertainty (Folding) Positions	126
	3.4.4 Extension of Grashof Criterion	128
3.5	Exercise Problems	131

4 DISPLACEMENT ANALYSIS   135

4.1	Graphical Method	135
4.2	Analytical Method	143
4.3	Input-Output Curves of 4R Linkages	151
4.4	Transmission Angle	159
4.5	Exercise Problems	162

5 VELOCITY AND ACCELERATION ANALYSIS   167

5.1	Introduction	167
5.2	Velocity Analysis (Graphical Methods)	168
	5.2.1 Some Fundamental Aspects	168
	5.2.2 Method of Instantaneous Centers	169

		5.2.3 Method of Velocity Difference	173
		5.2.4 Auxiliary Point Method	179
	5.3	Acceleration Analysis (Graphical Methods)	182
		5.3.1 Some Fundamental Aspects	182
		5.3.2 Method of Acceleration Difference	185
		5.3.3 Method of Normal Component	188
		5.3.4 Auxiliary Point Method	191
		5.3.5 Goodman's Indirect Method	198
		5.3.6 Method Using Euler-Savary Equation	210
	5.4	Kinematic Analysis (Analytical Method)	215
	5.5	Exercise Problems	219
6	**DIMENSIONAL SYNTHESIS**		**231**
	6.1	Classification of Synthesis Problems	231
	6.2	Exact and Approximate Synthesis	232
		6.2.1 Choice of Precision Points	233
	6.3	Graphical Methods (Three Positions)	237
		6.3.1 Motion Generation	237
		6.3.2 Path Generation	241
		6.3.3 Function Generation	243
		6.3.4 Relative Poles	247
	6.4	Dead-Center Problems	258
	6.5	Graphical Methods (Four Positions)	275
		6.5.1 Point-Position Reduction	275
		6.5.2 Application of Burmester Theory	280
	6.6	Analytical Methods (Three Positions)	283
		6.6.1 Bloch's Method	284
		6.6.2 Freudenstein's Method	287
	6.7	Analytical Methods (Four Positions)	293
		6.7.1 Bloch's Method	294
		6.7.2 Freudenstein's Method	295
	6.8	Optimization Method	298
	6.9	Branch and Order Defects	306
	6.10	Mechanical Error	308
		6.10.1 Deterministic Approach	309
		6.10.2 Stochastic Approach	311
	6.11	Exercise Problems	314

7 COUPLER CURVES — 327
- 7.1 Crunodes, Cusps, Symmetry 329
- 7.2 Cognate Linkages 338
- 7.3 Extension of Cognate Linkages 346
- 7.4 Parallel Motion Generator 350
- 7.5 Approximate Straight-Line Linkages 358
- 7.6 Exact Straight-Line Linkages 367
- 7.7 Exercise Problems 371

8 SPHERICAL AND SPATIAL LINKAGES — 375
- 8.1 Degrees of Freedom 376
- 8.2 Displacement Equation (Analytic Geometry) 378
- 8.3 Matrix Method 386
  - 8.3.1 Coordinate Transformation 387
  - 8.3.2 Link Coordinate System 389
  - 8.3.3 Derivation of Homogeneous Transformation Matrix 393
  - 8.3.4 Displacement Analysis 402
- 8.4 Velocity and Acceleration Analysis 410
- 8.5 Kinematic Synthesis 411
- 8.6 Mobility Analysis 419
  - 8.6.1 Type Identification of an R-S-S-R Linkage . . 419
  - 8.6.2 Limit Position Analysis 424
- 8.7 Principle of Transference 432
  - 8.7.1 Dual Number 432
  - 8.7.2 Dual Angle 434
  - 8.7.3 Displacement Equation of an R-C-C-C Linkage 435
- 8.8 Exercise Problems 437

9 CAM MECHANISMS — 443
- 9.1 Types and Fundamentals of Cam Mechanisms 443
  - 9.1.1 Types of Plane Cam Mechanisms 446
  - 9.1.2 Basic Features and Nomenclature 450
  - 9.1.3 Pressure Angle 453
- 9.2 Follower Movement, Displacement Diagram 454
  - 9.2.1 Displacement Functions 456
  - 9.2.2 Uniform Motion 459

	9.2.3	Simple Harmonic Motion	459
	9.2.4	Parabolic or Piecewise Uniform Acceleration Motion	460
	9.2.5	Cycloidal Motion	462
	9.2.6	Polynomial Functions	463
9.3	Determination of Basic Dimensions		466
	9.3.1	Translating Flat-Face Follower	468
	9.3.2	Translating Roller Follower	470
	9.3.3	Flocke's Method	480
9.4	Cam Synthesis (Graphical Approach)		490
	9.4.1	Radial Translating Follower	491
	9.4.2	Offset Translating Follower	493
	9.4.3	Oscillating Follower	495
9.5	Cam Synthesis (Analytical Approach)		497
	9.5.1	Flat-Face Translating Follower	497
	9.5.2	Translating Roller Follower	500
	9.5.3	Oscillating Flat-Face Follower	502
	9.5.4	Oscillating Roller Follower	504
9.6	Spatial Cam Mechanisms		506
	9.6.1	Basic Features of Cylindrical Cam Mechanisms	507
	9.6.2	Graphical Synthesis of Cylindrical Cams	509
	9.6.3	Analytical Synthesis of Cylindrical Cams	511
9.7	Exercise Problems		513

10 GEARS 519

10.1	Gear Tooth Action and The Law of Gearing		521
	10.1.1	Fundamental Law of Gearing	523
	10.1.2	Path of Contact, Pitch Circle and Line of Action	526
	10.1.3	Circular Pitch, Diametral Pitch and Module	526
10.2	Involute Spur Gears		528
	10.2.1	Involute Teeth Action	528
	10.2.2	Involute Nomenclature	532
	10.2.3	Involutometry	536
	10.2.4	Advantages of Involute Teeth	541
10.3	Characteristics of Involute Action		543
	10.3.1	Contact Ratio	543
	10.3.2	Interference and Undercutting	553

- 10.3.3 Standardization of Involute Gears 557
- 10.3.4 Minimum Number of Teeth 559
- 10.3.5 Backlash.................... 562
- 10.3.6 Sliding Phenomenon of Gear Teeth 568
- 10.4 Cycloidal (Trochoidal) Gears 572
  - 10.4.1 Trochoidal Curves 572
  - 10.4.2 Cycloidal (Trochoidal) Teeth 574
  - 10.4.3 Cycloidal Tooth Action 576
  - 10.4.4 Advantages and Disadvantages of Cycloidal Teeth 579
- 10.5 Determination of Conjugate Profile 582
  - 10.5.1 Graphical Approach 582
  - 10.5.2 Analytical Approach 584
- 10.6 Internal Gears...................... 591
  - 10.6.1 Advantages and Disadvantages 591
  - 10.6.2 Internal Gear Tooth Shape 592
  - 10.6.3 Contact Ratio, Interference, Secondary Interference and Undercutting 597
- 10.7 Noncircular Gears 605
  - 10.7.1 Determination of Pitch Curves 606
  - 10.7.2 Tooth Profile of Noncircular Gears 609
- 10.8 Helical Gears 612
  - 10.8.1 Geometry of Helical Gears........... 614
  - 10.8.2 Equivalent Spur Gear and Virtual Number of Teeth 617
  - 10.8.3 Parallel Helical Gears 619
  - 10.8.4 Crossed Helical Gears 622
- 10.9 Bevel Gears 627
- 10.10 Worm Gears...................... 633
- 10.11 Gear Trains 637
  - 10.11.1 Simple Gear Trains............... 638
  - 10.11.2 Compound Gear Trains 640
  - 10.11.3 Epicyclic Gear Trains 641
- 10.12 Exercise Problems 649

BIBLIOGRAPHY 657

INDEX 661

Chapter 1

INTRODUCTION

Devices have been used from time immemorial to augment human muscle power. Today we find devices in common use which supplement even human brain power and control systems. The most generalized name of all such devices is *machine*. Thus, a machine may include a wide range of devices starting from a simple lever or a chain-pulley to a complex robot or a computer. In this text, however, we shall consider only mechanical devices. Let us start our discussion with the definition of mechanisms and machines (a systematic study of mechanisms and machines is the objective of this text).

1.1 Mechanisms and Machines

In engineering, *mechanisms* and *machines* are two very common and frequently used terms. However, it is not easy to accurately define these in a few words[1]. Broadly, they refer to devices which transfer mechanical motions and forces from a source to an output member. Such a definition, recommended by IFToMM (International Federation for the Theory of Machines and Mechanisms), thus does not distinguish a mechanism from a machine. On the other hand, several texts on the subject suggest that if the idea of transferring motion rather than force predominates, the device should be called

---

[1]Reuleaux, F.: Theoretische Kinematik, Friedr.Vieweg & Sohn, Braunschweig, 1875 - gives 17 definitions.

a *mechanism* whereas, when substantial forces are also involved the device should be called a *machine*. For example, we can talk of a *wiper mechanism* for an automobile windshield (Fig. 1.1-1) and a hydraulically operated *dumping machine* (Fig. 1.1-2). In the former, the rotary motion of the motor (source) is converted to an oscillatory motion of the wiper-blades (output); the motor torque is transferred at the wiper to overcome the friction forces at the blade-screen interface. In the latter, the motions and forces available at the hydraulic cylinders, Z_1, Z_2 and Z_3 (sources), are transferred to the bin (output). It may be appreciated that all parts comprising such a device must resist deformation and to a good approximation can be treated as rigid bodies (in most of the applications). At this stage, a mechanism (and a machine) can be defined as a combination of rigid bodies so shaped and connected that they move upon each other with definite relative motion for the purpose of transferring motions and forces from a source (input) to an output. As discussed below, the word mechanism or machine is used depending on the objective and scope of the study of these devices.

It is well recognized that the investigations of the motions of a rigid body (or interconnected rigid bodies) can be conveniently separated into two parts. The first one, dealing only with the geometric aspects, is called *kinematics*. In kinematics, no reference is made to the forces causing the motions. Given the motions of one (or more) member(s), those of the others in the device are investigated in order to satisfy the geometric constraints. In the second part, called *kinetics*, the motion characteristics are investigated with reference to the forces involved. The inertias of different moving parts obviously play an important role in kinetics.

For the same system, the word mechanism is used if one is studying the kinematics, and the word machine is used while studying the kinetics. As an example, let us consider a commonly used device consisting of a crank, a connecting rod and a piston within a cylinder, shown in Fig. 1.1-3. Such a device is used for converting a rotary motion (of the crank) into a reciprocating motion (of the piston, as in an air compressor) or vice versa (as in an internal combustion engine). This device is called a slider (piston)-crank mechanism if we are studying the kinematics, e.g., the relationship between the linear

1.1. MECHANISMS AND MACHINES

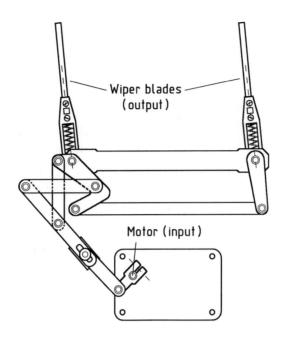

Figure 1.1-1

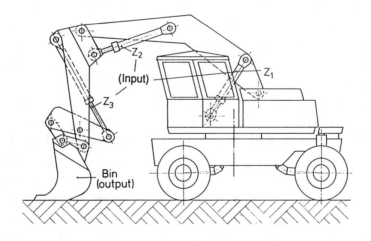

Figure 1.1-2

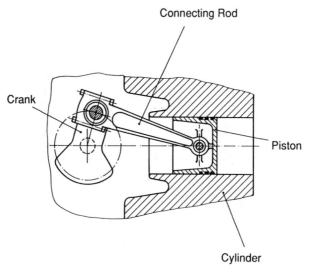

Figure 1.1-3

velocity of the slider and the angular velocity of the crank, which is governed entirely by geometric dimensions. However, the same device may be referred to as a machine while studying its kinetics, e.g., the relationship between the driving moment at the crank and the forces acting on the piston.

It is worthwhile to note that the separation of kinematics and kinetics is possible only for a rigid body and a system of interconnected rigid bodies. For a high-precision, high-speed device, the approximation of its members by rigid bodies no longer remains valid. In a nonrigid member, forces cause deformations and change the geometry. Consequently the study of force and motion must take place simultaneously, which significantly increases the complexity of the problem. This book deals only with kinematics of interconnected rigid bodies.

It may be noted at this stage that, as recognized by Ampere' in 1834, kinematics of interconnected rigid bodies forms a separate field of study vis-a-vis Newtonian dynamics. Since then, many specialized techniques, results and theorems have been derived towards a systematic study of kinematics of mechanisms. The study of kinematics not only forms the first step of designing a large number of mechanical devices, but very often this first step is also the most critical one if complex motion requirements are to be satisfied.

There are two aspects of the study of kinematics, namely *analysis* and *synthesis*. In *kinematic analysis*, the motion characteristics, such as displacement, velocity and acceleration, are investigated for given geometrical parameters. On the other hand, *kinematic synthesis* deals with the inverse problem, where the geometric parameters are to be determined so as to generate the desired motion characteristics. It is needless to say that both aspects are important to a designer, i.e., synthesis for coming up with a design and analysis for checking the performance of his design. A number of fundamental concepts in kinematics of rigid bodies are important for both analysis and synthesis. Before going into the details of the development and applications of these concepts, we will start with some basic definitions and nomenclature.

1.2 Kinematic Pairs

A mechanism is defined as a combination of rigid bodies connected so that each moves with respect to the other. A clue to the behaviour of the mechanism lies in the nature of the connections, called *kinematic pairs*, and in the type of relative motions they permit. If we consider the relative motion of two unconnected rigid bodies in a three-dimensional space, then six independent coordinates are required to describe this relative motion. In other words, this relative movement has six degrees-of-freedom, three of which are translations along three mutually perpendicular axes. The remaining three are rotations about each of these axes. As soon as the two bodies are connected, forming a kinematic pair, one or more (up to five) of these six degrees-of-freedom are curtailed. Broadly, kinematic pairs are classified on the basis of the particular degree(s)-of-freedom which is (are) retained, i.e., the nature of the relative movement which is still permitted. The degree-of-freedom of a kinematic pair is given by the number of independent coordinates required to completely specify the relative movement. These coordinates are commonly known as pair variables. Some typical kinematic pairs are discussed below. It may be mentioned that there are many different ways of physically constructing a kinematic pair, however, only the allowed relative

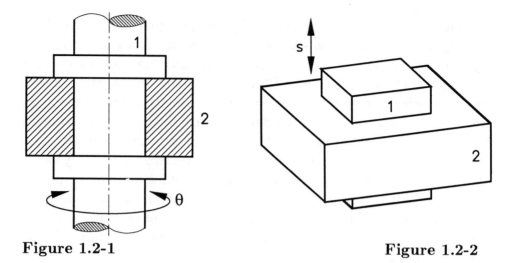

Figure 1.2-1   Figure 1.2-2

movement decides the type of the kinematic pair. In this section, only illustrative sketches of various kinematic pairs indicating the possible relative movements will be provided.

Revolute Pair

A *revolute pair*, illustrated in Fig. 1.2-1, allows only relative rotation between bodies 1 and 2 about one prescribed axis. Thus, a revolute pair has a single degree-of-freedom expressed by the pair variable θ.

Prismatic Pair

As shown in Fig. 1.2-2, a *prismatic pair* allows only a relative rectilinear translation between bodies 1 and 2. Thus, a prismatic pair has a single degree-of-freedom expressed by the pair variable s (think of sliding).

Screw Pair

As shown in Fig. 1.2-3, a *screw pair* also has a single degree-of-freedom, since the relative motion between bodies 1 and 2 can be expressed by a single pair variable θ or s. These two variables are related through the equation

1.2. KINEMATIC PAIRS

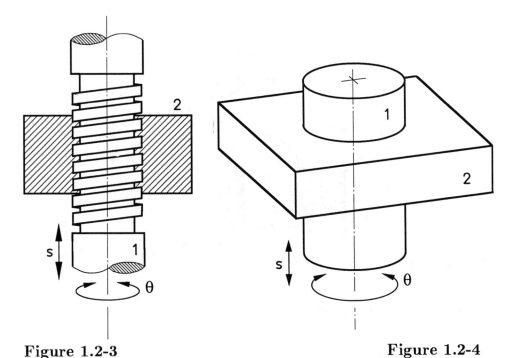

Figure 1.2-3 **Figure 1.2-4**

$$\frac{\Delta\theta}{2\pi} = \frac{\Delta s}{L} \qquad (1.2.1)$$

with L as the lead of the thread. It may be noted that the revolute and prismatic pairs are the limiting situations of the screw pair with $L = O$ and ∞, respectively. Furthermore, while revolute and screw pairs have prescribed axes (about which relative rotation takes place), a prismatic pair has only a prescribed direction (along which relative translation occurs). We should remember that a prescribed axis is given by a unique line in space whereas a prescribed direction is indicated by all lines parallel to that direction.

Cylindric Pair

As shown in Fig. 1.2-4, a *cylindric pair* has two degrees-of-freedom because it allows both rotation and translation, parallel to the axis of rotation, between the connected elements. The pair variables are θ and s. A cylindric pair can be thought of as a special combination of

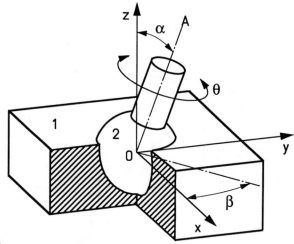

Figure 1.2-5

a revolute and a prismatic pair.

Spheric Pair

As shown in Fig. 1.2-5, a ball and socket joint forms a *spheric pair*.

The name originates from the fact that the pair allows a relative "spheric" motion between bodies 1 and 2. All points on moving body 2, assuming body 1 to be fixed (i.e., considering the relative motion), move on the surfaces of concentric spheres. The pair has three rotational degrees-of-freedom, since three independent angular coordinates are necessary to completely describe the relative motion. For example, as explained in Fig. 1.2-5, two angles (α and β) are required to specify the position of the axis OA and the third coordinate (θ) describes the rotation about the axis OA. It may be remembered that rotations about different axes do not commute and the most common convention of describing a spheric motion is through three Euler angles (pair variables), rather than through three rotations about mutually perpendicular axes[2]. For detailed discussion of this aspect, refer to Chapter 8. A spheric pair can be thought of as a combination of three revolute pairs with their axes intersecting at a point.

---

[2]Goldstein, H. : Classical Mechanics, Addison-Wesley Press,Inc. Reading, Mass., 1950.

1.2. KINEMATIC PAIRS

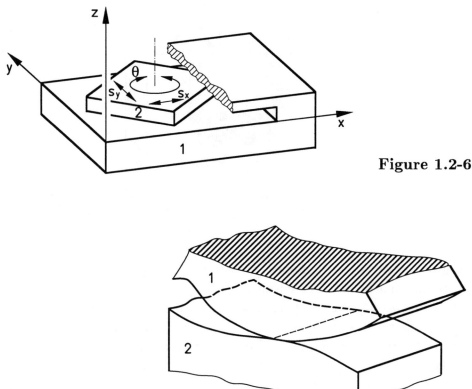

Figure 1.2-6

Figure 1.2-7

Planar Pair

As shown in Fig. 1.2-6, a *planar pair* also has three degrees-of-freedom; two of these are relative translations in the XY plane and the third is the rotation about the Z axis. The pair variables are s_x, s_y and θ, where s_x, s_y refer to relative translations of a point on body 2 along the X and Y directions, respectively, and θ to the relative rotation (between bodies 1 and 2) about an axis passing through this point and parallel to the Z axis. A planar pair can be thought of as a special combination of two prismatic and one revolute pairs.

The portions of rigid bodies that actually form a kinematic pair are referred to as *elements*. It may be noted that all six kinematic pairs described above are illustrated through sketches signifying area

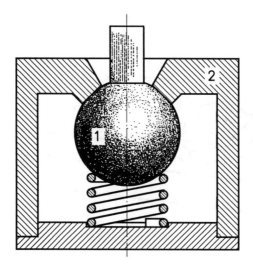

Figure 1.2-8

contact between the elements. However, in the physical construction there may not be any area contact, e.g., a prismatic pair can be formed between two flat surfaces by inserting a number of rollers between them, or a revolute pair between a straight shaft and its housing may be in the form of a number of ball bearings between them. Historically, these six types of kinematic pairs are called *lower pairs* to distinguish them from the type of connections between a pair of gear teeth or a cam-follower where the possible relative movement can be simulated only by a line or point contact. The latter type of connections illustrated in Fig. 1.2-7, is called a *higher pair*. In Figs. 1.2-1 to 1.2-6, the contact between the elements of a kinematic pair is shown to be maintained only by the geometric forms of the contact surfaces. These are known as *form-closed* pairs. If the contact is maintained by an external force (e.g., that of a spring) as shown in Fig. 1.2-8, then the pair is known as a *force-closed* pair. The lower pairs discussed thus far have degrees-of-freedom varying from one to three. For the sake of completeness, we can talk of connections having four and five degrees-of-freedom. These are illustrated in Figs. 1.2-9a and 1.2-9b, respectively. In Fig. 1.2-9a, a sphere is placed in either a vee (combination of a spheric and a prismatic pair) or a semi-cylinder of same radius; in Fig. 1.2-9b, a sphere is placed on a flat surface (combination of a spheric and two prismatic pairs).

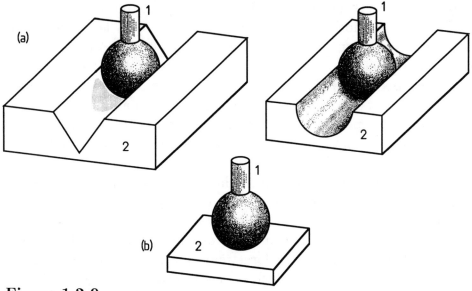

Figure 1.2-9

1.3 Kinematic Chains and Linkages

Each rigid body contributing the elements to form kinematic pairs is referred to as a link. A *kinematic chain* is a series of links connected by kinematic pairs. The chain is said to be *closed* if every link is connected to at least two other links; otherwise it is termed an *open chain*. A link which is connected to only one other link is known as a *singular link*. If it is connected to two other links, it is called a *binary link*. Similarly, if a link is connected to three other links, it is referred to as a *ternary link*. *Quaternary* and still higher order links are defined in the same manner. By definition, a closed chain cannot contain a singular link. A chain consisting only of binary links is termed a *simple chain*. Otherwise, it is called a *compound chain*.

A mechanism can be defined as a closed kinematic chain with one fixed link. The word fixed link implies the *frame of reference* and the relative motion, with respect to this frame, forms the subject matter of study of kinematics of the said mechanism. For example, if we are investigating the steering mechanism or the windshield wiper mechanism of an automobile, then the body of the automobile (which may not be fixed) serves as the fixed link. For the mechanisms under discussion, the motions of various parts relative to the automobile body are important and are treated as their absolute motions. This fixed link or the frame of reference is very often called a *frame*.

In this text, we shall restrict our discussions to those mechanisms that arise out of a closed kinematic chain. Robot manipulators, which are essentially open kinematic chains, are not considered, since in recent times kinematics of manipulators has grown as a seperate field of study.

The degrees-of-freedom of a mechanism are given by the number of independent pair variables needed to completely define the relative movements between all its links. A mechanism is said to be *constrained*, if the number of input motions equals the degrees-of-freedom of the mechanism. A large majority of the mechanisms have a single degree-of-freedom. Such a constrained mechanism obviously requires a single input motion to completely specify the motions of all other links. For a single degree-of-freedom, constrained mechanism, the paths along which the points on its various links can move are uniquely defined by the geometrical constraints. Here onwards, unless otherwise specified, when we refer to a mechanism, we shall mean only constrained mechanisms with a single input.

According to many kinematicians, a mechanism consisting of only the lower pairs may be called a *linkage*. However, in this book, the terms linkage and mechanism are used as synonyms. Furthermore, the theory involved in the study of linkages is different in many ways from that required for the study of higher pair mechanisms like gears and cams. The analysis and synthesis of gears and cams will be discussed separately.

1.4 Structure and Kinematic Diagrams

While studying different kinematic aspects of chains and linkages, we use simplified diagrams omitting all details that are irrelevant for the particular study. For example, a *structure diagram* of a kinematic chain indicates only the existence of interconnections among various links and is used to investigate how many basically different mecha-

1.4. STRUCTURE AND KINEMATIC DIAGRAMS

nisms can be generated from the same chain. In a structure diagram, a link is represented by a point, while every lower pair connecting two such points is represented by a solid line, and likewise every higher pair is represented by a dashed line. Graph-theoretic methods are widely used for investigating isomorphism problems of chains, i.e., to identify whether two apparently dissimilar chains are topologically so or not. The discussion on structure diagrams is beyond the scope of this text[3].

Kinematics of a mechanism is governed by (i) the type, (ii) the sequential order of connections and (iii) the relative locations and orientations of the kinematic pairs that are used to form the kinematic chain. All other parameters and physical dimensions of the actual device are immaterial so far as kinematics is concerned. Therefore, to focus the attention on the three important characteristics mentioned above, symbolic notations and skeleton diagrams (called *kinematic diagrams*) are used to describe and represent real-life mechanisms. The *symbolic notations* attempt to describe the type and sequencing of the kinematic pairs in a concise manner, while the kinematic diagram shows the essential geometry of a mechanism. The conventions, commonly followed in the literature for this purpose, are discussed here.

In a kinematic diagram, the links are numbered with the fixed link (indicated by hatched lines) given the number 1. Some typical conventions followed while drawing a kinematic diagram are illustrated in Fig. 1.4-1. In this figure, the axes of the revolute pairs are assumed to be perpendicular to the plane of the paper and pass through the centers of the small circles which are used to represent them. Also, the direction of the relative translation at the prismatic pair is assumed to be parallel to the plane of the paper. It should be noted, from Fig. 1.4-1d, that in a kinematic diagram, the positions of the axes of revolute pairs and the directions of sliding in the prismatic pairs are not altered from those existing in the actual device. Kinematic diagrams of the mechanisms shown in Figs. 1.1-3 and 1.1-1 are in Figs. 1.4-2

---

[3] See Manolescu,N.I.: "For a united point of view in the study of the structural analysis of kinematic chains and mechanisms." Journal of Mechanisms, Vol.3, pp. 149-169 (1968).

CHAPTER 1. INTRODUCTION

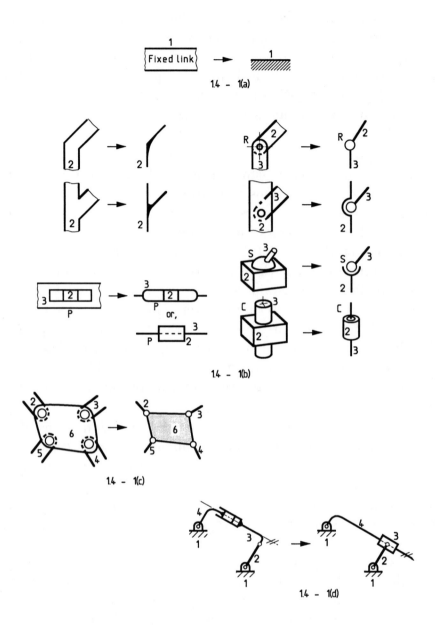

Figure 1.4-1

1.4. STRUCTURE AND KINEMATIC DIAGRAMS

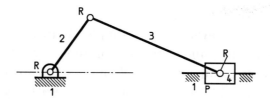

1-Frame 2-Crank 3-Connecting Rod 4-Piston

Figure 1.4-2

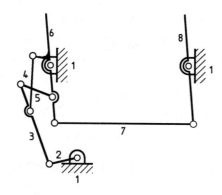

1-Frame 2-Input Crank 6 & 8 Wiper Blades

Figure 1.4-3

and 1.4-3, respectively. The symbols for the six lower pairs described in Section 1.2 are as follows:

Type of Pair	Symbol	Combinations
Revolute	R	
Prismatic	P	
Screw (Helix)	H	(R/P)
Cylindric	C	(RP)
Spheric	S (or G for Globular)	(R_3)
Planar	E	(RP_2)

In Section 1.2, we have already noted that all the lower pairs can be thought of as combinations of R and P pairs. This fact is indicated above for each symbol within parentheses. The numerical subscript refers to the number of the concerned pair. Following this convention, the kinematic pairs shown in Figs. 1.2-9a and 1.2-9b can be written as R_3P and R_3P_2, respectively.[4]

A mechanism originated from a simple chain can be easily identified by a closed loop, and can be represented by the sequence of the kinematic pairs occuring in the loop. For example, the mechanism shown in Fig. 1.4-2 can be symbolically written as an R-R-R-P (or 3R-1P) mechanism. Similarly, some other mechanisms can be referred to as R-S-S-R or R-C-C-C, etc. However, for a mechanism arising out of a compound chain (e.g., that shown in Fig. 1.4-3), there exists more than one closed loop. Consequently, more than one sequences of symbols are necessary for a complete description of such a mechanism. The choice of these sequences is also not unique. The reader is invited to verify the last two statements with reference to the multi-loop mechanism shown in Fig. 1.4-3. Thus, the symbolic notations are neither simple nor unambigious towards identifying the characteristics of these mechanisms.

1.5 Planar and Spatial Mechanisms

Mechanisms or linkages can be most broadly classified as planar and spatial. If all the points of a mechanism move in parallel planes, then it is defined as a *planar mechanism*. A *spatial mechanism* is one in which all points of the mechanism do not move in parallel planes. A special category of spatial mechanisms is called *spherical* where all points of the mechanism move on the surfaces of concentric spheres.

It may be noted that a planar linkage can have only revolute and prismatic pairs with all the revolute axes normal to the plane of motion and all the prismatic pair axes parallel to this plane. Of course, planar pairs can be included, but as noted in Section 1.2, a planar pair (RP_2) is nothing but a combination of 2 P and 1 R pairs satisfying the restriction just mentioned above. Among planar linkages, those consisting of only R pairs occupy a special place. This is mainly because of great ease of construction and maintenance of an R pair as compared to a P pair. The ease of maintenance is due to low frictional

---

[4]The letter P already represents a prismatic pair. Therefore, to avoid confusion, a planar pair is being symbolized by the letter E (In the German language Ebene means Plane).

1.5. PLANAR AND SPATIAL MECHANISMS

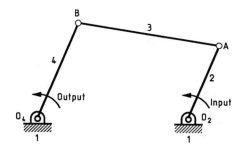

Figure 1.5-1

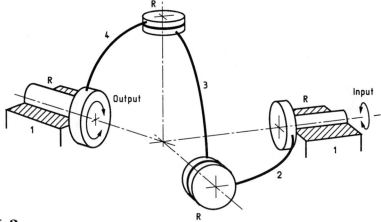

Figure 1.5-2

torque on the small pin of the R pair.

A spherical linkage consists of entirely revolute pairs with all the axes of R pairs intersecting at a point. A spatial linkage can have any type of lower pair. Figures 1.5-1, 1.5-2 and 1.5-3 show, respectively, a planar, a spherical and a spatial linkage. All these are 4R linkages. Therefore, it is realized that if a mechanism consists of only R or P or R and P pairs, it is better to add the suitable word, planar, spherical or spatial to the symbolic notation for more complete description of the linkage. Furthermore, the 4R spatial linkage with skewed revolute axes, shown in Fig. 1.5-3, is movable only for very special dimensions of the links (as will be seen in Chapter 8).

The vast majority of linkages in use today are "essentially" planar.

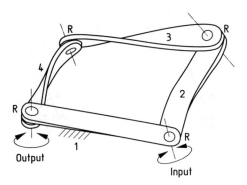

Figure 1.5-3

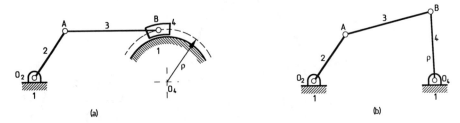

Figure 1.6-1

By using the word "essentially", attention is drawn to the fact that just like a rigid body, a planar mechanism is an idealization because in practice, for example, all R-pair axes can never be made exactly parallel. Significant developments have taken place in the techniques of analysis and synthesis of a planar linkage. For such a linkage, a single view perpendicular to the plane of motion reveals the true motions. Consequently, a large number of graphical methods are available that are based on the principles of Euclidean geometry. With the advent of digital computers, lately a shift towards analytical tools is taking place. Graphical methods, however, provide better insight and visualization and still occupy a predominant place in planar kinematics. A large portion of this book will be devoted to the analysis and synthesis of planar linkages through both graphical and analytical methods. Only one chapter exclusively discusses the spatial linkages.

1.6. LIMIT AND DISGUISE OF REVOLUTE PAIRS

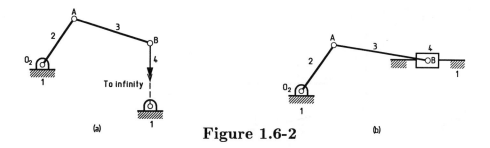

Figure 1.6-2

1.6 Limit and Disguise of Revolute Pairs

We have already seen that revolute and prismatic pairs are the basic building blocks of all lower pairs. All other lower pairs can be thought of as combinations of these two pairs. Moreover, a planar linkage consists of only these two types of kinematic pairs. In this section, it will be pointed out that a prismatic pair can always be thought of as the limit of a revolute pair. To demonstrate this, let us first consider a *curved slider* connection between two links 1 and 4 of a mechanism, as shown in Fig. 1.6-1a. A little thought would convince us that this connection is a revolute pair, since the pair variable describing the relative movement between 1 and 4 is still an angle. Hence, this mechanism can be represented by the kinematic diagram of a 4R planar linkage shown in Fig. 1.6-1b. If the radius of curvature (ρ) of the path of point B becomes infinitely large, the pair variable transforms from angular movement to linear displacement. Consequently, the connection between links 1 and 4 becomes a prismatic pair. Following this line of argument, as explained in Figs. 1.6-2a and 1.6-2b, a slider-crank mechanism is obtained as the limit of a 4R planar linkage when one revolute pair moves to infinity. Notice that the revolute pair moves to infinity along a direction perpendicular to that of the slider movement.

As discussed above, with reference to the curved slider, the form of a revolute pair may not always be apparent. Other forms of *disguise* of a revolute pair may be generated due to some practical design considerations. A typical example is in the eccentric-driven slider-crank mechanisms used in punching presses (Fig. 1.6-3a). Here, the relative motion between the eccentrically hinged disc 2 and the connecting

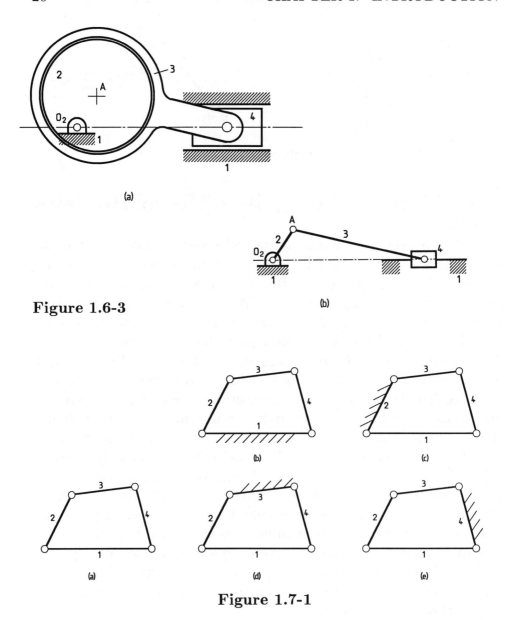

Figure 1.6-3

Figure 1.7-1

rod 3 is one of pure rotation about the center (A) of the disc. Consequently, the kinematic diagram of this mechanism will be as shown in Fig. 1.6-3b.

1.7. KINEMATIC INVERSION

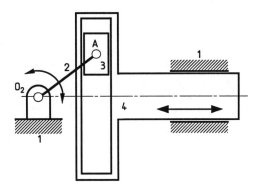

Figure 1.7-2

1.7 Kinematic Inversion

From the definition of mechanism given in Section 1.3, it is seen that by fixing the links of a closed chain one at a time, we get as many different mechanisms as the number of links in the chain. This process of fixing different links of the same kinematic chain to produce distinct mechanisms is called *kinematic inversion*. By distinct, it is meant that the input-output relation (as given by the absolute motions of the links connected to the frame) is different for these mechanisms. The most important point, however, is that the relative motions between various links remain unchanged under kinematic inversion. This invariance of relative motions under kinematic inversion is a very useful concept for both kinematic analysis and synthesis.

Let us consider a 4R kinematic chain shown in Fig. 1.7-1a. The four different mechanisms obtained by kinematic inversion from this chain are shown in Figs. 1.7-1b through 1.7-1e. Let us consider two of these mechanism, shown in Figs. 1.7-1b and 1.7-1d. In view of the statements made in the preceding paragraph, we may note that these two mechanisms are different so far as the relationship between the absolute motions of links 2 and 4 in them is concerned. In fact, the relationship between the absolute motions of links 2 and 4 in Fig. 1.7-1b is the same as that between the relative motions of links 2 and 4 with respect to link 1 in Fig. 1.7-1d. As we shall see later in Chapter 3, the movability characteristics of these mechanisms obtained by

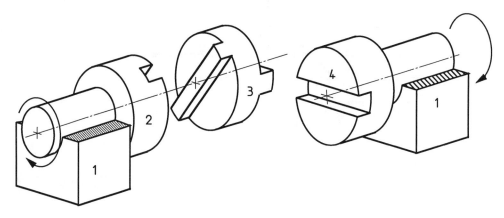

Figure 1.7-3

kinematic inversion are also diverse.

Physical constructions of mechanisms obtained from kinematic inversion of the same chain may widely differ. Their kinematic similarity becomes apparent only after drawing the kinematic diagrams. For example, let us consider two mechanisms, namely, the scotch-yoke mechanism and Oldham's coupling, shown in Figs. 1.7-2 and 1.7-3, respectively. The first mechanism is used to generate simple harmonic rectilinear motion (of link 4) from uniform rotary motion (of link 2). The latter is used for connecting two parallel shafts. This coupling transmits an angular velocity ratio of unity. It may be verified that these two mechanisms are obtained by kinematic inversions of the R-R-P-P chain, shown in Fig. 1.7-4a. The scotch-yoke mechanism is obtained by fixing link 1 of this chain while the Oldham's coupling is obtained by fixing link 2. The kinematic diagrams of the scotch-yoke mechanism and the Oldham's coupling are shown in Figs. 1.7-4b and 1.7-4c, respectively. In both these figures, links 2 and 4 (connected to the frame 1) are the input-output members.

1.8 Four-Link Planar Mechanisms

Four-link planar mechanisms have versatile applications. Over the last century, the kinematics of this class of mechanisms has been studied

1.8. FOUR-LINK PLANAR MECHANISMS

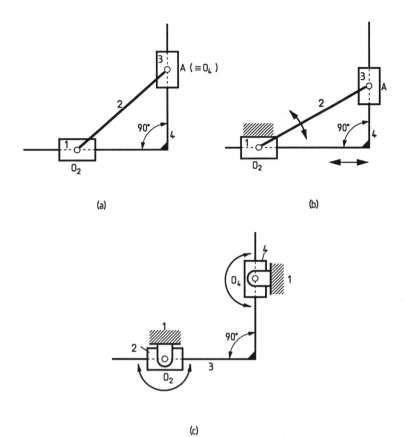

Figure 1.7-4

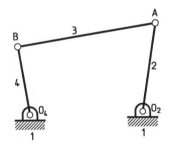

Figure 1.8-1

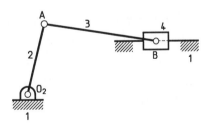

Figure 1.8-2

quite thoroughly. A large portion of this book is devoted to the details of such studies. In this section, however, several of the four-link planar mechanisms are briefly discussed in order to introduce some standard nomenclature. Let us consider various possible combinations of R and P pairs with a total of four links.

4R Mechanisms

A 4R planar mechanism is shown in Fig. 1.8-1. This mechanism is most commonly refered to as a *four-bar linkage*[5]. The two members connected to the frame are called the *crank* (input) and the *follower*(output). The intermediate moving member connecting the crank and follower is called the *coupler*. The revolute pairs connected to the frame, i.e., those at O_2 and O_4 are called *ground pivots* or *fixed hinges*. The other two revolute pairs at A and B are referred to as *moving hinges*. The line O_2O_4 is called the frame-line. It may be noted that, since in most cases the input motion is provided by an electric motor, at least one of the links connected to the frame should be able to make complete rotation for a mechanism to be practically useful. A link (connected to the frame) making complete rotation is called a crank. Thus, 4R planar mechanisms (depending on the rotatability of links 2 and 4) can be of the following three types:

(i) *Double-crank* - When both links 2 and 4 rotate completely.

(ii) *Crank-rocker* - When only one of the links 2 and 4 rotates while the other oscillates.

(iii) *Double-rocker* - When neither link 2 nor 4 rotates, both perform only oscillatory motion.

---

[5]We may note that the name *four-bar* is a misnomer because there are only three moving bars. Moreover, this name does not reveal how the four bars are connected, which is crucial to the behaviour of the relative motions between the four bars.

1.8. FOUR-LINK PLANAR MECHANISMS

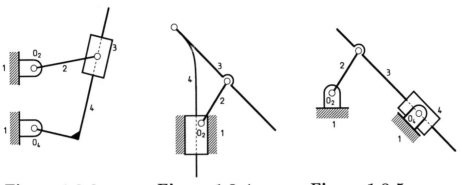

Figure 1.8-3   **Figure 1.8-4**   **Figure 1.8-5**

3R-1P Mechanisms

The most common 3R-1P mechanism, known as *slider-crank mechanism*, has already been shown in Fig. 1.4-2. A more generalized version of the same mechanism, known as *offset slider-crank*, is indicated in Fig. 1.8-2. Unlike in Fig. 1.4-2, in the present mechanism the axis of the prismatic pair through the point B does not pass through O_2. Some commonly used inversions of the same 3R-1P chain (with or without offset) are shown in Figs. 1.8-3 to 1.8-5. The mechanism shown in Fig. 1.8-3 is used for producing quick return of the cutting tool (during the idle stroke) in shapers and slotting machines. The mechanisms shown in Figs. 1.8-4 and 1.8-5 are used, respectively, in hand pumps and foot operated bicycle pumps.

R-R-P-P Mechanisms

Two mechanisms obtained by kinematic inversion of an R-R-P-P chain, namely, *scotch-yoke mechanism* and *Oldham's coupling*, have already been shown in Figs. 1.7-2 and 1.7-3, respectively. Another mechanism, obtained by kinematic inversion of the same chain, known as *elliptic trammel*, is shown in Fig. 1.8-6. It may be easily verified that as the mechanism moves, all points of link 3, except A, B and the midpoint C of AB, move on an elliptic path. The midpoint C moves on a circle (with center at O) whereas A and B are constrained to move along straight lines.

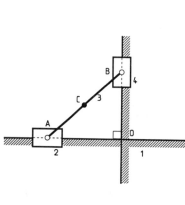
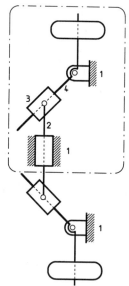

Figure 1.8-6   Figure 1.8-7

R-P-R-P Mechanisms

Figure 1.8-7 shows a part (indicated by circumscribing dashed lines) of *Davis automobile steering gear*. This part is seen to be an R-P-R-P mechanism.

A 3P-1R chain cannot give rise to a mechanism since no relative movement between various links can take place in such a chain. A little thinking should convince the reader that with three prismatic pairs in succession, all the links can only have relative translation whereas the revolute pair does not allow such a motion. A 4P-chain with one link fixed does not generate a constrained mechanism. These aspects will be discussed later in Chapter 3.

1.9 Equivalent Linkages

For the purpose of kinematic analysis, a planar higher-pair mechanism can sometimes be replaced by an *equivalent planar linkage* consisting of only lower pairs. It may be emphasized that this equivalence, in general, is valid only for studying instantaneous velocity and accel-

1.9. EQUIVALENT LINKAGES

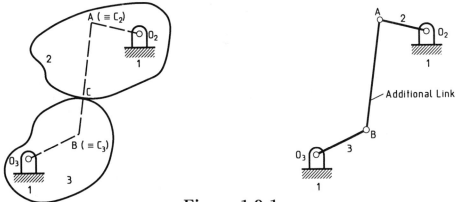

Figure 1.9-1

eration characteristics. For example, let us consider the higher-pair mechanism shown in Fig. 1.9-1a. Links 2 and 3, both connected to the fixed link through an R pair, are in contact with each other at C, forming a higher pair. This higher pair allows simultaneous rolling and sliding motion between links 2 and 3. Let $A(=C_2)$ and $B(=C_3)$ be the centers of curvature of surfaces 2 and 3, respectively, at the point C. The instantaneous equivalent lower-pair linkage shown in Fig. 1.9-1b, calls for an additional link AB, and the higher pair is replaced by two revolute pairs at A and B. It may noted that (due to second order contact of a curve with its circle of curvature) the relative positions between links 2 and 3 of the original and the equivalent mechanism are the same for three infinitesimally separated time intervals. Consequently, the instantaneous equivalence holds good only for velocity and acceleration analysis. Note that, in general, the centers of curvature A and B change as the mechanism moves. Thus, only for the special situations of circular or straight profiles, the equivalence may remain valid for the entire cycle of motion[6].

Another example of an equivalent lower-pair linkage for a cam-follower system is shown in Fig. 1.9-3. The sliding block 4 is the additional link and the higher pair is replaced by two lower pairs, one

---

[6]It may be further mentioned that the equivalent linkage is not unique. As explained in Fig. 1.9-2, one can place one of the moving hinges anywhere at A on link 2 with the other moving hinge B on link 3 at the center of curvature of the path of A traced out on link 3.

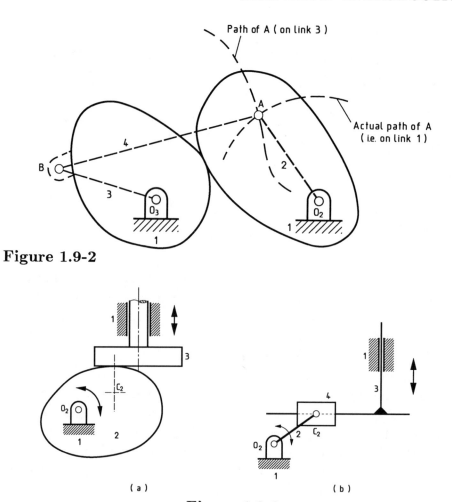

Figure 1.9-2

Figure 1.9-3

revolute and the other prismatic. The center of curvature of the cam surface at its point of contact with the follower is at C_2. The center of curvature of the flat-face follower is at infinity and, consequently, the additional lower pair becomes prismatic.

1.10 A Note on Subsequent Chapters

With the introductory materials provided in the preceding sections, we can ask ourselves at this stage the following question: "For kine-

1.10. A NOTE ON SUBSEQUENT CHAPTERS

matic design of mechanisms what should the designer study?"

In this section, we shall briefly discuss this question and the sequencing of topics that will be followed in this book.

The kinematic design of a mechanism, like any other design problem, has no unique solution. Consequently, the designer has to evaluate all possible designs that can satisfy the requirements. As a first step, the requirements have to be specified as precisely as possible, along with the level of accuracy to which each requirement is to be satisfied. The next step, known as *type synthesis*, is to determine which class of mechanisms or linkages can serve the purpose. Considerations in type synthesis, more often than not, go beyond the realm of kinematics. At the end of the type synthesis, the designer may have more than one solution. In the next step, each solution is optimized with regard to the involved parameters. Thereafter, each optimal solution is analysed to check how well a solution meets the requirements[7]. The final selection, again, may not be from kinematic considerations. Many other factors including aesthetics, economics, etc. play important roles towards the final choice. In this text, however, our discussions will be limited only to kinematic considerations without any regard to the other steps involved in the value analysis discussed above.

It is needless to mention that a thorough background of the basic features of planar kinematics of a rigid body and interconnected rigid bodies is a fundamental prerequisite for the study of kinematics of planar mechanisms. This background material, developed in Chapter 2, is useful for both analysis and synthesis.

Before entering the study of interconnected rigid bodies, we should appreciate that any number of rigid bodies connected through an arbitrary number of kinematic pairs may not give rise to a mechanism. The subject matter dealing with this aspect of kinematics (of interconnected rigid bodies) is normally called number synthesis or movability (or mobility) analysis. Another important aspect of a mechanism is concerned with the possible ranges of movement of various links. Movability, the range of movement and some associated results (for

---

[7]Dittrich, G.: Systematik der Bewegungsaufgaben und rundsätzliche Lösungsmöglichkeiten, VDI Berichte, Nr.VDI-Verlag, Düsseldorf (1988).

planar linkages) are discussed in Chapter 3.

It has already been mentioned that there are two phases in the design of mechanisms, namely analysis and synthesis. For the sake of convenience, we shall discuss various methods of kinematic analysis before going into the synthesis of planar linkages. Chapter 4 is devoted to displacement analysis, i.e., to the study of relative displacements between various links of a given mechanism. The methods required for velocity and acceleration analyses of various mechanisms are outlined in Chapter 5. Besides providing some general information regarding motion characteristics of mechanisms, these methods are useful to a designer for evaluating the performance of a mechanism designed by him (her).

Methods of dimensional synthesis of planar linkages are taken up in Chapter 6. These are used for determining the geometrical parameters of a chosen type of linkage so that it can fulfill the desired motion characteristics.

A knowledge of the nature of the curves generated by a point on a moving link is extremely useful to a mechanism designer[8]. This aspect, with special reference to the *coupler curves* (curves traced by points on the coupler) of 4R planar linkages, is discussed in Chapter 7.

Chapter 8 is exclusively devoted to the study of spatial linkages. Movability, mobility, kinematic analysis and synthesis of spatial linkages are discussed with examples.

Cams, used extensively for generating complicated motion characteristics (especially with dwell periods), form one of the most versatile groups of higher pair mechanisms. Methods of synthesis and analysis of different cam-follower systems are covered in Chapter 9.

Another most common higher pair mechanism is a geared system. The theory of various gears is so vast that it forms a subject by itself. In Chapter 10, only a brief introduction to the theory of gearing and gear trains is included. This is considered to be sufficient for a mechanism designer.

Besides the analytical and graphical methods discussed in this

---

[8]Artobolevsky, I.I.: Mechanisms for the generation of plane curves, Pergamon Press, Oxford (1964).

1.11. EXERCISE PROBLEMS

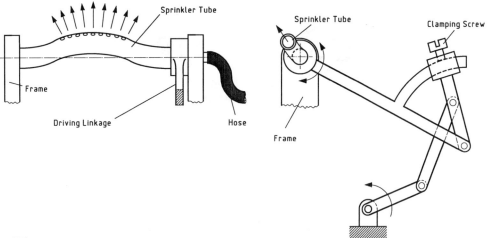

Figure 1.11-1

book, use of digital computers for both computation and graphics is necessary for most real-life designs of mechanisms. Very often fabrication of a scale model using cheap materials can bring to attention various shortcomings of a designed mechanism that are difficult to locate in the theory or drawing board. The designer, therefore, should attempt, whenever possible, to use both computers and models to make a design perfect.

1.11 Exercise Problems

1. Draw the kinematic diagrams of the linkages used in

    (a) The lawn sprinkler (Fig. 1.11-1) with variable swing. Consider both clamped and unclamped configurations.

    (b) The V-engine (Fig. 1.11-2).

    (c) The device (Fig. 1.11-3) used for obtaining the indicator (pressure versus time) diagram of an IC engine.

    (d) The minidrafter (Fig. 1.11-4). Consider both clamped and unclamped (at A) situations.

    (e) The earth-moving machinery (Fig. 1.1-2).

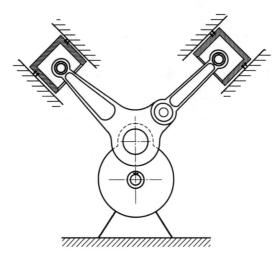

Figure 1.11-2

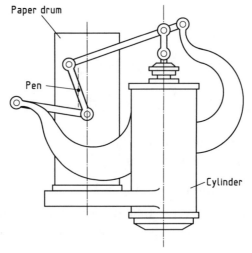

Figure 1.11-3

1.11. EXERCISE PROBLEMS

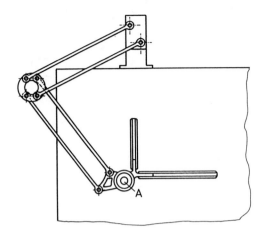

Figure 1.11-4

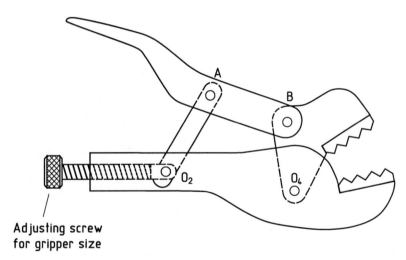

Adjusting screw
for gripper size

Figure 1.11-5

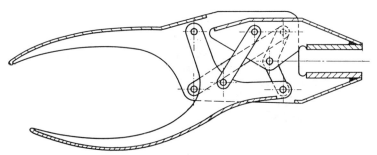

Figure 1.11-6

(f) The vice-grip pliers (Fig. 1.11-5).

(g) The parallel-jaw pliers (Fig. 1.11-6).

(h) The variable displacement engine (Fig. 1.11-7).

2. For each higher-pair mechanism shown in Fig. 1.11-8, draw the

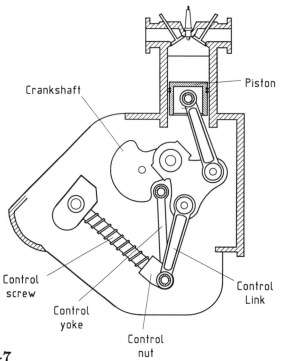

Figure 1.11-7

1.11. EXERCISE PROBLEMS

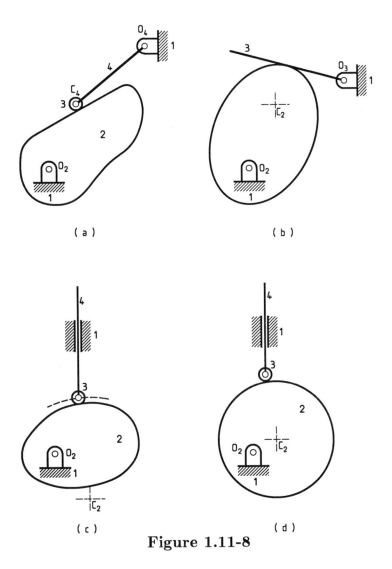

Figure 1.11-8

kinematic diagram of the corresponding equivalent lower-pair linkage. Identify the cases where the equivalence holds good for the entire cycle of motion.

Chapter 2

PLANAR KINEMATICS OF RIGID BODIES

A rigid body is a hypothetical solid that does not deform. In other words, it is a conglomeration of infinite number of particles where the distance between any two particles does not change. When all the points (particles) of a rigid body move in parallel planes, the rigid body is said to be in plane motion. In such a situation, a single view along the normal to these planes reveals the true motion of all the points. This view is drawn on paper and the plane of the paper is referred to as the plane of motion. The angular motion of a rigid body in plane motion has an axis normal to the plane of the paper.

For a discussion of plane motion, use of vectors is not essential. However, we shall use vectors for notational compactness. The vectors representing angular motion quantities (such as angular velocity, angular acceleration, etc.) will always be perpendicular to the plane of the paper. The vector pointing away from the paper (signifying counterclockwise direction [CCW]) is treated as positive.

2.1 Introduction

In this chapter, we assume the reader to be somewhat familiar with plane kinematics of a rigid body. The velocity and acceleration analysis of a rigid body, useful in Newtonian (Eulerian) dynamics, is

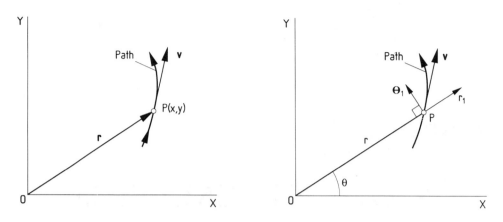

Figure 2.2-1   Figure 2.2-2

normally covered in any preliminary course on dynamics[1]. We shall briefly recapitulate the velocity and acceleration relationships between points on a rigid body. However, the concepts that are more useful in the context of kinematics of a planar linkage will be developed. Among others, these especially include the displacement analysis (not useful in dynamics) of a rigid body. It may be recalled that the motion of a rigid body is completely described by the translational motion of a point (particle) on the body together with the angular motion of a line attached to the body. The angular motion characteristics of this line are referred to as those of the rigid body itself.

2.2 Plane Motion of a Particle

In Fig. 2.2-1, the point P indicates the position of a particle (in plane motion) on its path at an instant, t. The position vector OP of the particle, in the rectangular XY coordinate system, is given by

$$\boldsymbol{r} = \boldsymbol{OP} = x\boldsymbol{i} + y\boldsymbol{j} \qquad (2.2.1)$$

where $\boldsymbol{i}$, $\boldsymbol{j}$ are unit vectors along X and Y directions, respectively. The coordinates of the point P at the instant are x and y, both

---

[1] For example, see Meriam, J.L.: Dynamics, John Wiley & Sons, Inc. New York, 1980.

2.2. PLANE MOTION OF A PARTICLE

being functions of time. The velocity (tangential to the path) and acceleration of the particle are given as follows:

$$\text{velocity} \quad \boldsymbol{v} = \dot{x}\boldsymbol{i} + \dot{y}\boldsymbol{j} \quad (2.2.2)$$
$$\text{acceleration} \quad \boldsymbol{a} = \ddot{x}\boldsymbol{i} + \ddot{y}\boldsymbol{j} \quad (2.2.3)$$

where (˙) denotes differentiation with respect to time.

If polar coordinates (r, θ) are used to define the position of the particle on its path (Fig. 2.2-2), then the position, velocity and acceleration of the particle are given as follows:

$$\text{position} \quad \boldsymbol{r} = r\boldsymbol{r}_1; \quad r \geq 0 \quad (2.2.4)$$
$$\text{velocity} \quad \boldsymbol{v} = \dot{r}\boldsymbol{r}_1 + (r\dot{\theta})\boldsymbol{\theta}_1 \quad (2.2.5)$$
$$\text{acceleration} \quad \boldsymbol{a} = (\ddot{r} - r\dot{\theta}^2)\boldsymbol{r}_1 + (r\ddot{\theta} + 2\dot{r}\dot{\theta})\boldsymbol{\theta}_1 \quad (2.2.6)$$

where $\boldsymbol{r}_1$ and $\boldsymbol{\theta}_1$ are unit vectors along r and θ directions, respectively, with $\boldsymbol{r}_1$ always directed along $\boldsymbol{OP}$. It may be noted that unlike $\boldsymbol{i}$ and $\boldsymbol{j}$, $\boldsymbol{r}_1$ and $\boldsymbol{\theta}_1$ are not constant vectors; their magnitudes are always unity but their directions are not fixed. For the special case of circular motion when $r = $ constant, we obtain the following expressions from Equations (2.2.4) to (2.2.6):

$$\text{position} \quad \boldsymbol{r} = r\boldsymbol{r}_1; \quad r \geq 0 \quad (2.2.7)$$
$$\text{velocity} \quad \boldsymbol{v} = (r\dot{\theta})\boldsymbol{\theta}_1 \quad (2.2.8)$$
$$\text{acceleration} \quad \boldsymbol{a} = (-r\dot{\theta}^2)\boldsymbol{r}_1 + (r\ddot{\theta})\boldsymbol{\theta}_1 \quad (2.2.9)$$
$$= (-v^2/r)\boldsymbol{r}_1 + \dot{v}\boldsymbol{\theta}_1 \quad (2.2.10)$$

where v is the speed (i.e., the magnitude of $\boldsymbol{v}$).

Another coordinate system using the path variables is most conveniently used for describing curvilinear motion of a particle. Here the two axes are oriented along the path normal ($\boldsymbol{n}$) and the path tangent ($\boldsymbol{t}$). The path *normal* is always directed *towards the center of path curvature* and the path *tangent along the direction of velocity* (Fig. 2.2-3). In Fig. 2.2-3, C is the center of path-curvature at

40   CHAPTER 2. PLANAR KINEMATICS OF RIGID BODIES

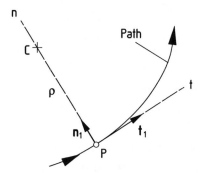

Figure 2.2-3

P and ρ is the radius of curvature. In this $n-t$ system, the velocity and the acceleration of the particle are given by

$$v = v t_1 = \rho \omega t_1; \qquad v > 0 \qquad (2.2.11)$$
$$a = (v^2/\rho) n_1 + \dot{v} t_1 \qquad (2.2.12)$$

where n_1, t_1 are unit vectors along the path normal and path tangent and ω is the magnitude of the angular velocity of the line CP. Again, for the special case of circular motion (of radius r), Equations (2.2.11) and (2.2.12) reduce, respectively, to

$$v = r\omega t_1 \qquad (2.2.13)$$
$$a = (v^2/r) n_1 + \dot{v} t_1 = (v^2/r) n_1 + r\dot{\omega} t_1 \qquad (2.2.14)$$

Comparison of Equations (2.2.10) and (2.2.14) indicates that in case of circular motion n_1 is directed opposite to r_1.

It is easy to verify from Equations (2.2.7) through (2.2.9) that for circular motion of P, we can write

$$v = \omega \times r \qquad (2.2.15)$$
$$a = \omega \times (\omega \times r) + \alpha \times r \qquad (2.2.16)$$

2.3. PLANE MOTION OF A RIGID BODY

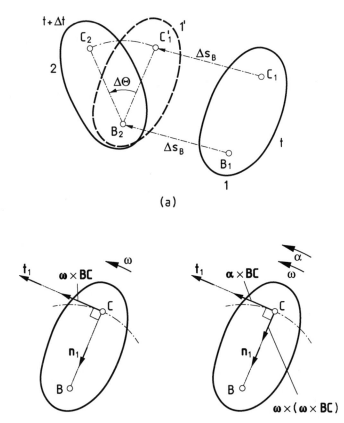

Figure 2.3-1

where $\boldsymbol{\omega}\ (=\dot{\theta}\,\boldsymbol{k})$ and $\boldsymbol{\alpha}\ (=\ddot{\theta}\,\boldsymbol{k})$ are, respectively, the angular velocity and acceleration of the line OP, with $\boldsymbol{k}$ representing a unit vector perpendicular (pointing outward) to the plane of motion and θ being measured in counterclockwise direction.

2.3 Plane Motion of a Rigid Body

Figure 2.3-1a shows the positions of a rigid body in plane motion at instants t and $t + \Delta t$. These two positions are indicated as 1 and 2 in the figure. Two arbitrarily chosen points on the body are labelled B and C with subscripts 1 and 2 indicating their positions at t and

$t + \Delta t$, respectively. The movement of the rigid body during Δt can be decomposed into a translation $s_B = \boldsymbol{B_1 B_2} = \boldsymbol{C_1 C'_1}$ (from 1 to 1′) and a rotation $\Delta \theta$ (from 1′ to 2), measured positive in counterclockwise direction, as indicated in the figure. During the translational part of the movement, all points on the body have identical motion. The difference in the motion of any two points is entirely due to the rotational part.

The angular velocity and acceleration of the rigid body (i.e., of the line BC) are given, respectively, by

$$\boldsymbol{\omega} = \omega \boldsymbol{k}$$
$$\boldsymbol{\alpha} = \alpha \boldsymbol{k} = \dot{\omega} \boldsymbol{k}$$
$$\text{with} \quad \omega = \dot{\theta} = \lim_{\Delta t \to 0} \frac{\Delta \theta}{\Delta t};$$

where θ is measured in counterclockwise direction and $\boldsymbol{k}$ is a unit vector perpendicular to the plane of motion (i.e., of the paper) and pointing outward.

Since the distance BC remains unchanged, the velocity difference between points C and B (caused entirely by the rotational motion) is easily obtained from Equation (2.2.15) as[2]

$$\boldsymbol{v}_{CB} = \boldsymbol{v}_C - \boldsymbol{v}_B = \boldsymbol{\omega} \times \boldsymbol{BC} \qquad (2.3.1)$$

Similarly, the acceleration difference between points C and B is obtained from Equation (2.2.16) as

$$\boldsymbol{a}_{CB} = \boldsymbol{a}_C - \boldsymbol{a}_B = \boldsymbol{\omega} \times (\boldsymbol{\omega} \times \boldsymbol{BC}) + \boldsymbol{\alpha} \times \boldsymbol{BC} \qquad (2.3.2)$$

The reader is advised to note the directions of the various terms appearing on the right-hand-side of Equations (2.3.1) and (2.3.2) as indicated in Figs. 2.3-1b and 2.3-1c. Moreover, it may also be pointed out that while using Equations (2.3.1) and (2.3.2) we must ensure that points B and C belong to the same rigid body.

---

[2]Note that Equation (2.3.1) implies that two points on a rigid body have the same component of velocities along the line joining them.

2.3. PLANE MOTION OF A RIGID BODY

The first term on the right-hand-side of Equation (2.3.2), called the normal component and completely determined by the angular velocity, is written as

$$a_{CB}^n = \boldsymbol{\omega} \times (\boldsymbol{\omega} \times \boldsymbol{BC}) \qquad (2.3.3)$$

The normal component is always directed from C towards B. The second term on the right-hand-side of Equation (2.3.2), called the tangential component and completely determined by the angular acceleration, is written as

$$a_{CB}^t = \boldsymbol{\alpha} \times \boldsymbol{BC} \qquad (2.3.4)$$

The tangential component is perpendicular to BC and directed in the sense of $\boldsymbol{\alpha}$.

Attention may be drawn to the following facts concerning the plane motion of a rigid body.

(i) Complete information regarding velocity involves three scalar unknowns, namely, two velocity components of any arbitrarily chosen point and the angular velocity of the rigid body. Thus, the velocities of two points on a rigid body (involving four scalar quantities) cannot be chosen arbitrarily.

(ii) Complete information regarding acceleration involves four scalar unknowns, namely, two acceleration components of an arbitrarily chosen point and the angular velocity and angular acceleration of the rigid body.

(iii) In view of the above statements, it is obvious that knowledge about the motion of two points of a rigid body definitely provides complete information about the motion of that body.

CHAPTER 2. PLANAR KINEMATICS OF RIGID BODIES

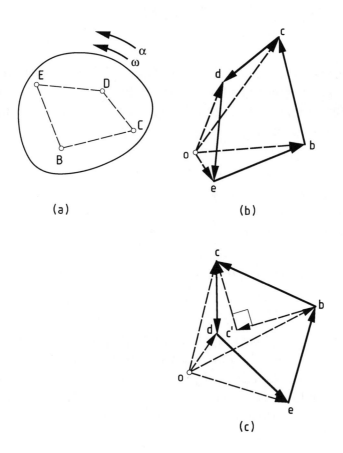

Figure 2.4-1

2.4 Velocity and Acceleration Images

As mentioned in the previous section, if the velocities and accelerations of any two points of a rigid body are known, then the velocity and acceleration of any other point on the same rigid body can be determined. A graphical approach to such a solution is facilitated through the concept of velocity and acceleration images. This concept, explained in this section, is especially useful in the kinematic analysis of mechanisms having ternary and higher order links.

Referring to Fig. 2.4-1a, let $B, C, D, E, \ldots$ be a number of points on a rigid body that is in planar motion. Let ***ob, oc, od, oe,*** ... represent, respectively, the velocities of the points $B, C, D, E, \ldots$ as shown in Fig. 2.4-1b. From Equation (2.3.1), we get

$$\boldsymbol{bc} = \boldsymbol{oc} - \boldsymbol{ob} = v_C - v_B = \omega \times \boldsymbol{BC}$$

Similarly, $\boldsymbol{cd} = \omega \times \boldsymbol{CD}, \boldsymbol{de} = \omega \times \boldsymbol{DE}, \boldsymbol{eb} = \omega \times \boldsymbol{EB}$, and so on. From the above relations, we can write

$$|\omega| = \frac{bc}{BC} = \frac{cd}{CD} = \frac{de}{DE} = \frac{eb}{EB} = \frac{bd}{BD} = \ldots$$

Hence, it is readily seen the polygon $bcde$ in Fig. 2.3-1b, called the velocity image of $BCDE$, is geometrically similar to the polygon $BCDE$ in Fig. 2.4-1a. The velocity image is also seen to be rotated through a 90° angle in the sense of ω from the corresponding polygon in the rigid body. Thus, given the velocities of any two points, (say B and C, by ***ob*** and ***oc***, respectively) the velocity of any other point (say D) can be obtained as ***od*** where the point d is located by drawing a triangle bcd similar to BCD. It should be noted that the letters b, c, d identifying the tips of the velocity vectors are in the same sequence (clockwise or counterclockwise) as the corresponding points B, C, D in the rigid body.

Referring to Fig. 2.4-1c, let ***ob, oc, od*** and ***oe*** represent the accelerations of the points B, C, D and E, respectively. From Equations (2.3.2) to (2.3.4), we get

$$\boldsymbol{bc} = \boldsymbol{oc} - \boldsymbol{ob} = \boldsymbol{a}_C - \boldsymbol{a}_B = \boldsymbol{a}_{CB}^n + \boldsymbol{a}_{CB}^t$$

$$= \omega \times (\omega \times BC) + \alpha \times BC$$
$$= bc' + c'c$$

where
$$bc = (bc'^2 + c'c^2)^{1/2} = (\omega^4 + \alpha^2)^{1/2} BC.$$

Similarly,

$$cd = (\omega^4 + \alpha^2)^{1/2} CD, \; de = (\omega + \alpha)^{1/2} DE \quad \text{and}$$
$$eb = (\omega^4 + \alpha^2)^{1/2} EB.$$

From the above relations we can again write

$$\frac{bc}{BC} = \frac{cd}{CD} = \frac{de}{DE} = \frac{eb}{EB}$$

Thus, the polygon $bcde$ in Fig. 2.4-1c, called the acceleration image of $BCDE$, is a scale drawing of the polygon $BCDE$. The geometrically similar acceleration image is rotated by an angle $[\pi - \tan^{-1}(\alpha/\omega)]$ in the counterclockwise direction[3] from the corresponding polygon in the rigid body. All other statements made above regarding the velocity image are also valid for the acceleration image.

2.5 Instantaneously Coincident Points

In mechanisms, very often a point of a rigid body is guided along a path prescribed in another moving rigid body. The kinematic analysis in such situations is facilitated by considering the motion relationships between two instantaneously coincident points belonging to these two rigid bodies. These relationships are discussed in this section for ready reference in future sections.

Figure 2.5-1 shows two moving rigid bodies, numbered 2 and 3. A fixed reference frame is denoted by OXY. A point *of body 3*

---

[3]Note that rotational parameters in the CCW direction have been treated as positive. Thus, a clockwise α should be treated as negative.

2.5. INSTANTANEOUSLY COINCIDENT POINTS

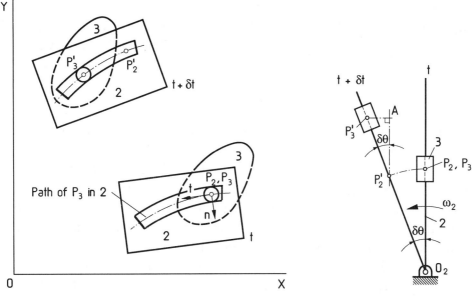

Figure 2.5-1 **Figure 2.5-2**

is guided along a path *prescribed in body 2*. The configurations at two different instants, t and $t + \delta t$, are indicated. The points P_2 and P_3, belonging respectively to bodies 2 and 3, are coincident at t. The displacements of these two points during δt are given by $\boldsymbol{P_2 P'_2}$ (for P_2) and $\boldsymbol{P_3 P'_3}$ (for P_3). From Fig. 2.5-1, we can write $\boldsymbol{P_3 P'_3} = \boldsymbol{P_2 P'_2} + \boldsymbol{P'_2 P'_3}$. Dividing this equation by δt and taking the limit $\delta t \to 0$, we get the following relation between the velocities of P_3 and P_2:

$$v_{P_3} = v_{P_2} + v_{P_3}|_2 \qquad (2.5.1)$$

where $v_{P_3}|_2$ indicates the velocity of P_3 as seen by an observer attached to body 2. The direction of $v_{P_3}|_2$ is obviously tangential to the path prescribed in body 2.

From Equation (2.5.1) it may appear that the accelerations of P_3 and P_2 can also be related by replacing $\boldsymbol{v}$'s by $\boldsymbol{a}$'s in Equation (2.5.1). However, *this is incorrect*. The correct relationship is expressed as

$$a_{P_3} = a_{P_2} + a_{P_3}|_2 + 2\omega_2 \times v_{P_3}|_2 \qquad (2.5.2)$$

where ω_2 is the angular velocity of body 2. The last term on the right-hand-side of Equation (2.5.2) is known as *Coriolis' acceleration*, given by

$$a_{Cor} = 2\omega_2 \times v_{P_3}|_2 \qquad (2.5.3)$$

The existence of Coriolis' acceleration is explained below with the help of a simple example[4].

Referring to Fig. 2.5-2, block 3 slides along link 2, which is rotating about O with angular velocity ω_2. Let the sliding velocity of 3 on 2, i.e., $v_{P_3}|_2$, be constant (of magnitude $v_{P_3}|_2$), when $a_{P_3}|_2$ is zero. Since $v_{P_3}|_2$ is along link 2, the transverse displacements (perpendicular to link 2) of P_2 and P_3 should be the same in the absence of Coriolis' acceleration (i.e., with $a_{P_3} = a_{P_2}$). From the figure, however, it is evident that the transverse displacement of P_3 exceeds that of P_2 by AP'_3. Assuming δt to be small, we can write

$$AP'_3 = P'_2 P_3 . \delta\theta = v_{P_3}|_2 \delta t . \omega_2 \delta t.$$

Since AP'_3 is proportional to $\frac{1}{2}(\delta t)^2$, it can be attributed to an acceleration of magnitude $2v_{P_3}|_2 \omega_2$ which is the Coriolis acceleration. The reader is advised to check that the directions indicated by vector Equation (2.5.3) are in conformity with the above discussion.

Returning to Equation (2.5.2) we can further write this equation as

$$a_{P_3} = a_{P_2} + a_{P_3}^n|_2 + a_{P_3}^t|_2 + a_{Cor} \qquad (2.5.4)$$

where

$$a_{P_3}^n|_2 = [v_{P_3}^2|_2/\rho]\,n_1 \qquad (2.5.5)$$

with ρ as the radius of curvature of the prescribed path.

It may be mentioned that while studying the kinematics of a mechanism, Equations (2.5.1) and (2.5.2) are useful even for coincident points belonging to two links (rigid bodies) that may not be directly

---

[4] For a rigorous derivation of Equations (2.5.1) and (2.5.2), see any standard text on dynamics.

2.6. INSTANTANEOUS CENTER OF VELOCITY

connected. Thus, if two coincident points are denoted by P_i and P_j, belonging, respectively, to bodies numbered i and j, the kinematic relationship between these two points can be written following the above equations. A note of caution should be taken for Equation (2.5.1), and especially Equation (2.5.2). To render useful information, first we must ensure that the path of P_i in body j or that of P_j in body i is known. These two paths are not *necessarily* the same (only their tangents at the coincident point are necessarily the same). This statement is independent of whether the bodies i and j are directly connected or not. Finally, *if the path of P_i in body j is known*, then the following relations will be found very useful in the kinematic analysis of mechanisms:

$$v_{P_i} = v_{P_j} + v_{P_i}|_j \quad (2.5.6)$$

with

$$v_{P_i}|_j = v_{P_i}|_j \, t_1 \quad (2.5.7)$$

and

$$a_{P_i} = a_{P_j} + a_{P_i}^n|_j + a_{P_i}^t|_j + 2\omega_j \times v_{P_i}|_j \quad (2.5.8)$$

where

$$a_{P_i}^n|_j = [v_{P_i}^2|_j/\rho] n_1 \quad (2.5.9)$$

with ρ as the radius of curvature of the path of P_i in body j and n_1, t_1 as the unit vectors associated with this path.

2.6 Instantaneous Center of Velocity

For a rigid body in plane motion, there exists a point in the plane of motion (either within, at or outside the physical boundary of the rigid body) whose instantaneous velocity is zero. This point is called the *instantaneous center of velocity*. The word **centro** is also used

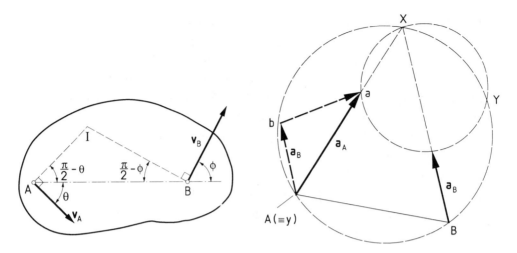

Figure 2.6-1 **Figure 2.6-2**

in the literature. Obviously, the rigid body can be thought of as momentarily under pure rotation about this center (so far as only the velocities are concerned). To prove the existence of the instantaneous center of velocity, let us consider two points, A and B, on a moving rigid body as shown in Fig. 2.6-1. The points A and B are chosen so that their velocities are nonparallel. Let I be the point of intersection of the normals to v_A and v_J drawn at A and B, respectively. Point I on the rigid body cannot have any velocity component along IA because both the velocity of A along IA and the velocity difference between points I and A along IA (both I and A being on the same rigid body) are zero. Similar arguments, starting from point B, show that the velocity of I along IB is also zero. Since the velocity of I along both IA and IB is zero, the only conclusion that can be drawn is that the velocity of I is zero, i.e., I is the instantaneous center of velocity. With the point I having zero velocity, both A and B are momentarily in circular motion with respect to I. So, the following relation must hold good :

$$\frac{v_A}{IA} = \frac{v_B}{IB} \qquad (2.6.1)$$

where the above ratio is the magnitude of the instantaneous angular velocity of the rigid body. The proof of relation (2.6.1) can be given

2.6. INSTANTANEOUS CENTER OF VELOCITY

simply as follows:
Referring to Fig. 2.6-1, we can write

$$v_A cos\theta = v_B cos\phi \qquad (2.6.2)$$

since the component of v_{BA} along AB must be zero (A and B being two points on the same rigid body). Again from the triangle IAB, applying sine rule

$$IA/sin(\frac{\pi}{2} - \phi) = IB/sin(\frac{\pi}{2} - \theta)$$

or,

$$IAcos\theta = IBcos\phi. \qquad (2.6.3)$$

Dividing Equation (2.6.2) by Equation (2.6.3) we get the relation (2.6.1).

The above discussion assumed that all points of the rigid body cannot have velocities that are parallel. In situations where velocities of all points on a rigid body are parallel, the velocity magnitudes are necessarily the same and the rigid body is said to be under pure translation. For a translating rigid body, the instantaneous center of velocity lies at infinity in the direction perpendicular to the direction of translation.

The concept of instantaneous center of velocity can be extended to the relative motion between two moving rigid bodies. If two rigid bodies, numbered 2 and 3, are in motion, we can define a *relative instantaneous center*, I_{23}, as a point on 3 having zero relative velocity with respect to (i.e., having the same absolute velocity as) a coincident point on 2. Consequently, the instantaneous relative motion of 3 with respect to 2 appears to be a pure rotation about I_{23}. It is evident from the definition that I_{23} and I_{32} are identical.

Relative instantaneous centers are very useful in the study of kinematics of planar mechanisms. The (absolute) instantaneous center of velocity of a rigid body, defined earlier, is the relative instantaneous center of velocity with respect to the fixed link (numbered 1) of the mechanism. A mechanism consisting of N links has a total of $N(N-1)/2$ relative instantaneous centers. In some situations, the relative instantaneous centers are easily identified by inspection of the geometry of the permitted relative motion. Some examples in this category are as follows:

(i) If two bodies are connected by an R pair, then the location of the axis of the R pair is the relative instantaneous center because one body is in pure rotation with respect to the other about that point.

(ii) If two bodies are connected by a P pair, then the relative instantaneous center is at infinity in a direction normal to the direction of the P pair (the relative motion is one of pure translation).

(iii) If one body is rolling without slipping over the other, then the point of contact is the relative instantaneous center.

(iv) If a body rolls and simultaneously slides over the other, then the relative instantaneous center lies on the common normal to the contacting surfaces at the point of contact. The exact location cannot be determined by inspection only (without the application of more information).

Finally, it may be mentioned that the acceleration of the instantaneous center of velocity is generally not zero. The instantaneous center of acceleration also exists and is defined as a point on a rigid body having instantaneously zero acceleration. Being of little use, it will not be discussed here.[5]

---

[5]See Sandor, G.N. & Erdman, A.G.: "Advanced Mechanism Design: Analysis & Synthesis", Vol. 2, p. 337. Prentice Hall Inc. New Jersey, 1984.

Rosenauer, N. & Willis, A.H.: "Kinematics of Mechanism", Associated General Publications, Sydney, 1953.

Several graphical methods of locating the instantaneous center of acceleration are given in the second reference. Given the accelerations of two points A and B of a rigid body, the instantaneous center of acceleration, not withstanding special situations, is located at Y as shown (without proof) in Fig. 2.6-2.

In this figure, X is the point of intersection of a_A and a_B drawn, respectively, at A and B. Two circles are drawn, one through X and tips of a_A and a_B and the other through X, A and B. The second point of intersection (one being at X) of these two circles is at Y. We may note by drawing a_B also at A, that $\Delta A(\equiv y)ab$ is the acceleration image of ΔYAB.

2.7. ARONHOLD-KENNEDY THEOREM

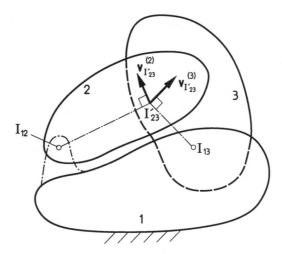

Figure 2.7-1

2.7 Aronhold-Kennedy Theorem

The Aronhold-Kennedy theorem states that *if three rigid bodies are in relative motion* (in a plane) *with respect to one another, the three relative instantaneous centers of velocity are collinear*. This theorem holds true regardless of whether the bodies are connected or not.

A simple proof of the theorem follows. Figure 2.7-1 shows three bodies, numbered 1, 2 and 3, in relative motion. Let I_{12} and I_{13} be two of the three relative instantaneous centers of velocity. It may be noted that these points may or may not lie within the physical boundaries of the bodies involved. Without any loss of generality, the relative motion between these three bodies can be discussed by assuming one of them, say 1, is fixed. Under this assumption, both I_{12} and I_{13} have zero velocities. Thus, momentarily the body 2 is in pure rotation about I_{12} and the body 3 about I_{13}. By definition of the relative instantaneous center of velocity, the point I_{23} must have the same velocity whether considered as a point of body 2 or of body 3.

Let us assume, in contradiction to the Aronhold-Kennedy theorem, that I_{23} lies outside the line $I_{12}\ I_{13}$ at I'_{23} as indicated in Fig. 2.7-1. When considered a point of body 2, the direction of velocity of the point I'_{23} ($v_{I'_{23}}^{(2)}$) is perpendicular to the line $I_{12}\, I'_{23}$. Similarly, if I'_{23} is considered as a point of body 3, then the direction of its velocity ($v_{I'_{23}}^{(3)}$) is perpendicular to the line $I_{13}\, I'_{23}$. The directions of ($v_{I'_{23}}^{(2)}$) and ($v_{I'_{23}}^{(3)}$) can never be the same (Fig. 2.7-1) unless I'_{23} lies on the line $I_{12}\, I_{13}$. Therefore, I_{12}, I_{13} and I_{23} must be collinear. This completes the proof of the Aronhold-Kennedy theorem of three centers.[6]

This theorem is immensely useful for locating the relative instanta-

---

[6]The proof can also be completed without assuming body 1 to be fixed, as shown below:
Using Equation (2.3.1) we can write

$$\mathbf{v}_{I_{23}}^{(2)} - \mathbf{v}_{I_{12}}^{(2)} = \omega_2 \times \mathbf{I}_{12}\mathbf{I}_{23} \tag{i}$$

and

$$\mathbf{v}_{I_{23}}^{(3)} - \mathbf{v}_{I_{13}}^{(3)} = \omega_3 \times \mathbf{I}_{13}\mathbf{I}_{23} \tag{ii}$$

By definition

$$\mathbf{v}_{I_{23}}^{(2)} = \mathbf{v}_{I_{23}}^{(3)},\ \mathbf{v}_{I_{12}}^{(2)} = v_{I_{12}}^{(1)},\ \text{and}\ \mathbf{v}_{I_{13}}^{(3)} = \mathbf{v}_{I_{13}}^{(1)} \tag{iii}$$

Subtracting Equation (ii) from Equation (i) and using Equation (iii)

$$\mathbf{v}_{I_{13}}^{(1)} - \mathbf{v}_{I_{12}}^{(1)} = \omega_2 \times \mathbf{I}_{12}\mathbf{I}_{23} - \omega_3 \times \mathbf{I}_{13}\mathbf{I}_{23}$$

or

$$\omega_1 \times \mathbf{I}_{12}\mathbf{I}_{13} = \omega_2 \times \mathbf{I}_{12}\mathbf{I}_{23} - \omega_3 \times \mathbf{I}_{13}\mathbf{I}_{23}$$

or,

$$\omega_1 \times (\mathbf{I}_{12}\mathbf{I}_{23} - \mathbf{I}_{13}\mathbf{I}_{23}) = \omega_2 \times \mathbf{I}_{12}\mathbf{I}_{23} - \omega_3 \times \mathbf{I}_{13}\mathbf{I}_{23}$$

With $\omega_1 = \mathbf{k}\omega_1$, $\omega_2 = \mathbf{k}\omega_2$ and $\omega_3 = \mathbf{k}\omega_3$, the above equation results in

$$(\omega_1 - \omega_2)\mathbf{k} \times \mathbf{I}_{12}\mathbf{I}_{23} = (\omega_1 - \omega_3)\mathbf{k} \times \mathbf{I}_{13}\mathbf{I}_{23}$$

Hence,

$$\frac{(I_{12}I_{23})_x}{(I_{13}I_{23})_x} = \frac{(I_{12}I_{23})_y}{(I_{13}I_{23})_y} = \frac{\omega_1 - \omega_3}{\omega_1 - \omega_2} \tag{iv}$$

where the subscripts x and y refer to the in-plane x and y components, respectively. Equation (iv) obviously implies that I_{12}, I_{23} and I_{13} must be collinear. Whether I_{23} lies inside or outside $I_{12}\, I_{13}$ depends on the sign of the quantity $(\omega_1 - \omega_3)/(\omega_1 - \omega_2)$.

2.7. ARONHOLD-KENNEDY THEOREM

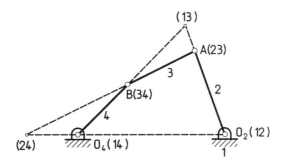

Figure 2.7-2

neous centers of a constrained mechanism. It may be mentioned, that for a single degree-of-freedom linkage (i.e., consisting only of lower pairs) having up to six links, all the relative instantaneous centers can be obtained geometrically by using the theorem of three centers. However, for a linkage consisting of eight or more links, it may or may not be possible to locate all the relative instantaneous centers by such geometrical means. For a mechanism having higher pair(s), it may not be possible to geometrically locate all the relative instantaneous centers even if the mechanism has only four links. For linkages with more than one degree-of-freedom, it is not possible to geometrically locate all the relative instantaneous centers.

Problem 2.7-1

Figure 2.7-2 shows a 4R planar linkage. Locate all the relative instantaneous centers of the linkage at this configuration.

Solution

With 4 links, there are (4x3/2 =) 6 relative instantaneous centers. Of these, 4 are at the revolute joints as indicated in the figure as $O_2 = 12$, $A = 23$, $B = 34$ and $O_4 = 14$. (From here foreward the centers I_{ij} are simply written as ij). The last two, namely 13 and 24, are determined by applying of the theorem of three centers as explained below. Considering three links 1, 2 and 3, we can say 12, 23 and 13 are collinear. Again considering links 1, 4 and 3 we know 14, 34 and 13 are collinear. Thus, 13 lies at the intersection of two lines, one joining 12 and 23 and the other joining 14 and 34. Similarly, 24 is located at the intersection of two lines, one joining 12 and 14 and the other joining 23 and 34.

CHAPTER 2. PLANAR KINEMATICS OF RIGID BODIES

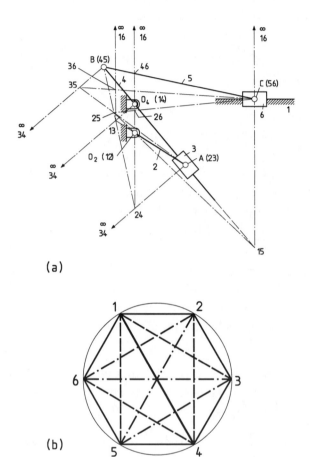

(a)

(b)

Figure 2.7-3

2.7. ARONHOLD-KENNEDY THEOREM

Problem 2.7-2

For the Whitworth quick-return mechanism, shown in Fig. 2.7-3a, obtain all the relative instantaneous centers of velocity at the given configuration.

Solution

This mechanism has 6 links with five R and two P pairs. Therefore, there are (6x5/2 =) 15 relative instantaneous centers with 7 of them clearly discernible by the kinematic pairs. For such a large number of instantaneous centers, it is better to proceed in a systematic manner as followed in this example. First, all the centers are listed in a triangular array as follows :

$$
\begin{array}{ccccc}
\times & \checkmark & \times & \checkmark & \times \\
12 & 13 & 14 & 15 & 16 \\
\times & \checkmark & & & \\
23 & 24 & 25 & 26 & \\
\times & & & & \\
34 & 35 & 36 & & \\
\times & \checkmark & & & \\
45 & 46 & & & \\
\times & & & & \\
56 & & & &
\end{array}
$$

Each link is identified by a point and numbered according to the link number, as shown in Fig. 2.7-3b. For visual clarity, these points are marked equidistant on a circle and such a drawing is referred to as a circle diagram.

Thus, for the present problem, in the first step, the relative instantaneous centers 12, 14, 23, 45, 56 (at the 5 R pairs) and 34, 16 (for the P pairs) are located in Fig. 2.7-3a. These are designated by a cross mark (X) in the list and the corresponding lines are drawn, indicated by solid lines, in Fig. 2.7-3b. We may note, as shown in Fig. 2.7-3a, that for a prismatic pair, the instantaneous center lying at infinity can be thought to be the point of intersection of parallel lines which are perpendicular to the direction of relative sliding.

In the second step, all the lines (in Fig. 2.7-3b) constituting the common side of two triangles, whose other sides have been drawn in the first step, are identified. The relative instantaneous centers designated by these lines are determined in this step by the application of the theorem of three centers. As an example, let us consider line 24, which is the common side of the triangles 124 and 234, where the lines 12, 14 and 23, 34 have been drawn in the first step. Consequently, as explained in Fig. 2.7-3a, the center 24 is located at the intersection of the lines joining the two pairs of centers, viz., (12, 14) and (23, 34). The other centers similarly located are 13, 15 and 46. The four centers located in this step are designated by a check mark ($\sqrt{}$) in the list and are indicated by dashed lines in Fig. 2.7-3b.

By successive application of the procedure, using all the lines drawn thus far, we can determine the remaining instantaneous centers, which are indicated by chain lines in Fig. 2.7-3b.

Problem 2.7-3

Figure 2.7-4a shows a 10R mechanism with 8 links. Investigate whether all the relative instantaneous centers of velocity for this mechanism at the given configuration can be obtained.

Solution

The total number of relative instantaneous centers for this mechanism is $8 \times 7/2 = 28$, of which 10 are easily located at the R pairs. Figure 2.7-4b shows the corresponding circle diagram, explained in Problem 2.7-2, where the lines corresponding to the 10 R pairs are indicated by solid lines. At this stage we cannot identify any quadrilateral in this figure, i.e., any pair of triangles with a common side yet to be drawn but with all other sides already indicated by solid lines. Consequently, no further relative instantaneous center can be located by the application of the theorem of three centers.[7]

2.8 Fixed and Moving Centrodes

The instantaneous center of velocity of a moving rigid body is, as the name implies, applicable at an instant, i.e., for a given position or configuration of the moving body. As the rigid body moves, the locus

---

[7]It should be mentioned that there exist other 10R-8 link single degree-of-freedom linkages for which all the relative instantaneous centers can be located by applying the theorem of three centers.

2.8. FIXED AND MOVING CENTRODES

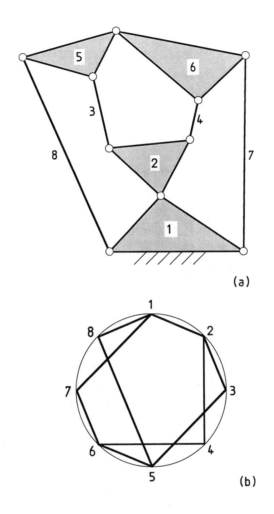

Figure 2.7-4

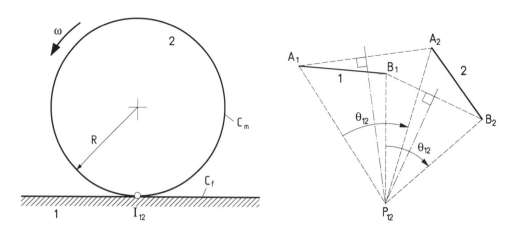

Figure 2.8-1 **Figure 2.8-2**

of the instantaneous center on the fixed reference plane is termed the *fixed centrode*. The locus of the instantaneous center on the moving body is called the *moving centrode*. We shall denote the fixed and moving centrodes by C_f and C_m, respectively. Consider the motion of a circular wheel that rolls without slipping on a fixed, horizontal surface (Fig. 2.8-1). The instantaneous center of velocity I_{12} is always at the point of contact between the wheel and the surface (refer to Section 2.6). Thus, as the wheel moves on, the locus of I_{12} on the fixed body 1 is obviously the line C_f, and that on wheel 2 is the periphery of the wheel, C_m. For better appreciation of the role of centrodes in the kinematic analysis and synthesis of planar linkages, we shall first develop some concepts associated with finite displacements of a rigid body.

Let us consider two finitely separated positions 1 and 2 of a rigid body (Fig. 2.8-2), where the positions of two fixed points on the body are denoted by A_1, A_2 and B_1, B_2, respectively. The movement of the body from position 1 to 2 can be thought of as a pure rotation about a point P_{12}. This is true only where the initial (1) and final (2) positions are concerned, in complete disregard of the intermediate positions. The point P_{12}, located at the intersection of the midnormals of A_1A_2 and B_1B_2, is called the pole. As indicated in Fig. 2.8-2, the finite movement is represented by a rotation of the rigid triangle $P_{12}A_1B_1$ about P_{12} through an angle θ_{12}. When $\theta_{12} \to 0$, the pole P_{12} becomes

2.8. FIXED AND MOVING CENTRODES

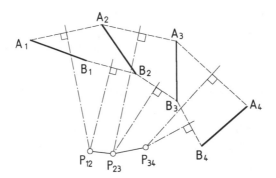

Figure 2.8-3

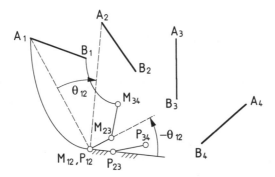

Figure 2.8-4

the instantaneous center of velocity for the position 1. An important use of the pole concept can be explained by considering more than two positions of the moving rigid body (Fig. 2.8-3)[8]. Using the definition of a pole, the poles P_{12}, P_{23}, P_{34} are located corresponding to the movements of the body successively through four positions, viz., 1 to 2, 2 to 3 and 3 to 4. The polygon joining such successive poles $P_{i(i+1)}$ for $i = 1, 2, 3, -, n, -$ is referred to as the *fixed pole polygon*.

Let us identify the point M_{23} (Fig. 2.8-4) in the first position of the moving body which moves into coincidence with P_{23} in the second position. As shown in Fig. 2.8-4, M_{23} can be located by rotating $P_{12}P_{23}$ about P_{12} through an angle $-\theta_{12}$. Similarly, the point M_{34}

---

[8]More discussion on poles in general for a number of finite displacements of a rigid body will be presented in Section 2.11.

denotes the point in the first position of the rigid body, which moves into coincidence with P_{34} in the third position. It can easily be seen that $\triangle A_2B_2P_{23} \cong \triangle A_1B_1M_{23}$ and $\triangle A_3B_3P_{34} \cong \triangle A_1B_1M_{34}$ and so on. The polygon joining the points $M_{12}(\equiv P_{12})$, M_{23}, M_{34} and so on, is referred to as the *moving pole polygon*. To emphasize that the points M_{12}, M_{23}, M_{34}, etc., belong to the first position of the moving body, these points are shown (Fig. 2.8-4) as rigidly connected to A_1B_1. It can now be visualized that by rolling the moving pole polygon over the fixed pole polygon, the rigid body moves through positions 2,3,4, starting from position 1. If we consider a limiting case of the motion discussed above, when all successive rotations, θ_{12}, θ_{23}, etc., are infinitesimally small, then the following statements can be made:

(i) The successive poles P_{12}, P_{23}, etc., move infinitesimally close to each other (indicating the locations of the instantaneous centers of velocity on the fixed plane) and generate a smooth curve (in general) instead of a polygon. This curve is the fixed centrode.

(ii) The points M_{12}, M_{23}, etc., also generally generate a smooth curve, which is nothing more than the moving centrode attached to the moving body.

Since the point of contact between the fixed and moving centrodes is the instantaneous center (with zero velocity), we can conclude that the continuous motion of a rigid body can be represented by rolling (without slipping) the moving centrode over the fixed centrode.[9]

Just as the concept of instantaneous center of velocity, the concept of the fixed and moving centrodes can also be easily extended to the situation of two moving rigid bodies. The two centrodes, one attached to each body, are called the *relative centrodes* with the relative instantaneous center of velocity as their point of contact.

---

[9] A logical extension of the concept of centrodes for spatial motion of a rigid body has been attempted in terms of two ruled surfaces, known as fixed and moving axodes. However, thus far the application of the concept of axodes has met with very little success. For details see Hunt, K.H.: Kinematic Geometry of Mechanisms, Oxford University Press, Oxford, 1978.

2.8. FIXED AND MOVING CENTRODES

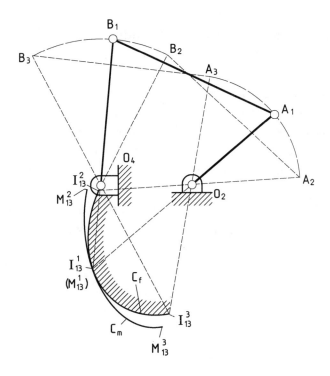

Figure 2.8-5

Problem 2.8-1

A 4R linkage is shown in Fig. 2.8-5. Obtain portions of the fixed and moving centrodes of the coupler around the configuration $O_2 A_1 B_1 O_4$ indicated in this figure.

Solution

Using the theorem of three centers, we locate I_{13} (see Problem 2.7-1) for the given configuration and indicate it as I_{13}^1, where the superscript 1 refers to the first configuration indicated by the subscripts of A and B.

Two more configurations, around the given one, are indicated as 2 and 3 and the corresponding locations of the relative instantaneous center are denoted by I_{13}^2 and I_{13}^3. Considering a number of such configurations (not shown in the figure), we draw a smooth curve passing through the relative instantaneous centers to obtain the fixed centrode, C_f.

The points M_{13}^2 and M_{13}^3 are most conveniently obtained with the help of a transparent paper by using the following congruence relations:

$$\Delta A_2 B_2 I_{13}^2 = \Delta A_1 B_1 M_{13}^2$$

and

$$\Delta A_3 B_3 I_{13}^3 = \Delta A_1 B_1 M_{13}^3$$

For example, mark the three points A_2, B_2 and I_{13}^2 on tracing paper and place the tracing paper on the drawing (Fig. 2.8-5) such that A_2 and B_2 coincide, respectively, with A_1 and B_1. The location of I_{13}^2 then gives M_{13}^2. Similar constructions are done for many more configurations and the smooth curve through $M_{13}^1 (\equiv I_{13}^1), M_{13}^2, M_{13}^3$, gives the moving centrode, C_m, in configuration 1.

Another method of obtaining C_m is to make a kinematic inversion of the linkage with the coupler held fixed in its first position (Fig. 2.8-6). The fixed centrode for this inverted mechanism is the moving centrode C_m obtained in Fig. 2.8-5. The fixed centrode of the mechanism shown in Fig. 2.8-6 is obtained in the manner explained with reference to Fig. 2.8-5.

2.8. FIXED AND MOVING CENTRODES

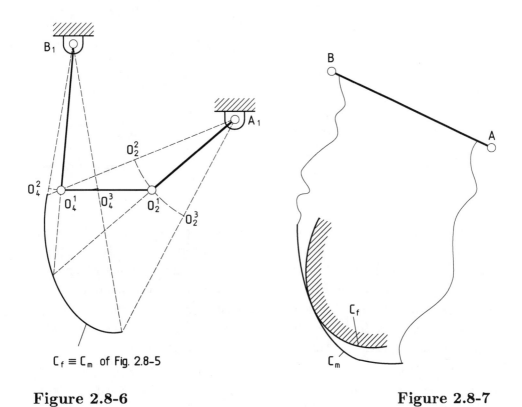

$C_f \equiv C_m$ of Fig. 2.8-5

Figure 2.8-6

Figure 2.8-7

66 CHAPTER 2. PLANAR KINEMATICS OF RIGID BODIES

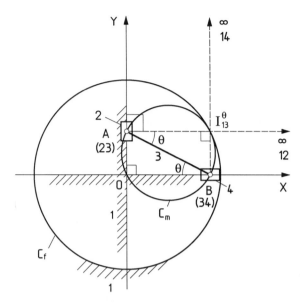

Figure 2.8-8

The moving centrode is rigidly connected to the coupler. The constrained motion of the coupler plane generated by the 4R linkage is exactly duplicated by rolling C_m over C_f in the absence of the 4R linkage, as explained in Fig. 2.8-7. The remarkable property that *the plane motion of one rigid body relative to another is completely equivalent to the rolling motion of their relative centrodes* is useful information for kinematic synthesis of linkages.

Problem 2.8-2

Determine the fixed and moving centrodes for link 3 (of length l) of the elliptic trammel mechanism shown in Fig. 1.8-6.

Solution

The elliptic trammel is drawn again in Fig. 2.8-8, with a chosen XY coordinate axis. The length $AB = l$. The configuration of this constrained mechanism is expressed by the angle θ. For this configuration, the instantaneous center 13 is determined by using the theorem of three centers (See Fig. 2.8-8) and is denoted by I_{13}^θ. It is easy to see that the x and y coordinates of I_{13}^θ are $x_I = l\cos\theta$ and $y_I = l\sin\theta$. Eliminating θ from these two equations, the locus of I_{13} (as θ changes)

2.9. IC VELOCITY

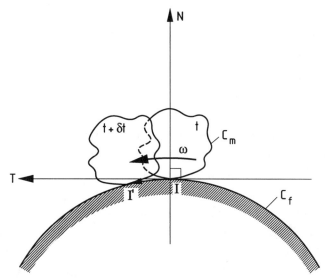

Figure 2.9-1

can be written as

$$x_I^2 + y_I^2 = l^2. \qquad (a)$$

Thus, the locus of I_{13}, i.e., the fixed controde C_f, is obtained as a circle of radius l $(= AB)$ with its center at O.

Since AB always subtends a right angle at I_{13}, the locus of I_{13} in the moving plane of AB is a circle with AB as its diameter. Thus, the moving centrode C_m is a circle with AB as its diameter.[10]

2.9 IC Velocity

When a rigid body moves with an instantaneous angular velocity, ω, then its motion can be equivalently represented by rolling C_m (attached to the body) over C_f with an angular velocity ω. We consider

---

[10]The two circles representing C_f and C_m are known as *Cardan circles*. In Cardan motion, a circle rolls inside another circle of radius twice as large. Any point on the circumference of the rolling circle then generates a straight line, which is a diameter of the fixed circle (degenerate case of a hypocycloid). Note that in Fig. 2.8-8, the points A and B lying on C_m are in fact generating diameters of C_f.

68   CHAPTER 2. PLANAR KINEMATICS OF RIGID BODIES

a time interval, δt, between instants t and $t + \delta t$, during which the point of contact between C_m and C_f (i.e., the instantaneous center of velocity) shifts from I to I' (Fig. 2.9-1). The time rate change of this shift expressed as $\lim_{\delta t \to 0} \frac{II'}{\delta t}$ is called the *IC velocity*, v_I. The direction of IC velocity is along the common tangent (between C'_m and C_f) at I, denoted by IT. The line IT is referred to as the centrode tangent and the line IN (perpendicular to IT) as the centrode normal. The centrode tangent is important in the theory of path curvature, to be developed in subsequent sections.

It should be emphasized that, by definition, the velocity of the instantaneous center is zero. Hence, IC velocity should not be confused as the velocity of the instantaneous center. In fact, IC velocity does not refer to the velocity of any material particle or point; it refers to the time rate of shift of the instantaneous center along the fixed centrode. Sometimes the determination of IC velocity is important just because its direction gives the centrode tangent (without going through the determination of the centrodes).[11]

Let us consider the rolling wheel shown in Fig. 2.8-1. It is obvious that in this example, v_I is toward the left along C_f and has a magnitude ωR, with R as the radius of the wheel. Next, let us consider the motion of the coupler of a 4R linkage discussed in Problem 2.8-1. Figure 2.9-2a shows a particular configuration of the linkage when the angular velocity of the coupler is ω_3. Our objective is to determine v_I.

For determination of v_I, we first "materialize" the instantaneous center I_{13} by the center of the pin P, which connects the two forks rigidly attached to links 2 and 4 (see Fig. 2.9-2a). It is readily seen that as the linkage moves, the pin center always occupies the position of the instantaneous center I and, thus, the path of the pin center is nothing more than the fixed centrode C_f. Therefore, the velocity (v_p) of the pin center (a "material" point) is v_I.[12]

For a prescribed value of ω_3, the IC velocity ($v_I \equiv v_P$) can now be

---

[11] IC velocity is also used to determine the inflection circle to be discussed in Section 2.10.

[12] A reader totally unfamiliar with simple velocity analysis of linkages should skip the rest of this section and return to this point after going through Chapter 5.

2.9. IC VELOCITY

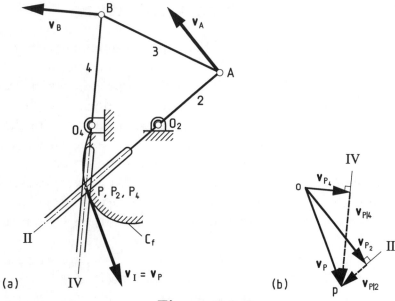

Figure 2.9-2

determined by considering three instantaneously coincident points P, P_2 and P_4, where P_2 is attached to link 2 and P_4 to link 4. The path of P in link 2 is along the fork center-line II and that in link 4 is along the fork center-line IV. The velocity analysis is explained in Figs. 2.9-2a and 2.9-2b, assuming a counterclockwise rotation of coupler 3 with an angular velocity $\omega_3 = 1$ rad/s. The steps required for completing the velocity diagram in Fig. 2.9-2b are as follows:

(i) $v_A = \omega_3 \times PA = \omega_2 \times O_2A$ determines ω_2

(ii) $v_B = \omega_3 \times PB = \omega_4 \times O_4B$ determines ω_4

(iii) $v_P = v_{P_2} + v_P|_2 = \omega_2 \times O_2P_2 + v_P|_2$

(iv) $v_P = v_{P_4} + v_P|_4 = \omega_4 \times O_4P_4 + v_P|_4$

From steps (iii) and (iv) with directions of $v_P|_2$ (along II) and $v_P|_4$ (along IV) being known, v_P is obtained as op in Fig. 2.9-2b.

2.10 Theory of Path Curvature

The constrained motion of a rigid body implies a unique path for each point on the body. The knowledge of instantaneous curvature (applicable for a particular configuration) of such a path is useful for both kinematic analysis and synthesis. For example, during synthesis of a 4R planar mechanism we may need to know which points of the coupler can generate an approximate circular arc (i.e., a path of almost constant radius of curvature), or an approximate straight line (i.e., a path of infinite radius of curvature), around a given configuration. We may note from Equation (2.2.12) that for a moving point P, its instantaneous radius of curvature, ρ_P, may be obtained from the relation $(\boldsymbol{a}_P)^n = (v_P^2/\rho_P)\,\boldsymbol{n}_1$, where v_P and $(\boldsymbol{a}_P)^n$ are, respectively, its speed and normal component of acceleration. For a constrained motion, however, the path and its curvature are decided entirely by the geometric constraints and do not depend intrinsically on the velocity and acceleration of the point; only the rate of traversing the constrained path is governed by the velocity and acceleration. The theory of path curvature and, more specifically, the Euler-Savary equation, provide a direct (geometric) method of determining the radius of curvature of a moving point (of a constrained rigid body) without going through its velocity and acceleration analyses. Keeping in view the applications in the field of mechanisms, most of the theory will be illustrated through the motion of the coupler of a 4R planar linkage (also see Chapter 7). It may be emphasized, however, that the theory of path curvature is equally applicable to any point of a rigid body under constrained plane motion.

2.10.1 The Inflection Circle

Before deriving the radius of path curvature of a specific point (on a constrained rigid body in plane motion), we shall first locate the points having infinite instantaneous radius of path curvature. It will be seen that all such points (for a given configuration) lie on a circle called the *inflection circle*.

Figure 2.10-1 shows the configuration of a moving plane (indicated by number 2) at an instant when its angular velocity and acceleration

2.10. THEORY OF PATH CURVATURE

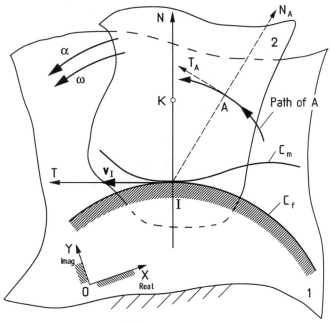

Figure 2.10-1

are ω and α, respectively. Both ω and α are taken positive in CCW direction. The fixed plane is numbered as 1. OXY is a fixed, Cartesian coordinate system. All vectorial quantities will be expressed as complex numbers with the real part denoting the X-component and the imaginary part denoting the Y-component. The motion of plane 2 is completely represented by rolling the moving centrode, C_m, over the fixed centrode, C_f, with angular velocity and acceleration as ω and α, respectively. Consider a point A on the moving plane. In Fig. 2.10-1, I is the instantaneous center of velocity, IT and IN are, respectively, the centrode tangent and centrode normal, AT_A and AN_A are, respectively, the path tangent and path normal (for the point A). Since the angular velocity ω is counterclockwise, we can write the velocity of the point A,

$$v_A = i\omega \boldsymbol{IA}; \quad (i = \sqrt{-1}) \qquad (2.10.1)$$

It should be noted that Equation (2.10.1) is an equation correlating two complex numbers representing the vectors v_A and $\boldsymbol{IA}$. The same (bold-face) symbol is being used for a vector and the complex number representing it. From here foreward, a little effort on the part of the reader will easily resolve this duplicity of symbol. If Equation (2.10.1) is interpreted as a vector equation, then $(i\omega)$ is not a scalar factor. The real number ω is a scalar factor, which leaves the direction of the vector $\mathbf{IA}$ unaltered and only scales its magnitude by a factor $|\omega|$. On the other hand, the factor $i\ (=\sqrt{-1})$ leaves the magnitude of the vector $\mathbf{IA}$ unaltered but rotates it through 90° in the sense of ω.

Differentiating both sides of the above equation with respect to time, t, we obtain the acceleration of the point A,

$$\begin{aligned} \mathbf{a}_A &= i\alpha \mathbf{IA} + i\omega(\frac{d}{dt}(\mathbf{IA})) \\ &= i\alpha \mathbf{IA} + i\omega(\frac{d}{dt}(\mathbf{OA} - \mathbf{OI})) \\ &= i\alpha \mathbf{IA} + i\omega \mathbf{v}_A - i\omega \mathbf{v}_I \end{aligned}$$

(since $\mathbf{v}_A = \frac{d}{dt}(\mathbf{OA})$ and IC velocity $\mathbf{v}_I = \frac{d}{dt}(\mathbf{OI})$).

Using Equation (2.10.1) in the above equation, we get

$$\begin{aligned} \mathbf{a}_A &= i\alpha \mathbf{IA} - \omega^2 \mathbf{IA} - i\omega \mathbf{v}_I \\ &= i\alpha \mathbf{IA} - \omega^2(\mathbf{IA} + i\mathbf{v}_I/\omega) \end{aligned} \quad (2.10.2)$$

Let us introduce a point K, such that $\mathbf{IK} = -i\mathbf{v}_I/\omega$. Since $\mathbf{v}_I$ is along $\mathbf{IT}$, note that $\mathbf{IK}$ must be along $\mathbf{IN}$ (i.e., at 90° clockwise from the direction of $\mathbf{v}_I$). We may now rewrite Equation (2.10.2) as

$$\mathbf{a}_A = i\alpha \mathbf{IA} - \omega^2(\mathbf{IA} - \mathbf{IK}) = i\alpha \mathbf{IA} - \omega^2 \mathbf{KA}. \quad (2.10.3)$$

The first term on the right-hand-side of this equation represents an acceleration component along the path tangent. Therefore, at a given instant if the point A is so chosen that $\angle IAK = 90°$, implying $\mathbf{KA}$ along the path tangent, then $\mathbf{a}_A$ is entirely tangential. In other words, for such a choice of A, the normal component of acceleration of A, $(\mathbf{a}_A)^n$, is 0, i.e., $\rho_A \to \infty$. Since the only condition for $\rho_A \to \infty$ is

2.10. THEORY OF PATH CURVATURE

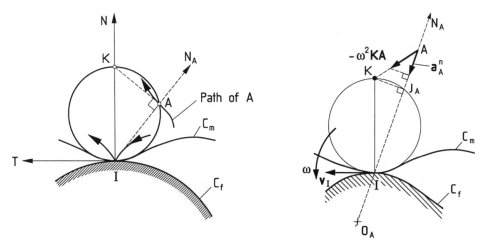

Figure 2.10-2 **Figure 2.10-3**

$\angle IAK = 90°$, all the points lying on the circle on the circle of diameter IK have infinite instantaneous radius of curvature (Fig. 2.10-2). The circle of diameter IK is called the inflection circle and the point K is known as the *inflection pole*.

It should be remarked that for every instant or configuration, there exists a unique inflection circle. Thus, the curvature of the points lying on the inflection circle is infinite only for the configuration under consideration. The paths of these points are not exact straight lines, rather they have an approximately flat segment around this configuration. Further, if a point on the inflection circle is in the vicinity of the instantaneous center, then the flat segment is very small. In fact, at the instantaneous center, the path has a cusp (Fig. 2.10-2). For a point to generate an exact straight line, it must lie on the inflection circles for all configurations as happens in case of a Cardan motion (see Problem 2.10-3).

2.10.2 The Euler-Savary Equation

Referring to Fig. 2.10-3, consider a point (on the moving plane with angular velocity ω) that does not lie on the inflection circle. The positive direction of path normal AN_A is taken along the vector $\boldsymbol{IA}$. It is now known that the normal component of acceleration of the point

74 CHAPTER 2. PLANAR KINEMATICS OF RIGID BODIES

A, $(a_A)^n$, lies along the path normal. The magnitude $|a_A^n| = (v_A^2/AO_A)$ where O_A is the center of path curvature and v_A is the speed of A. If the magnitude and sense (sign) of a_A^n is represented by a real number $a_A^n = v_A^2/AO_A$, then a_A^n is positive when AO_A is positive, i.e., directed in the same sense as IA. It may be noted that in Fig. 2.10-3, AO_A is negative. Further, from Equation (2.10.3)

$$\begin{aligned}(a_A)^n &= \text{Projection of } (-\omega^2 \mathbf{KA}) \text{ along } \mathbf{IA} \\ &= -\omega^2 [\text{Projection of } (\mathbf{KA}) \text{ along } \mathbf{IA}] \\ &= -\omega^2 [\text{Projection of } (\mathbf{IA} - \mathbf{IK}) \text{ along } \mathbf{IA}] \\ &= -\omega^2 (IA - IJ_A) \\ &= -\omega^2 J_A A \end{aligned} \quad (2.10.4)$$

where J_A is the point of intersection of IA and the inflection circle (Fig. 2.10-3).

Since $v_A = \omega IA$, we can also write

$$(a_A)^n = \frac{\omega^2 IA^2}{AO_A}. \quad (2.10.5)$$

Recognizing, $AO_A = -O_A A$, from Equations (2.10.4) and (2.10.5), we finally get the Euler-Savary equation in the form

$$O_A A = \frac{IA^2}{J_A A}. \quad (2.10.6)$$

We shall refer to the above equation as the *first form of the Euler-Savary equation*. It may be emphasized that this equation correlates three directed (algebraic) quantities, viz., IA, $J_A A$ and $O_A A$, where the sign convention mentioned earlier implies $O_A A$ and $J_A A$ must be in the same sense along the line IA. When point A lies outside the inflection circle (Fig. 2.10-3), IA, $J_A A$ and $O_A A$ all have the same sense. On the otherhand, if point A lies within the inflection circle (Fig. 2.10-4), $J_A A$ and $O_A A$ again have the same sense, which is opposite to that of IA, and Equation (2.10.6) still holds good. The reader is advised to note the shift of point O_A as the location of A crosses the inflection circle.

2.10. THEORY OF PATH CURVATURE

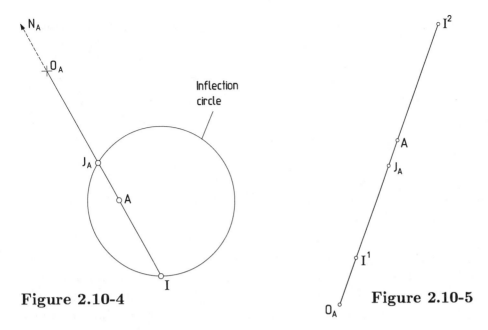

Figure 2.10-4

Figure 2.10-5

By using the Euler-Savary equation, we can obtain one of the four points, I, A, J_A and O_A, on a line when the other three are known. We should note that for given locations of A, J_A and O_A there exist two possible locations of I (as is evident from IA^2 appearing in Equation (2.10.6)). Figure 2.10-5 graphically depicts the two possible solutions as I^1 and I^2. Further, if point A has to be determined with the locations of the other three points given, then Equation (2.10.6) cannot yield the solution. In such a situation, Equation (2.10.6) is recast as explained below.

From Equation (2.10.6), we get

$$O_A A . J_A A = IA^2.$$

Remembering the sign convention, the above equation is written as

$$(IA - IO_A)(IA - IJ_A) = IA^2.$$

On simplification, we obtain

$$\frac{1}{IA} - \frac{1}{IO_A} = \frac{1}{IJ_A}. \qquad (2.10.7)$$

We shall call this equation the *second form of the Euler-Savary equation*. In this form, point A can be determined when the locations of I, J_A and O_A are known.

There is one more useful application of Equation (2.10.7). This is when point A is at infinity. In such a situation, Equation (2.10.6) fails to yield any meaningful information, but Equation (2.10.7) implies that J_A and O_A lie on opposite sides of I and are equidistant from it (see Problem 2.10-2).

The practical application of the Euler-Savary equation is often made via the inflection circle and known curvatures of two points of the given rigid body (see Problem 2.10-1).

The importance of the adopted sign convention for correct application of the Euler-Savary equation has already been explained. Not much difficulty is encountered in following the sign convention when a graphical method is used to solve a particular problem. However, special care is needed to maintain the sign convention while solving a problem numerically, using a computer. Sandor and Erdman[13] suggest the use of vector notation in complex exponential form, as detailed below.

Consider the four collinear vectors along IN_A, namely $\boldsymbol{A} = \boldsymbol{IA}$, $\boldsymbol{J} = \boldsymbol{IJ_A}$, $\boldsymbol{O} = \boldsymbol{IO_A}$ and $\boldsymbol{\rho} = \boldsymbol{O_A A}$, all of which are expressed in the complex exponential form with their arguments measured from a reference line. The Euler-Savary equation can then be put in the following "sign-proof" forms depending on which three points (out of I, A, J_A and O_A) are given and which one is to be determined.

Case (i): Given I, A and J_A, determine O_A:

$$\boldsymbol{\rho} = \frac{A^2}{|\boldsymbol{A} - \boldsymbol{J}|} e^{i \, arg(\boldsymbol{A} - \boldsymbol{J})} \qquad (2.10.8)$$

where "arg" stands for argument.

Case (ii): Given I, O_A and A, determine J_A:

$$\boldsymbol{J} = \boldsymbol{A} - \frac{A^2}{\rho^2}\boldsymbol{\rho}. \qquad (2.10.9)$$

---

[13] Sandor, G.N. and Erdman, A.G.: Advanced Mechanism Design - Analysis and Synthesis, Vol. 2 - Prentice Hall, Englewood Cliffs, New Jersey, 1984.

2.10. THEORY OF PATH CURVATURE

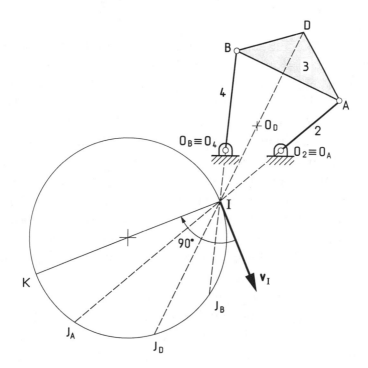

Figure 2.10-6

Case (iii): Given, I, O_A and J_A, determine A:

$$A = \frac{JO}{J+O}. \qquad (2.10.10)$$

Case (iv): Given O_A, J_A and A, determine I:

$$A = \pm \text{Abs}[(|A - J|\rho)^{\frac{1}{2}}] e^{i\, arg(\rho)}. \qquad (2.10.11)$$

The choice of I between the two possible locations given by Equation (2.10.11) is to be determined from other information or requirements.

Problem 2.10-1

A 4R planar linkage is shown in Fig. 2.10-6. Determine the radius of curvature of the path of the coupler point D at the configuration indicated.

Solution

It may be observed that the centers of curvature of two coupler points, viz., A and B are, respectively, at O_2 ($\equiv O_A$) and O_4 ($\equiv O_B$). The instantaneous center of velocity of the coupler is at I. Using Equation (2.10.6), J_A on IA and J_B on IB are located. Note that J_AA and O_AA are in the same sense and likewise, J_BB and O_BB are also in the same sense to satisfy the sign convention. Thus, three points, I, J_A and J_B, all lying on the inflection circle for this given configuration, are located. Draw the inflection circle passing through these points. Extend ID to intersect the inflection circle at J_D. Locate O_D on the line ID so as to satisfy Equation (2.10.6) and the sign convention. Hence, the radius of curvature of the path of point D is O_DD.

We may note that the inflection circle could have been obtained in another way explained below.

Obtain the IC velocity (v_I) with an assumed counterclockwise angular velocity (ω_3) of the coupler as explained in Fig. 2.9-2. Then the diameter of the inflection circle IK can be obtained at $90°$ in the clockwise direction from v_I with $IK = v_I/\omega_3$.[14]

Problem 2.10-2

A 3R-1P mechanism is shown in Fig. 2.10-7. Determine for the given configuration, the radius of curvature of the path of A (a point on link 3 instantaneously coincident with the axis of the R-pair between links 2 and 4) traced out on the moving plane of link 2.

Solution

Since the path of A_3 on link 2 remains invariant under kinematic inversion, we first make a kinematic inversion of the given mechanism (at the prescribed configuration) with link 2 as the fixed link as shown in Fig. 2.10-8a. Thus, we are now required to find out the radius of path curvature of A_3 on the fixed link of this kinematically inverted mechanism. Recalling that a P pair is a limiting case of an R pair (see Section 1.6), the inverted mechanism reduces to a 4R planar linkage with link 3 as the coupler. One R pair (D) of the coupler is at infinity in a direction perpendicular to CO_4 (Fig. 2.10-8b). The instantaneous center of this coupler link is at I, the intersection of O_2C and O_4D. The center of curvature of the path of D is at O_4.

---

[14]For an application of this approach see Problem 5.3-8.

2.10. THEORY OF PATH CURVATURE

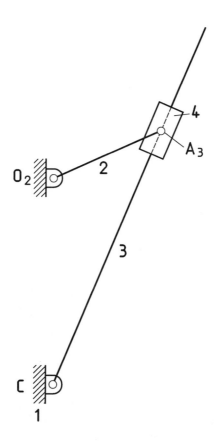

Figure 2.10-7

80 CHAPTER 2. PLANAR KINEMATICS OF RIGID BODIES

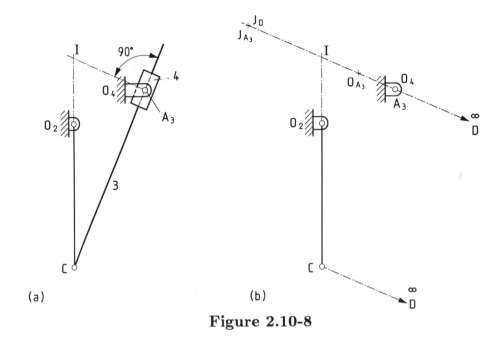

Figure 2.10-8

Hence, using Equation (2.10.7) for the point D, we can locate J_D as shown in Fig. 2.10-8b, with $IJ_D = IO_4$. Now using Equation (2.10.6) for A_3 and keeping in mind $J_D \equiv J_{A_3}$ (since A_3 lies on the line ID), we get

$$O_{A_3}A_3 = \frac{IA_3^2}{J_{A_3}A_3} = \frac{IA_3^2}{J_DO_4} = \frac{IA_3}{2},$$

since

$$IA_3 = \frac{1}{2}J_DO_4.$$

Thus, the center of curvature of the path of A on link 2 is the midpoint of IA_3.

2.10. THEORY OF PATH CURVATURE

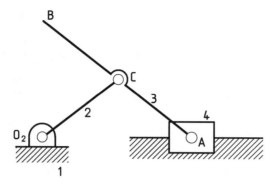

Figure 2.10-9

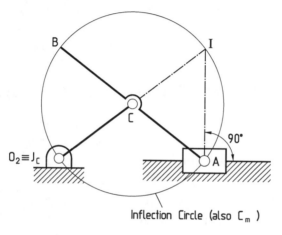

Figure 2.10-10

Problem 2.10-3

Figure 2.10-9 shows a 3R-1P mechanism. Prove that as the mechanism moves, the point B generates an exact straight line if $O_2C = AC = BC$.

Solution

We shall first solve the problem by showing that B always lies on the inflection circle of link 3. Referring to Fig. 2.10-10, the instantaneous center of velocity of link 3 is at I, since the velocity of C is perpendicular to O_2C and that of A is horizontal. Point A must lie on the inflection circle of link 3 because the path of A has infinite radius of curvature. From simple geometry, since $O_2C = AC$ and

$\angle IAO_2 = 90°$, AC must be equal to CI. Applying Equation (2.10.6) for point C, we get $J_C \equiv O_2$. So we get three points, namely I, A and O_2 on the inflection circle. Thus, the inflection circle of link 3 is a circle with its center at C and radius equal to AC. Again from simple geometry, since $AC = BC$, point B must lie on this inflection circle. It is obvious that the above reasoning holds good for all configurations and, hence, B generates an exact straight line.

The reader may verify that for the mechanism shown in Fig. 2.10-9, the inflection circles for all configurations coincide with the moving centrode (C_m) of link 3. Further, we may point out, by referring to Fig. 1.8-6, that the midpoint (C) of link AB in this figure moves on a circle with center at O. Thus, we can add a link OC (with R pairs) and remove link 4 without altering the motion of link 3, i.e., the point B continues to move along a straight line as in Fig. 2.10-9.

2.10.3 The Cubic of Stationary Curvature

For a given configuration, the points of the moving plane (where $\frac{d\rho}{ds} = 0$, with s being measured along the respective paths) lie on a *third order curve* called the *cubic of stationary curvature*, from here forward abbreviated as CSC. We may note that $\frac{d\rho}{ds} = 0$ does not imply a constant radius of curvature. However, at the locations where $\frac{d\rho}{ds} = 0$, we may expect, in general (unless prejudiced by some other considerations) that the curvature changes rather slowly, i.e., the paths of the points on the CSC should approximate circular arcs. In fact, the CSC identifies points where the path will have at least a fourth order contact with the respective osculating circles (i.e., the circles of curvature). It may be further mentioned that just like the inflection circle, the CSC applies to a given configuration (instant of motion). The method of obtaining the CSC is explained below.

A constrained motion of a rigid body is represented by rolling C_m over C_f, as depicted in Fig. 2.10-11. At the instant under consideration, ω is the angular velocity of the moving rigid body, the inflection pole is at K and the instantaneous center of velocity is at I. Thus, the inflection circle is denoted by k. Consider a point A on the moving body at this instant whose center of path curvature is at O_A and the path normal is along $\boldsymbol{AN_A}$. The orientation of $\boldsymbol{AN_A}$ is given by the

2.10. THEORY OF PATH CURVATURE

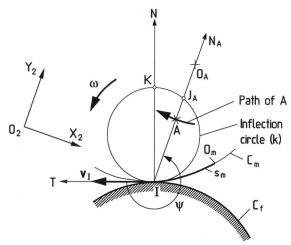

Figure 2.10-11

angle ψ (measured from $\boldsymbol{IT}$ in the counterclockwise direction).

With the inflection circle diameter $IK = D$ (always positive), $IA = r$ and $O_A A = \rho$, we obtain from Fig. 2.10-11

$$J_A A = IA - IJ_A = r + D \sin \psi \tag{2.10.12}$$

and using Equation (2.10.12) in Equation (2.10.6) we can write

$$\rho = \frac{r^2}{r + D \sin \psi}. \tag{2.10.13}$$

Now we introduce a new variable s_m, which defines the position of I along C_m (from some arbitrary origin O_m), such that s_m increases with time (Fig. 2.10-11). Applying chain-rule of differentiation we get

$$\frac{d\rho}{ds} = \frac{d\rho}{ds_m} \frac{ds_m}{ds}. \tag{2.10.14}$$

Since, in general $\frac{ds_m}{ds} \neq 0$, from Equation (2.10.14) $\frac{d\rho}{ds} = 0$ implies $\frac{d\rho}{ds_m} = 0$. Thus, the equation of the CSC can be obtained as $\frac{d\rho}{ds_m} = 0$. Differentiating Equation (2.10.13) with respect to s_m, we get

$$\frac{d\rho}{ds_m} = \frac{2r(r+D\sin\psi)\frac{dr}{ds_m} - r^2(\frac{dr}{ds_m} + D\cos\psi\frac{d\psi}{ds_m} + \sin\psi\frac{dD}{ds_m})}{(r+D\sin\psi)^2}.$$
(2.10.15)

The derivatives $\frac{dr}{ds_m}$ and $\frac{d\psi}{ds_m}$, appearing on the right-hand-side of Equation (2.10.15), are evaluated in a manner explained below. Let the coordinate system $O_2 X_2 Y_2$ be embedded in the moving plane and a unit vector along IT be defined as $\boldsymbol{\sigma} = \boldsymbol{v}_I/|\boldsymbol{v}_I|$. Now the vector IA can be expressed in this coordinate system as

$$IA = re^{i\psi}\boldsymbol{\sigma}.$$

From Fig. 2.10-11, we can now write

$$IA = O_2A - O_2I = re^{i\psi}\boldsymbol{\sigma}.$$

Differentiating the above expression with respect to s_m, we get

$$\frac{d}{ds_m}(O_2A) - \frac{d}{ds_m}(O_2I) = \frac{dr}{ds_m}e^{i\psi}\boldsymbol{\sigma} + re^{i\psi}\frac{d\boldsymbol{\sigma}}{ds_m} + ire^{i\psi}\frac{d\psi}{ds_m}\boldsymbol{\sigma}. \qquad (2.10.16)$$

Since the coordinate system $O_2X_2Y_2$ and the point A are both attached to the moving body, $\frac{dO_2A}{ds_m} = 0$. Further, referring to Fig. 2.10-12, $\frac{d\boldsymbol{\sigma}}{ds_m} = i\boldsymbol{\sigma}/R_m$, where R_m is the radius of curvature of the moving centrode. The following sign convention for R_m has been used while deriving the above relation. If the curvature of C_m is such that the rotation of R_m (with time) is in the same sense of ω, then R_m is taken as positive. Note that in Fig. 2.10-12, R_m is negative. It is obvious to realize from the definition that $\frac{d}{ds_m}(O_2I) = \boldsymbol{\sigma}$. Substituting these three derivatives in Equation (2.10.16), we finally obtain

$$-\boldsymbol{\sigma} = (\frac{dr}{ds_m} + i\frac{r}{R_m} + ir\frac{d\psi}{ds_m})e^{i\psi}\boldsymbol{\sigma}$$

or,

2.10. THEORY OF PATH CURVATURE

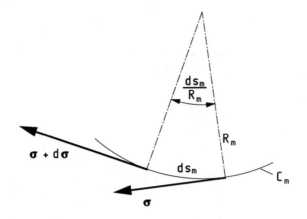

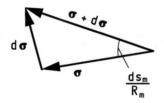

Figure 2.10-12

$$-1 = (\frac{dr}{ds_m} + i\frac{r}{R_m} + ir\frac{d\psi}{ds_m})e^{i\psi}$$

or,

$$-\cos\psi + i\sin\psi = \frac{dr}{ds_m} + i(\frac{r}{R_m} + r\frac{d\psi}{ds_m})$$

Equating real and imaginary parts of both sides of the above equation, we get the desired derivatives as

$$\frac{dr}{ds_m} = -\cos\psi$$

and

$$\frac{d\psi}{ds_m} = \frac{\sin\psi}{r} - \frac{1}{R_m}. \qquad (2.10.17)$$

Using Equation (2.10.17) in Equation (2.10.15) followed by algebraic manipulations, the equation for the CSC is obtained from $\frac{d\rho}{ds_m} = 0$ as

$$r^2 \sin\psi \cos\psi (\frac{1}{M'\sin\psi} + \frac{1}{N'\cos\psi} - \frac{1}{r}) = 0 \qquad (2.10.18)$$

where

$$\frac{1}{M'} = \frac{1}{3}(\frac{1}{R_m} - \frac{1}{D}) \quad \text{and} \quad 1/N' = -\frac{1}{3}\frac{1}{D}\frac{dD}{ds_m}. \qquad (2.10.19)$$

In a Cartesian coordinate system XY, with its origin at I and X axis along IT, Equation (2.10.18) reduces, after substituting $\cos\psi = x/r$ and $\sin\psi = y/r$, to

$$xy[\frac{\sqrt{x^2+y^2}}{M'y} + \frac{\sqrt{x^2+y^2}}{N'x} - \frac{1}{\sqrt{x^2+y^2}}] = 0$$

or,

$$(x^2 + y^2)(\frac{x}{M'} + \frac{y}{N'}) - xy = 0. \qquad (2.10.20)$$

The above equation is a cubic one, justifying the name cubic of stationary curvature.

2.10. THEORY OF PATH CURVATURE

Referring to Equation (2.10.18), we may note that the equation for the CSC in polar form is commonly written as

$$\frac{1}{M'\sin\psi} + \frac{1}{N'\cos\psi} - \frac{1}{r} = 0. \qquad (2.10.21)$$

There are, however, special situations when Equation (2.10.21) fails to yield a branch of CSC, because the term $r^2 \sin\psi \cos\psi$ in Equation (2.10.18) is ignored. For example, with circular centrodes, one can show that $\frac{1}{N'}$ becomes zero[15], when the straight line, $\cos\psi = 0$ (i.e., the Y axis in the Cartesian system), along with the circle $M'\sin\psi = r$, locate the points on the CSC. Note that the straight line is of first degree and the circle is of second degree so that the product still remains cubic.

While using Equation (2.10.21), it is difficult to obtain the values of M' and N' from Equation (2.10.19). These values are obtained by plugging in the coordinates of two points, which are known to be on the CSC, into Equation (2.10.21) (see Problem 2.10-4).

2.10.4 Ball's Point

The points of the moving plane lying on the inflection circle have infinite radius of path curvature, i.e., the circle of curvature for a point on the inflection circle degenerates into a straight line, which is the path tangent. Thus, we can say that the path of such a point coincides with the path tangent for three infinitesimally separated positions. The path of a point on the CSC, on the other hand, coincides with the circle of curvature for four infinitesimally separated positions. Therefore, we can conclude that the path of the point of intersection (which, in general, exists) of the inflection circle and the CSC coincides with the path tangent for four infinitesimally separated positions. This point is generally given the symbol U (undulation point) and is called *Ball's point* (named after its discoverer Sir Robert J. Ball)[16]. The path of a Ball's point is expected to have a flat portion. The curve

---

[15]Sandor, G.N. and Erdman, A.G., Op.cit.

[16]For a detailed discussion on Ball's point and the characteristic of its path, see Hunt., K.H.: Kinematic Geometry of Mechanisms, Oxford University Press, Oxford, 1978.

88 CHAPTER 2. PLANAR KINEMATICS OF RIGID BODIES

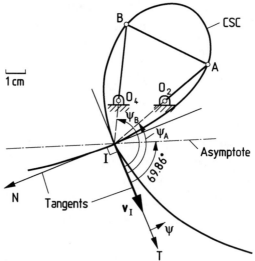

Figure 2.10-13

obtained as the locus of Ball's point, when the moving plane takes up different configurations, is called the Ball's point curve. The reader is advised that for the special case of circular centrodes with $1/N' = 0$ (when the CSC has a straight line branch) the inflection pole K is the Ball's point.

Problem 2.10-4

Obtain the cubic of stationary curvature for the coupler of the 4R planar linkage at the configuration shown in Fig. 2.9-2.

Solution

The linkage configuration, the instantaneous center of velocity of the coupler and the direction of v_I obtained in Fig. 2.9-2, are reproduced in Fig. 2.10-13. We may note that the direction of v_I could have also been obtained by drawing the inflection circle (see Problem 2.10-1), if necessary, to a reduced scale, and identifying its diameter as IK. The direction of v_I is at 90° CCW from $\mathbf{IK}$.

The moving R pairs at A and B are definitely on the CSC since these two points move on circular paths. For the points A and B, from measurements, we get

$$r_A = 6.5 \text{ cm}, \ \psi_A = 106°; \ r_B = 6.28 \text{ cm}, \ \psi_B = 150.5°.$$

2.11. FINITE MOVEMENTS

Using these values in Equation (2.10.21) we obtain the following equations

$$\frac{1}{.9613M'} + \frac{1}{N'}\left(-\frac{1}{.2756}\right) = \frac{1}{6.5}$$

and

$$\frac{1}{.4924M'} + \frac{1}{N'}\left(-\frac{1}{.8704}\right) = \frac{1}{6.28}$$

whose solution yields

$$\frac{1}{M'} = .0649 \quad \text{and} \quad \frac{1}{N'} = -.0238.$$

Hence, the equation for the CSC is finally obtained from Equation (2.10.21) as

$$\frac{.0649}{\sin\psi} - \frac{.0238}{\cos\psi} - \frac{1}{r} = 0. \tag{a}$$

Taking the limit $r \to \infty$ in Equation (a), the equation of the asymptote to the CSC is obtained as $\tan\psi = 2.7269$, i.e., a line at an angle $69.86°$ with the direction of v_I (See Fig. 2.10-13). Since r becomes zero for both $\psi = 90°$ and $\psi = 180°$, there is a self-crossing or double point (called crunode) of the CSC at I. The centrode normal IN and the centrode tangent IT are the two tangents to the CSC at I. Solving for r with various values of ψ from Equation (a) we draw the CSC as shown in Fig. 2.10-13. Negative values of r are obtained for some values of ψ which imply $\psi = \psi + 180°$.

2.11 Finite Movements

In Sections 2.6 - 2.10 we developed various concepts like instantaneous center of velocity, path-radius of curvature of a point and the cubic of stationary curvature. We can say that these concepts, associated with a moving lamina refer, to infinitesimally separated positions of the lamina, e.g., the instantaneous center is connected to two such positions, the radius of path curvature to three such positions, and the

CHAPTER 2. PLANAR KINEMATICS OF RIGID BODIES

CSC to four positions. In this section we shall discuss analogous concepts, viz., pole, pole triangle and opposite-pole quadrilateral, which are connected to finitely separated positions of a moving lamina. It will be seen that poles are associated with two positions of the moving lamina. Likewise, pole triangles are associated with three positions and opposite-pole quadrilaterals are associated with four positions of the moving lamina. The analogy between CSC and the circle-point curve (associated with four positions) will also be discussed. We shall restrict our discussion only to the results that will be used later in the synthesis of planar mechanisms (Chapter 6). For proof of these results and other details, the reader is referred to the classic text of R.S. Hartenberg and J. Denavit: Kinematic Synthesis of Linkages, McGraw-Hill Book Co. (New York), 1964.

2.11.1 Poles

In Section 2.8, we noted that corresponding to two finitely separated positions, say 1 and 2, of a moving lamina, we can define a pole P_{12}, a pure rotation about which moves the lamina from position 1 to position 2. The amount of rotation was indicated by θ_{12}. It was also pointed out (Fig. 2.8-2) that when in the limit $\theta \rightarrow 0$, the pole P_{12} reduces to the instantaneous center of velocity at position 1.

2.11.2 Pole Triangles

Referring to Fig. 2.11-1, let us consider three finitely separated positions 1, 2 and 3 of the moving lamina (M). The moving lamina is represented by the line AB attached to it; the suffixes 1, 2 and 3 refer to the corresponding positions. We can identify three poles, namely P_{12}, P_{23} and P_{13}, associated with the given three positions. A rotation of θ_{12} about P_{12} brings the lamina from position 1 to 2, and a further rotation of θ_{23} about P_{23} moves it from position 2 to 3, whereas a rotation of θ_{13} ($= \theta_{12} + \theta_{23}$) about P_{13} moves the lamina directly from position 1 to 3. The triangle $P_{12}P_{23}P_{13}$, associated with the three positions 1, 2 and 3, is called the *pole triangle*. The pole triangle is drawn on the fixed plane (F). It can be shown that the angle (of the pole triangle) at the vertex P_{ij} is either equal to or differs by 180°

2.11. FINITE MOVEMENTS

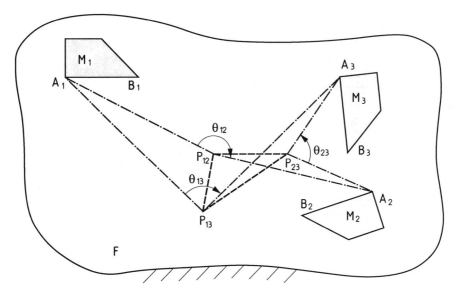

Figure 2.11-1

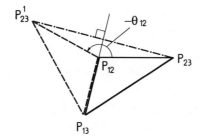

Figure 2.11-2

from $\theta_{ij}/2$, where θ_{ij} denotes the rotation of M as it moves from the i-th to the j-th position (see Exercise Problem 2.13). Thus, the pole triangle completely defines two successive movements.

2.11.3 Image Poles

The pole triangle drawn in F defines two successive movements of M with respect to F. Sometimes it is necessary to consider an inverse situation, i.e., to consider M fixed and F moving, in order to describe

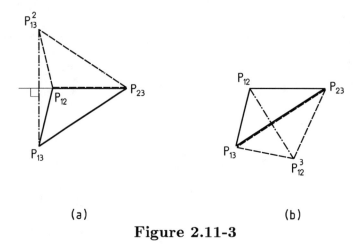

Figure 2.11-3

the same relative movements between M and F. When F moves, the pole triangle $P_{12}P_{23}P_{13}$ moves along with it. If we want to draw the pole triangle in M in its first position (i.e., in M_1) then P_{12} and P_{13} stay where they are in F (because both these poles involve the first position M_1) but P_{23} moves to, say, P_{23}^1, which is called the *image pole* of P_{23} (see Fig. 2.11-2). Recalling the relative movement between F and M, it is obvious that P_{23}^1 is obtained by rotating the line P_{12} P_{23} about P_{12} through an angle $-\theta_{12}$. Using the relationship between the angles of a pole triangle and the rotations described by the pole triangle, it can be shown that P_{23}^1 is the mirror image of P_{23}, with the mirror placed along P_{12} P_{13}. The pole triangle $P_{12}P_{23}^1P_{13}$ drawn in M_1 defines the same two relative movements between M and F as described by $P_{12}P_{23}P_{13}$ drawn in F.

The pole triangles drawn in M_2 and M_3 can now be obtained by rotating $P_{12}P_{23}^1P_{13}$, respectively, about P_{12} through θ_{12} and about P_{13} through θ_{13}. The pole triangle in M_2 is obtained as $P_{12}P_{23}P_{13}^2$ and in M_3 as $P_{12}^3P_{23}P_{13}$, as shown in Figs. 2.11-3a and 2.11.3b. The pole with a superscript indicates image pole, with the superscript identifying the configuration of the moving lamina.

2.11. FINITE MOVEMENTS

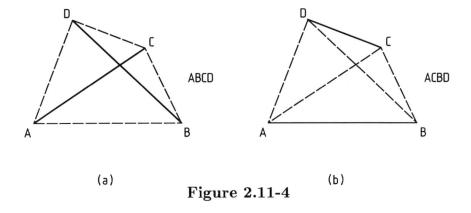

Figure 2.11-4

2.11.4 Opposite-pole Quadrilateral

Let us now consider three successive movements of the lamina M, i.e., four finitely separated configurations M_1, M_2, M_3 and M_4. We can define six poles, viz., P_{12}, P_{23}, P_{34}, P_{13}, P_{14} and P_{24}, associated with these three finite movements. *Opposite poles* are defined as a pair of poles carrying different subscripts. Thus, there are three pairs of opposite poles, as listed below:

$$(P_{12}, P_{34}) \quad (P_{23}, P_{14}) \quad (P_{13}, P_{24})$$

An *opposite-pole quadrilateral* is defined as one whose diagonals connect two pairs of opposite poles. The definitions of the sides and diagonals to be used in the present context warrant special attention because these are quite different from those we are familiar with in plane geometry.

The sides and diagonals of a quadrilateral are defined depending only on the order in which the four vertices of the quadrilateral are named (and with complete disregard to the actual geometric shape of the quadrilateral). If a quadrilateral is named $ABCD$ (Fig. 2.11-4a), then its diagonals are AC and BD, whereas if the quadrilateral with the same vertices is named $ACBD$ (Fig. 2.11-4b), then its diagonals are AB and CD. Thus, the lines joining alternate vertices are called diagonals. The lines joining the two vertices in any other order are called the sides of the quadrilateral. The sides are referred to as adjacent if they have one vertex in common and as opposite if they

94   CHAPTER 2. PLANAR KINEMATICS OF RIGID BODIES

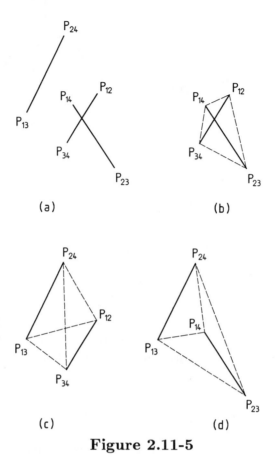

Figure 2.11-5

do not have any common vertex.

With the definitions of diagonals given above, we can construct three opposite-pole quadrilaterals using the six poles. These are

$$(P_{12}P_{23}P_{34}P_{14}), \ (P_{12}P_{13}P_{34}P_{24}) \ and \ (P_{23}P_{13}P_{14}P_{24}).$$

For the six poles shown in Fig. 2.11-5a, (where each pair of opposite poles are connected by a line) the three opposite-pole quadrilaterals are shown in Figs. 2.11-5b to 2.11-5d. In Figs. 2.11-4 and 2.11-5, the diagonals are indicated by solid lines and the sides by dashed lines. In Fig. 2.11-5b, the two pairs of opposite sides are $(P_{12}P_{23}, P_{14}P_{34})$ and $(P_{23}P_{34}, P_{12}P_{14})$. Note the order of the subscripts in each pair of opposite sides. The common subscripts, in a pair, e.g., 1 and 3 in

2.11. FINITE MOVEMENTS 95

the first and 2 and 4 in the second, occur in same sequence. In other words, we should not write the first pair of opposite sides as $P_{12}\ P_{23}$ and $P_{34}\ P_{14}$, whereas we can write it as $P_{23}\ P_{12}$ and $P_{34}\ P_{14}$. The identification of opposite sides in Figs. 2.11-5c and 2.11-5d is left as an exercise for the reader.

2.11.5 Circle-Point Curve and Center-Point Curve

If we consider any three positions of a moving lamina (say M_1, M_2 and M_3), then any point A on it takes up three positions (say A_1, A_2 and A_3). Obviously, it is always possible to construct a circle through A_1, A_2 and A_3. In the special situation where A_1, A_2 and A_3 are collinear, we can consider the straight line joining these points as a degenerate case of a circle with infinite radius. A question may be raised as to what happens if we consider a fourth position of the lamina (say M_4) when A moves to A_4. For given M_1, M_2, M_3 and M_4 and with an arbitrary choice of A we do not expect A_1, A_2, A_3 and A_4 to lie on a circle. But are there any special locations of A for which A_1, A_2, A_3 and A_4 would lie on a circle? The answer to this question is yes; there exists a unique (for specified M_1, M_2, M_3 and M_4) curve on M, the points of which occupy positions lying on a circle as M takes up the specified four positions. This curve is called the *circle-point curve* and the points on this curve are referred to as *circle points*. The center of the circle, on which a circle point lies in its four positions, is called a *center point*. Thus a center point is obviously a point on the fixed plane F. The curve obtained in F by joining the center points corresponding to all the circle points is called the *center-point curve*. It may be worthwhile to emphasize that the circle-point curve (drawn in M) and the corresponding center-point curve (drawn in F) are unique for the three specified successive movements of M with respect to F. Furthermore, the roles of these two curves are interchanged if the same relative movements (between M and F) are maintained in an inverse situation, i.e., by moving F and keeping M fixed at one of the configurations where the circle-point curve has been drawn.

A little thought would convince us that the poles P_{12}, P_{13} and P_{14} must be on the circle-point curve drawn in M_1. This is so because P_{12} in M_2 does not move as M goes from M_1 to M_2. Thus, four positions

96   CHAPTER 2.   PLANAR KINEMATICS OF RIGID BODIES

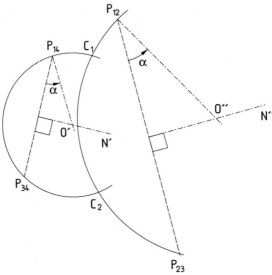

Figure 2.11-6

of P_{12} (as M moves through M_1 to M_4 via M_2 and M_3) essentially reduce to three different locations through which a circle can always be drawn. Similar arguments hold good for P_{13} (as M goes from M_1 to M_3) and P_{14} (as M goes from M_1 to M_4). The image poles P_{23}^1, P_{24}^1 and P_{34}^1 also lie on the circle-point curve drawn in M_1 since only three different locations exist for each of these points, as M takes up four specified configurations. For example, P_{23}^1 does not move as M moves from M_2 to M_3. Similarly, P_{24}^1 does not move when M moves from M_2 to M_4 and P_{34}^1 does not move when M moves from M_3 to M_4. Recalling the analogous roles of the circle-point curve and the center-point curve and following the arguments presented in this section, it is easy to see that the center-point curve passes through all six poles associated with the three successive movements.

The method of locating a center point (other than the poles) uses the following theorem:

Two opposite sides of any opposite-pole quadrilateral subtend angles at a center point that are either equal or differ by 180°. Conversely, at any point where two opposite sides of any opposite-pole quadrilateral subtend angles that are either equal or differ by 180° is a center point. Use of the above theorem for constructing two center

2.11. FINITE MOVEMENTS

points is illustrated in Fig. 2.11-6[17] with the pair of opposite sides P_{12} P_{23} and P_{14} P_{34}. In this figure, N' and N'' are the midnormals of the two opposite sides. A point O' is arbitrarily chosen on N', then the point O'' is located on N'' such that $\angle O'' P_{12} P_{23} = \angle O' P_{14} P_{34} = $ (say α). Two circles are drawn with O' and O'' as centers and $O' P_{14}$ and $O'' P_{12}$ as their respective radii. The points of intersection of these two circles locate two center points C_1 and C_2. The angles subtended by the two opposite sides at C_1 are as follows:

$$\angle P_{14} C_1 P_{34} = (\frac{\pi}{2} - \alpha) \text{ CCW}$$

$$\angle P_{12} C_1 P_{23} = (\alpha + \frac{\pi}{2}) \text{ CW} = [(\frac{\pi}{2} - \alpha) - \pi] \text{ CCW}$$

Therefore, these two angles differ by π as required for C_1 to be a center point.

Similar constructions can be carried out with different values of α in the range $-90° \leq \alpha \leq 90°$, and also with any pair of opposite sides (of any opposite-pole quadrilateral). For obtaining the circle points in M_1, we have to consider the six poles as P_{12}, P_{23}^1, P_{34}^1, P_{13}, P_{14} and P_{24}^1 and then proceed just as in the case of the center points (see Problem 2.11-1).

To each circle point, A_m, there exists a unique center point C_m, which can be determined from another result that states the angle subtended by the line A_m C_m at the pole P_{ij} is either equal to, or differs by 180° from, $\theta_{ij}/2$, where θ_{ij} is the rotation of M from the i-th to the j-th position (see Problem 2.11-1).

Finally, we may note that the circle-point curve relates to four finitely separated positions exactly the same way as the CSC relates to four infinitesimally separated positions. In fact, the center-point curve in general (i.e., the circle-point curve on the fixed plane F), like the CSC, drawn on F is a third order curve (see Problem 2.11-2). In some situations, the center-point curve splits into two branches, one extending to infinity and the other being closed. Under special

---

[17]For computational method of obtaining center points using the same theorem, See Luck, K. and Modler, K.H.: Konstruktionslehre der Getriebe, Akademie-Verlag, Berlin (1990).

98   CHAPTER 2. PLANAR KINEMATICS OF RIGID BODIES

conditions the center-point curve again like the CSC, splits into two branches, one of which is a straight line and the other a circle (second order curve) (see Problem 2.11-3).

Problem 2.11-1

Six poles, corresponding to three successive movements of a lamina, are shown in Fig. 2.11-7a. The magnitudes of two rotations are known to be $\theta_{12} = 48°$ (CW) and $\theta_{13} = 116°$ (CW). Obtain a point A on the circle-point curve in M_1. Locate the corresponding center-point C and determine the magnitude of rotation θ_{34}.

Solution

Referring to Fig. 2.11-7b, the image pole P_{23}^1 is the mirror image of P_{23}, with the mirror placed along $P_{12} P_{13}$. Similarly, P_{24}^1 is the mirror image of P_{24}, with the mirror placed along $P_{12} P_{14}$ and P_{34}^1 is the mirror image of P_{34}, with the mirror placed along $P_{13} P_{14}$. Referring to Fig. 2.11-7c, a circle point A is obtained by the method explained in Fig. 2.11-6. The only difference between these two figures is that in Fig. 2.11-7c the opposite-pole quadrilateral (involving image poles) has the pair of opposite sides as $P_{12} P_{23}^1$ and $P_{14} P_{34}^1$. The determination of the center point C corresponding to the circle point A is explained in Fig. 2.11-7d. To locate C, we draw two lines l_1 and l_2 such that l_1 is at an angle $\theta_{12}/2$ (CW) from AP_{12} and l_2 is at an angle $\theta_{13}/2$ (CW) from AP_{13}. The point of intersection of l_1 and l_2 determines the center point (C) corresponding to the circle point A. Repeating the constructions explained in Figs. 2.11-7c and 2.11-7d with different values of α and using different opposite-pole quadrilaterals we can construct the circle- point curve and the center-point curve shown in Fig. 2.11-7d.

Now $\angle AP_{14}C = \theta_{14}/2$, so θ_{14} is obtained as 84° (CW). Thus, $\theta_{34} = \theta_{14} - \theta_{13} = [84° - 116°](CW) = 32°$ (CCW), which can also be seen as double the angle $\angle P_{13}P_{34}P_{14}$ (Fig.2.11-7a).

Problem 2.11-2

Using the theorem regarding the angles subtended by the pair of opposite sides of an opposite-pole quadrilateral at the center point, show that the center-point curve is a cubic.

Solution

Referring to Fig. 2.11-8, let us consider the opposite-pole quadrilateral designated by $P_{12}P_{23}P_{34}P_{14}$, where a pair of opposite sides are

2.11. FINITE MOVEMENTS

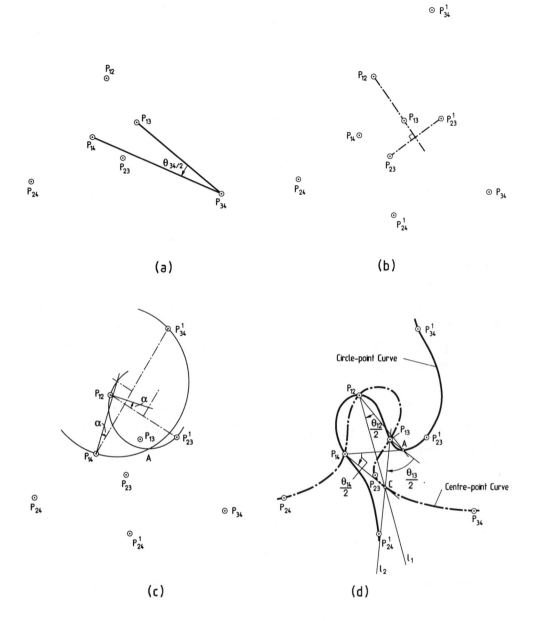

Figure 2.11-7

100   CHAPTER 2. PLANAR KINEMATICS OF RIGID BODIES

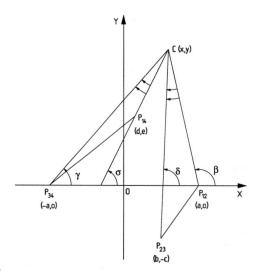

Figure 2.11-8

$P_{12}P_{23}$ and $P_{14}P_{34}$. The X-axis of a Cartesian coordinate system fixed in F is taken along $P_{34}P_{12}$ with the origin of the coordinate system at the midpoint of $P_{12}P_{34}$. In this coordinate system, let the coordinates of the four poles be

$$P_{12} : (a, o),\ P_{34} : (-a, o),\ P_{23} : (b, -c),\ P_{14} : (d, e).$$

The coordinates of a center point C are denoted by (x, y). Since the concerned theorem suggests that either

$$\angle P_{12}CP_{23} = \angle P_{14}CP_{34} \qquad (a)$$

or

$$\angle P_{12}CP_{23} = 180° + \angle P_{14}CP_{34} \qquad (b)$$

we can write the condition for C to be a center point as

$$\tan \angle P_{12}CP_{23} = \tan \angle P_{14}CP_{34} \qquad (c)$$

which takes care of both conditions (a) and (b). From Fig. 2.11-8, condition (c) can be rewritten as

$$\tan(\beta - \delta) = \tan(\sigma - \gamma) \qquad (d)$$

2.11. FINITE MOVEMENTS

where

$$\tan\delta = \frac{y+c}{x-b}, \tan\beta = \frac{y}{x-a}$$

$$\tan\sigma = \frac{y-e}{x-d}, \quad \text{and} \quad \tan\gamma = \frac{y}{x+a}.$$

From Equation (d) we get

$$\frac{\tan\beta - \tan\delta}{1 + \tan\beta \tan\delta} = \frac{\tan\sigma - \tan\gamma}{1 + \tan\sigma \tan\gamma}.$$

Substituting for tangents of various angles in the above equation and simplifying, we obtain

$$\frac{y(x-b) - (y+c)(x-a)}{(x-a)(x-b) + y(y+c)} = \frac{(y-e)(x+a) - y(x-d)}{(x+a)(x-d) + y(y-e)}$$

or,

$$\frac{y(a-b) - c(x-a)}{(x-a)(x-b) + y(y+c)} = \frac{y(a+d) - e(x+a)}{(x+a)(x-d) + y(y-e)}$$

or,

$$(x^2 - y^2 - a^2)((c-e)x - (b-d)y + be + cd) + 2xyg = 0 \quad (e)$$

where,

$$g = (b-d)x + (c-e)y - (a^2 + ce - bd)$$

Equation (e) is a cubic equation, since the highest power of the polynomial involved is three.

Problem 2.11-3

Three special situations for the locations of various poles and image poles are indicated in Figs. 2.11-9a to 2.11-9c. Verify that, for Figs. 2.11-9a and 2.11-9b the center-point curve splits into a straight

102   CHAPTER 2. PLANAR KINEMATICS OF RIGID BODIES

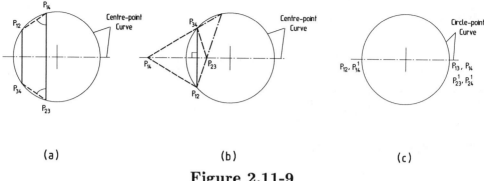

Figure 2.11-9

line and a circle and for Fig. 2.11-9c the circle-point curve splits into a straight line and a circle.[18]

Solution

In Fig. 2.11-9a, four poles, viz., $P_{12}, P_{23}, P_{34}, P_{14}$ constitute a trapezium with a pair of equal angles, i.e., $\angle P_{12}P_{14}\ P_{23} = \angle P_{14}P_{23}P_{34}$. Hence, the circumcircle of this trapezium and the midnormal of $P_{12}P_{34}$ and $P_{14}P_{23}$ (a diameter of the circumcircle) are the center-point curves. This is so because all the points on the circumcircle and this diameter subtend equal angles at the pair of opposite sides.

In Fig. 2.11-9b, P_{12} and P_{34} are located symmetrically on a line perpendicular to the line $P_{14}P_{23}$. Here again, the center-point curve splits into a circle (passing through P_{12}, P_{34} and the points of intersection of the opposite sides of the opposite-pole quadrilateral $P_{12}\ P_{23}\ P_{34}\ P_{14}$) and a straight line (joining P_{14} and P_{23}). Verification of this fact is left for the reader to complete.

In Fig. 2.11-9c, some of the poles and image poles are coincident. This situation occurs when M_1 and M_2 are infinitesimally separated. So are M_3 and M_4, whereas M_2 and M_3 are finitely separated. It is easy to verify that the circle-point curve splits into the circle and straight line as indicated in the figure.

---

[18] For many more such special situations, see Getriebetechnik Lehrbuch - J. Volmer Autorenkollektiv - VEB Verlag Technik Berlin, 1969, p. 468.

2.11.6 Burmester Points

We have already noted that all the ∞^2 points (since there are ∞ lines on a plane and ∞ points on each line) of the moving plane M occupy positions lying on a circle when M occupies three arbitrary positions. When four arbitrary positions of M are considered, only ∞^1 points lying on the circle-point curve (corresponding to these four positions) occupy positions lying on a circle. Extending this trend, while considering five arbitrary positions of M (i.e., four successive movements of M), it is expected that only ∞^0, i.e., a finite number of points on M, occupy positions lying on a circle. These points are called *Burmester points*. Obviously, if a sixth position of M is added, we conclude that, in general, no point (∞^{-1} points) on M occupies positions through which a circle can be drawn.

The determination of Burmester points by geometric methods is not easy[19]. Consider two sets of four arbitrary positions, such as 1, 2, 3, 4 and 1, 2, 3, 5. The circle-point curves for these two sets are two cubic curves and at most can have 9 points of intersection (not necessarily all real). Three of these points of intersection are obviously P_{12}, P_{13} and P_{23}^1, which lie on both circle-point curves (see Exercise Problem 2.16). Out of the remaining six, two are always imaginary. Thus, the number of real intersection points, which are the Burmester points, can be zero, two or four (since imaginary roots occur in pairs).

2.12 Exercise Problems

2.1 A slider-crank mechanism is shown in Fig. 2.12-1a. The absolute velocity and acceleration vectors for the end-points A and B of the connecting rod are indicated in Figs. 2.12-1b and 2.12-1c, respectively, for the configuration shown in Fig. 2.12-1a. Determine the absolute velocity and acceleration of the midpoint C of the connecting rod AB. Also determine the magnitude and direction of the angular velocity and angular acceleration of the connecting rod at this instant.

---

[19] For an analytic approach, See Freudenstein, F. and Sandor, G.N.: On the Burmester Points of a Plane, J. of Appl. Mech. 28, Trans. ASME (1961) pp.41-49.

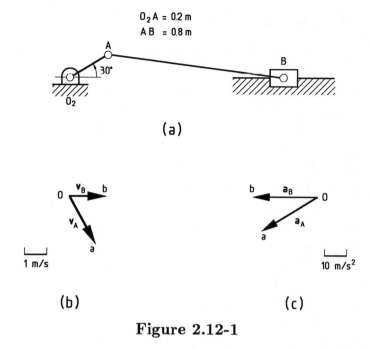

Figure 2.12-1

2.2 Locate the points on the plane of the connecting rod of Problem 2.1 that, at the instant under consideration, have (i) zero velocity and (ii) zero acceleration, respectively.

2.3 In the mechanism shown in Fig. 2.12-2, the point P moves in the circular slot in link 3 as shown. Considering two instantaneously coincident points P_2 and P_3, determine the angular velocity and angular acceleration of link 3 at the given configuration, if the disc 2 rotates with constant angular velocity of 30 rad/s (CW).

2.4 Determine all the relative instantaneous centers for the six-link mechanism shown in Fig. 2.12-3.

2.5 Figure 2.12-4 shows a mechanism where links 4 and 5 are always kept in contact by an external force. It is desired to have a pure rolling contact between 4 and 5 at the instant shown. To realize this, determine the location along AB of the R pair between links 1 and 2.

2.6 Figure 2.12-5 shows schematically a planar approximation of the

2.12. EXERCISE PROBLEMS

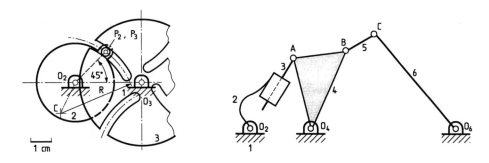

Figure 2.12-2

Figure 2.12-3

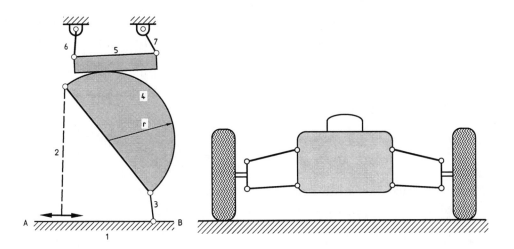

Figure 2.12-4

Figure 2.12-5

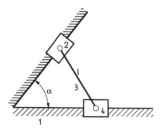

Figure 2.12-6

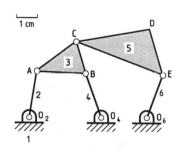

Figure 2.12-7

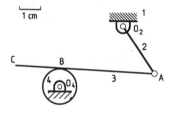

Figure 2.12-8

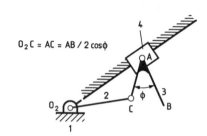

Figure 2.12-9

front suspension of an automobile. The roll center refers to the point about which the body of the automobile seems to rotate with respect to the road. Assuming no slip between the tires and the road, determine the location of the roll center.

2.7 Determine the fixed and moving centrodes of link 3 of the mechanism shown in Fig. 2.12-6. Note that the centrodes turn out to be Cardan circles.

2.8 Locate the center of curvature of the path of the point D of link 5 at the configuration shown in Fig. 2.12-7.

2.9 Determine the radius of curvature of the path of the point C on link 3, at the position indicated in Fig. 2.12-8.

2.10 Figure 2.12-9 shows a 3R-1P mechanism. Prove that as this mechanism moves, the point B generates an exact straight line if $O_2C = AC = AB/(2\cos\phi)$.

2.11 Show that Bobillier construction, described below with reference to Fig. 2.12-10, satisfies the Euler-Savary equation:

2.12. EXERCISE PROBLEMS

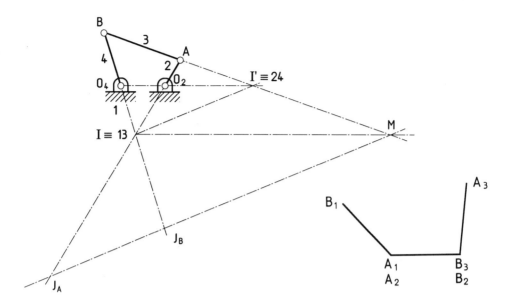

Figure 2.12-10   **Figure 2.12-11**

(a) Locate the relative instantaneous centers 13 and 24, respectively at I and I' for the 4R linkage O_2ABO_4.

(b) Draw IM, parallel to $I'O_2O_4$, to intersect $I'AB$ at M.

(c) The line drawn through M and parallel to II' intersects O_2A and O_2B at J_A and J_B, respectively.

2.12 The line II' in Fig. 2.12-10 is called the collineation axis. Prove that for the 4R linkage the ratio ω_2/ω_4 reaches an extremum value when II' is perpendicular to the coupler AB. This is known as Freudenstein's Theorem.

2.13 Three positions of a moving lamina are indicated in Fig. 2.12-11 by three positions of a line (AB) on it, which are indicated as A_iB_i, with $i = 1, 2, 3$. Locate the pole triangle describing the two successive movements of the lamina. Show that the angles at the vertex P_{ij} of the pole triangle is either $\theta_{ij}/2$ or $\pi + \theta_{i,j}/2$, where $\theta_{i}j$ denotes the rotation from i-th to j-th position.

2.14 Four poles are indicated in Fig. 2.12-12. Three points, C_1, C_2

108 CHAPTER 2. PLANAR KINEMATICS OF RIGID BODIES

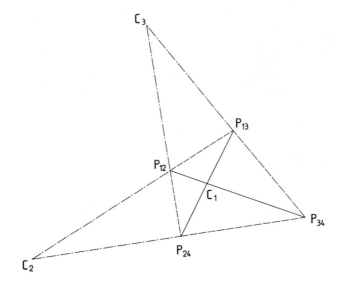

Figure 2.12-12

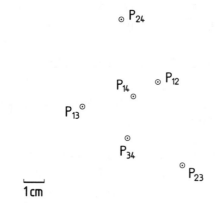

Figure 2.12-13

2.12. EXERCISE PROBLEMS

and C_3, are located as shown in this figure. Which of these are center points? Justify your answer.

2.15 Six poles corresponding to four positions of a moving plane are indicated in Fig. 2.12-13. Obtain at least 20 circle points and 20 center points corresponding to these movements.

2.16 Though the circle-point curves for two sets of four positions (1, 2, 3, 4 and 1, 2, 3, 5) intersect at P_{12}, P_{13} and P_{23}^1, these points cannot be considered as Burmester points. Justify this statement.

Chapter 3

MOBILITY AND RANGE OF MOVEMENT

We have already stated that an arbitrary number of rigid bodies connected through a number of kinematic pairs in an arbitrary fashion does not give rise to a mechanism. In this Chapter, we will discuss some requirements and criteria that must be met in order for an assembly of rigid bodies to serve as a useful mechanism. First we shall discuss some mobility criteria, which are governed solely by the number of links and kinematic pairs. Later, we shall see how the relative link-lengths govern the possible ranges of movements of various links in simple four-link mechanisms.

3.1 Kutzbach Equation and Grübler Criterion

In Section 1.3 we defined the degrees-of-freedom of a mechanism as the number of independent pair variables required to completely specify the relative movements between various links of the mechanism. It is generally possible to determine the degrees-of-freedom of a mechanism by counting the number of links and the number of types of kinematic pairs constituting the mechanism. Certain exceptions to this rule will be discussed in Section 3.3.

In order to develop the relationship between the degrees-of-freedom of a planar mechanism and its number of links and kinematic pairs, let n be the number of links, out of which one is fixed, and let j be the number of simple R pairs (i.e., those that connect only two links). An unconnected rigid body in plane motion has three degrees-of-freedom; two of these represent translatory motions (in the plane of motion) and the third represents a rotation about an axis normal to the plane of motion. So, when $(n-1)$ links of the mechanism move, in the absence of any connection, the total degrees-of-freedom are $3(n-1)$. Once we connect two links by an R pair, there cannot be any relative translation between them. Thus, two degrees-of-freedom (translational) are lost at every R pair and only one degree-of-freedom (rotational) remains. Hence, the number of degrees-of-freedom of the mechanism

$$F = 3(n-1) - 2j. \qquad (3.1.1)$$

The above equation is known as *Kutzbach equation* .

If $F = 1$, the mechanism is said to be of single degree-of-freedom, where only one input motion is required to generate a unique output motion. With $F = 2$, the mechanism needs two independent input motions to specify a unique output motion, and the mechanism is referred to as a two degrees-of-freedom mechanism. If $F = 0$, then the assembly is referred to as a structure where no relative movement between its various members can take place. Finally, $F = -1$ or less implies redundant constraints giving rise to what are known as statically indeterminate structures. Most of the mechanisms used in machinery are of single degree-of-freedom; for this condition, putting $F = 1$ in Equation (3.1.1), we have

$$2j - 3n + 4 = 0. \qquad (3.1.2)$$

The simple estimate of constrained movement expressed by Equation (3.1.2) is known as *Grübler criterion* for plane mechanisms.

To have a closed chain with simple R pairs, we must have a minimum of three links with three R pairs (Fig. 3.1-1). Using Equation (3.1.1), we get

$$F = 3 \times (3-1) - 2 \times 3 = 0.$$

3.1. KUTZBACH EQUATION AND GRÜBLER CRITERION

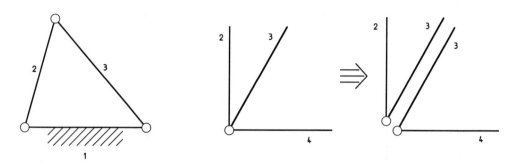

Figure 3.1-1 **Figure 3.1-2**

It can be easily seen from Fig. 3.1-1 that there cannot be any relative movement between the links and the assembly is really a structure.

As noted in Section 1.6 a prismatic (P) pair having one degree-of-freedom is a limiting case of an R pair. Hence, j in Equations (3.1.1) and (3.1.2) can also include prismatic pairs. If a mechanism includes higher order R pairs, j in Equations (3.1.1) and (3.1.2) should be replaced by

$$j = j_1 + 2j_2 + 3j_3 + \ldots + ij_i \qquad (3.1.3)$$

where j_i is the number of R pairs, each of which connects $(i+1)$ links (i.e., j_1 is the number of simple R pairs and so on). Equation (3.1.3) is justified as follows. Figure 3.1-2 explains why an R pair in the category j_2 is equivalent to two simple R pairs. Similarly, an R pair in the category j_i is equivalent to i numbers of simple R pairs. A higher pair has two degrees-of-freedom (one translational and one rotational) and curtails one translational degree-of-freedom (along the common normal at the point of contact). Following the same argument as before, the degrees-of-freedom of a mechanism having higher pairs can be written as

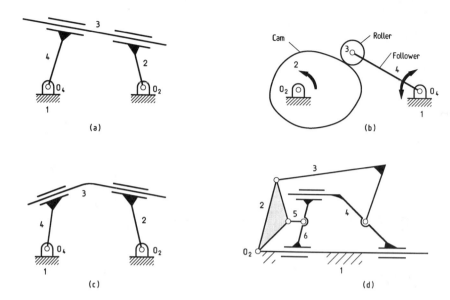

Figure 3.1-3

$$F = 3(n-1) - 2j - h \qquad (3.1.4)$$

where h is the number of higher pairs. We can also use Equation (3.1.1) after replacing each higher pair with an additional link and two R pairs (see Section 1.9). If a higher pair contact prevents slipping due to high friction, i.e., maintains a rolling contact, then such a higher pair curtails two degrees-of-freedom and should be included in j_1.

Quite often, one or more links of a mechanism may have a redundant degree-of-freedom. If a link can be moved without causing any movement in the rest of the mechanism, then the link is said to have a redundant degree-of-freedom. For example, consider the mechanisms shown in Figs. 3.1-3a and 3.1-3b. Link 3 in Fig. 3.1-3a and roller 3 in Fig. 3.1-3b can, respectively, slide and rotate without causing any other movement. Thus, each represents one redundant degree-of-freedom. The effective degree-of-freedom of a mechanism can be expressed as

$$F_e = 3(n-1) - 2j - h - F_r \qquad (3.1.5)$$

where F_r is the number of redundant degrees-of-freedom. It is interesting to note that if the P pairs of Fig. 3.1-3a are not parallel (as shown in Fig. 3.1-3c) then $F_r = 0$. Thus, the mechanism shown in Fig. 3.1-3c is mobile (with $F_e = 1$) whereas, that shown in Fig. 3.1-3a is locked with no mobility ($F_e = 0$, $F_r = 1$).

Care should be taken while using Equation (3.1.5), in order to ensure that the constraints imposed by various kinematic pairs are independent. For instance, consider the three parallel P pairs shown in Fig. 3.1-3d. Any two of these are enough to ensure the desired translational motion (without relative rotation) between links 1, 4 and 6 and the third pair is kinematically redundant. Therefore, for this mechanism, j in Equation (3.1.5) is 7 (5R + 2P) and not 8.

3.2 Applications of Grübler Criterion

In this section, Grübler criterion given by Equation (3.1.2) will be used for deriving certain useful generalized results applicable to single degree-of-freedom mechanisms consisting of simple pairs.

3.2.1 Minimum Number of Binary Links

In a mechanism, there cannot be any singular link. Let n_2 = number of binary links, n_3 = number of ternary links, n_4 = number of quaternary links and so on. Then, the total number of links

$$n = n_2 + n_3 + n_4 + \ldots + n_i. \tag{3.2.1}$$

Each simple pair consists of two elements as shown in Fig. 3.2-1a (1^+ and 1^- or 2^+ and 2^-). Thus, the total number of elements in the mechanism

$$e = 2j \tag{3.2.2}$$

where j is the number of simple pairs. As explained in Fig. 3.2-1a, a binary link has two elements (1^+ and 2^+). Similarly, a ternary link, connected at three pairs, has three elements, and so on. Thus,

$$e = 2n_2 + 3n_3 + 4n_4 + \ldots + in_i$$

116   CHAPTER 3. MOBILITY AND RANGE OF MOVEMENT

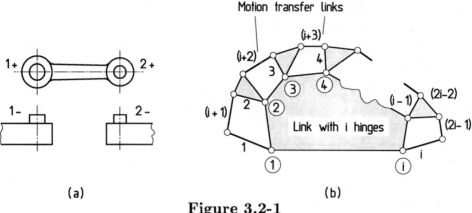

(a)   (b)

Figure 3.2-1

or, using Equation (3.2.2) we get

$$2j = 2n_2 + 3n_3 + 4n_4 + \ldots + in_i. \tag{3.2.3}$$

Substituting Equations (3.2.1) and (3.2.3) in Grübler criterion given by Equation (3.1.2), we have

$$2n_2 + 3n_3 + 4n_4 + \ldots + in_i - 3(n_2 + n_3 + \ldots + n_i) + 4 = 0$$

$$\text{or,} \quad n_2 = 4 + \sum_{p=4}^{i}(p-3)n_p. \tag{3.2.4}$$

Therefore, the minimum number of binary links is four.

3.2.2 Highest Order Link in an n-Link Mechanism

Let the highest order link be of ith order, i.e., the maximum number of pairs to which a link can be connected is i, where the total number of links is n. We approach the problem in an indirect manner, i.e., by finding the minimum number of links n required for closure (and satisfying Equation (3.1.2)) when one link is connected to i number of kinematic pairs as shown in Fig. 3.2-1b. To obtain a closed chain in the simplest manner, we connect a link, denoted by 1,2,3, ... i at each of these kinematic pairs. To transfer motion from one of these i

3.2. APPLICATIONS OF GRÜBLER CRITERION

links to the next one, we add $(i-1)$ number of motion transfer links, which are numbered $(i+1), (i+2), (i+3), \ldots (2i-1)$. Thus, the closure is obtained with a total number of $2i$ links. In other words, the highest order link is the ith with

$$i = n/2. \qquad (3.2.5)$$

We can now check whether the mechanism so obtained satisfies Equation (3.1.2). For the mechanism of Fig. 3.2-1b,

$$n = 2i, \quad j = i + 2 + 2(i-2) = 3i - 2,$$
$$\text{so,} \quad 2j - 3n + 4 = 2(3i - 2) - 3(2i) + 4 = 0.$$

Thus, Grübler criterion is satisfied. We conclude that for a single degree-of-freedom planar mechanism with n links, the highest order link is of the order $n/2$.

3.2.3 Number Synthesis

By number synthesis, we shall refer to the determination of the numbers and types (order) of various links and the number of simple pairs that give rise to single degree-of-freedom planar linkages. From Equation (3.1.2) we see that $3n = 2j + 4$. Therefore, the total number of links n must be even. Further, the minimum number of binary links is 4, i.e., the four-link mechanisms are the simplest (see Section 1.8). The next higher order mechanism starts with $n = 6$. From Equation (3.2.5) we get $i = n/2 = 3$, i.e., the highest order of the links in a six-link mechanism is ternary. Thus, from Equation (3.2.1) we can write

$$n_2 + n_3 = 6.$$

Again, from Equation (3.2.4) with $i = 3$, we have

$$n_2 = 4.$$

The above two equations imply $n_2 = 4$ and $n_3 = 2$. From Equation (3.1.2) with $n = 6$, j turns out to be 7. Thus, a six-link mechanism must have 4 binary and 2 ternary links, connected by 7 simple pairs.

118   CHAPTER 3. MOBILITY AND RANGE OF MOVEMENT

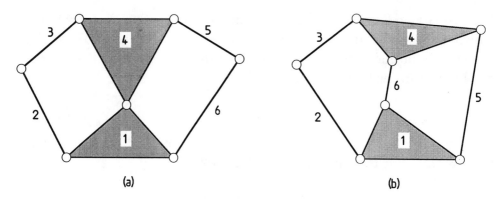

Figure 3.2-2

Two different arrangements are possible. In one arrangement (Fig. 3.2-2a), two ternary links are directly connected to each other, which is known as *Watt's chain* . In the other arrangement (Fig. 3.2-2b), two ternary links are connected by a binary link, which is called *Stephenson's chain.*

By kinematic inversion, two different types of mechanisms are obtained from a *Watt's chain*. If one of the two ternary links is fixed (1 or 4 in Fig. 3.2-2a), we get a so-called Watt's mechanism of type-I. This mechanism is easily seen to be equivalent to two four-link mechanisms connected in series where the input of the second is the output of the first four-link mechanism. If any of the four binary links (2, 3, 5, 6 in Fig. 3.2-2a) is fixed, then we get what is known as Watt's mechanism of type-II. If the two members connected to the frame are used as input and output, then it is obvious that the input/output relationship of this mechanism is identical to that of a four-link mechanism. We should note, however, that the present mechanism has three floating links as compared to one (the coupler) in a four-link mechanism. For example, with link 2 (Fig. 3.2-2a) as the frame, link 4 moves as the coupler of a four-link mechanism. However, the two other floating links 5 and 6 have more complicated motion characteristics and can be used for path generation and rigid-body guidance (see Chapter 6), difficult to achieve by the coupler of a simple four-link mechanism.

Three different types of mechanisms are obtained by kinematic in-

3.3. FAILURE OF MOBILITY CRITERIA

versions of a Stephenson's chain. Stephenson's mechanism of type-I is obtained by fixing one of the ternary links (1 or 4 in Fig. 3.2-2b). Stephenson's mechanism of type-II is obtained by fixing one of the binary links that are connected to each other (2 or 3 in Fig. 3.2-2b). Stephenson's mechanism of type-III is obtained by fixing one of the binary links that are connected to the two ternary links at both ends (5 or 6 in Fig. 3.2-2b). We may note the similarity between Watt's mechanism of type-II and Stephenson's mechanism of type-III. Thus, a properly designed six-link mechanism can produce complicated motion characteristics that may not be satisfactorily generated by a simple four-link mechanism.

The number of different possible chains and, consequently, the number of different mechanisms increases rather rapidly as the total number of links n increases. For example, if we consider $n = 8$, then $i = 4$. Therefore, the highest order link is quaternary. Thus, Equations (3.2.1) and (3.2.4) yield, respectively,

$$n_2 + n_3 + n_4 = 8$$
$$\text{and} \quad n_2 = 4 + n_4.$$

Three possible solutions to the above equations are given below:

$$n_2 = 4, \quad n_3 = 4, \quad n_4 = 0$$
$$n_2 = 5, \quad n_3 = 2, \quad n_4 = 1$$
$$n_2 = 6, \quad n_3 = 0, \quad n_4 = 2$$

The different arrangements possible with each of these solutions are left as an exercise for the reader[1].

3.3 Failure of Mobility Criteria

Special situations exist where the mobility criteria discussed in Section 3.1 fail to yield correct result. For example, consider three rigid bodies, shown in Fig. 3.3-1a, forming a closed loop with three prismatic

---
[1] See Rosenauer, N. and Willis, A.H.: Kinematics of Mechanisms, Dover, N.Y., 1967.

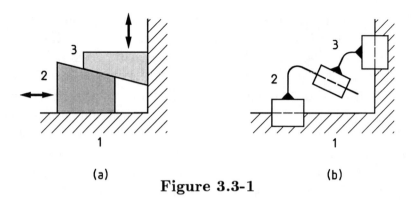

Figure 3.3-1

pairs. It is obvious that body 2 can be moved horizontally, causing definite relative motion between all three members. Fig. 3.3-1b represents the kinematic diagram of the arrangement shown in Fig. 3.3-1a. In this assembly, $n = 3$ and $j = 3$. Thus, Equation (3.1.1) suggests $F = 0$ i.e., the arrangement is a structure with no mobility, which is obviously wrong. The degree-of-freedom of this loop $F = 1$. Thus, in any mechanism if there exists a closed loop with three successive prismatic pairs, then the degree-of-freedom obtained from Equation (3.1.1) should be increased by one for each such loop.

Mobility criteria discussed in Section 3.1 are completely independent of the kinematic dimensions of the mechanism. Consequently, there exist mechanisms with special dimensions that violate these mobility criteria. A mechanism, immobile according to these criteria, may have mobility due only to its special geometric dimensions. Such a mechanism is called an *overclosed linkage*. Some examples of overclosed linkages are given below.

Consider a 4R mechanism, shown in Fig. 3.3-2a, in the form of a parallelogram. If another binary link parallel to the coupler is added, as indicated in Fig. 3.3-2b or 3.3-2c, then the resulting mechanism, with 5 links and 6 R pairs should be immobile according to Equation (3.1.1). However, it is intuitively obvious that the additional link does not put any extra constraint and the degree-of-freedom of the mechanism remains at $F = 1$.

Next, let us discuss a *crossed-slider trammel*, shown in Fig. 3.3-3. The special geometric relation required for this mechanism to have $F = 1$ is $OC = CA = CB$. According to Equation (3.1.1),

3.3. FAILURE OF MOBILITY CRITERIA

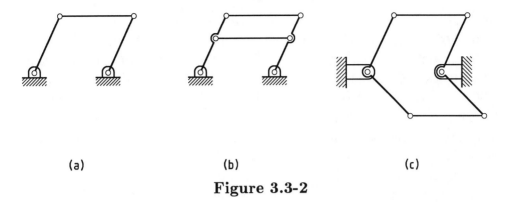

(a) (b) (c)

Figure 3.3-2

this mechanism with $n = 5$ and $j = 6$ should again be immobile. Comparing Fig. 3.3-3 with Fig. 1.8-6, we note that the crossed-slider trammel is obtained from the elliptic trammel by kinematic inversion after adding the link OC with R pairs at O and C and keeping this link OC fixed. Since the point C in Fig. 1.8-6 moves along a circle of radius $AB/2$, i.e., the distance between O and C remains constant, joining these two points by a rigid link does not impose any extra constraint. In other words, the degree-of-freedom of the crossed-slider trammel is one. It is interesting to note that in the crossed-slider trammel, $\omega_3/\omega_5 = 2$ at all configurations (just like a pair of internal gears where the ratio of the number of teeth is 2). Proof of this result, by the use of instantaneous centers is left as an exercise for the reader. Where is the relative instantaneous center 24? How do you physically explain the location of 24?[2]

A curious overclosed linkage, where the redundant constraint is not obvious, known as *Kempe-Burmester focal mechanism*, is shown in Fig. 3.3-4. This eight-link mechanism has $j_1 = 8$ and $j_3 = 1$. Thus, from Equations (3.1.1) and (3.1.3) we get

$$F = 3 \times (8 - 1) - 2 \times (8 + 3 \times 1) = -1.$$

---

[2]Application of Aronhold-Kennedy theorem of three centers fails to locate 24 uniquely. However, links 2 and 4 providing same constraints, are in relative translation along a direction perpendicular to AB. Thus, 24 is at ∞ along AB. Further, one of the links 2 and 4 can be removed without altering the motion characteristics of the mechanism. Once one of the sliders (along with the associated P and R pairs) is removed, we get $n = 4$, $j = 4$, yielding $F = 1$.

But the linkage has $F = 1$, with suitable choices of the points T, P (on AB), Q (on O_4B), R (on O_2O_4) and S (on O_2A), for a given arbitrary starting 4R linkage O_2ABO_4. If T is within O_2ABO_4, then the angles subtended at T by the pair of opposite sides (i.e., O_2O_4, AB and O_2A, O_4B) of the quadrilateral O_2ABO_4 should add up to π, i.e., $\angle O_2TO_4 + \angle BTA = \angle O_4TB + \angle ATO_4 = \pi$. Each pair of opposite internal 4R linkages are geometrically related to each other, as indicated in Fig. 3.3-4. The points P, Q, R and S lie on a circle. The point T may also lie outside the quadrilateral O_2ABO_4. In that case, the pair of opposite sides of the quadrilateral O_2ABO_4 should subtend equal angles at T.

In Fig. 3.3-4 the R pair at O_4 can be removed and still the point O_4 on link 4 remains stationary because of the constrained motion. Thus, link 4 (of the eight-link mechanism) can undergo pure rotation without providing a rotation center to this link. For many other interesting details and applications of the Kempe-Burmester focal mechanism, the reader is referred to *Motion Geometry of Mechanisms* by E.A. Dijksman, Cambridge University Press, 1976.

3.4 Grashof Criterion

The mobility criterion discussed thus far tells nothing about the ranges of possible movements of various links of a mechanism, which depend on the relative link lengths. As already stated in Section 1.8, for a mechanism to be driven by a rotary input (not oscillatory) there must be a link, connected to the frame, which should be capable of making a complete revolution. Generalized results, so far as the ranges of movement of various links are concerned, are available only for four-link mechanisms. The most important result towards this end is known as *Grashof criterion*, which can be stated as follows:

In a 4R planar kinematic chain, the shortest link can make complete rotation with respect to all other links, and vice versa, if

$$l_{min} + l_{max} < l' + l'' \qquad (3.4.1)$$

where l_{min} = length of the shortest link, l_{max} = length of the longest link and l', l'' = lengths of the two other links.

3.4. GRASHOF CRITERION

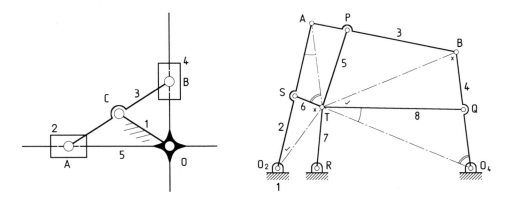

Figure 3.3-3 **Figure 3.3-4**

A 4R chain satisfying condition (3.4.1) is called a *Grashof chain* and the linkages obtained by fixing a link of such a chain are called *Grashof linkages*. The proof of complete rotatability of the shortest link, with respect to all other links, and vice versa, in a Grashof chain, is given below.

Consider a 4R mechanism with link-lengths l_1, l_2, l_3 and l_4, where we assume the shortest link of length l_1 to be the frame[3]. If link 2 has to make a complete rotation, then obviously it must be able to attain the two extreme positions shown in Figs. 3.4-1a and 3.4-1b. This implies that the presence of links 3 and 4 should not prevent the maximum and the minimum possible separations between the points O_4 and A. Thus, for the $\triangle O_4 AB$ to exist in both extreme situations, the link-lengths should satisfy the following conditions:

$$l_1 + l_2 < l_3 + l_4, \quad l_4 < l_3 + (l_2 - l_1), \quad l_3 < l_4 + (l_2 - l_1)$$

These inequalities can be rewritten as

$$l_1 + l_2 < l_3 + l_4, \qquad (3.4.2a)$$

$$l_1 + l_4 < l_3 + l_2, \quad \text{and} \qquad (3.4.2b)$$

---

[3]This particular choice of fixed link does not affect the relative motion between various links.

124   CHAPTER 3. MOBILITY AND RANGE OF MOVEMENT

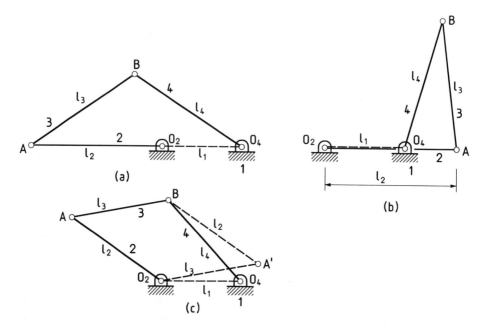

Figure 3.4-1

$$l_1 + l_3 < l_4 + l_2. \tag{3.4.2c}$$

Thus, when the conditions (3.4.2a, b and c) are satisfied, link 2 can make complete rotation with respect to link 1. If link 4 has to make complete rotation, then the necessary conditions are similarly obtained which are nothing more than those given by conditions (3.4.2a, b and c) after interchanging l_2 and l_4. It is easy to see that the set of conditions (3.4.2a, b and c) remains unaltered by interchanging l_2 and l_4. To prove that the set (3.4.2a, b and c) also constitutes the conditions for complete rotation of link 3, we proceed as follows. Construct a 4R linkage $O_2A'BO_4$ by interchanging l_2 and l_3, as shown in Fig. 3.4-1c. The rotatability condition of O_2A' in this linkage can be obtained from (3.4.2a, b and c) by interchanging l_2 and l_3. It can again be easily verified that the set of conditions (3.4.2a, b and c) remains invariant under interchanging of l_2 and l_3. Now, overlapping the original linkage O_2ABO_4 on to $O_2A'BO_4$, we get a constrained six-link mechanism (with $j = 7$), where $O_2A'BA$ is a parallelogram. Thus, in this six-link mechanism, O_2A' and AB always remain paral-

3.4. GRASHOF CRITERION

lel to each other, i.e., they undergo identical motion. Therefore, the conditions for rotatability of O_2A' are also those for the rotatability of AB. Hence, the set of conditions (3.4.2a, b and c) constitutes the requirement for the rotatability of all three links 2, 3 and 4 with respect to the shortest link 1, and vice versa. Adding the condition (3.4.2a) to (3.4.2b), (3.4.2b) to (3.4.2c) and (3.4.2c) to (3.4.2a), in turn, we get, respectively, $l_1 < l_3$, $l_1 < l_2$ and $l_1 < l_4$, to confirm that l_1 is the shortest link-length ($\equiv l_{min}$). Further, it can be verified from the set of conditions (3.4.2a, b and c) that the sum of l_{min} and l_{max} (any one of l_2, l_3 and l_4) is smaller than the sum of the remaining two link-lengths.

3.4.1 Inversions of a Grashof Chain

Before proceeding to the discussion of various inversions of a Grashof chain, we shall first verify that in a Grashof chain (Fig. 3.4-1a), link 3 cannot make complete rotation with respect to either link 2 or 4.[4] If link 3 has to make complete rotation with respect to link 2, then links 2 and 3 must be able to fall in a line as shown in Fig. 3.4-2. This configuration implies $l_1 + l_4 > l_2 + l_3$, which violates the Grashof criterion, since l_1 is the shortest link-length. Similarly, if link 3 has to make complete rotation with respect to link 4 then, the required condition is $l_1 + l_2 > l_3 + l_4$, which again violates the Grashof criterion.

In view of the complete rotatability of the shortest link and the oscillatory movements between the adjacent links among the other three, we conclude that by kinematic inversions from a Grashof chain, one can obtain all the three possible variations of a 4R linkage (see Section 1.8), viz., crank-rocker, double-crank and double-rocker, as given below:

(i) A double-crank results if the shortest link is fixed, since all other links can completely rotate with respect to the shortest one.

(ii) A double-rocker results if the shortest link is made the coupler, i.e., the link not connected to the shortest one is made the frame.

---

[4]In fact, in a Grashof chain, the links other than the shortest can only oscillate with respect to each other through an angle less than 180°.

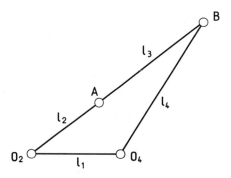

Figure 3.4-2

This is so because then both of the links connected to the fixed link can only oscillate.

(iii) Two crank-rockers with the shortest link as the crank, result if any of the links adjacent to the shortest one is fixed.

3.4.2 Inversions of a Non-Grashof Chain

If the link-lengths of a 4R chain satisfy the following condition

$$l_{min} + l_{max} > l' + l'' \tag{3.4.3}$$

then the chain is called a *non-Grashof chain*. All four inversions of a non-Grashof chain yield double-rocker linkages.

3.4.3 Inversions of Chains with Uncertainty (Folding) Positions

Let us consider the threshold situation between conditions (3.4.1) and (3.4.3), i.e., when

$$l_{min} + l_{max} = l' + l''. \tag{3.4.4}$$

If Equation (3.4.4) is satisfied, then in general, the four inversions result in mechanisms similar to those obtained from a Grashof chain.

3.4. GRASHOF CRITERION

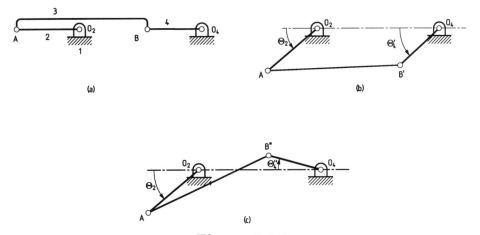

Figure 3.4-3

However, in all these mechanism, there exists configuration(s) when the four links become collinear. The collinear configuration is referred to as *uncertainty position* or *folding position*, since from this configuration the mechanism, unless properly guided, can go into more than one different configuration. This is illustrated in Figs. 3.4-3a to 3.4-3c for a crank-rocker linkage. From the uncertainty configuration indicated in Fig. 3.4-3a, the linkage, unless guided, can move into one of the configurations indicated in Figs. 3.4-3b and 3.4-3c.[5] Similar investigations of the double-crank and double-rocker linkages at the uncertainty configuration are left as exercises for the reader.

The equality condition (3.4.4) is also true when a chain has two pairs of equal links. This results in two special chains, namely,

(i) The parallelogram configuration (in which the equal links are not adjacent) shown in Fig. 3.3-2a. All four inversions of a parallelogram chain yield double-crank mechanisms if the linkage is properly guided at the uncertainty configurations. The addition of an extra coupler, shown in Figs. 3.3-2b and 3.3-2c, provides a means for ensuring a parallelogram configuration to be preserved while crossing through the uncertainty configuration. In

---

[5]It may be noted that the relative instantaneous center 24 is indeterminate at the uncertainty configuration, implying non-unique (uncertain) input-output relationship at this configuration.

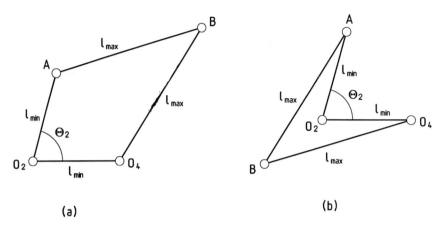

Figure 3.4-4

Fig. 3.3-2c, all five links never become collinear, which helps in maintaining a good motion transmission characteristic when four links become collinear.

(ii) The *deltoid* or *kite* configuration (in which the equal links are adjacent) is shown in Fig. 3.4-4a. Here, when any of the longer links (l_{max}) is fixed, two crank-rocker mechanisms are obtained. On the other hand, when any of the shorter links (l_{min}) is fixed, two double-crank mechanisms result. In the latter, one revolution of the longer crank causes two revolutions of the shorter link (crank). Such a mechanism is known as the *Galloway mechanism*. With O_2O_4, in Fig. 3.4-4a, as the frame, one revolution of O_2A from this position yields the configuration indicated in Fig. 3.4-4b.

3.4.4 Extension of Grashof Criterion

(i) 3R - 1P Mechanisms

In Section 1.6 we saw that a prismatic pair can be thought of as a limiting case of a revolute pair with the axis of the pair at infinity (along a direction perpendicular to the relative sliding at the P pair). Accordingly, as explained in Figs. 3.4-5a to 3.4-5c, we can consider a

3.4. GRASHOF CRITERION

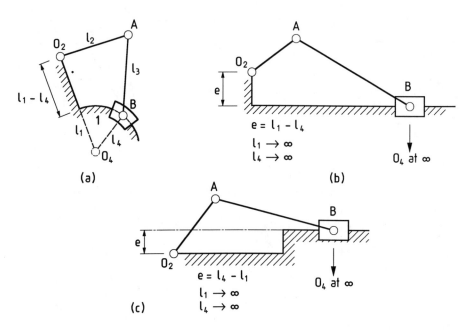

Figure 3.4-5

3R - 1P mechanism as a 4R mechanism where two link-lengths (say l_1 and l_4) are infinite (with O_4 at infinity) and the eccentricity e of the 3R - 1P mechanism is given by $e = |l_1 - l_4|$. Thus, Grashof criterion (3.4.1), can now be written as

$$l_{min} + e < l', \qquad (3.4.5)$$

when l_{min} can make complete rotation.

If $l_{min} + e > l'$, then no link can make complete rotation and with $l_{min} + e = l'$, the linkage behaves, as before, like a Grashof 3R-1P mechanism with uncertainty configurations.

(ii) RRPP Mechanisms

Extending the concept outlined in the previous paragraph, we can readily see that an RRPP mechanism (Fig. 1.8-6) can be thought of as a 4R mechanism with two of the four R pairs at infinity. Consequently, three link-lengths become infinity and the link with R pairs at its

130 CHAPTER 3. MOBILITY AND RANGE OF MOVEMENT

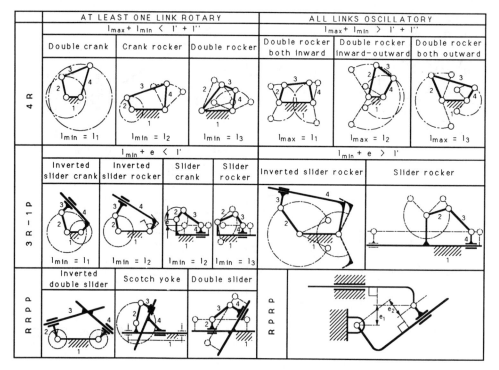

e, e_1, e_2, offset; l_{max}, longest link length; l_{min}, shortest link length; l', l'', other link lengths

Figure 3.4-6

ends is the shortest (and of finite length). Thus, Grashof's criterion is always satisfied and the link with R pairs at its ends can always make complete rotation with respect to the other three links.

(iii) RPRP Mechanisms (Fig. 1.8-7)

In this case, all link lengths of the equivalent 4R linkage become infinity. Hence, there exists nothing like a shortest link and no link can make complete rotation

The general characteristics of movement ranges for four-link mechanisms are summarized in Fig. 3.4-6. It may be noted that the mechanisms with uncertainty configurations are omitted here. Further, we may note that, in general, an RPRP mechanism has two characteristic kinematic dimensions indicated as e_1 and e_2 (Fig. 3.4-6) whereas in Fig. 1.8-7, one of these was zero.

3.5. EXERCISE PROBLEMS

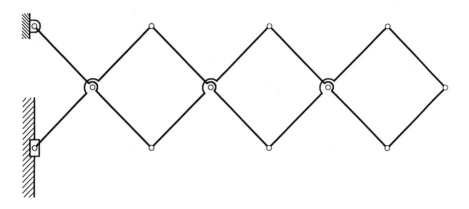

Figure 3.5-1

3.5 Exercise Problems

3.1 Determine the effective degrees-of-freedom of the mechanisms shown in Figs. 3.5-1 to 3.5-6.

3.2 How many input motions are required in the earth-moving ma-

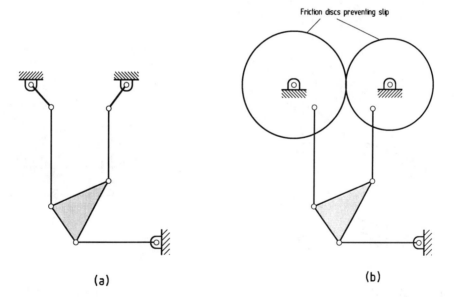

Figure 3.5-2

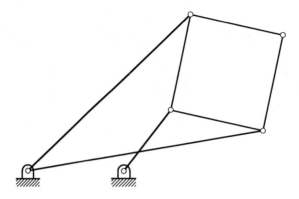

Figure 3.5-3

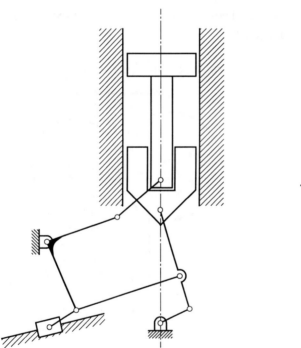

Figure 3.5-4

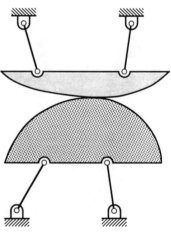

Figure 3.5-5

3.5. EXERCISE PROBLEMS

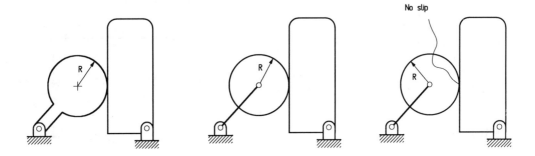

Figure 3.5-6

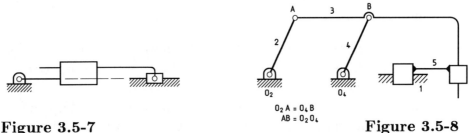

Figure 3.5-7

Figure 3.5-8

chinery shown in Fig. 1.1-2, in order to generate a constrained motion of the dumping bin?

3.3 What is the effective degree-of-freedom of the RPRP mechanism with $e_1 = e_2 = 0$, as shown in Fig. 3.5-7?

3.4 An assortment of five links of lengths 5 cm, 8 cm, 15 cm, 19 cm and 28 cm is available for constructing a 4R crank-rocker mechanism. Sketch the crank-rocker mechanism indicating the crank and all other link lengths.

3.5 Prove that in a Grashof linkage the rocker can never cross the line of frame.

3.6 Sketch a parallelogram linkage at its uncertainty configuration when (a) one of the longer links is fixed, (b) one of the shorter links is fixed. Drawing the two possible configurations just beyond the uncertainty configurations for each of the two cases (a)

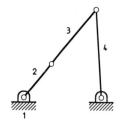 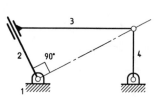

Figure 3.5-9 **Figure 3.5-10**

and (b), observe the difference in the changeover characteristics for the two cases.

3.7 Determine the effective degrees-of-freedom of the mechanism shown in Fig. 3.5-8.

3.8 (a) Consider the dead-center configuration of a crank-rocker 4R linkage, as shown in Fig. 3.5-9. Locate all the relative instantaneous centers. Note that the subscripts 1 and 4 are interchangeable implying $\omega_4 \equiv 0$ (for any value of ω_2). This means that the mechanism gets locked at this configuration if 4 is the driving (input) link.

(b) Consider the configuration of a 3R-1P mechanism shown in Fig. 3.5-10. Locating all the relative instantaneous centers, comment on the speciality of this configuration.

Chapter 4

DISPLACEMENT ANALYSIS

By displacement analysis of a mechanism the following problem is solved: Given the kinematic dimensions and input movements of a mechanism, determine the movements of all the other links.

Both graphical and analytical methods can be employed for displacement analysis. So far as the graphical method is concerned, no generalized discussion can be provided. Rather, we shall demonstrate the method through a few examples.

4.1 Graphical Method

In the graphical method of displacement analysis, the mechanism is drawn to a convenient scale and the unknown quantities are determined through suitable geometrical constructions.

Problem 4.1-1

Figure 4.1-1a shows a slotted-lever quick-return mechanism used in shapers. Determine the stroke length of the tool and the quick-return ratio (defined as the ratio of the times taken for the forward cutting stroke and the backward return stroke), assuming constant angular speed of the input link 2.

The stroke length of the tool is adjusted by varying the length O_2A. If the length O_2A is reduced to 8 cm, determine the new stroke

136 CHAPTER 4. DISPLACEMENT ANALYSIS

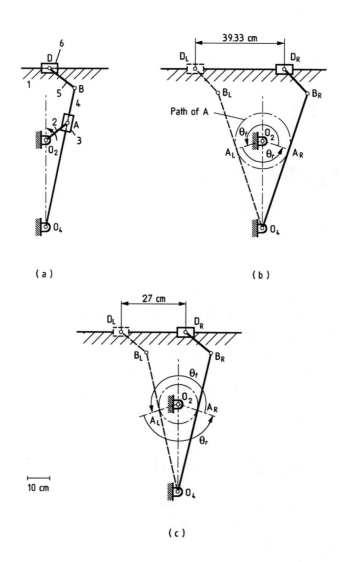

Figure 4.1-1

4.1. GRAPHICAL METHOD

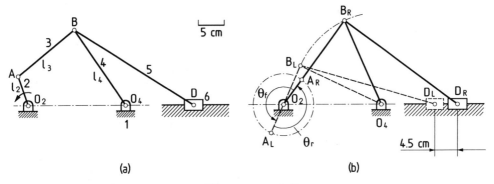

Figure 4.1-2

length and the corresponding quick-return ratio (all other dimensions remain unchanged).
Solution

Referring to Fig. 4.1-1b, the positions of various links, corresponding to the extreme right and extreme left positions of the tool are indicated by subscripts R and L, respectively. It may be noted that at these extreme positions, the axis of the slotted-lever (O_4B) is tangential to the path of the point A. Using the scale of the diagram, the stroke length of the tool $= D_R D_L = 39.33$ cm. The quick-return ratio is given by θ_f/θ_r, since link 2 rotates at a constant speed. From measurement, $\theta_f/\theta_r = 217°/143° = 1.52$.

When the length of O_2A is reduced to 8 cm (Fig. 4.1-1c), the solution is repeated yielding stroke length $= 27$ cm and the quick-return ratio $= 208°/152° = 1.37$.

It may be noted that the quick-return ratio is reduced when the stroke length of the tool is decreased. A slight variation of this mechanism, used in slotting machines, provides a quick-return ratio independent of the stroke length of the tool (see Exercise Problem 4.1).

Problem 4.1-2

Referring to Fig. 4.1-2a, the output sliding link 6 is driven by the oscillating output link of a 4R crank-rocker mechanism, where the input crank rotates at a constant angular speed. Determine the travel of the output link 6 and the quick-return ratio.
Solution

The extreme positions of the slider correspond to the extreme

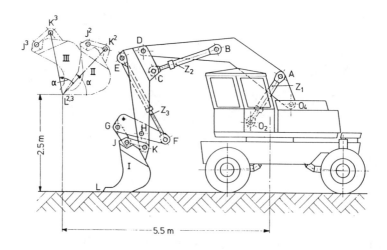

Figure 4.1.3a

positions of the rocker (Fig. 4.1-2b), which are obtained from the fact that at these instants the crank and the coupler become collinear, i.e., $O_2B_R = l_2 + l_3$ and $O_2B_L = l_3 - l_2$. Further, the rocker does not cross the line of frame. From measurements, the travel of link 6 $D_R D_L = 4.5$ cm; the quick-return ratio $\theta_f/\theta_r = 193°/167° = 1.16$.

Problem 4.1-3

Consider the earth-moving machine shown in Fig. 1.1-2 to the scale indicated in Fig. 4.1-3a. The hinge K of the bin is positioned by using the hydraulic actuators Z_1 and Z_2 whereas the actuator Z_3 is used mainly to tilt the bin. The ranges of movement of Z_1 and Z_2 are such that

$$(O_2A)_{\min} = 1100 \text{ mm}, \quad (O_2A)_{\max} = 1800 \text{ mm},$$
$$(CB)_{\min} = 1500 \text{ mm} \quad \text{and} \quad (CB)_{\max} = 2400 \text{ mm}.$$

Determine

(i) the positions of all the revolute joints corresponding to position II of the bin, indicated in Fig. 4.1-3a

(ii) the zone in which the hinge K lies when the full ranges of move-

4.1. GRAPHICAL METHOD

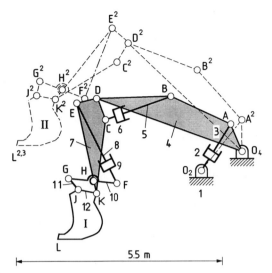

Figure 4.1-3b

ments of Z_1 and Z_2 are utilized

(iii) the maximum value of the tilt angle α, keeping the lip L at the location indicated for position II

Solution

(i) The locations of the revolute joints J and K, corresponding to positions II are denoted by J^2 and K^2, respectively, in Fig. 4.1-3b. O_2 and O_4 are two fixed points. Since the distances KD and O_4D remain constant, the location of the R pair at D, corresponding to configuration II, is obtained at the intersection of two circular arcs (one of radius KD with center at K^2 and the other of radius O_4D with center at O_4) and denoted by D^2 in Fig. 4.1-3b. Since two points (namely O_4 and D) of link 4 are now located, the points A^2 and B^2 on the same link can be easily located. Similarly, since K^2 and D^2 on link 7 are located, the other points, viz., C^2, E^2 and H^2 on the same link can be easily located. (Use of a transparent paper to locate the points on the same rigid body with the known locations of two points is probably most convenient.) Thereafter, G^2 is located, using the distances GH and GJ which remain constant. Finally, with G^2 and H^2 known, F^2 is located since these three points belong to link 10.

140 CHAPTER 4. DISPLACEMENT ANALYSIS

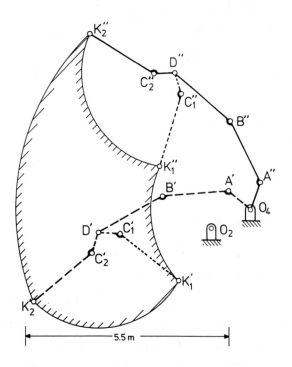

Figure 4.1-3c

4.1. GRAPHICAL METHOD

(ii) Referring to Fig. 4.1-3c, A' and A'' indicate the locations of A when O_2A is, respectively, minimum (= 1100 mm) and maximum. It may be noted that A lies on the circle with O_4 as center and O_4A as radius. The locations of D, corresponding to A' and A'', are indicated by D' and D''. As indicated in Fig. 4.1-3c, C'_1 is the position of C with both O_2A and CB at their respective minimum values. Similarly, C'_2 is the position of C with O_2A at its minimum and CB at its maximum value. The point C''_1 indicates the position of C with O_2A at its maximum and CB at its minimum value. Finally, the point C''_2 shows the position of C with both O_2A and CB, at their respective maximum values. The four locations of K, corresponding to four different combinations of the minimum and the maximum values of O_2A and CB, are likewise denoted by K'_1, K''_1 and K'_2, K''_2. With both the minimum and the maximum values of CB, links 4 and 8 get rigidly connected. This rigid configuration rotates about O_4. Therefore, K'_1 and K''_1 (both correspond to the minimum value of CB) are joined by a circular arc with O_4 as the center. Similarly, K'_2 and K''_2 are also joined by a circular arc with O_4 as the center. When O_2A is at any one of its extreme values, link 4 is immovable (gets rigidly connected to the frame 1); hence, the point K moves on a circle with D as its center. Therefore, K''_1 and K''_2 (both correspond to the maximum value of O_2A) are joined by a circular arc with D'' as its center. Similarly, K'_1 and K'_2 are joined by a circular arc with D' as its center.

Thus, the zone, in which the hinge K lies when the full ranges of the movements of Z_1 and Z_2 are utilized, is obtained. This zone, enclosed by the four circular arcs mentioned above, is indicated by the hatched boundary in Fig. 4.1- 3c.

(iii) Referring to Fig. 4.1-3d, link 12 (i.e., the bin) is rotated about its tip L (corresponding to position II) until the point K reaches its limit, represented by the circular arc (reproduced from Fig. 4.1-3c). From measurement, the angle α is found out to be $46°$.

To start the displacement analysis of a single degree-of-freedom linkage by graphical method, we need a closed loop containing only four links, two of which are the frame and input. The reader may appreciate this point better by attempting graphical displacement analysis of Stephenson's mechanisms of type I (with various combinations of input and output links) and type II. It is easy to verify that graph-

CHAPTER 4. DISPLACEMENT ANALYSIS

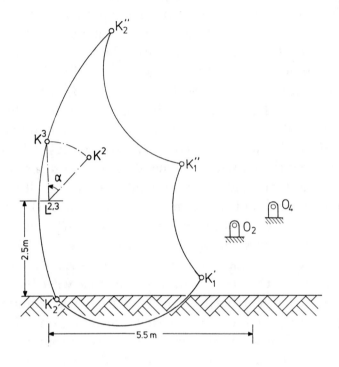

Figure 4.1-3d

ical displacement analysis of the mechanism shown in Fig. 3.5-2b is impossible, since no closed loop consisting of only four links can be identified. In fact, the existence of adequate numbers of closed loops containing only four links holds the key to the success of graphical displacement analysis.

4.2 Analytical Method

Whenever a high level of accuracy is desired or the analysis has to be repeated for a large number of configurations, an analytical approach, amenable to computer programming, is preferred. Furthermore, analytical method has to be used when graphical displacement analysis is not possible, as discussed in the previous section. In most cases the resulting algebraic equations are nonlinear and are solved by a numerical procedure. In some simple, but commonly used, mechanisms the resulting nonlinear algebraic equations can be solved analytically. The most versatile method, using the loop closure equations, is discussed below.

First, all the independent closed loops (see Section 1.4) are identified in the kinematic diagram of the given mechanism. Every link length and slider displacement (from a convenient reference point) are represented by suitable planar vectors. The vectors are expressed through complex exponential notation. A vector equation (complex equation) corresponding to each independent loop is established. Each such vector (complex) equation gives rise to two scalar (real) equations by treating the real and the imaginary parts separately. The solution to these equations yields the movements of all the links, when the input motion(s) is prescribed. The method is illustrated through some examples.

Problem 4.2-1

Figure 4.2-1 shows a 4R linkage with link lengths as l_1, l_2, l_3 and l_4. Obtain the angles θ_3 and θ_4 as functions of the input movement θ_2 and the link-lengths.

Solution

Referring to Fig. 4.2-1, all links are denoted as vectors, viz., l_1, l_2, l_3 and l_4. All the angles are measured counterclockwise from the

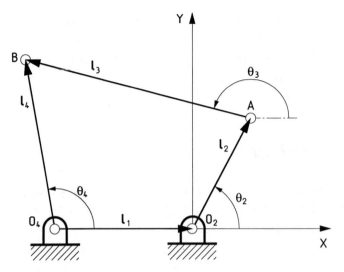

Figure 4.2-1

X-axis, which is along the fixed vector l_1. Thus, $\theta_1 = 0$.

Considering the closed loop $O_2ABO_4O_2$, we can write

$$\boldsymbol{l_1} + \boldsymbol{l_2} + \boldsymbol{l_3} - \boldsymbol{l_4} = \boldsymbol{o}.$$

Using complex exponential notation, with $\theta_1 = 0$, the above equation can be written as

$$l_1 + l_2 e^{i\theta_2} + l_3 e^{i\theta_3} - l_4 e^{i\theta_4} = 0.$$

Equating the real and imaginary parts of this equation separately to zero, we get

$$l_1 + l_2 \cos\theta_2 + l_3 \cos\theta_3 - l_4 \cos\theta_4 = 0 \qquad (a)$$

and

$$l_2 \sin\theta_2 + l_3 \sin\theta_3 - l_4 \sin\theta_4 = 0. \qquad (b)$$

For given values of the link-lengths and prescribed θ_2, Equations (a) and (b) constitute a set of nonlinear equations in two unknowns,

4.2. ANALYTICAL METHOD

namely θ_3 and θ_4. The solution of these two equations completes the displacement analysis.

In the case of this simple mechanism under discussion, it is possible to have a closed form solution of Equations (a) and (b). This 4R linkage being very common and versatile we shall present the closed form solution in detail. In most cases, the existence of the closed form solution also implies the success of the graphical method of displacement analysis.

Rearranging Equations (a) and (b),

$$l_3 \cos \theta_3 = (l_4 \cos \theta_4 - l_1) - l_2 \cos \theta_2 \qquad (c)$$

and

$$l_3 \sin \theta_3 = l_4 \sin \theta_4 - l_2 \sin \theta_2. \qquad (d)$$

Squaring both sides of the above equations and adding, we get

$$\begin{aligned} l_3^2 &= l_1^2 + l_2^2 + l_4^2 - 2l_4 \cos \theta_4 (l_1 + l_2 \cos \theta_2) \\ &\quad - 2l_4 \sin \theta_4 (l_2 \sin \theta_2) + 2l_1 l_2 \cos \theta_2 \end{aligned}$$

or,

$$a \sin \theta_4 + b \cos \theta_4 = c \qquad (e)$$

where

$$a = \sin \theta_2, \quad b = \cos \theta_2 + (l_1/l_2) \quad \text{and}$$
$$c = (l_1/l_4) \cos \theta_2 + \left((l_1^2 + l_2^2 + l_4^2 - l_3^2)/(2l_2 l_4) \right). \qquad (f)$$

It should be noted that for given values of θ_2 and the link-lengths, the coefficients a, b and c of Equation (e) are known. To solve for θ_4 from Equation (e) we substitute

$$\sin \theta_4 = 2 \tan(\theta_4/2)/[1 + \tan^2(\theta_4/2)]$$

and

$$\cos \theta_4 = [1 - \tan^2(\theta_4/2)]/[1 + \tan^2(\theta_4/2)] \qquad (g)$$

in Equation (e), to yield

$$(b+c)\tan^2(\theta_4/2) - 2a\tan(\theta_4/2) + (c-b) = 0$$

which gives

$$\tan(\theta_4/2) = (a \pm \sqrt{a^2+b^2-c^2})/(b+c). \qquad (h)$$

Thus, for each given value of θ_2 and the link-lengths, we get, in general, two distinct values of θ_4 as follows:

$$\theta_4^{(1)} = 2\tan^{-1}[(a+\sqrt{a^2+b^2-c^2})/(b+c)] \qquad (i)$$

$$\theta_4^{(2)} = 2\tan^{-1}[(a-\sqrt{a^2+b^2-c^2})/(b+c)] \qquad (j)$$

These two values correspond to the two different ways in which the 4R - linkage can be closed for any given value of θ_2, as explained in Fig. 4.2-2 (solved by graphical method)[1].

To solve for the coupler orientation θ_3, we can eliminate θ_4 from Equations (a) and (b) in a manner similar to that described above. The following results are obtained:

$$\theta_3^{(1)} = 2\tan^{-1}[(a+\sqrt{a^2+b^2-c'^2})/(b+c')] \qquad (k)$$

and

$$\theta_3^{(2)} = 2\tan^{-1}[(a-\sqrt{a^2+b^2-c'^2})/(b+c')] \qquad (l)$$

where

$$c' = -\left\{(l_1/l_3)\cos\theta_2 + \left[(l_1^2+l_2^2+l_3^2-l_4^2)/(2l_2l_3)\right]\right\}. \qquad (m)$$

The reader may note the following alternative method for determining the input-output relationship given by Equation(e). Referring to Fig. 4.2-1, the coordinates of the points A and B can be written as

$$x_A = l_2\cos\theta_2 \quad \text{and} \quad y_A = l_2\sin\theta_2, \qquad (n)$$

$$x_B = -l_1 + l_4\cos\theta_4 \quad \text{and} \quad y_B = l_4\sin\theta_4. \qquad (o)$$

---

[1]These two configurations for the same position of link 2 are referred to as geometric inversions.

4.2. ANALYTICAL METHOD

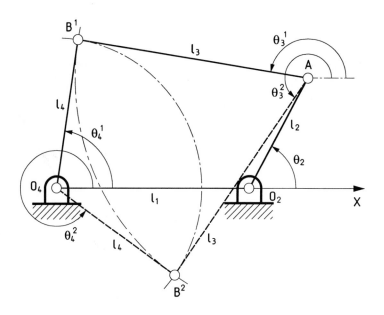

Figure 4.2-2

Now
$$(x_A - x_B)^2 + (y_A - y_B)^2 = AB^2 = l_3^2. \qquad (p)$$

Substituting for the coordinates and simplifying, we get

$$l_3^2 = l_1^2 + l_2^2 + l_4^2 - 2l_4 \cos\theta_4 (l_1 + l_2 \cos\theta_2)$$

$$-2l_4 \sin\theta_4 (l_2 \sin\theta_2) + 2l_1 l_2 \cos\theta_2 \qquad (q)$$

which is same as Equation (e).

Problem 4.2-2

Figure 4.2-3a shows the quick-return shaper mechanism. Given the link-lengths l_1, l_2, l_4, l_5, l_6 and the input angle θ_2, obtain all the other variables, viz., θ_4, s_4, θ_5 and s_6, of this mechanism.

Solution

The vector diagram (Fig. 4.2-3b) is drawn by representing the link-lengths and slider displacements by vectors. Two independent closed loops, viz., $O_2 O_4 A O_2$ and $O_2 ABCDO_2$, are identified in this

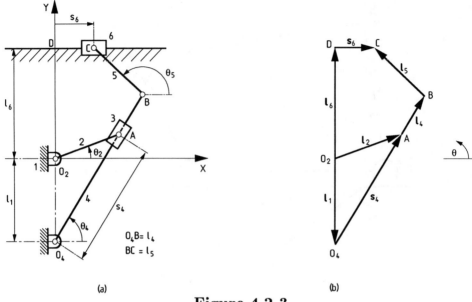

Figure 4.2-3

diagram. The vector equations, corresponding to these two loops, are written, respectively, as

$$l_1 + s_4 = l_2 \tag{a}$$

and

$$l_2 + (l_4 - s_4) + l_5 = l_6 + s_6. \tag{b}$$

Using complex exponential notations and substituting proper values of the angles, the above two equations are rewritten as

$$l_1 e^{i\frac{3\pi}{2}} + s_4 e^{i\theta_4} - l_2 e^{i\theta_2} = 0 \tag{c}$$

and

$$l_2 e^{i\theta_2} + (l_4 - s_4) e^{i\theta_4} + l_5 e^{i\theta_5} - l_6 e^{i\frac{\pi}{2}} - s_6 e^{i \cdot 0} = 0. \tag{d}$$

Equating the real and imaginary parts of Equations (c) and (d) separately to zero, we obtain

$$s_4 \cos\theta_4 = l_2 \cos\theta_2, \tag{e}$$

4.2. ANALYTICAL METHOD

$$s_4 \sin\theta_4 = l_1 + l_2 \sin\theta_2, \qquad (f)$$

$$(l_4 - s_4)\cos\theta_4 + l_5 \cos\theta_5 - s_6 = -l_2 \cos\theta_2 \quad \text{and} \qquad (g)$$

$$(l_4 - s_4)\sin\theta_4 + l_5 \sin\theta_5 = l_6 - l_2 \sin\theta_2. \qquad (h)$$

Equations (e) to (h) constitute a set of four nonlinear equations in four unknowns, namely, θ_4, θ_5, s_4 and s_6. However, as in Problem 4.2-1, in this case, also closed form solution of this set of equations is possible, as shown below.

From Equations (e) and (f) we get

$$s_4 = (l_1^2 + l_2^2 + 2l_1 l_2 \sin\theta_2)^{1/2} \qquad (i)$$

and

$$\theta_4 = \tan^{-1}\left[(l_1 + l_2 \sin\theta_2)/(l_2 \cos\theta_2)\right]. \qquad (j)$$

The quadrant in which θ_4 lies can be determined from the signs of the numerator and the denominator of the argument of $\tan^{-1}$ in Equation (j). Once s_4 and θ_4 are known, θ_5 and s_6 can be obtained from Equations (g) and (h), after substituting for $\sin\theta_5$ and $\cos\theta_5$ in terms of $\tan(\theta_5/2)$. We may note that two roots of θ_5 and s_6 are resulted.

At this stage, it is worthwhile to recognize that for complex mechanisms, the set of (nonlinear) loop closure equations needs to be solved numerically. The most commonly used method for solving such equations is the Newton-Raphson method. For details on this method, the reader may refer to any standard text on numerical analysis. Generalized computer programs are also available for kinematic analysis of planar linkages. Some such available programs are KINSYN, DRAM, ADAMS, IMP and LINKAGES.4. Finally, we may note that if a mechanism needs n number of quantities to describe its configuration and there exists L number of independent loops, then the loop-closure equations generate $2L$ constraint equations (involving these n quantities), and if $(n-2L)$ quantities (equal to the degrees-of-freedom of the mechanism) are specified as inputs, then the remaining $2L$ number of quantities can be obtained to complete the displacement analysis.

CHAPTER 4. DISPLACEMENT ANALYSIS

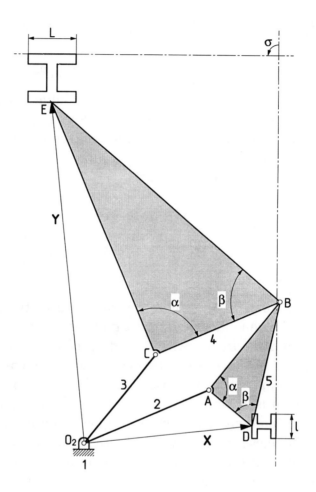

Figure 4.2-4

Problem 4.2-3

Figure 4.2-4 shows a plagiograph (copying) mechanism. Determine the scale ($\equiv L/l$) and the relative orientation ($\equiv \sigma$) of the reproduction. All link-lengths and angles are indicated in the figure. Note that O_2ABC is a parallelogram.

Solution

Let the locations of the points D and E be represented by vectors $\boldsymbol{X}$ and $\boldsymbol{Y}$, respectively (Fig. 4.2-4). From the figure,

$$\boldsymbol{X} = \boldsymbol{O_2A} + \boldsymbol{AD}$$

and

$$\boldsymbol{Y} = \boldsymbol{O_2C} + \boldsymbol{CE} = (AB/AD)\boldsymbol{AD}e^{i\alpha} + (CE/CB)\boldsymbol{O_2A}e^{i\alpha}$$

(since O_2ABC remains a parallelogram)

$$= (\boldsymbol{O_2A} + \boldsymbol{AD})e^{i\alpha}(AB/AD)$$

(As $CE/CB = AB/AD$ from similar triangles)

$$= \boldsymbol{X}(AB/AD)e^{i\alpha}.$$

Hence, $(d\boldsymbol{Y}/d\boldsymbol{X}) = (AB/AD)e^{i\alpha} = $ constant. Thus, the figure is copied to a scale (AB/AD) and at an orientation α from the original figure.

4.3 Input-Output Curves of 4R Linkages

In Section 3.4 we noted that there are basically two different types of 4R linkages, viz., Grashof and non-Grashof. A Grashof linkage can be a crank-rocker, a double-crank or a double-rocker, whereas a non-Grashof linkage can be only a double-rocker. Figures 4.3-1a to d show the typical input (θ_2)-output (θ_4) characteristics of various types of 4R linkages. It should be noted that, as expected, except at the ends of the rocker-swings, for every value of θ_2 there are two values of θ_4 (if

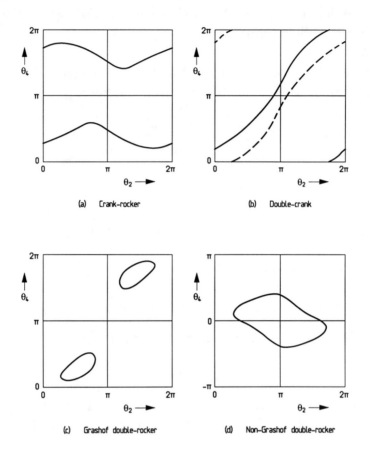

Figure 4.3-1

4.3. INPUT-OUTPUT CURVES OF 4R LINKAGES

any real value for θ_4 exists) and vice-versa. The following distinctive features of various plots shown in Figs. 4.3-1a to d should be noted.

(i) For a Grashof linkage (of all types), there are two disconnected branches of θ_4 vs. θ_2 characteristics, implying that they refer to different ways the linkage (of given link-lengths) can be assembled. Once assembled, the input-output relationship is governed by the curve corresponding to that assembly only. To produce the θ_4 vs. θ_2 characteristics indicated by the other curve, the linkage has to be dismantled and reassembled in the other configuration. It is left to the reader to be convinced about the typical characteristics shown in Figs. 4.3-1a to c. It is helpful to remember the result stated in Exercise Problem 3.5 and that two mirror-image configurations (with the mirror along the line of frame) exist for every 4R linkage.

(ii) For a non-Grashof linkage, the θ_4 vs. θ_2 curve is a single closed loop, implying that the linkage, once assembled, need not be dismantled to generate the entire curve.

(iii) For a rocking link (connected to the frame), there exists a real value of the angle made by that link (with the line of frame), for which there exists only one real value of the angle made by the other link connected to the frame.

(iv) For the θ_4 vs θ_2 curves of a double-crank linkage, there exist points of inflection, but no maxima or minima.

Problem 4.3-1

Assume that a 4R linkage can be assembled with link-lengths l_1, l_2, l_3 and l_4, where links 2 and 4 are connected to the frame 1. Derive the condition for complete rotatability of link 2.

Solution

From feature (iii) mentioned above, for link 2 to make comple rotation, there should not exist any real value of θ_2 for which θ_4 has a single value. From Equation (h) of Problem 4.2-1, the condition for θ_4 to have a single value is

$$a^2 + b^2 = c^2 \qquad (a)$$

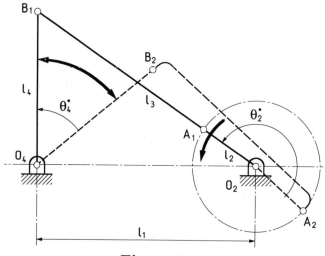

Figure 4.3-2

where

$$a = \sin\theta_2, \quad b = \cos\theta_2 + (l_1/l_2) \quad \text{and}$$
$$c = (l_1/l_4)\cos\theta_2 + \left[(l_1^2 + l_2^2 - l_3^2 + l_4^2)/(2l_2l_4)\right].$$

Substituting for a, b and c in Equation (a), we get a quadratic equation in $\cos\theta_2$, the two roots of which are obtained, after simplification, as

$$\cos\theta_2 = \left[l_1^2 + l_2^2 - (l_3^2 + l_4^2 \pm 2l_3l_4\right]/(2l_1l_2). \tag{b}$$

Now, for link 2 to make complete rotation, the roots of θ_2 from Equation (b) should be nonreal, i.e., $\cos\theta_2$ should not lie in the range $-1 \leq \cos\theta_2 \leq 1$. Hence, the conditions for complete rotatability of link 2 can be written as

$$\left[l_1^2 + l_2^2 - (l_3^2 + l_4^2) - 2l_3l_4\right]/(2l_1l_2) < -1 \tag{c}$$

and

$$\left[l_1^2 + l_2^2 - (l_3^2 + l_4^2) + 2l_3l_4\right]/(2l_1l_2) > 1. \tag{d}$$

From Equation (c) we get

$$l_2 + l_1 < l_3 + l_4. \tag{e}$$

4.3. INPUT-OUTPUT CURVES OF 4R LINKAGES

With $l_3 > l_4$, from Equation (d) we get

$$l_1 - l_2 > l_3 - l_4$$

or

$$l_2 + l_3 < l_1 + l_4. \qquad (f)$$

Since $l_3 > l_4$, from Equation (f) we can also write

$$l_2 + l_4 < l_1 + l_3. \qquad (g)$$

Similarly, with $l_4 > l_3$, again conditions (f) and (g) are resulted from the condition (d). Thus, Equations (e), (f) and (g) together give the condition for complete rotatability of link 2. It may be noted that Equations (e), (f) and (g) imply that link 2 is the shortest link of a Grashof linkage (compare these conditions with Equations 3.4.2a to c.

Of the three possible variations of a 4R linkage, crank-rocker is most commonly used in practice. It may be worthwhile to note (at this stage) a useful relation between the link-lengths which govern the quick-return ratio (assuming uniform crank speed) of such a linkage. As explained in Problem 4.1-2, the extreme positions of the rocker are attained when the crank and the coupler become collinear (see Fig. 4.3-2). In this figure, as the crank rotates through an angle θ_2^*, the rocker moves from position O_4B_2 to O_4B_1 through an angle θ_4^* and, during a rotation $(2\pi - \theta_2^*)$ of the crank, the follower swings back from O_4B_1 to O_4B_2. Note that θ_2 and θ_4 are measured in the same sense. As is evident from Fig. 4.3-2, if $\theta_2^* = \pi$, then the line B_1B_2 passes through O_2, i.e., $\angle B_1O_2O_4 = \angle B_2O_2O_4$. Under this condition, the rocker takes equal time during its forward and backward motion and there is no quick-return effect. Thus, from triangles $B_1O_4O_2$ and $B_2O_4O_2$ we can write

$$l_4^2 = (l_2 + l_3)^2 + l_1^2 - 2l_1(l_2 + l_3)\cos \angle B_1O_2O_4$$

and

$$l_4^2 = (l_3 - l_2)^2 + l_1^2 - 2l_1(l_3 - l_2)\cos \angle B_2O_2O_4.$$

When $\angle B_1O_2O_4 = \angle B_2O_2O_4$, the above two equations, after simplification, result in

$$l_1^2 + l_2^2 = l_3^2 + l_4^2. \tag{4.3.1}$$

Thus, Equation (4.3.1) constitutes the condition for the rocker to take equal time for its forward and backward motion with uniform crank speed. Further, it is easy to see that, with A_1, A_2, B_1 and B_2 on the same line,

$$l_2 = l_4 \sin(\theta_4^*/2) \tag{4.3.2}$$

where θ_4^* is the swing angle of the rocker.

If $\theta_2^* > \pi$ (as shown in Fig. 4.3-2), then the line $B_1 B_2$ intersects the line of frame outside $O_4 O_2$ and the reader can prove that (Hint: Note that $\angle B_2 O_2 O_4 > \angle B_1 O_2 O_4$, hence $\cos \angle B_1 O_2 O_4 > \cos \angle B_2 O_2 O_4$)

$$l_3^2 + l_4^2 > l_1^2 + l_2^2.$$

Similarly, with $\theta_2^* < \pi$, the line $B_1 B_2$ intersects the line of frame within $O_4 O_2$ and

$$l_3^2 + l_4^2 < l_1^2 + l_2^2.$$

So, we see that the quick-return ratio depends on the ratio

$$(l_1^2 + l_2^2)/(l_3^2 + l_4^2).$$

Problem 4.3-2

Figure 4.3-3a shows a windshield wiper mechanism where the wiper blade 5 is the coupler of a parallelogram linkage $O_4 CDO_6$. The input member $O_2 A$ rotates at a constant angular speed.

(a) Obtain the field being wiped.

(b) It is desired that the field of wiping be made symmetrical about the vertical line through O_4 by changing only the angle δ. Determine the required value of δ.

(c) A symmetrical wiping field of 450 mm width is desired, with a further constraint of equal time for forward and backward motion of the wiper, by changing only δ, $O_2 A$ and AB, with all other dimensions unchanged. Determine the required values of δ, $O_2 A$ and AB.

4.3. INPUT-OUTPUT CURVES OF 4R LINKAGES

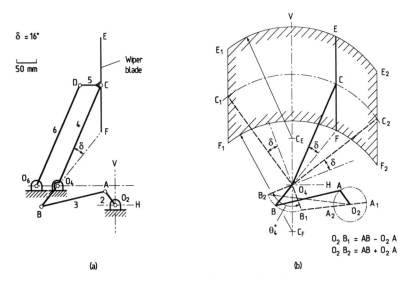

Figure 4.3-3a & b

Solution

(a) Since link 5 is the coupler of a parallelogram linkage, the wiper blade (5) always remains parallel to itself, i.e., link 5 undergoes curvilinear translation. Points E and F move on circular paths, with their centers on the vertical line through O_4. In Fig. 4.3-3b the extreme positions of link 4 are obtained in a manner explained in Problem 4.1-2. The corresponding locations of EF are indicated by E_1F_1 and E_2F_2. Two circles are drawn, one through E_1, E_2 and E, and the other through F_1, F_2 and F. We may note that $O_4C_E = CE$ and $O_4C_F = CF$. The field of wiping (bounded by these two circles and the vertical lines E_1F_1 and E_2F_2) is indicated by hatching in Fig. 4.3-3b.

(b) The swing angle of link 4 is seen to be θ_4^* (Fig. 4.3-3b). Thus, $\angle B_1O_4B_2 = \angle C_1O_4C_2 = \theta_4^*$. It is desired that $\angle C_1O_4V = \angle C_2O_4V = \theta_4^*/2$ when the field of wiping will be symmetrical about the line OV. Accordingly, Fig. 4.3-3c is drawn with $\angle C_1O_4V = \angle C_2O_4V = \theta_4^*/2$. From this figure the required δ is found to be $20°$.

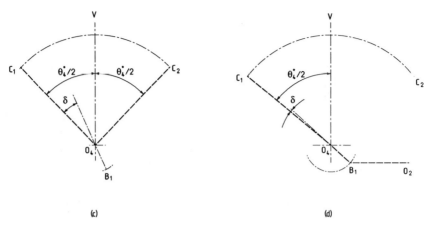

Figure 4.3-3c & d

(c) From measurements (Fig. 4.3-3a), $l_1 = O_2O_4 = 165$ mm, $O_4C = 288$ mm and $l_4 = O_4B = 70$ mm. Referring to Fig. 4.3-3c we note that when the field of wiping is symmetrical, its width $C_1C_2 = 2O_4C \sin(\theta_4^*/2)$. Hence, for $C_1C_2 = 450$ mm we get

$$\theta_4^* = 2\sin^{-1}(450/576) = 102.75°. \quad (a)$$

With equal time for forward and backward motion of the wiper, i.e., of link 4 as well, from Equations (4.3.1) and (4.3.2) we can write

$$165^2 + (O_2A)^2 = (AB)^2 + 70^2 \quad (b)$$

and

$$O_2A = 70\sin(\theta_4^*/2) = 54.69 \text{ mm}. \quad (c)$$

Using Equations (c) in (b) we get

$$AB = 159.11 \text{ mm}. \quad (d)$$

Figure 4.3-3d is drawn with θ_4^* given by Equation (a) and $O_2B_1 = (AB - O_2A)$ with AB and O_2A given by Equations (d) and (c), respectively. From measurements, the required value of δ is found to be $5°$.

4.4. TRANSMISSION ANGLE

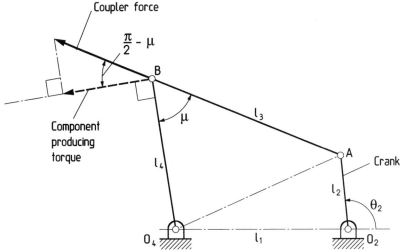

Figure 4.4-1

4.4 Transmission Angle

For smooth functioning of a mechanism in real life we should ensure that, besides satisfying the kinematic requirements, the mechanism moves freely. No kinematic criterion can be found that ensures this free movement. A complete dynamic force analysis including inertia, gravity and friction forces is necessary to check the free or smooth running of a mechanism. However, at the stage of the kinematic design we would like to ensure that, at least under static conditions the output member receives a large component of the force (or torque from the member driving it) in the direction of its movement. Neglecting gravity and friction, under static conditions all binary links become two-force members. Consequently, the force in any such link is along the link vector and the force analysis becomes simple. Based on these conditions, for 4R, 3R-1P and such simple mechanisms, an index called the *transmission angle* is used to measure the free-running quality.

For a 4R linkage, the transmission angle μ is defined as the acute angle between the coupler and the follower so that $0 \leq \mu \leq \pi/2$. As explained in Fig. 4.4-1, when $\mu = \pi/2$ the entire coupler force is utilized to produce the torque, which drives the follower. For good

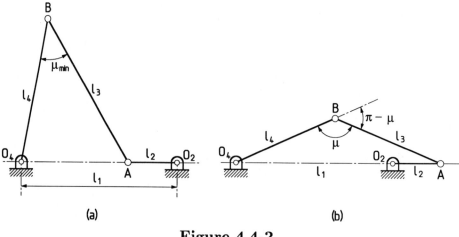

Figure 4.4-2

performance of a mechanism it is suggested that $\mu_{\min} > 30° - 35°$. For a double-rocker, obviously there are instants when $\mu = 0$ (i.e. when the coupler and the follower become collinear). As shown below, if the driving member (crank) rotates completely, then the transmission angle takes its minimum value when the crank is collinear with the line of the frame.

Referring to Fig. 4.4-1,

$$O_4 A^2 = l_1^2 + l_2^2 + 2l_1 l_2 \cos \theta_2 = l_3^2 + l_4^2 - 2l_3 l_4 \cos \mu$$

or,

$$\cos \mu = \left[(l_3^2 + l_4^2) - (l_1^2 + l_2^2) - 2l_1 l_2 \cos \theta_2 \right] / (2l_3 l_4). \qquad (4.4.1)$$

If $\mu < \pi/2$, then μ is minimum when $\cos \mu$ is maximum (with $\cos \mu > 0$). From Equation (4.4.1) this occurs with $(l_3^2 + l_4^2) > (l_1^2 + l_2^2)$ when $\theta_2 = \pi$, as indicated in Fig. 4.4-2a. It may be remembered that for a crank-rocker the condition $(l_3^2 + l_4^2) > (l_1^2 + l_2^2)$ implies, as shown in Fig. 4.3-2, that $B_1 B_2$ intersects the line of frame outside $O_4 O_2$, with $\theta_2^* > \pi$.

Likewise, if $\mu > \pi/2$, then the transmission angle is defined as $(\pi - \mu)$ and $\cos \mu < 0$. In such a situation the transmission angle is minimum when μ is maximum. This occurs, as can be seen from

4.4. TRANSMISSION ANGLE

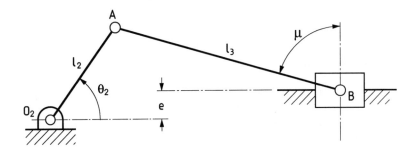

Figure 4.4-3

Equation (4.4.1), with $(l_1^2 + l_2^2) > (l_3^2 + l_4^2)$, when $\theta_2 = 0$. This configuration is indicated in Fig. 4.4-2b. Again it may be noted that, for a crank-rocker, the condition $(l_1^2 + l_2^2) > (l_3^2 + l_4^2)$ implies that θ_2^*, in Fig. 4.3-2, is less than π, with $B_1 B_2$ intersecting the line of frame within $O_4 O_2$.

For a crank-rocker without quick-return ($\theta_2^* = \pi$), i.e., when $(l_1^2 + l_2^2) = (l_3^2 + l_4^2)$, the minimum transmission angle occurs for both $\theta_2 = 0$ and π.

For a slider-crank mechanism (Fig. 4.4-3), the transmission angle is defined as the acute angle between the connecting rod and the normal to the slider movement, as indicated in the figure. The reader may find that the minimum transmission angle is given by[2]

$$\mu_{\min} = \cos^{-1}\left[(l_2 + e)/l_3\right]. \tag{4.4.2}$$

Problem 4.4-1

A crank-rocker 4R linkage is without any quick-return effect (i.e. unit quick-return ratio). Given that the rocker swing angle is θ_4^* and the minimum transmission angle is $\mu_{\min}$, determine the link-length ratios in terms of θ_4^* and $\mu_{\min}$.

Solution

For the desired crank-rocker mechanism, from Equations (4.3.1) and (4.3.2) we get

$$l_1^2 + l_2^2 = l_3^2 + l_4^2 \tag{a}$$

---

[2] For transmission angle of six-link mechanisms see Tao, D.C.: Applied Linkage Synthesis, Addison-Wesley, Reading, Massachusetts, (1964).

and
$$l_2 = l_4 \sin(\theta_4^*/2). \tag{b}$$

From Fig. 4.4-2 we can write

$$(l_1 - l_2)^2 = l_3^2 + l_4^2 - 2l_3 l_4 \cos \mu_{\min}. \tag{c}$$

Algebraic manipulation of Equations (a), (b) and (c) finally yield the link-length ratios as

$$l_3/l_1 = \sin(\theta_4^*/2)/\cos \mu_{\min},$$

$$l_4/l_1 = \left[\left\{1 - (l_3/l_1)^2\right\} / \left\{1 - \sin^2(\theta_4^*/2)\right\}\right]^{1/2}$$

and

$$l_2/l_1 = \left[(l_3/l_1)^2 + (l_4/l_1)^2 - 1\right]^{1/2}.$$

We shall see in Chapter 6 that for a prescribed value of θ_4^*, with $\theta_2^* = \pi$, the value of $\mu_{\min}$ has to be less than a limiting value. This limiting value can be obtained from the condition that the link-length ratios obtained above satisfy Grashof criterion for a crank-rocker linkage. It is left for the reader to show that $\mu_{\min} < (\pi/2 - \theta_4^*/2)$.

It will be further seen, in Section 6.10, that the input-output relationship of a 4R linkage with poorer transmission angle is also more sensitive to manufacturing tolerances and clearances in the joints (called mechanical errors). Thus, a satisfactory transmission angle is a very important design constraint for all real-life mechanisms in order to ensure quiet and jerk-free operation and less sensitivity to manufacturing defects.

When the transmission angle is zero under static condition, it is implied that infinite force is required to move the output member. In case of a vice-grip plier (Fig. 1.11-5), the two links O_2A and AB are almost in one line at the gripping position. Thus, with O_4B as the driving (input) member, the transmission angle is almost equal to zero, which implies self-locking.

4.5 Exercise Problems

4.1 Figure 4.5-1 shows the quick-return mechanism used in slotting machines. The stroke-length of the tool is adjusted by changing

4.5. EXERCISE PROBLEMS

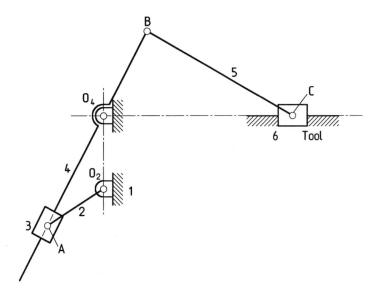

Figure 4.5-1

the link-length O_4B. Determine the stroke-length and the quick-return ratio for uniform angular speed of link 2.

4.2 The crank and the connecting rod of a slider-crank mechanism are of lengths 10 cm and 16 cm, respectively. Determine the maximum possible quick-return ratio for the slider, that can be achieved with constant angular velocity of the crank.

4.3 Figure 4.5-2 shows a mechanism for laying cloth in a textile machine. Crank 2 rotates at a constant speed. The dimensions of the mechanism are as follows:

$a = 550$ mm, $\quad b = O_2A = 170$ mm, $\quad c = 550$ mm,
$d = 80$ mm, $\quad e = 400$ mm, $\quad$ swing-arm length $l_s = 900$ mm.

(a) Determine the width of the layer (l_w) being laid, the eccentricity (m) and the quick-return ratio of the swinging arm.

(b) It is desired to have $w = 1000$ mm, with $m = 0$ and a further requirement of equal time for the forward and the

CHAPTER 4. DISPLACEMENT ANALYSIS

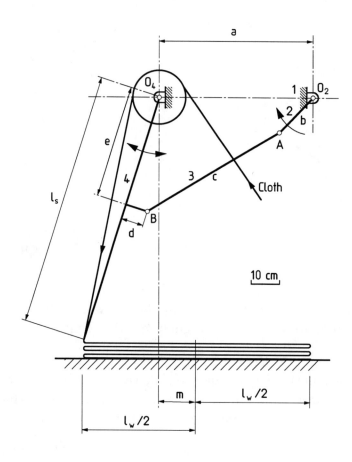

Figure 4.5-2

4.5. EXERCISE PROBLEMS

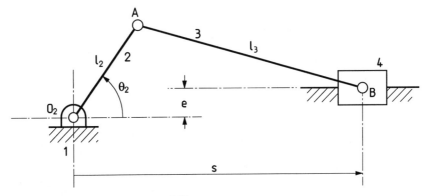

Figure 4.5-3

backward motions of the swinging arm. Determine the required values of c, d and e with all other dimensions unchanged.

4.4 Figure 4.5-3 shows an offset slider-crank mechanism. Determine the input-output $(\theta_2 - s)$ relationship using the loop closure equation.

4.5 Following the procedure outlined in Problem 4.3-1, obtain the condition for complete rotatability of link 2 in Fig. 4.5-3.

4.6 For a crank-rocker 4R linkage without quick-return what is the maximum value of the transmission angle? At, what value of the crank angle θ_2 does this maximum occur?

4.7 Figure 4.5-4 shows a six-link mechanism. Using complex exponential notation for link-vectors, show that the point E moves along a straight line. What is the inclination of this line to the axis of the slider-movement? Also, obtain the ratio of the velocities of the points E and B given $O_2A = CD$ and $AC = O_2D$.

4.8 For a crank-rocker 4R linkage, the centrodes for the coupler cannot be closed curves since there exist two configurations in the cycle when the crank and the rocker become parallel, signifying the instantaneous center at infinity. Justify this statement.

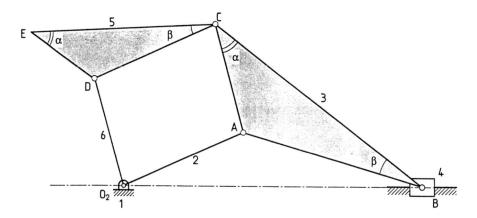

Figure 4.5-4

4.9 For a 4R crank-rocker linkage the rocker swings through an angle θ_4^*. Show that the minimum and the maximum possible values of crank rotation during the forward stroke of the follower are given, respectively, by

$$(\theta_2^*)_{\min} = (\pi/2) + (\theta_4^*/2) \text{ and } (\theta_2^*)_{\max} = (3\pi/2) + (\theta_4^*/2).$$

4.10 Consider the modified crank-shaper mechanism shown in Fig. 5.5-11, with $O_2A = 4$ cm, $O_5B = 5$ cm and $BC = 25$ cm. Determine the quick return ratio for the slider (6) with constant angular velocity of O_2A.

4.11 Consider the mechanism shown in Fig. 3.5-10. Prove that all the links in this mechanism can have only oscillatory motion. Also obtain the extreme positions of links 2 and 4 by a graphical procedure.

Chapter 5

VELOCITY AND ACCELERATION ANALYSIS

5.1 Introduction

The quality of the design of a mechanical system depends on the proper use of knowledge in the fields of dynamics, stress analysis and some associated areas of mechanical engineering, including material properties. Kinematic analysis, dealing with the displacement, velocity and acceleration of the members of a system, is the first step towards dynamic analysis, which ultimately provides information about forces, moments, etc.

A large number of methods exist for velocity and acceleration analysis, which help a designer to either critically examine an already existing system or predict the motion characteristics of a proposed design. Displacement analysis, being substantially different in nature from velocity and acceleration analysis, was discussed in Chapter 4. In this chapter, various methods for velocity and acceleration analysis of planar mechanisms are presented.

The kinematics of a mechanism can be studied either by graphical method or by analytical procedure. The former approach is usually preferred by the designers, as it yields results quickly and provides an overall idea about the motion characteristics of all members of the system under consideration. On the other hand, analytical approach is suitable when highly accurate results are desired and/or a similar analysis has to be repeated for a large number of configurations of a mechanism. One should also remember that there are problems that cannot be solved by graphical method. It should be noted that to conduct kinematic analysis only the kinematic dimensions of a mechanism are necessary.

To carry out velocity analysis three basically different approaches will be presented. These are (i) method of instantaneous centers, (ii) velocity difference method and (iii) auxiliary point method. Acceleration analysis can be carried out by using the acceleration difference approach. But, for a complex mechanism, special techniques are needed. The techniques for acceleration analysis of complex mechanisms that will be discussed in this chapter are (i) auxiliary point method, (ii) method of normal component, (iii) Goodman's indirect method and (iv) use of Euler-Savary equation. At the end, velocity and acceleration analysis using analytical approach will be presented.

5.2 Velocity Analysis (Graphical Methods)

5.2.1 Some Fundamental Aspects

Before starting the discussions on velocity analysis, a few important points of fundamental importance will be presented in order to facilitate a smoother presentation of the subsequent sections. Transfer of motion by a hinge connection between two bodies i and j leads to some interesting results that can be used in velocity analysis. Figure 5.2-1 shows two rigid bodies i and j connected by a revolute pair at B. If v_A is the velocity of a point A on body i, then v_A^I is the component of v_A along AB (labelled as I). Now, the component of the velocity of the point B_i on i along I must be equal to v_A^I because the body i is rigid. We also know that the velocity of the point B_j on j will be identical to that of B_i. Hence, $v_{B_j}^I = v_{B_i}^I = v_A^I$. Since for a hinged connection at B, $v_{B_i} \equiv v_{B_j}$, from here forward, we shall drop the

5.2. VELOCITY ANALYSIS (GRAPHICAL METHODS)

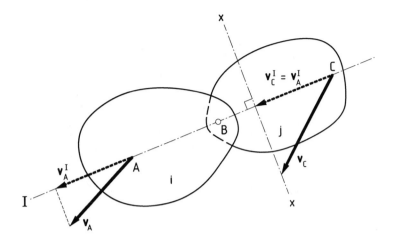

Figure 5.2-1

subscript for such points and write only v_B. Now if C is any point on j on the line obtained by joining A and B, the component of its velocity along I will be equal to v_B^I ($\equiv v_A^I$). Thus, we may conclude that the tip of the vector v_C must be on the line xx (Fig. 5.2-1).

When motion is transferred from body i to body j by physical contact at a point P, as shown in Fig. 5.2-2, the components of the velocities of P_i and P_j along nn (common normal at P) are equal. So if v_{P_i} is known, the line xx on which the tip of v_{P_j} lies can be determined, as indicated in Fig. 5.2-2.

5.2.2 Method of Instantaneous Centers

The concept of instantaneous center of velocity can be very conveniently used for velocity analysis of those mechanisms for which the relative instantaneous centers can be geometrically determined. The most commonly used technique is to find out the location of relative instantaneous center ij, where i is the input link (whose velocity is prescribed) and j is the link whose velocity must be determined. The velocity of the particle on body i at the location of ij can be obtained. The velocity of the particle on body j at the location of ij, being same as that of the particle on i, is thus known. Once the velocity of this

170 CHAPTER 5. VELOCITY AND ACCELERATION ANALYSIS

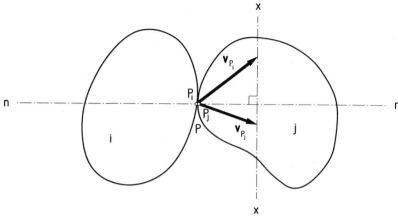

Figure 5.2-2

point on body j is known, the velocity of link j can be determined, if it is a hinged or a sliding link. The procedure will be further clarified with the help of a few examples.

Problem 5.2-1

Link 2 of the 4R linkage shown in Fig. 2.7-2 is rotating at a speed of 1 rad/s in the CW direction. Find out the angular velocity of link 4, given $O_2O_4 = 19.5$ cm.

Solution

Since the velocity of link 2 is prescribed and that of link 4 must be determined, first we locate 24, as shown in Fig. 2.7-2. The distance of 24 from O_2 is 26.25 cm as per the given scale. Thus, a particle on link 2 at the location of 24 has a velocity of (26.25 x 1 =) 26.25 cm/s, in the vertically upward direction. This must also be the velocity of a particle on link 4 at the location of 24. Since link 4 is hinged to the frame at O_4, the distance of which, from the instantaneous center 24 is 6.7 cm, the angular velocity of 4 must be equal to (26.25/6.7 =) 3.92 rad/s in the CW direction.

When a mechanism has many links, it is advisable to follow a systematic method in order to determine the required instantaneous center. Since the total number of relative instantaneous centers are large in such cases, it should be attempted to determine the required instantaneous center with minimum effort. Use of the circle diagram is quite suitable for this purpose.

5.2. VELOCITY ANALYSIS (GRAPHICAL METHODS)

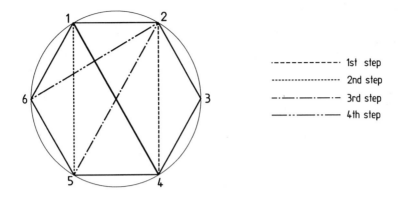

Figure 5.2-3

Problem 5.2-2

Link 2 of the quick-return mechanism shown in Fig. 2.7-3a rotates at 3 rad/s in the CW direction. Determine the velocity of the slider 6 at the instant shown, given $O_2O_4 = 11$ cm.

Solution

Since there are 6 links, the total number of all the relative instantaneous centers is 15. Out of these, 7 are directly identified by the lower pairs (as indicated by the solid lines in Fig. 2.7-3b). We have to determine 26. First, 24 is determined from the intersection of the lines (12-14) and (23-34). Next, we determine 15 from the intersection of lines (16-65) and (14-45). Then, 25 is found out as the intersection of lines (12-15) and (24-45). Once 25 is known, 26 is determined by the lines (12-16) and (25-56) (see Fig. 2.7-3a). A particle on link 2 at the location of 26 (8.5 cm above O_2) moves towards the right with a velocity of $(3 \times 8.5 =) 25.5$ cm/s. As link 6 is a translating member, all points on it (including the one at the location of 26) move with the same velocity. Hence, the slider 6 moves towards the right with a velocity of 25.5 cm/s. Note that we had to find out only 4 relative instantaneous centers out of the 8 unknown ones. The steps are indicated in the circle diagram shown in Fig. 5.2-3.

Mechanisms with higher pairs can also be easily analysed for velocity using the concept of instantaneous center.

172 CHAPTER 5. VELOCITY AND ACCELERATION ANALYSIS

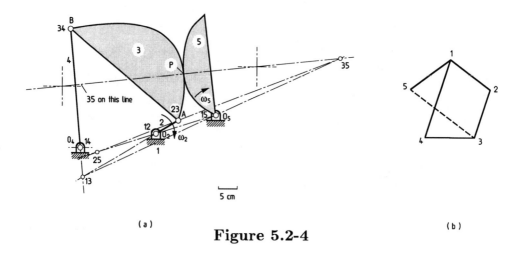

Figure 5.2-4

Problem 5.2-3

Figure 5.2-4a shows a five-link mechanism in which the follower 5 is driven by the cam 3, which is the coupler link of the 4R linkage. If the input link 2 rotates with an angular velocity $\omega_2 = 10$ rad/s (CW), determine the angular velocity of the output link 5.

Solution

First, all the relative instantaneous centers, determined by the lower pairs, are located as indicated in Fig. 5.2-4a. Since motion is transferred to link 5 by link 3 through a higher pair contact at P, the relative instantaneous center 35 lies on the common normal to the contacting surfaces at P. Next, the circle diagram is drawn (as shown in Fig. 5.2-4b), in which the lower pairs are represented by solid lines and the higher pair by a broken line. Our interest is to determine 25. As can be seen from Fig. 5.2-4b, 35 must be known if 25 is to be determined. Since 1-2-3-4-1 is a quadrilateral with the sides as solid lines (implying known or already determined instantaneous centers) the diagonal 13 can be obtained. Once 13 is known, location of 35 on the common normal is determined when line 35 in Fig. 5.2-4b is replaced by a solid line. Finally, the instantaneous center represented by the diagonal 25, is found out and its distances from O_2 and O_5 are measured to be 17 cm and 34 cm, respectively. Hence $\omega_5 = (17/34) \times 10$ rad/s (CW) $= 5$ rad/s (CW).

5.2. VELOCITY ANALYSIS   (GRAPHICAL METHODS)

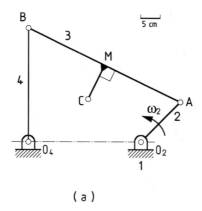

(a)

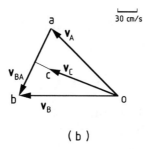

(b)

Figure 5.2-5

5.2.3 Method of Velocity Difference

The most straightforward way to carry out velocity analysis is to construct a velocity diagram starting from the input link. In these diagrams, the fixed link is represented by a point called the pole of the velocity diagram. While solving the problems, we shall explain the method using velocity difference.

Problem 5.2-4

Figure 5.2-5a shows a 4R linkage in which link 2 is rotating in the CCW direction at a speed of 10 rad/s. C is a point on the coupler AB at a distance of 10 cm from the midpoint of AB. Determine the velocity of point C.

174 CHAPTER 5. VELOCITY AND ACCELERATION ANALYSIS

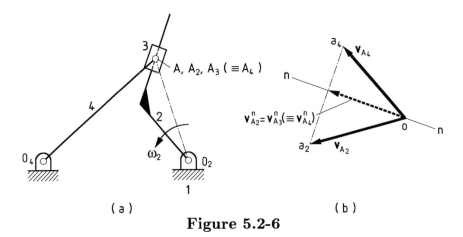

Figure 5.2-6

Solution

The fixed link is represented by the velocity diagram pole o and the velocity of point A is given by

$$v_A = \omega_2 \times O_2 A. \tag{a}$$

The magnitude of v_A is 150 cm/s and is perpendicular to O_2A. This is represented by oa in the velocity diagram (Fig. 5.2-5b). Thus, from Equation (2.3.1) we can write

$$v_B = v_A + v_{BA}. \tag{b}$$

We also know that v_{BA} is perpendicular to AB, as both points A and B are on the same rigid link 3. It is further known that, as link 4 is hinged to the frame at O_4, the velocity of B must be perpendicular to O_4B. Hence, the intersection of the line from the tip of v_A and $\perp^r$ [1] to AB and the line from o $\perp^r$ to O_4B represents b, implying ab is v_{BA} and ob is v_B. The point c is located so that $\triangle ABC \cong \triangle abc$ (see Section 2.4). The required velocity v_C is given by oc as indicated in Fig. 5.2-5b. Measurement yields v_C to be equal to 117 cm/s.

In the above example, the motion transfer points A and B are well identified and do not change with configuration. However, when a mechanism consists of a prismatic or higher pair, two coincident points on the two links at the location of the contact must be considered. As

---

[1] The symbols $\|^l$ and $\perp^r$ stand for parallel and perpendicular, respectively.

5.2. VELOCITY ANALYSIS (GRAPHICAL METHODS)

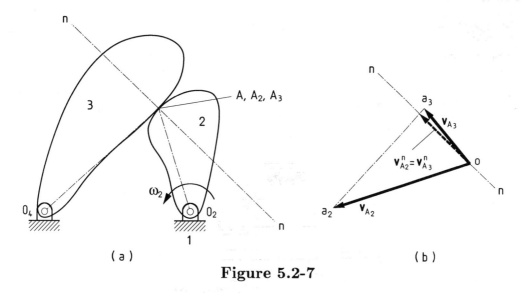

Figure 5.2-7

explained in Section 5.2.1, the components of the velocities of these two points (at the instant under consideration) in the direction of common normal nn (Fig. 5.2-2) are equal. For example, Fig. 5.2-6a shows a four-link mechanism with three R pairs and one P pair. These particles A_2, A_3 and A_4, can be chosen at A but attached to bodies 2, 3 and 4, respectively. As links 3 and 4 are connected by an R pair at A, A_2 and A_4 are always the same. Now, the velocity of point A, $v_{A2} = \omega_2 \times O_2 A$. The velocity of A_4, $v_{A_4} (= v_{A_2} + v_{A_4 A_2})$ must be $\perp^r$ to $O_4 A$ and at the same time its component along nn should be equal to that of v_{A_2} [2] (Fig. 5.2-6b). When a mechanism consists of a higher pair, as shown in Fig. 5.2-7a, the procedure remains as described above. To solve for the output velocity ω_3, the condition that v_{A_2} and v_{A_3} have equal components along nn must be used, along with the condition that v_{A_3} is perpendicular to $O_3 A$ (see Fig. 5.2-7b). The concept of coincident points is sometimes very useful in kinematic analysis, as will be depicted through the following example.

Problem 5.2-5

Figure 5.2-8a shows a four-link mechanism consisting of three R pairs and one P pair. The input link 2 is rotating with an angular

---

[2] It also implies that $v_{A_4 A_2}$ should be along the common tangent which is $\perp^r$ to nn.

176 CHAPTER 5. VELOCITY AND ACCELERATION ANALYSIS

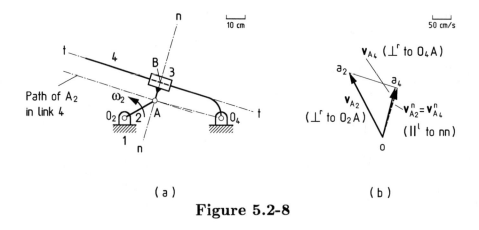

Figure 5.2-8

velocity (ω_2) equal to 10 rad/s in the CCW direction, as indicated. Find out the angular velocity of link 4.

Solution

To start with, we conceive two points A_2 and A_4 at the location of the hinge A, attached to bodies 2 and 4, respectively. It should be noted that the path of A in link 4 is parallel to the axis of the prismatic pair. Therefore, the components of v_{A_2} and v_{A_4} along nn should be equal. Figure 5.2-8b shows the velocities v_{A_2} and v_{A_4} satisfying this condition (which also implies that $v_{A_4 A_2}$ will be parallel to the axis of the prismatic pair). From the diagram, v_{A_4} comes out to be 135 cm/s. Since the distance OA is equal to 37 cm, ω_4 is 3.69 rad/s in the CW direction.

The problem can also be solved by considering two coincident points B_3 and B_4 (at the intersection of the P pair axis and the normal drawn to the axis from A) attached to bodies 3 and 4, respectively. First, we can write

$$v_A = v_{B_4} + v_{AB_3} + v_{B_3 B_4} = \omega_4 \times O_4 B + \omega_3 \times BA + v_{B_3 B_4}.$$

Now let us assume $\omega_4 = 10$ rad/s in the CW direction and note that $\omega_3 = \omega_4$[3]. We can then construct the velocity diagram representing

---

[3]Note that whenever two links are connected by a prismatic pair, their angular velocities (and also angular accelerations) are equal.

5.2. VELOCITY ANALYSIS (GRAPHICAL METHODS)

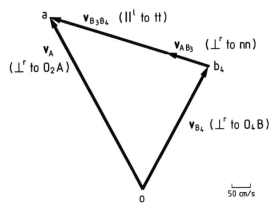

Figure 5.2-9

the above equation since, the direction of v_A is known ($\perp^r$ to O_2A). This is shown in Fig. 5.2-9. The magnitude of v_A comes out as 515 cm/s and, consequently, ω_2 is 27.1 rad/s in the CCW direction. Since in the problem, ω_2 is given as 10 rad/s in the CCW direction, $\omega_4 = 10 \times 10/27.1$ rad/s (CW) $= 3.69$ rad/s (CW). If the sliding velocity of link 3 on link 4 ($v_{B_3B_4}$) is to be determined, from Fig. 5.2-9, it comes out as 330 cm/s (towards left). Thus, the actual value of $v_{B_3B_4} = 330 \times 10/27.1$ cm/s $= 122$ cm/s (towards left).

One should note that since ω_4 and ω_3 (equal to ω_4) are unknown, the problem was solved (the second approach) by considering the output link 4 to be the input link with some *assumed* velocity, and determining the *corresponding* ω_2. Since the ratio of velocities of different members of a mechanism depends only on the configuration, such a procedure is possible in the case of velocity analysis.

Problem 5.2-6

In the slotted-lever quick-return mechanism shown in Fig. 5.2-10a, the crank O_2A rotates at a speed of 30 r.p.m.(CCW). For the position shown, determine the velocity of the ram (i.e., of the point C).

Solution

The magnitude of the angular velocity of link 2 is $\omega_2 = 2\pi (30/60)$ rad/s (CCW) $= 3.14$ rad/s (CCW). The fixed hinge points O_2 and O_4 of the mechanism are represented as o in the velocity diagram (Fig. 5.2-10b). Let us consider two coincident points $A_2(\equiv A)$ and A_4 on

178 CHAPTER 5. VELOCITY AND ACCELERATION ANALYSIS

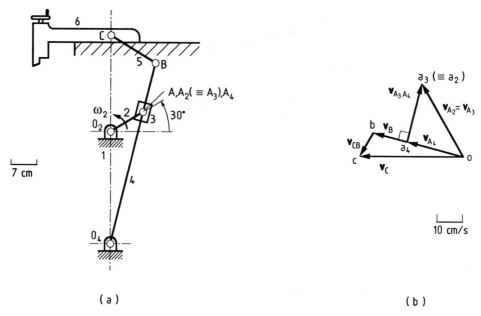

Figure 5.2-10

links 2 and 4, respectively. We can now write[4]

$$v_{A_2} = \omega_2 \times O_2 A = v_{A_4} + v_{A_3 A_4}.$$

v_{A_2} is completely known, the direction of v_{A_4} is known ($\perp^r$ to $O_4 A$) and we also know the direction of $v_{A_3 A_4}$, which is the direction of sliding ($\|^l$ to $O_4 A$). Hence, the velocity triangle, $oa_4 a_3$, can be drawn as indicated in Fig. 5.2-10b. As both the points A_4 and B are on the same link 4, v_B is easily obtained (from the concept of velocity image) by extending oa_4 to b so that $O_4 B / O_4 A = ob/oa_4$ (Fig. 5.2-10b). Again, we can write

$$v_C = v_B + v_{CB}$$

---

[4]Note that to be able to draw the vector polygon representing a vector algebraic equation at the most two quantities (i.e., two magnitudes or two directions or one magnitude and one direction) can be unknown.

5.2. VELOCITY ANALYSIS   (GRAPHICAL METHODS)

in which v_B is completely known and the direction of v_{CB} ($\perp^r$ to BC) and v_C (along the sliding direction of C) are also known. Now we can complete the velocity triangle obc as shown in Fig. 5.2-10b. From measurements, we get

$$v_C = 40 \text{ cm/s} \quad \text{(towards left)}.$$

5.2.4 Auxiliary Point Method

Most of the velocity analysis problems can be solved by using either the concept of instantaneous center or method of velocity difference. However, there are cases where all the instantaneous centers cannot be determined geometrically (see Section 2.7), or the method of velocity difference fails because some velocity equations contain more than two unknowns. These situations may arise in both single and multiple degree-of-freedom mechanisms. The auxiliary point approach can be conveniently used to solve such problems. This approach is, thus, more powerful as compared to the direct applications of the velocity-difference and instantaneous-center methods.

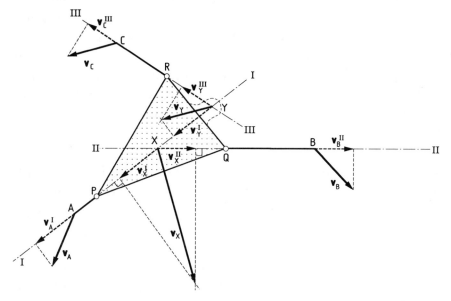

Figure 5.2-11

In this method, *auxiliary lines* are drawn through the *motion-transfer points* of a floating link in the directions along which the velocity components of these points can be determined. *Auxiliary points* attached to the link are defined as the points of intersection of the auxiliary lines. The basic idea behind the method of using the auxiliary points is illustrated below.

Figure 5.2-11 shows a ternary link whose motion-transfer points are P, Q and R. Letting the velocities of points A on link AP (v_A), B on link BQ (v_B) and C on link CR (v_C), already be known, we must obtain complete information regarding the velocity of the ternary link PQR. Referring to Section 5.2.1, we note that the velocity component of P along AP, being same as that of A, is known. Thus, we take the first auxiliary direction along AP and label it as I (Fig. 5.2-11). From similar considerations, we identify auxiliary directions II through Q along BQ and III through R along CR. The auxiliary directions I and II intersect at the auxiliary point X. The velocity component of X along I, v_X^I ($= v_P^I = v_A^I$), is now known. Similarly, the velocity component of X along II, v_X^{II} ($= v_Q^{II} = v_B^{II}$), is also known. Therefore, the components of velocity of the (auxiliary) point X on the ternary link are known in two different directions, viz., I and II. Hence, v_X determined as indicated in the figure. Similarly, another auxiliary point Y (on the ternary link) is obtained at the intersection of I and III, and v_Y can also be determined. Thus, the velocity of the ternary link is completely known. It should be remembered that the location of an auxiliary point can be outside the physical boundary of the link under consideration but, it should be always considered to be physically connected to the link.

Problem 5.2-7

Figure 5.2-12 shows a mechanism in which link 2 rotates at 1.25 rad/s (CW) and the hydraulic cylinder-piston actuator O_7D expands at a rate of 10 cm/s. Determine the angular velocity of link 5 at this instant.

Solution

It can be easily verified that this problem cannot be solved either by the velocity-difference method or by the method of instantaneous centers. Thus, we have to adopt the method using the concept of auxiliary point. From the information given, point A of link 3 moves

5.2. VELOCITY ANALYSIS (GRAPHICAL METHODS)

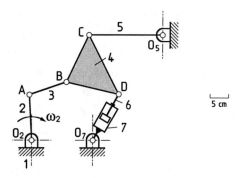

Figure 5.2-12

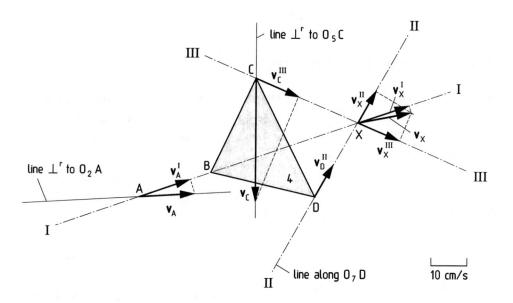

Figure 5.2-13

with a velocity of approximately 15 cm/s in a direction perpendicular to O_2A (Fig. 5.2-13). The component of v_A along AB is determined as shown in Fig. 5.2-13, and we take auxiliary direction I along AB as indicated. We also know that the component of the velocity of point D along O_7D is nothing but the expansion rate of the actuator, i.e., 10 cm/s. Therefore, we take auxiliary direction II along O_7D. Hence, if X (at the intersection of I and II) is a point on link 4, the components of its velocity along I and II are v_A^I and 10 cm/s, respectively. Thus, v_X is obtained as shown in Fig. 5.2-13. Once v_X is known, the velocity of point C on link 4 can be obtained as follows. Points X and C are joined by a line marked III. Then, v_C^{III} ($= v_X^{III}$) is obtained as shown in the figure. Finally, v_C is determined as a vector, perpendicular to O_5C, whose component along III is v_C^{III}. From v_C, the angular velocity of link 5 (ω_5) is obtained as 1.6 rad/s (CCW).

5.3 Acceleration Analysis (Graphical Methods)

5.3.1 Some Fundamental Aspects

As was done in the case of velocity analysis, in the case of acceleration analysis it is also desirable to discuss a few important and fundamental aspects in order to facilitate smooth presentation of the subsequent sections on acceleration analysis.[5]

Let us again consider links i and j, hinged at B, as shown in Fig. 5.3-1a. If the component of acceleration of any point A on i along AB (line I as indicated in Fig. 5.3-1a) is a_A^I, then

$$a_B^I = a_A^I + \omega_i^2 . BA. \qquad (5.3.1)$$

Similarly, considering B and C to be points on link j, a_C^I and a_B^I are related as

---

[5] One must always remember that while starting acceleration analysis it should be ensured that the corresponding velocity analysis is completed and all the velocities are known.

5.3. ACCELERATION ANALYSIS (GRAPHICAL METHODS)

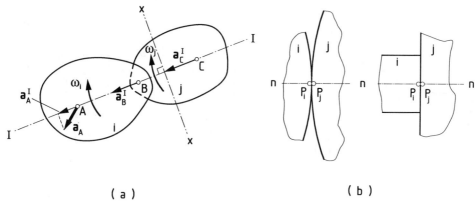

(a)   (b)

Figure 5.3-1

$$a_C^I = a_B^I + \omega_j^2.CB. \tag{5.3.2}$$

From Equations (5.3.1) and (5.3.2)

$$a_C^I = a_A^I + \omega_i^2.BA. + \omega_j^2.CB. \tag{5.3.3}$$

Thus, if a_A^I is known, then the line XX, on which the tip of the vector a_C should lie, can be determined using Equation (5.3.3) as shown in Fig. 5.3-1a. This is useful in acceleration analysis using the technique of auxiliary point.

When motion is transferred between bodies i and j only through physical contact (Fig. 5.3-1b), then the acceleration difference between the coincident points P_i and P_j has components, not only in the direction of sliding but also along the common normal nn at the contact point. However, the velocity difference between P_i and P_j can only be along the direction of sliding.

Unlike velocities, the ratios of various accelerations in a mechanism is not dependent on the configuration only; therefore the technique of interchanging the input output links is not applicable. The basic relations between the accelerations of two points on a moving body, discussed in Section 2.3, are used in this section. The concept of coincident points will also be used often. The acceleration difference between such points has been discussed in Section 2.5. The most straightforward approach towards solving acceleration analysis

184 CHAPTER 5. VELOCITY AND ACCELERATION ANALYSIS

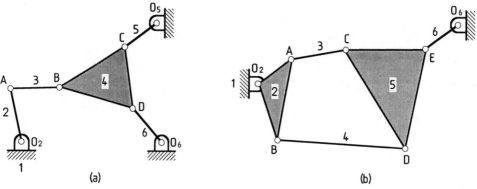

Figure 5.3-2

problems is the method of acceleration difference. However, such an approach is feasible when the radii of curvature of the paths of all the motion-transfer points are known. But if a mechanism has a ternary or a higher-order floating link (i.e., one which is not connected to the frame), the direct successive application of the acceleration difference equation of the form $a_B = a_A + a_{BA}$ (where A and B are the motion transfer points of a link) may fail to complete the acceleration analysis. Such a mechanism is considered to be kinematically complex. Figure 5.3-2 shows two such mechanisms.

In the mechanism shown in Fig. 5.3-2a, if 2 is the input link, a_B cannot be determined from a_A as the radius of path curvature of B is not known. However, if in the same mechanism, 6 (or 5) is chosen as the input link, then a_C and a_D can be determined, and, thereafter, we can find out a_B and a_A. Such mechanisms are called *complex mechanisms with low degree of complexity*. This transformation from a complex to a simple mechanism by changing the input link from 2 to 6, is not possible for the mechanism shown in Fig. 5.3-2b, and such a mechanism is called a *complex mechanism with high degree of complexity*. If the radii of path curvature of at least two motion transfer points on each of the floating links are known, the mechanism has a low degree of complexity. Otherwise, the mechanism has a high degree of complexity. Special methods need to be adopted for analysing the acceleration of a complex mechanism, a few of which are discussed in this section. It is also possible to adopt the direct method

5.3. ACCELERATION ANALYSIS (GRAPHICAL METHODS)

of acceleration difference if the unknown radius of path curvature can be determined. Euler-Savary equation may, sometimes, be found useful in this connection.

Acceleration analysis of mechanisms involving higher pairs is also quite involved, and the most straightforward way to handle such problems is to construct an equivalent linkage consisting of lower pairs only (see Section 1.9). Of course, there are methods for acceleration analysis of higher pair mechanisms without usage of an equivalent linkage.

5.3.2 Method of Acceleration Difference

As in the case of velocity analysis using the method of velocity difference, in the case of acceleration also, a vector equation is formed that can be solved by drawing an acceleration diagram. The problem is simple when all the motion transfer points are R pairs only. The procedure will be illustrated with the help of the following example.

Problem 5.3-1

If the input link O_2A of the four-link mechanism of Problem 5.2-4 rotates with an angular velocity of 10 rad/s in the CCW direction and an angular acceleration of 75 rad/s in the CW direction, find out the acceleration of point C.

Solution

Before beginning any acceleration analysis, it is necessary to carry out the velocity analysis. Since the data are the same the results of Problem 5.2-4 can be directly used. The acceleration of point B can be written as

$$a_B = a_A + a_{BA}. \qquad (a)$$

Since a_B and a_A can be written as a_{BO_4} and a_{AO_2}, respectively, Equation (a) can be written as follows:

$$a_{BO_4} = a_{AO_2} + a_{BA} \qquad (b)$$

Now, using Equations (2.3.2), (2.3.3) and (2.3.4) the above equation can be written as follows:

186 CHAPTER 5. VELOCITY AND ACCELERATION ANALYSIS

$$\omega_4 \times (\omega_4 \times O_4 B) + \alpha_4 \times O_4 B = \omega_2 \times (\omega_2 \times O_2 A) + \alpha_2 \times O_2 A$$
$$+ \omega_3 \times (\omega_3 \times AB) + \alpha_3 \times AB \quad (c)$$

where

$$\omega_4 \times (\omega_4 \times O_4 B) = a^n_{BO_4},$$

$$\alpha_4 \times O_4 B = a^t_{BO_4},$$

$$\omega_2 \times (\omega_2 \times O_2 A) = a^n_{AO_2},$$

$$\alpha_2 \times O_2 A = a^t_{AO_2},$$

$$\omega_3 \times (\omega_3 \times AB) = a^n_{BA} \quad \text{and}$$

$$\alpha_3 \times AB = a^t_{BA}.$$

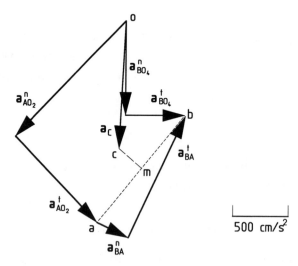

Figure 5.3-3

5.3. ACCELERATION ANALYSIS (GRAPHICAL METHODS)

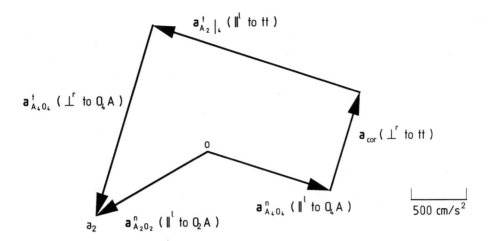

Figure 5.3-4

We note that $a^n_{BO_4}$, $a^n_{AO_2}$, $a^t_{AO_2}$ and a^n_{BA} are completely known and the directions of $a^t_{BO_4}$ and a^t_{BA} are known. Hence, the acceleration polygon can be completed.[6] Figure 5.3-3 shows the completed polygon representing Equation (c). From Fig. 5.2-5b, ω_4 and ω_3 come out as 5.4 rad/s (CCW) and 2.6 rad/s (CCW), respectively, which have been used in constructing the acceleration diagram. Once points a and b (the tips of the vectors $a_A = a^n_{AO_2} + a^t_{AO_2}$ and $a_B = a^n_{BO_2} + a^t_{BO_2}$) are obtained, point c is determined so that $\triangle ABC \cong \triangle abc$ (see Section 2.4). Then, oc represents a_C, which comes out as 1133 cm/s² in the direction of oc (Fig. 5.3-3).

The concept of coincident points is useful while solving an acceleration analysis problem of a mechanism having prismatic pairs.

Problem 5.3-2

If the angular velocity of the input link 2 of the inverted slider-crank mechanism described in Problem 5.2-5 is constant, determine the angular acceleration of link 4.

---

[6] Recall that a vector equation is equivalent to two scalar equations. In other-words, a vector equation can be solved (graphically in this case) if only two unknowns (two directions or two magnitudes or one magnitude and one direction) are involved.

188 CHAPTER 5. VELOCITY AND ACCELERATION ANALYSIS

Solution

As we did when solving Problem 5.2-5, we consider coincident points A_2 and A_4 at A attached to bodies 2 and 4, respectively. The following quantities were determined in Problem 5.2-5:

$$\omega_4 = 3.69 \text{ rad/s} \quad (CW),$$

$$v_{A_2}|_4 = v_{A_3}|_4 = v_{B_3}|_4 = v_{B_3 B_4} = 125 \text{ cm/s}$$

(towards left). Note that link 3 being a sliding member, A and B have the same relative motion with respect to body 4. Now, from Equation (2.5.5)

$$\boldsymbol{a}_{A_2} = \boldsymbol{a}_{A_4} + \boldsymbol{a}^n_{A_2}|_4 + \boldsymbol{a}^t_{A_2}|_4 + \boldsymbol{a}_{Cor} \qquad (a)$$

where $\boldsymbol{a}^n_{A_2}|_4$ is zero, since the radius of curvature of the path of A_2 in link 4 is infinity. The Coriolis acceleration $\boldsymbol{a}_{Cor}$ is given by (2.5.4) as follows: $\boldsymbol{a}_{Cor} = 2\boldsymbol{\omega}_4 \times \boldsymbol{v}_{A_2}|_4 = 922$ cm/s in a direction $\perp^r$ to tt (Fig. 5.2-8a), i.e., along $\boldsymbol{AB}$. Equation (a) can be expanded as follows:

$$\boldsymbol{a}^n_{A_2 O_2} + \boldsymbol{a}^t_{A_2 O_2} = \boldsymbol{a}^n_{A_4 O_4} + \boldsymbol{a}^t_{A_4 O_4} + \boldsymbol{a}^t_{A_2}|_4 + \boldsymbol{a}_{Cor} \qquad (b)$$

in which $\boldsymbol{a}^n_{A_2 O_2}, \boldsymbol{a}^t_{A_2 O_2}, \boldsymbol{a}^n_{A_4 O_4}$ and $\boldsymbol{a}_{Cor}$ are completely known and the directions of $\boldsymbol{a}^t_{A_4 O_4}$ and $\boldsymbol{a}^t_{A_4}|_4$ are known. Hence, the acceleration polygon can be completed as indicated in Fig. 5.3-4. It should be remembered that the order of addition of the vectors in Equation (b) is unimportant. From Fig. 5.3-4, the magnitude of $\boldsymbol{a}^t_{A_4 O_4}$ comes out as 1738 cm/s.

Since $\boldsymbol{a}^t_{A_4 O_4} = \boldsymbol{\alpha}_4 \times \boldsymbol{O}_4 \boldsymbol{A}, \alpha_4 = 47$ rad/s^2 in the CCW direction.

5.3.3 Method of Normal Component

This method is applicable to mechanisms with low degree of complexity only. It is also useful as a supplement to the auxiliary point method (presented in Section 5.3.4) for certain mechanisms with high degree of complexity, when the latter method alone is not sufficient. The underlying principle of this method is that the acceleration component of a point P on a link, in a direction perpendicular to its velocity (called the normal component), is independent of the angular acceleration of the link. Thus, $a^n_p = v^2_p/\rho_p$, where ρ_p is the radius of

5.3. ACCELERATION ANALYSIS (GRAPHICAL METHODS)189

curvature of the path of point P. The steps to be followed in applying this method are as follows:

(i) Transform the mechanism into a simple one by changing the input link.

(ii) Carry out the velocity analysis[7] with this alternative input link and determine the true velocities.

(iii) Draw an auxiliary acceleration diagram based on the true velocities and zero acceleration of the alternative input link. Determine the normal component of acceleration of the motion-transfer point (on the floating link) whose radius of path curvature is unknown.

(iv) Construct the true acceleration diagram with the actual input acceleration, using the known normal component of acceleration of the motion-transfer point with unknown radius of path curvature.

Problem 5.3-3

In the mechanism shown in Fig. 5.3-5a, link 2 rotates with a constant angular velocity $\omega_2 = 10$ rad/s, as indicated (CW). Determine α_6.

Solution

With link 2 as the driving link, the mechanism is complex since the path curvature of B is unknown. However, if link 6 (or 5) is considered to be the input link, the mechanism is transformed into a simple one.

To start with, the velocity diagram (Fig. 5.3-5b) is drawn to some scale assuming link 6 to be the input link. When the diagram is completed, its scale is determined by $v_A = \omega_2.O_2A = 10 \times 7.5$

---

[7]It should be noted that the velocity diagram can be completed without any reference to actual magnitude of the velocities. Only the scale of the diagram will vary according to the magnitude of the input velocities. This is because the ratios of the velocities depend only on the geometric configuration as already mentioned earlier.

190 CHAPTER 5. VELOCITY AND ACCELERATION ANALYSIS

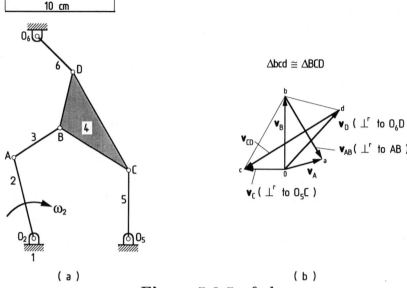

Figure 5.3-5a & b

cm/s = 75 cm/s in the direction obtained in the diagram.[8] Having determined the scale, all velocities can be obtained as follows: $v_D = 150$ cm/s, $\omega_6 = 32$ rad/s (CCW), $\omega_3 = 27$ rad/s (CCW), $\omega_4 = 21$ rad/s (CW) and $\omega_5 = 13$ rad/s (CCW).

Next, an auxiliary acceleration diagram is constructed, with link 6 as the input member and taking $\alpha_6 = 0$, based on the actual velocities already obtained (see Fig. 5.3-5c). The auxiliary acceleration of the point B, $\boldsymbol{a}_{\underset{o}{B}}$ [9], is obtained from this diagram. From $\boldsymbol{a}_{\underset{o}{B}}$ and $\boldsymbol{a}_{\underset{o}{D}}$, $\boldsymbol{a}^n_{\underset{o}{B}}$ and $\boldsymbol{a}^n_{\underset{o}{D}}$ are determined, as shown in Fig. 5.3-5d, where $\boldsymbol{a}^n_{\underset{o}{B}}$ is the component of $\boldsymbol{a}_B \perp^r$ to v_B and $\boldsymbol{a}^n_{\underset{o}{D}} = \boldsymbol{a}_{\underset{o}{D}}$, as α_6 is taken to be zero. Since $\boldsymbol{a}^n_{\underset{o}{B}} = \boldsymbol{a}^n_B$ and $\boldsymbol{a}^n_{\underset{o}{D}} = \boldsymbol{a}^n_D$, we can draw the true acceleration diagram with 2 as the input link; this is shown in Fig. 5.3-5e. From this figure, $\alpha_6 = a^t_D/O_6 D = 320$ rad/s^2 (CW).

---

[8] The scale could have been negative with a wrong choice of the direction of ω_6, when v_A from the diagram would have been opposite to the actual sense.

[9] Symbol o at the bottom will now refer to values corresponding to zero input acceleration.

5.3. ACCELERATION ANALYSIS (GRAPHICAL METHODS)

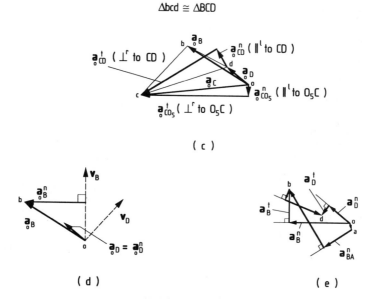

Figure 5.3-5c, d & e

5.3.4 Auxiliary Point Method

As in the case of velocity analysis, the auxiliary point method is also useful for acceleration analysis. It is applicable to all mechanisms with low degree of complexity and most mechanisms with high degree of complexity. In certain mechanisms with high degree of complexity, this method alone may not be sufficient and must be used in conjunction with the method of normal components described in the previous section.

In this method, auxiliary points are determined on the higher-order floating link at the intersection of auxiliary lines drawn through the motion-transfer points of the link along directions in which components of accelerations can be determined. Two such auxiliary points are sufficient. Components of acceleration of each of these auxiliary points in two directions are known and hence, their accelerations can be determined. Once the accelerations of two points of the floating link are determined, the concept of acceleration image can be employed to complete the analysis.

192 CHAPTER 5. VELOCITY AND ACCELERATION ANALYSIS

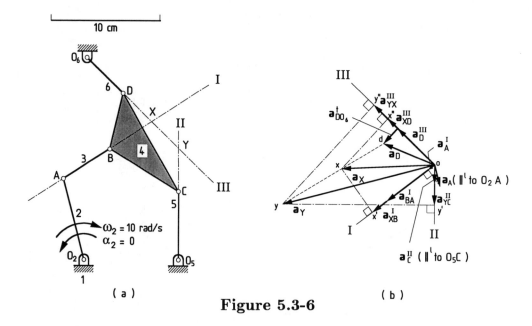

Figure 5.3-6

Problem 5.3-4

Solve Problem 5.3-3 by the auxiliary point method.

Solution

Velocities are to be determined first and we will assume the values already obtained. From Fig. 5.3-6a, we see that the acceleration of point A is completely known, whereas for points C and D, components of their accelerations along CO_5 and DO_6, respectively, are known (as these depend only on ω_5 and ω_6, which have already been determined). Thus, three lines I, II and III are drawn through B, C and D in the directions AB, O_5C and O_6D, respectively. X and Y are identified as the two auxiliary points (on link 4) at the intersection of I and III and at the intersection of II and III, respectively.[10]

Using, Equation (5.3.3) we can write

$$a_X^I = a_A^I + a_{BA}^I + a_{XB}^I$$
$$= a_A^I + \omega_3^2.BA + \omega_4^2.XB = ox' \quad \text{(Fig.5.3} - 6\text{b)}.$$

---

[10]To solve this particular problem, only one auxiliary point would have been enough, but we are using X and Y to demonstrate the procedure in general.

5.3. ACCELERATION ANALYSIS (GRAPHICAL METHODS)

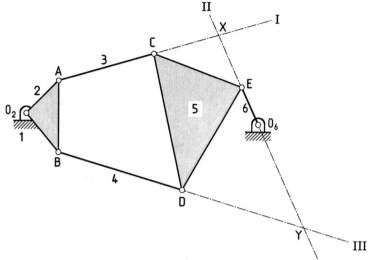

Figure 5.3-7

Similarly

$$a_X^{III} = a_D^{III} + a_{XD}^{III}$$
$$= a_D^{III} + \omega_4^2.XD = ox''.$$

Finally, a_X is found to be ox (see Fig. 5.3-6b).
To determine a_Y, we first determined

$$a_Y^{II} = a_C^{II} + a_{YC}^{II} = a_C^{II} + \omega_4^2.YC = oy'$$

and

$$a_Y^{III} = a_D^{III} + a_{YD}^{III} = a_D^{III} + a_{XD}^{III} + a_{YX}^{III} = a_D^{III} + \omega_4^2.YD = oy''.$$

Hence, a_Y ($= oy$) is determined as shown. Once a_X and a_Y are known, a_D is determined using the concept of acceleration image. Then, α_6 can be obtained, as $a_{DO_6}^t/O_6D = 320$ rad/s^2 (CW).

In the above example, it was not necessary to consider two auxiliary points. However, in case of mechanisms with high degree of complexity, it is necessary to identify two auxiliary points on the higher order floating link. One such mechanism is shown in Fig. 5.3-7. When the motion of link 2 is prescribed (i.e., ω_2 and α_2 are given), v_A, v_B, a_A and a_B are known. To solve the problem, two auxiliary points X and Y are identified, as indicated in Fig. 5.3-7. First, the velocity analysis is carried out using the following relations:

$$\begin{aligned} v_X^I &= v_A^I, & v_X^{II} &= 0, \\ v_Y^{III} &= v_B^{III} & \text{and} \quad v_Y^{II} &= 0. \end{aligned}$$

Once the velocities are determined, accelerations of the auxiliary points (on link 5) can be obtained using the following relations:

$$\begin{aligned} a_X^I &= a_A^I + a_{CA}^I + a_{XC}^I = a_A^I + \omega_3^2.AC + \omega_3^2.CX \quad \text{(all towards } A\text{)}, \\ a_X^{II} &= a_E^{II} + a_{XE}^{II} = \omega_6^2.O_6E + \omega_5^2.XE \quad \text{(all towards } O_6\text{)}, \\ a_Y^{III} &= a_B^{III} + a_{DB}^{III} + a_{YD}^{III} = a_B^{III} + \omega_4^2.DB + \omega_5^2.DY \quad \text{(all towards } B\text{) and} \\ a_Y^{II} &= a_E^{II} + a_{YE}^{II} = \omega_6^2.O_6E - \omega_5^2.YE \quad \text{(all towards } O_6\text{)} \end{aligned}$$

When a_X and a_Y are known, the rest of the problem is quite straightforward.

In the two examples discussed thus far, all the motion transfer points were hinges. A similar procedure is followed for doing kinematic analysis of mechanisms in which a floating link contains a prismatic pair(s). Consider the ternary floating link shown in Fig. 5.3-8. Motions of points A and D are prescribed, i.e., v_A, v_D, a_A and a_D are known. The acceleration analysis can be completed as explained below.

We have already seen that a prismatic pair is a limiting case of a revolute pair. Thus, the relative motion between point D (on the slider) and the coincident point E (on the ternary link) remains the same when D is connected to the ternary link 3 by a link hinged to the ternary link at ∞ in the direction perpendicular to the slot axis. Hence, the corresponding auxiliary line is a line through D (and E) perpendicular to the slot axis. The other two auxiliary lines are along AB and O_2C as indicated in Fig. 5.3-8. Any two of the three auxiliary points X, Y and Z can be considered. Let us take X and Y as the auxiliary points for kinematic analysis. It is assumed that the

5.3. ACCELERATION ANALYSIS (GRAPHICAL METHODS)

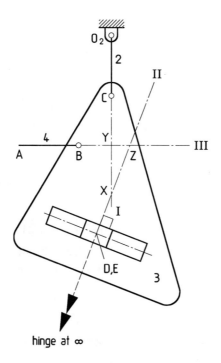

Figure 5.3-8

196 CHAPTER 5. VELOCITY AND ACCELERATION ANALYSIS

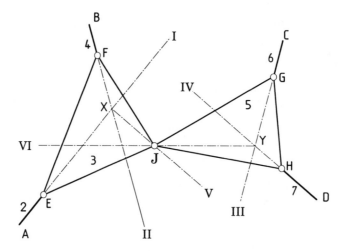

Figure 5.3-9

velocity analysis has already been performed and only the acceleration analysis is presented below. The following quantities are known: a_A^{III}, a_C^I ($= \omega_2^2 . O_2 C$ towards O_2), a_D^{II} and a_{DE}^{II} (Coriolis component given by $2\omega_3 \times v_{DE}$).[11] Therefore,

$$a_X^I = a_C^I + a_{XC}^I = \omega_2^2 . CO_2 + \omega_3^2 . XC \quad \text{and}$$
$$a_X^{II} = a_D^{II} + a_{ED}^{II} + a_{XE}^{II} = a_D^{II} - a_{DE}^{II} + a_{XE}^{II}$$
$$= a_D^{II} - 2\omega_3 \times v_{DE} + \omega_3^2 . XE.$$

When the components of a_X in two different directions I and II are known, a_X can be determined. To determined a_Y, the following equations are used:

$$a_Y^I = a_C^I + a_{YC}^I = \omega_2^2 . CO_2 + \omega_3^2 . YC$$
$$a_Y^{III} = a_A^{III} + a_{BA}^{III} + a_{YB}^{III} = a_A^{III} + \omega_4^2 . BA + \omega_3^2 . YB$$

---

[11] Use of vector notation here is convenient to take care of the Coriolis acceleration. Otherwise, algebraic equations would have been adequate, as in other cases. Further, note that $v_{DE} = v_{D|_3}$.

5.3. ACCELERATION ANALYSIS (GRAPHICAL METHODS)

If two ternary links (say 3 and 5) are connected as shown in Fig. 5.3-9, the analysis must be carried out in two steps with the help of six auxiliary directions. This is illustrated as follows. Let a_A, a_B, a_C and a_D be given (what should be known, at least, are their components along AE, BF, CG and DH, respectively, which are identified as the auxiliary directions I, II, III and IV). The vector a_X can be determined from a_X^I and a_X^{II} using the following equations:

$$a_X^I = a_A^I + a_{EA}^I + a_{XE}^I = a_A^I + \omega_2^2.EA + \omega_3^2.XE$$
$$a_X^{II} = a_B^{II} + a_{FB}^{II} + a_{XF}^{II} = a_B^{II} + \omega_4^2.FB + \omega_3^2.XF$$

Similarly, a_Y is determined from a_C and a_D using the following relations:

$$a_Y^{III} = a_C^{III} + a_{GC}^{III} + a_{YG}^{III} = a_C^{III} + \omega_6^2.GC + \omega_5^2.YG$$
$$a_Y^{IV} = a_D^{IV} + a_{HD}^{IV} + a_{YH}^{IV} = a_D^{IV} + \omega_7^2.HD + \omega_5^2.YH$$

Once a_X and a_Y are known, a_J can be obtained using the following relations (directions V and VI are obtained by joining XJ and YJ, respectively):

$$a_J^V = a_X^V + a_{JX}^V = a_X^V + \omega_3^2.JX \quad \text{and}$$
$$a_J^{VI} = a_Y^{VI} + a_{JY}^{VI} = a_Y^{VI} + \omega_5^2 JY.$$

Now by using the concept of acceleration image and the accelerations of points X, Y and J, accelerations of all other points can be determined.

It has already been noted that sometimes the direct use of auxiliary points may not be successful. In such cases, the auxiliary point method has to be used along with the technique of normal component. If link 6 (instead of link 2) is the input link with known motion, in the mechanism shown in Fig. 5.3-7, the acceleration analysis is performed by applying the method of normal component. An auxiliary acceleration diagram is constructed on the basis of true velocities, treating link 2 as the input link with zero acceleration. This is done

CHAPTER 5. VELOCITY AND ACCELERATION ANALYSIS

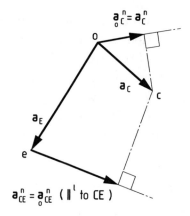

Figure 5.3-10

using the auxiliary point method, as explained earlier in this section. From this auxiliary acceleration diagram, the normal components of the auxiliary accelerations a_{oC} and a_{oD} are obtained. Now $a^n_{oC} = a^n_C$ and $a^n_{oD} = a^n_D$. Since in the actual problem a_E is known, a_C can be determined using the relation

$$a_C = a_E + a_{CE} = a_E + a^n_{CE} + a^t_{CE}$$

and the known a^n_C (in the direction perpendicular to v_C), as shown in Fig. 5.3-10. Once a_C is known, the motion of link 5 is determined as a_E is given.

5.3.5 Goodman's Indirect Method

Goodman's indirect approach to the acceleration analysis of complex mechanisms is based on the following two properties of a constrained mechanism:

(i) The angular velocities and accelerations of links are linear functions of the respective input quantities.

5.3. ACCELERATION ANALYSIS (GRAPHICAL METHODS)

(ii) The relative angular velocities and accelerations between different links of a mechanism remain unaffected by kinematic inversion.

It is also relevant to mention at this stage that this approach is not really a method by itself. When the acceleration analysis of a given problem cannot be carried out by any of the graphical methods directly, but an auxiliary analysis is possible with a change of input link and/or the frame, Goodman's approach can be used to derive the actual results from the values obtained through the auxiliary analysis.

Let the subscript i denote quantities related to the input link, and l denote quantities related to any other link of a constrained mechanism. If θ is the angular position of a link from a fixed reference line, then

$$\omega_l = d\theta_l/dt = (d\theta_l/d\theta_i)d\theta_i/dt = C_l\omega_i \quad (5.3.4)$$

where C_l is a quantity depending only on the geometric configuration of the mechanism (except at dead-center locations, when two links coincide with each other) and is independent of the velocities of different members. The above equation is the mathematical statement of the obvious fact that the velocity polygons of a mechanism at any instant, with different input velocities, are scale drawings of one another. The angular acceleration of link l

$$\begin{aligned} \alpha_l &= d\omega_l/dt = d(C_l\omega_i)/dt = \omega_i(dC_l/dt) + C_l\dot{\omega}_i \\ &= \omega_i \frac{d}{d\theta_i}(\frac{d\theta_l}{d\theta_i})\frac{d\theta_i}{dt} + C_l\alpha_i \\ &= C_l'\omega_i^2 + C_l\alpha_i \end{aligned} \quad (5.3.5)$$

where C_l' also represents a geometrical property. The first term on the RHS (right-hand side) of Equation (5.3.5) represents the angular acceleration of link l with input link acceleration $\alpha_i = 0$, and the velocity being equal to the actual velocity ω_i. Thus, Equation (5.3.5) can be written as

$$\alpha_l = \underset{\circ l}{\alpha} + C_l\alpha_i.$$

200 CHAPTER 5. VELOCITY AND ACCELERATION ANALYSIS

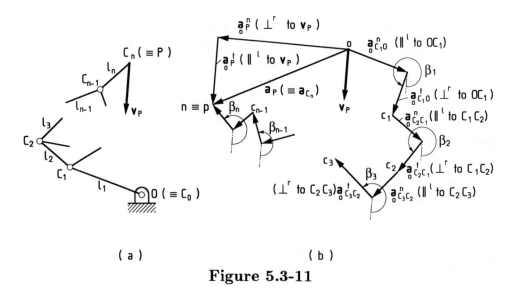

(a) (b)

Figure 5.3-11

Now $C_l = d\theta_l/d\theta_i = (d\theta_l/dt)/(d\theta_i/dt) = \omega_l/\omega_i$. Hence, finally the above equation takes the following form:

$$\alpha_l = \underset{o\,l}{\alpha} + (\omega_l/\omega_i)\alpha_i \qquad (5.3.6)$$

where $\underset{o\,l}{\alpha}$ denotes the angular acceleration of link l obtained from an auxiliary acceleration diagram drawn with true velocities, but with zero input acceleration. When the input link is a sliding member, Equation (5.3.6) takes the following form:

$$\alpha_l = \underset{o\,l}{\alpha} + (\omega_l/v_i)a_i \qquad (5.3.7)$$

where v_i and a_i denote the velocity and acceleration of the input link.

Now let us consider a mechanism in which a point P is connected to the fixed hinge O through a number of links and hinges, as shown in Fig. 5.3-11a. The other portion of the linkage has not been shown. The velocity of point P is v_P, as shown. Figure 5.3-11b shows the portion of the auxiliary acceleration diagram (relevant to the portion of the mechanism under consideration) when the velocities are the true velocities but the acceleration of the input link is zero. If we draw the true acceleration diagram, the normal components will remain unchanged. Therefore, the difference between a_P^t (in the true

5.3. ACCELERATION ANALYSIS (GRAPHICAL METHODS)

acceleration diagram) and a^t_{oP} (shown in Fig. 5.3-11b) is entirely due to the difference between the various tangential components (parallel to the corresponding velocities). Hence,

$$\begin{aligned}
a^t_P - a^t_{oP} &= (a^t_{C_1} - a^t_{oC_1})\cos\beta_1 + (a^t_{C_2 C_1} - a^t_{oC_2 C_1})\cos\beta_2 \\
&\quad + \ldots + (a^t_{C_n C_{n-1}} - a^t_{oC_n C_{n-1}})\cos\beta_n \\
&= (\alpha_1 - \alpha_{o1})l_1\cos\beta_1 + (\alpha_2 - \alpha_{o2})l_2\cos\beta_2 \\
&\quad + \ldots + (\alpha_n - \alpha_{on})l_n\cos\beta_n
\end{aligned}$$

where ($\beta_i - 90°$) is the angle of link $C_{i-1}C_i$ (with $C_o = 0$) measured from the direction of v_P in the CCW direction (Fig. 5.3-11b). Using Equation (5.3.6) in the above equation, we get (for rotary input link)

$$\begin{aligned}
a^t_P - a^t_{oP} &= l_1\frac{\omega_1}{\omega_i}\alpha_i\cos\beta_1 + l_2\frac{\omega_2}{\omega_i}\alpha_i\cos\beta_2 + \ldots + l_n\frac{\omega_n}{\omega_i}\alpha_i\cos\beta_n \\
&= (v_{C_1 O}\cos\beta_1 + v_{C_2 C_1}\cos\beta_2 + \ldots + v_{C_n C_{n-1}}\cos\beta_n)\frac{\alpha_i}{\omega_i}.
\end{aligned}$$

Since β's are measured from $v_{C_n}(\equiv v_P)$, it is obvious that

$$v_{C_1 O}\cos\beta_1 + v_{C_2 C_1}\cos\beta_2 + \ldots + v_{C_n C_{n-1}}\cos\beta_n = v_{C_n} \equiv v_P.$$

Using this relation in the previous equation, we get

$$a^t_P - a^t_{oP} = v_P\frac{\alpha_i}{\omega_i}$$

or

$$a^t_P = a^t_{oP} - v_P\frac{\alpha_i}{\omega_i} \quad \text{(for a rotary input link)}. \tag{5.3.8}$$

Similarly, for a sliding input link

$$a^t_P = a^t_{oP} - v_P\frac{a_i}{v_i}. \tag{5.3.9}$$

For a sliding output link, the total acceleration is in the direction of velocity. Hence,

$$a_s = a_{os} + v_s\frac{\alpha_i}{\omega_i} \quad \text{(for rotary input link)} \qquad (5.3.10)$$

and

$$a_s = a_{os} + v_s\frac{a_i}{v_i} \quad \text{(for sliding input link)}. \qquad (5.3.11)$$

The absolute velocities and accelerations mean those with respect to the fixed link g (ground link).[12] Thus, Equations (5.3.6) and (5.3.7) can be written as

$$\alpha_{lg} = \alpha_{olg} + \frac{\omega_{lg}}{\omega_{ig}}\alpha_{ig} \qquad (5.3.12)$$

and

$$\alpha_{lg} = \alpha_{olg} + \frac{\omega_{lg}}{v_{ig}}a_{ig}. \qquad (5.3.13)$$

In this form, the equations are applicable to any inversion of the mechanism where the use of g is not restricted to the frame. The subscript i denotes any alternative input link (not necessarily the actual input link) with assumed zero acceleration, on the basis of which the auxiliary acceleration diagram should be drawn.

This indirect approach can be applied to mechanisms with low degree of complexity in the following manner:

(i) Choose an alternative input link to transform the mechanism into a simple one. The auxiliary analysis is carried out with zero acceleration of this alternative input link.

(ii) Actual values are found by using Equations (5.3.6) to (5.3.11). For mechanisms with high degree of complexity, a direct kinematic inversion is made to transform the mechanism into a simple one.[13] The auxiliary analysis is conducted first assuming a suitable link as the input link with zero acceleration. Thereafter, using Equations (5.3.12) and (5.3.13), the desired result can be determined.

---

[12] Subscript f is often used to denote quantities related to floating link, so subscript g is used to denote the frame.

[13] One should note that such a conversion may not be always possible.

5.3. ACCELERATION ANALYSIS (GRAPHICAL METHODS)

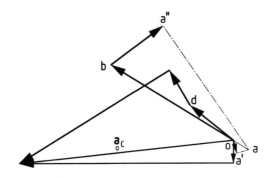

Figure 5.3-12

(iii) Other cases, in which kinematic inversion cannot lead to a simple mechanism and direct applications of other methods also fail, can be treated as follows. First, it should be examined whether acceleration analysis is possible by some method when a suitable link is assumed as the input link. If the answer is yes, then an auxiliary analysis is carried out assuming zero acceleration of the chosen input link. Actual results are obtained from the auxiliary results using Equations (5.3.6) to (5.3.11). When acceleration analysis of the mechanism (by any graphical method discussed thus far) is not possible by changing the input link only, it should be examined whether a kinematic inversion, along with a change of the input link, makes the analysis possible. If the answer is yes, then this auxiliary analysis is conducted first and, using Equations (5.3.12) and (5.3.13), the actual results can be determined.

It should be remembered that in deriving the foregoing equations, rotational quantities were considered positive in the CCW direction. A few examples are presented below.

Problem 5.3-5

Solve Problem 5.3-3 using the acceleration difference method along with Goodman's indirect approach.

Solution

This mechanism has a low degree of complexity and is transformed into a simple mechanism by considering link 6 as the alternative input

204 CHAPTER 5. VELOCITY AND ACCELERATION ANALYSIS

link. The velocity analysis, considering link 6 as the driver, has been carried out completely in Problem 5.3-3 and the true velocities of all the links have been obtained.

An auxiliary acceleration diagram with 6 as the input link, based on $\alpha_6 = 0$, and with true velocities, is drawn as shown in Fig. 5.3-12, a part of which was also obtained in Fig. 5.3-5c. In this diagram, $oa' = \omega_2^2 . O_2 A$ (parallel to O_2A) towards O_2, $ba'' = \omega_3^2 . AB$ (parallel to AB) towards B. The point a (so that $oa = a_A$) is determined by the intersection of two lines drawn from a'' (perpendicular to ba'') and from a' (perpendicular to oa'). In this auxiliary diagram $a'a = a^t_{oA}$, whose magnitude is 850 cm/s² (positive, as it is in the same direction as v_A). However, the actual input condition states $\alpha_A = 0$, i.e., $a^t_A = 0$. Using Equation (5.3.8) we can write

$$a^t_A = a^t_{oA} - v_A \frac{\alpha_6}{\omega_6}.$$

However, if a^t_A has to be equal to zero as prescribed (ω_2 is constant),

$$\alpha_6 = -a^t_{oA} \frac{\omega_6}{v_A} = -850 . \frac{28}{75} \text{ rad/s}^2 = -318 \text{ rad/s}^2.$$

Since the sign of α_6 has come out as negative it must be in the CW direction.

Problem 5.3-6

Investigate how the acceleration of link 2 of the six-link mechanism shown in Fig. 5.3-7 can be determined with prescribed ω_6 and α_6, using the concept of auxiliary point and Goodman's indirect approach.

Solution

It has already been stated that the problem cannot be solved by the direct application of only auxiliary point method when link 6 is the driver. However, if the ternary link 2 is the driver, kinematic analysis is possible by considering the two auxiliary points X and Y (shown in Fig. 5.3-7). The velocity analysis is first completed by considering link 2 as the input with any assumed velocity, as discussed earlier and the true velocities are determined. Next, α_2 is assumed to be zero and an auxiliary diagram is drawn on the basis of the true velocities with $\alpha_2 = 0$, and a_{oE} is determined as indicated qualitatively in Fig.

5.3. ACCELERATION ANALYSIS (GRAPHICAL METHODS)

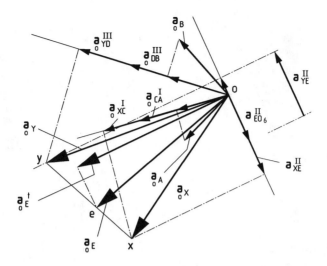

Figure 5.3-13

5.3-13. From this, $a^t_{o_E E}$ (in the direction of v_E i.e., perpendicular to $O_6 E$) is determined. Now, using Equation (5.3.8),

$$a^t_E = a^t_{o_E E} + v_E \frac{\alpha_2}{\omega_2}.$$

However, link 6 being the actual input link with ω_6 and α_6 prescribed, a_E and, therefore, a^t_E is known. the angular acceleration of link 2 can now be determined as follows:

$$\alpha_2 = \frac{\omega_2}{v_E}(a^t_E - a^t_{o_E E}).$$

Problem 5.3-7

Find out the acceleration of link 6 of the mechanism shown in Fig. 5.3-7, when ω_2 is 1 rad/s and constant, using Goodman's indirect approach along with kinematic inversion.

Solution

The mechanism is again drawn to scale, as shown in Fig. 5.3-14a. Inspection of the figure reveals that if either 3 or 4 is fixed and one of the links 2 and 5 is the input link, the mechanism becomes a simple one, for which kinematic analysis can be conducted by the

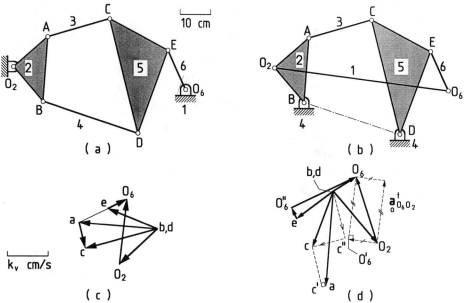

Figure 5.3-14

direct velocity-difference and acceleration-difference method. Let us fix link 4 as shown in Fig. 5.3-14b. Next, the velocity diagram of the inverted mechanism is drawn to some scale, as indicated in Fig. 5.3-14c. The velocity diagram is constructed assuming some angular velocity of link 2 with respect to link 4 (which is now the frame). The angular velocities of the various links with respect to link 4 are determined as given below:

$$\omega_{24} = 0.1k_v \text{ rad/s}, \; \omega_{14} = -0.028k_v \text{ rad/s}, \; \omega_{34} = -0.032k_v \text{ rad/s},$$

$$\omega_{54} = 0.053k_v \text{ rad/s} \quad \text{and} \quad \omega_{64} = 0.041k_v \text{ rad/s}.$$

We should note that as is conventional, the CCW direction is considered to be positive. In the actual problem, link 1 is fixed and ω_{21} is prescribed as 1 rad/s. Hence,

$$\omega_2 = \omega_{21} = \omega_{24} - \omega_{14} = (0.1 + 0.028)k_v = 0.128k_v = 1 \text{ rad/s}$$

5.3. ACCELERATION ANALYSIS (GRAPHICAL METHODS)

Therefore, $k_v = 7.81$.

Using this we determined $\omega_4, \omega_3, \omega_5$ and ω_6 as follows:

$$\omega_4 = \omega_{41} = -\omega_{14} = 0.028 k_v = 0.028 \times 7.81 \text{ rad/s} \equiv 0.22 \text{ rad/s (CCW)}$$
$$\omega_3 = \omega_{31} = \omega_{34} - \omega_{14} = (-0.032 + 0.028) k_v \text{ rad/s} \equiv 0.03 \text{ rad/s (CW)}$$
$$\omega_5 = \omega_{51} = \omega_{54} - \omega_{14} = (0.053 + 0.028) k_v \text{ rad/s} \equiv 0.63 \text{ rad/s (CCW)}$$
$$\omega_6 = \omega_{61} = \omega_{64} - \omega_{14} = (0.041 + 0.028) k_v \text{ rad/s} \equiv 0.54 \text{ rad/s (CCW)}$$

The actual angular velocities of the inverted mechanism are also determined as

$$\omega_{24} = 0.78 \text{ rad/s (CCW)},$$
$$\omega_{14} = 0.22 \text{ rad/s (CW)},$$
$$\omega_{34} = 0.25 \text{ rad/s (CW)},$$
$$\omega_{54} = 0.41 \text{ rad/s (CCW), and}$$
$$\omega_{64} = 0.32 \text{ rad/s (CCW)}.$$

Now, the auxiliary acceleration diagram, based on these velocities and $\alpha_{24} = 0$, can be drawn as shown in Fig. 5.3-14d. From this diagram, we get

$$\alpha_{o14} = \frac{a^t_{oO_6O_2}}{O_6O_2} = 0.113 \text{ rad/s}^2 \text{ (CCW) and}$$

$$\alpha_{o64} = \frac{a^t_{oO_6E}}{O_6E} = 0.69 \text{ rad/s}^2 \text{ (CCW)}.$$

Using Equation (5.3.12)

$$\alpha_{14} = \alpha_{o14} - \frac{\omega_{14}}{\omega_{24}} \alpha_{24} \qquad (a)$$

and

$$\alpha_{64} = \alpha_{o64} + \frac{\omega_{64}}{\omega_{24}} \alpha_{24}. \qquad (b)$$

Our objective is to determine α_{64}. Since $\alpha_2 = \alpha_{21} = 0$, using Equation (a)

$$\alpha_{21} = \alpha_{24} - \alpha_{14} = \alpha_{24} - \underset{\circ 14}{\alpha} + \frac{\omega_{14}}{\omega_{24}}\alpha_{24} = 0.$$

Substituting the values already known in the above equation,

$$\alpha_{24} - 0.113 + \frac{0.22}{0.78}\alpha_{24} = 0.$$

Solving the above equation, we get $\alpha_{24} = 0.09 \text{rad/s}^2$ (CCW).
Using, this value of α_{24} in Equations (a) and (b),

$$\alpha_{14} = (0.113 - 0.22/0.78 \times 0.09)\text{rad/s}^2 = 0.09 \text{rad/s}^2 \text{ (CCW) and}$$
$$\alpha_{64} = (0.69 + 0.32/0.78 \times 0.09)\text{rad/s}^2 = 0.73 \text{rad/s}^2 \text{ (CCW)}.$$

Finally,

$$\alpha_6 = \alpha_{61} = \alpha_{64} - \alpha_{14} = (0.73 - 0.09) \text{ rad/s}^2 = 0.64 \text{ rad/s}^2 \quad \text{(CCW)}.$$

Sometimes no kinematic inversion of a complex mechanism results in a simple mechanism. However, it may be possible to analyse the inverted mechanism using a powerful method, viz., auxiliary point method (which may not be possible for the actual mechanism). Once the results of the inverted problem are obtained, the actual results can be found out using Goodman's indirect approach. To illustrate this, let us consider the eight-link mechanism shown in Fig. 5.3-15a. It can be seen that the kinematic analysis of the given system is not possible by any graphical method, nor is it possible to tackle the problem by changing the input link. To solve the problem, first we invert the linkage by considering link 2 as the frame (see Fig. 5.3-15b). Then, ω_{12} and α_{12} will be respectively equal to $-\omega_{21}$ and $-\alpha_{21}$, which are prescribed. However, the kinematic analysis of this problem (with 2 as the frame and 1 as the input link) also is not possible. However, we can assume that link 3 is the input link of the inverted mechanism. Assume any ω_{32} and the velocity analysis can be completed, with the help of two auxiliary points X (on link 5) and Y (on link 8). Once v_X and v_Y (with assumed ω_{32}) are obtained, v_H can be determined as discussed in Section 5.3.4. With v_H and v_X known, v_{O_5} can be

5.3. ACCELERATION ANALYSIS (GRAPHICAL METHODS)

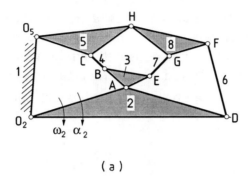

(a)

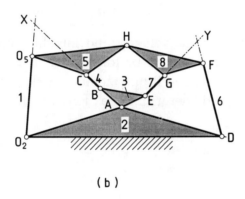

(b)

Figure 5.3-15

determined, from which ω_{12} (with assumed ω_{32}) can be obtained. But true $\omega_{12}(\equiv -\omega_{21} = -\omega_2)$ is known, and all the true velocities can be obtained with 2 as the frame and ω_{12} known.

After obtaining the true velocities of the inverted mechanism, an auxiliary acceleration analysis is conducted based on the true velocities and $\alpha_{32} = 0$. This is also carried out by first determining the accelerations of points X and Y and then finding out $\boldsymbol{a}_{oH}$. Once $\boldsymbol{a}_{oH}$ and $\boldsymbol{a}_{oX}$ are known, $\boldsymbol{a}_{oO_5}$ can be determined, which can yield α_{o12}. Using Equation (5.3.12),

$$\alpha_{12} = \alpha_{o12} + \frac{\omega_{12}}{\omega_{32}}\alpha_{32}.$$

However, since α_{12} is nothing but $-\alpha_{21}$, which is given, we get

$$\alpha_{32} = -(\alpha_{21} + \alpha_{o12})\frac{\omega_{32}}{\omega_{12}}.$$

In this way the angular velocity and angular acceleration of any link, with respect to link 2, can be found. The true values (i.e., with 1 as the frame and 2 as the input member with prescribed velocity and acceleration) can be determined as follows:

$$\omega_j = \omega_{j1} = \omega_{j2} - \omega_{12}$$
$$\alpha_j = \alpha_{j1} = \alpha_{j2} - \alpha_{12}$$

When no graphical method works, one must follow the analytical procedures to be discussed later in this chapter.

5.3.6 Method Using Euler-Savary Equation

We have already discussed, in Section 2.10, how the Euler-Savary equation provides a direct (geometric) method of obtaining the radius of path curvature of a point on a moving rigid lamina. In this section, we discuss how the Euler-Savary equation can be used profitably for acceleration analysis of complex mechanisms and mechanisms having higher pairs. For mechanisms having higher pairs, of course, the concept of equivalent lower-pair linkage (see Section 1.9) is also a very powerful technique.

5.3. ACCELERATION ANALYSIS (GRAPHICAL METHODS)

We may recall that the application of the Euler-Savary equation requires, as a first step, the determination of the inflection circle. Therefore, the use of the Euler-Savary equation for acceleration analysis is advisable when the required inflection circle can be easily determined. As discussed in Problem 2.10-1, the inflection circle of a moving body is most easily determined if the path curvature of any two points of the body are known. Otherwise, the inflection circle is obtained by determining the IC velocity v_I and using the relation $\boldsymbol{IK} = -i\boldsymbol{v}_I/\omega$, where I is the instantaneous center, K is the inflection pole and ω is the angular velocity. The line segment IK is the diameter of the inflection circle.[14] The IC velocity v_I is determined by considering a suitable auxiliary linkage with a pin at I connecting two forks, as demonstrated in Figs. 2.9-2a and 2.9-2b. The determination of the required IC velocity (and hence of the inflection circle) itself may become a difficult task for a mechanism with high degree-of-complexity, e.g., Problem 5.3-6.[15] We shall now demonstrate the use of the Euler-Savary equation for aceleration analysis with the help of examples.

Problem 5.3-8

Solve Problem 5.3-3 using the Euler-Savary equation.

Solution

We may note that the direct application of the relative acceleration method fails to solve this problem because the radius (ρ_B) of path-curvature of B is not known. However, ρ_B can be easily determined using the Euler-Savary equation, considering the 4R linkage 1-6-4-5, since ρ_C and ρ_D are known (see Problem 2.10-1). Referring to Fig. 5.3-16a, ρ_B is determined as $O_B B$. Once ρ_B is known, we get

$$a_B^n = v_B^2/\rho_B \quad \text{(parallel to } \boldsymbol{BO_B}\text{)}.$$

Thereafter, the acceleration analysis is completed as in Fig. 5.3-5e, which is reproduced in Fig. 5.3-16b. From this figure, $\boldsymbol{\alpha}_6$ is again

---

[14] This method of obtaining the inflection circle is normally referred to as Hartmann's construction.

[15] For solution of this problem using v_I and Euler-Savary equation, see Hirschhorn, J.: Kinematics and Dynamics of Plane Mechanisms, McGraw-Hill Book Co., New York, 1962.

CHAPTER 5. VELOCITY AND ACCELERATION ANALYSIS

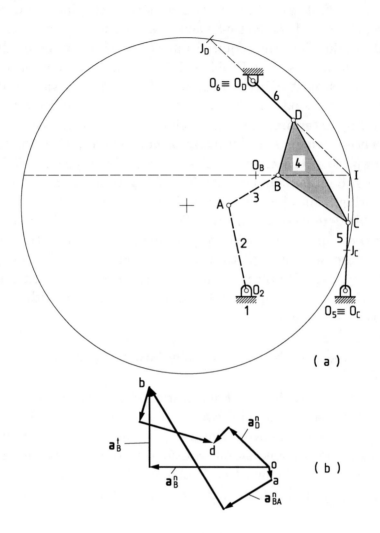

Figure 5.3-16

5.3. ACCELERATION ANALYSIS (GRAPHICAL METHODS)

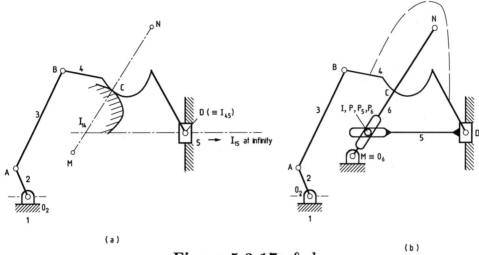

Figure 5.3-17a & b

obtained as 320 rad/s (CW). Note that in Problem 5.3-3, a_B^n was obtained through an auxiliary analysis shown in Figs. 5.3-5c and 5.3-5d, whereas here a_B^n is obtained directly, once O_B is located.

Problem 5.3-9

Figure 5.3-17a shows a five-link mechanism with a higher pair between 1 and 4 at C. With link 2 as the input link, perform the acceleration analysis assuming all the velocities (which can be easily determined using the method of instantaneous centers) are known. The centers of curvature, at the contact point C, for links 1 and 4 are at M and N, respectively.

Solution

We should first note that the acceleration analysis can be performed either by converting the mechanism into an equivalent lower-pair linkage or by finding ρ_B through the Euler-Savary equation. We shall discuss only the second approach.

The instantaneous center I_{14} lies on the line MN. Applying the theorem of three centers with I_{14}, I_{45} and I_{15}, the instantaneous center I_{14} is located, as explained in Fig. 5.3-17a.

In Fig. 5.3-17b, I_{14} is written as only I and our objective is to determine the IC velocity v_I with an assumed counter clockwise angular velocity ω_4. Towards this end, we draw the auxiliary linkage

214 CHAPTER 5. VELOCITY AND ACCELERATION ANALYSIS

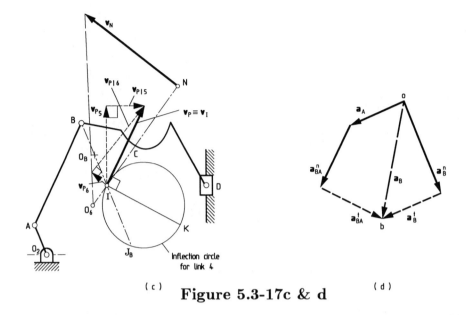

(c) Figure 5.3-17c & d (d)

shown in Fig. 5.3-17b, where the two forks, attached rigidly to link 5 and an imaginary link 6, are connected by a pin P at I. Link 6 is hinged to link 4 at N and to link 1 at M. The velocity of the pin P at this instant is v_I.

Let us consider two points, P_5 on link 5 and P_6 on link 6, instantaneously coincident with P. Assuming $\omega_4 = 1$ rad/s (CCW), we get $v_N = \omega_4 \times PN$ (recall $P \equiv I_{14}$) which is drawn in Fig. 5.3-17c. Since link 6 is rotating about $M(\equiv O_6)$, we obtain v_{P_6} from v_N, as $v_{P_6} = \frac{MP_6}{MN} v_N$, which is also indicated in Fig. 5.3-17c. Since link 5 is a slider, $v_{P_5} = v_D$, which is also known for $\omega_4 = 1$ rad/s (CCW). We use the following two equations

$$v_P = v_{P_5} + v_P|_5 \qquad (a)$$

and $\quad v_P = v_{P_6} + v_P|_6 \qquad (b)$

to determine $v_P(\equiv v_I)$. In the above two equations, $v_P|_5$ is along the fork in link 5 and $v_P|_6$ is along the fork in link 6. The determination of v_P, using these two equations, is graphically explained in Fig. 5.3-17c. Once v_P (i.e., v_I) is obtained, we get the inflection circle for link 4 with diameter equal to $IK = -iv_I/\omega_4$. Thereafter, using the Euler-Savary equation, we get the center of curvature O_B for the path of B.

The acceleration analysis can now be conducted as follows. From the input data, we obtain $\boldsymbol{a}_A$. From $\boldsymbol{v}_B$ and ρ_B we get $a_B^n = v_B^2/\rho_B$, parallel to $\boldsymbol{BO}_B$. The acceleration of the point B is determined as explained below. First we write

$$\boldsymbol{a}_B = \boldsymbol{a}_B^n + \boldsymbol{a}_B^t \qquad (c)$$

and

$$\boldsymbol{a}_B = \boldsymbol{a}_A + \boldsymbol{a}_{BA}^n + \boldsymbol{a}_{BA}^t. \qquad (d)$$

From Equations (c) and (d), we get

$$\boldsymbol{a}_B^n + \boldsymbol{a}_B^t = \boldsymbol{a}_A + \boldsymbol{a}_{BA}^n + \boldsymbol{a}_{BA}^t. \qquad (e)$$

Since Equation (e) involves only two unknowns, viz., the magnitudes of $\boldsymbol{a}_B^t$ ($\perp^r$ to $O_B B$) and $\boldsymbol{a}_{BA}^t$, it is solved graphically, as indicated in Fig. 5.3-17d (not to scale). Once $\boldsymbol{a}_B$ is determined, the rest of the acceleration analysis is straightforward and can be carried out using the method of acceleration difference.

5.4 Kinematic Analysis (Analytical Method)

It is evident from the previous sections that in a mechanism with a large number of links, it may not be possible to solve the kinematic analysis problem by graphical methods. Furthermore, it has also been mentioned that when highly accurate results are necessary or the kinematic analysis of a mechanism has to be repeated for a large number of configurations, analytical approach is more suitable. Analytical approach has also gained importance because of the present trend of computer applications in mechanism design.

Analytical approach for displacement analysis (which is the most difficult part of kinematic analysis) has already been presented in

Section 4.2. Using the results of displacement analysis, velocity and acceleration analyses can be carried out in a straightforward manner. The method is illustrated through the following examples.

Let us consider the 4R linkage discussed in Problem 4.2-1. To determine the velocities $\dot{\theta}_3$ and $\dot{\theta}_4$ and accelerations $\ddot{\theta}_3$ and $\ddot{\theta}_4$ for given values of θ_2, $\dot{\theta}_2$ and $\ddot{\theta}_2$ we proceed as follows:

First the displacement analysis is completed and the values for θ_3 and θ_4 are determined as discussed in Problem 4.2-1. Next let us consider the loop closure equation

$$l_1 + l_2 e^{i\theta_2} + l_3 e^{i\theta_3} - l_4 e^{i\theta_4} = 0 \tag{5.4.1}$$

derived while solving Problem 4.2-1. Differentiating both sides of Equation (5.4.1) with respect to time we get

$$il_2 \dot{\theta}_2 e^{i\theta_2} + il_3 \dot{\theta}_3 e^{i\theta_3} - il_4 \dot{\theta}_4 e^{i\theta_4} = 0. \tag{5.4.2}$$

Expanding Equation (5.4.2) and equating the real and imaginary parts to zero, we get

$$-l_2 \dot{\theta}_2 \sin\theta_2 - l_3 \dot{\theta}_3 \sin\theta_3 + l_4 \dot{\theta}_4 \sin\theta_4 = 0$$

and

$$l_2 \dot{\theta}_2 \cos\theta_2 + l_3 \dot{\theta}_3 \cos\theta_3 - l_4 \dot{\theta}_4 \cos\theta_4 = 0.$$

The above equations are linear algebraic equations in the velocities and with prescribed $\dot{\theta}_2$ the unknown velocities are determined as

$$\dot{\theta}_3 = \frac{l_2 \dot{\theta}_2 \sin(\theta_2 - \theta_4)}{l_3 \sin(\theta_4 - \theta_3)} \tag{5.4.3}$$

and

$$\dot{\theta}_4 = \frac{l_2 \dot{\theta}_2 \sin(\theta_2 - \theta_3)}{l_4 \sin(\theta_4 - \theta_3)}. \tag{5.4.4}$$

Further differentiation of Equation (5.4.2) with time yields the complex acceleration equation, which can also be split into real and imaginary parts. Thus, we get

5.4. KINEMATIC ANALYSIS (ANALYTICAL METHOD)

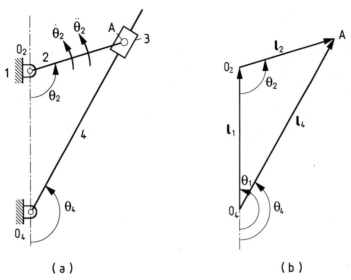

Figure 5.4-1

$$l_2 \sin \theta_2 . \ddot{\theta}_2 + l_3 \sin \theta_3 . \ddot{\theta}_3 - l_4 \sin \theta_4 . \ddot{\theta}_4$$

$$= l_4 \dot{\theta}_4^2 \cos \theta_4 - l_2 \dot{\theta}_2^2 \cos \theta_2 - l_3 \dot{\theta}_3^2 \cos \theta_3 \quad \text{and}$$

$$l_2 \cos \theta_2 \ddot{\theta}_2 - l_3 \cos \theta_3 \ddot{\theta}_3 + l_4 \cos \theta_4 \ddot{\theta}_4$$

$$= l_4 \dot{\theta}_4^2 \sin \theta_4 - l_2 \dot{\theta}_2^2 \sin \theta_2 - l_3 \dot{\theta}_3^2 \sin \theta_3.$$

Hence, if $\ddot{\theta}_2$ is prescribed, $\ddot{\theta}_3$ and $\ddot{\theta}_4$ can be determined (as a matter of fact, if any one of the three accelerations is known, the other two can be obtained) as follows:

$$\ddot{\theta}_3 = \frac{l_3 \dot{\theta}_3^2 \cos(\theta_3 - \theta_4) + l_2 \dot{\theta}_2^2 \cos(\theta_2 - \theta_4) + l_4 \dot{\theta}_4^2 + l_2 \ddot{\theta}_2 \sin(\theta_2 - \theta_4)}{l_3 \sin(\theta_4 - \theta_3)}$$

$$\ddot{\theta}_4 = \frac{l_2 \dot{\theta}_2^2 \cos(\theta_2 - \theta_3) + l_2 \dot{\theta}_3^2 + l_4 \dot{\theta}_4^2 \cos(\theta_3 - \theta_4) + l_2 \ddot{\theta}_2 \sin(\theta_2 - \theta_3)}{l_4 \sin(\theta_3 - \theta_4)}$$

In case of the 4R linkage presented above, all the vectors are of constant magnitudes. However, there may be situations where the length of a vector, while representing a mechanism by a vector polygon, depends on the configuration. Consider the quick-return mechanism shown in Fig. 5.4-1a. The input θ_2, $\dot{\theta}_2$ and $\ddot{\theta}_2$ are prescribed. Once the displacement analysis is over, θ_4 and l_4 are known. From the link-vector diagram shown in Fig. 5.4-1, we can write the loop equation

$$\mathbf{l}_1 + \mathbf{l}_2 - \mathbf{l}_4 = \mathbf{o}. \tag{5.4.5}$$

In complex notation, this can be written as

$$l_1 e^{i\theta_1} + l_2 e^{i\theta_2} - l_4 e^{i\theta_4} = 0. \tag{5.4.6}$$

Successive differentiation of Equation (5.4.6) with respect to time results in the following velocity and acceleration equations:

$$il_2\dot{\theta}_2 e^{i\theta_2} - il_4\dot{\theta}_4 e^{i\theta_4} - \dot{l}_4 e^{i\theta_4} = 0 \tag{5.4.7}$$

$$il_2\ddot{\theta}_2 e^{i\theta_2} - l_2\dot{\theta}_2^2 e^{i\theta_2} - il_4\ddot{\theta}_4 e^{i\theta_4} + l_4\dot{\theta}_4^2 e^{i\theta_4}$$

$$- \ddot{l}_4 e^{i\theta_4} - i\dot{l}_4\dot{\theta}_4 e^{i\theta_4} = 0 \tag{5.4.8}$$

since $\theta_1 = \pi$ (constant). However, if link 2 rotates with a constant angular velocity, i.e., $\ddot{\theta}_2 = 0$, the unknown velocities and acceleration quantities are found to be as follows:

$$\dot{\theta}_4 = \frac{l_2\dot{\theta}_2}{l_4} \cos(\theta_4 - \theta_2)$$

$$\dot{l}_4 = l_2\dot{\theta}_2 \sin(\theta_4 - \theta_2)$$

$$\ddot{\theta}_4 = \frac{l_2\dot{\theta}_2^2}{l_4} \sin(\theta_4 - \theta_2) - \frac{\dot{l}_4\dot{\theta}_4}{l_4}$$

$$\ddot{l}_4 = l_4\dot{\theta}_4^2 - l_2\dot{\theta}_2^2 \cos(\theta_2 - \theta_4)$$

5.5. *EXERCISE PROBLEMS* 219

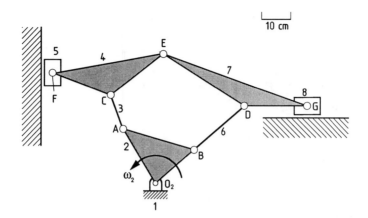

Figure 5.5-1

5.5 Exercise Problems

5.1 Determine ω_6 in the mechanism shown in Fig. 2.12-3, when link 2 rotates at 1 rad/s (CCW), using the method of instantaneous centers.

5.2 Determine ω_2 when link 6 of the six-link mechanism shown in Fig. 5.3-14a rotates in the CW direction with a speed of 2.5 rad/s, using the method of instantaneous center.

5.3 Link 2, of the eight-link mechanism shown in Fig. 5.5-1, rotates at 10 rad/s in the CCW direction. Determine the velocities of the sliders 5 and 8 using graphical method.

5.4 The kinematic arrangement of a machine is schematically shown in Fig. 5.5-2. The slotted lever and the crank O_2C are being rotated with angular velocities of 1000 rpm and 100 rpm, respectively, both in the CW direction. Determined the velocity of the midpoint P of link AB.

5.5 Figure 5.5-3 shows a simplified version of the feed-dog drive in a sewing machine. Circular cam 5 rotates at a constant speed of 200 rpm in the CCW direction as shown. Determine the velocity of point P at the instant shown.

CHAPTER 5. VELOCITY AND ACCELERATION ANALYSIS

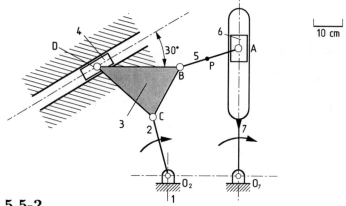

Figure 5.5-2

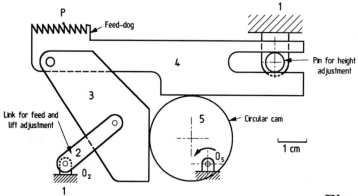

Figure 5.5-3

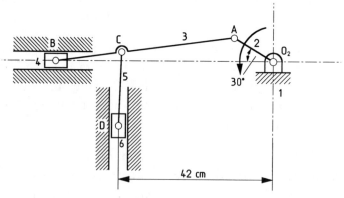

Figure 5.5-4

5.5. EXERCISE PROBLEMS

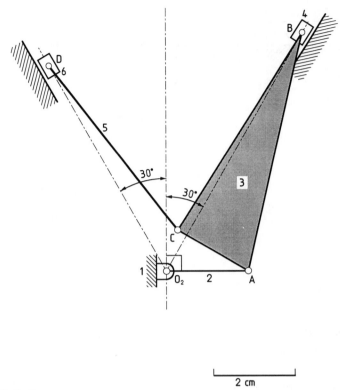

Figure 5.5-5

5.6 In the mechanism shown in Fig. 5.5-4, the crank O_2A rotates at 20 rpm in the CCW direction. For the given configuration, determine (i) the velocities of sliding at B and D, and (ii) the linear acceleration of D.

5.7 In the Whitworth quick-return mechanism shown in Fig. 2.7-3a, link 2 rotates at a constant speed of 120 rpm in the CW direction. Determine the velocity and acceleration of the slider D.

5.8 Figure 5.5-5 shows the mechanism used in a two-cylinder 60° V-engine with an articulated connecting rod. Crank 2 rotates in a clockwise direction at a speed of 2000 rpm. Determine the velocities and accelerations of the sliders at B and D.

5.9 The guide D in Fig. 5.5-6 has an upward velocity of 24 cm/s

222 CHAPTER 5. VELOCITY AND ACCELERATION ANALYSIS

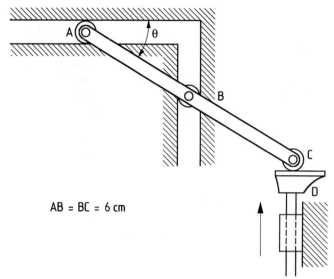

AB = BC = 6 cm

Figure 5.5-6

and an acceleration of 48 cm/s². Determine the corresponding values of v_A and a_A, given $\theta = 30°$ and $AB = BC = 6$ cm.

5.10 Two mechanisms are shown in Figs. 5.5-7a and 5.5-7b. Determine ω_3 and α_3 for the mechanism shown in Fig. 5.5-7a, at the instant indicated with constant $\omega_2 = 10$ rad/s (CW). Also, determine ω_4, α_4, ω_3 and α_3 for the mechanism shown in Fig. 5.5-7b with constant $\omega_2 = 10$ rad/s (CW) assuming no slip between links 3 and 4.

5.11 Determine the acceleration of link 6 of the mechanism shown in Fig. 2.12-3, when the sliding velocity of link 3 with respect to link 2 to is constant and equal to 10 cm/s in the upward direction.

5.12 A slotted disc, labelled 2 in Fig. 5.5-8, rotates in the CCW direction with a constant speed of 1000 rpm. Blocks 4 and 5 slide in the slots in the disc. The slots are at right angles. Block 6 slides in a fixed slot as indicated. A straight rigid rod, labelled 3, is pinned to the blocks at P, Q and R so that $PQ = 20$ cm and $PR = 60$ cm. Determine the velocity and acceleration of

5.5. EXERCISE PROBLEMS

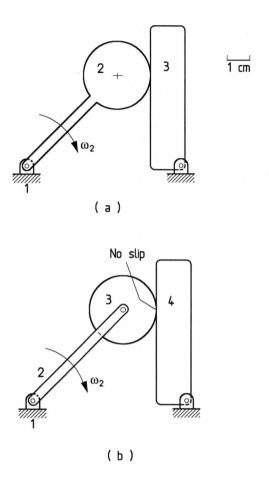

Figure 5.5-7

224 CHAPTER 5. VELOCITY AND ACCELERATION ANALYSIS

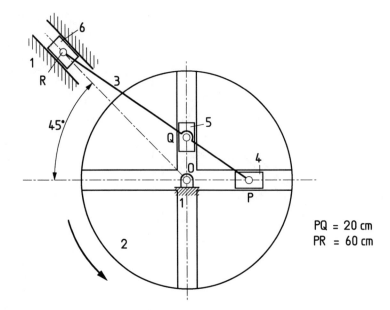

PQ = 20 cm
PR = 60 cm

Figure 5.5-8

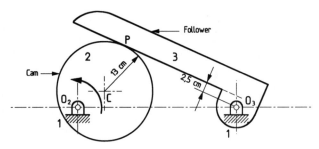

Figure 5.5-9

block 6 using graphical procedure.

5.13 Find out the acceleration of point P in Exercise Problem 5.5-4.

5.14 A circular disc cam with an oscillating follower is shown in Fig. 5.5-9. If the cam rotates at a constant speed of 1000 rpm in the CCW direction, determine the angular velocity and acceleration of the follower by graphical method, given $O_2C = 8.5$ cm, O_2O_3

5.5. EXERCISE PROBLEMS

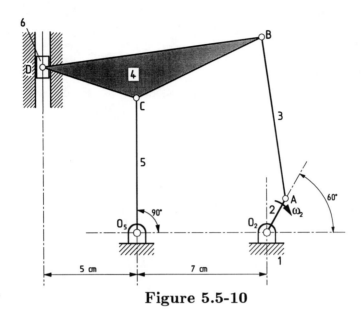

Figure 5.5-10

= 43 cm, and $\angle O_3 O_2 C = 30°$.

5.15 A Watt walking-beam mechanism is shown in Fig. 5.5-10. Determine the velocity and acceleration of the slider if the crank $O_2 A$ rotates with a constant speed of 1 rad/s in the CW direction, given $O_2 A = 2$ cm, $AB = 8.5$ cm, $BC = 7.5$ cm, $CD = 5.25$ cm and $O_5 C = 7$ cm.

5.16 If the speed of link 2 in Exercise Problem 5.3, is constant determine the accelerations of the sliders 5 and 8.

5.17 In the modified crank-shaper mechanism shown in Fig. 5.5-11, $\omega_2 = 10$ rad/s (CCW) and $\alpha_2 = 0$. Determine the velocity and acceleration of the slider at C, given $O_2 A = 4$ cm, $O_5 B = 5$ cm and $BC = 25$ cm.

5.18 In the Atkinson engine mechanism shown in Fig. 5.5-12, the velocity and acceleration of the slider at D are given as 40 cm/s and 450 cm/s (both towards right), respectively. Determine ω_2, α_2, ω_3 and α_3, given $O_2 A = 4.5$ cm, $AB = 12$ cm, $O_4 B = 6$ cm, $BC = 2$ cm, $AC = 13$ cm and $CD = 14$ cm.

226 CHAPTER 5. VELOCITY AND ACCELERATION ANALYSIS

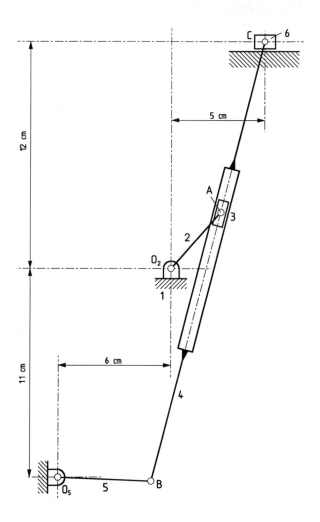

Figure 5.5-11

5.5. EXERCISE PROBLEMS

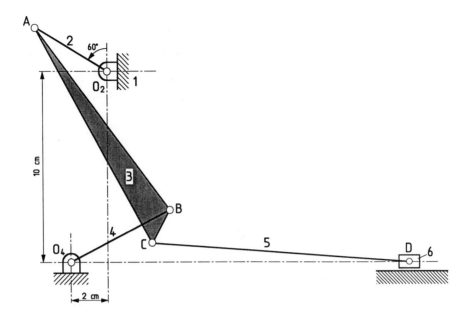

Figure 5.5-12

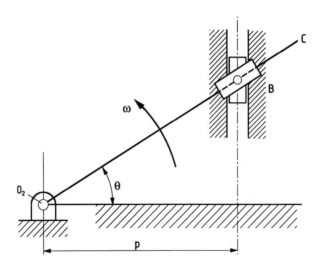

Figure 5.5-13

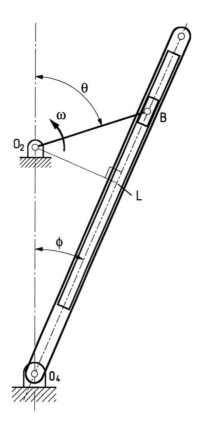

Figure 5.5-14

5.19 In Fig. 5.5-13 the block center B is free to move along AC and is constrained to move in the vertical direction. Show that

$$v_B = (p\omega/\cos^2\theta)j, \quad a_B = (2\omega^2 p \sin\theta/\cos^3\theta)j,$$

where j is a unit vector directed vertically upward.

5.20 For the slotted-lever quick-return mechanism shown in Fig. 5.5-14, show that

$$\ddot{\phi} = \omega^2 \frac{O_2 L}{O_4 B}\left(\frac{2BL}{O_4 B} - 1\right)$$

where $O_2 L$ is perpendicular to $O_4 B$ and ω is the constant angular velocity of the crank $O_2 B$.

5.5. EXERCISE PROBLEMS

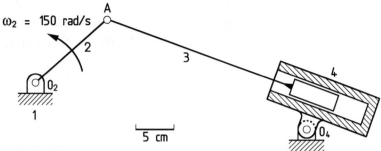

Figure 5.5-15

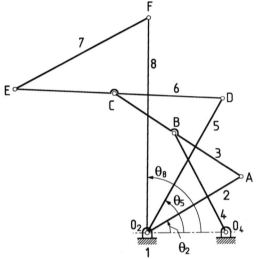

Figure 5.5-16

5.21 An oscillating engine mechanism is shown in Fig. 5.5-15. With 2 as the input link, perform the kinematic analysis, using an analytical method.

5.22 In Exercise Problem 5.5, determine the acceleration of the point P.

5.23 A curious linkage originally devised by Kempe for trisecting an angle, is shown in Fig. 5.5-16. This linkage consists of three anti-parallelograms, viz., O_2ABO_4, O_2DCA and O_2FED, all having the same ratio of their longer to shorter sides. Show that the angular motions of links 8, 5 and 2 bear the following relationships: $\alpha_8 : \alpha_5 : \alpha_2 = \omega_8 : \omega_5 : \omega_2 = 3 : 2 : 1$ for all configurations. (Hint: Show geometrically that $\theta_8 = 3\theta_2$ and $\theta_5 = 2\theta_2$ for all values of θ_2, which is measured from the transitional configuration with all links collinear).

Chapter 6
DIMENSIONAL SYNTHESIS

Dimensional synthesis involves the determination of link-lengths of a given linkage so as to fulfill the prescribed motion (kinematic) characteristics. Like kinematic analysis, dimensional synthesis can also be accomplished by both graphical and analytical methods. The choice of the proper method is very often decided by the class of synthesis problem. Therefore, before going into the details of various graphical and analytical techniques, we shall classify the various types of synthesis problems that are normally encountered. Most of the discussions in this chapter will be restricted to four-link mechanisms.

6.1 Classification of Synthesis Problems

Depending on the required kinematic characteristics to be satisfied by the designed linkage, dimensional synthesis problems can be broadly classified as listed below:

(i) *Motion Generation* - By motion generation we mean that a rigid body (i.e., one link of the given mechanism) has to be guided in a prescribed manner. The guidance may or may not be coordinated with the input motion.

(ii) *Path Generation* - If a point on a floating link (i.e., a link not connected to the frame) of the mechanism has to be guided along a prescribed path, then such a problem is classified as a problem of path generation. The generation of a prescribed path may or may not be coordinated with the input motion. Some problems of path generation will also be discussed in Chapter 7.

(iii) *Function Generation* - In this class of problems, the motion parameters (displacement, velocity, acceleration, etc.) of the output and input links are to be coordinated so as to satisfy a prescribed functional relationship.

(iv) *Dead-Center Problems* - In this class of problems, linkage dimensions are determined for fulfilling the prescribed dead-center configurations.

6.2 Exact and Approximate Synthesis

While generating a function or a path, or guiding a rigid body by a linkage, in most cases it is not possible to maintain the prescribed characteristics over the desired range of movements. In other words, the function or the path cannot be generated in an exact manner and one must attempt for an approximate synthesis. To achieve an acceptable level of approximation, two different approaches can be adopted. In the first one, known as the *precision* (or *accuracy*) *point approach*, the desired characteristics are exactly satisfied only at a finite number of configurations (corresponding to the precision or accuracy points) within the desired range. The difference between the desired and generated characteristics, which is nonzero except at the precision points, is called the structural error. The precision points are chosen so that the maximum structural error is as small as possible. In the second approach, the structural error, which may not be zero anywhere, is minimized in an overall sense in the entire range. This approach, which is analytical in nature, is discussed in Section 6.8, under the heading *Optimization Methods*.

6.2. EXACT AND APPROXIMATE SYNTHESIS

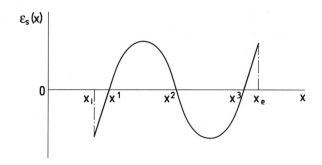

Figure 6.2-1

6.2.1 Choice of Precision Points

The choice of precision points is discussed in this section with reference to function generation problems. Basically, the same approach is applicable to other classes of problems, whenever a choice of precision points is necessary. Let the input movement (displacement) represent the independent variable x and the output movement is desired to correspond to a prescribed function $f(x)$ in the interval $x_i \le x \le x_e$, where the subscripts i and e refer to the initial and end values. Normally, a linear relationship is assumed between x and the input movement variable. Likewise, $f(x)$ and the output movement variable are also linearly related. These linearity constants are called the *scale factors*. The function generated by a mechanism (i.e., the output movement) will depend on x (i.e., the input movement) and the *link parameters*. Thus, the generated function is written as $F(x, D_1, D_2, \ldots D_n)$, where $D_1, D_2, \ldots D_n$ are the *design variables* (functions of link parameters). The difference between $f(x)$ and $F(x, D_1, D_2, \ldots D_n)$ has been defined as the structural error $\epsilon_s(x)$. Thus,

$$\epsilon_s(x) = f(x) - F(x, D_1, D_2, \ldots D_n). \qquad (6.2.1)$$

At precision points, $\epsilon_s(x) = 0$. The number of such precision points is equal to n, i.e., the number of design variables to be determined. A typical plot of $\epsilon_s(x)$ in the interval $x_i \le x \le x_e$ is shown in Fig. 6.2-1. In this figure, x^1, x^2 and x^3 represent three precision points where $\epsilon_s(x) = 0$. Intuitively, it appears that the maximum value of $|\epsilon_s(x)|$ will be minimum when all the maxima and minima of $\epsilon_s(x)$ are equal in magnitude, which in turn, is also equal to the value of $|\epsilon_s(x)|$ at $x = x_i$ and $x = x_e$. Such a solution, with n accuracy points, exists if $\epsilon_s(x)$ is proportional to *Chebyshev's polynomial* of degree n. Defining $a = (x_i + x_e)/2$ and $h = (x_e - x_i)/2$, the Chebyshev's polynomial of degree n is written as[1]

$$C_n(x) = \frac{h^n}{2^{n-1}} \cos\left[n \cos^{-1}\left(\frac{x-a}{h}\right)\right]. \qquad (6.2.2)$$

Thus, the points where $C_n(x) = 0$ implying $\epsilon_s(x) = 0$, are given by

$$\cos\left[n \cos^{-1}\left(\frac{x-a}{h}\right)\right] = 0$$

or

$$n \cos^{-1}\left(\frac{x-a}{h}\right) = \frac{(2k-1)\pi}{2}; k = 1, 2, \ldots n$$

or

$$x_k = a + h \cos\frac{(2k-1)\pi}{2n}; k = 1, 2, 3, \ldots n.$$

Thus, the accuracy points $x^1, x^2, \ldots x^n$ can be written as

$$x^j = a - h \cos\frac{(2j-1)\pi}{2n}; j = 1, 2, \ldots n. \qquad (6.2.3)$$

The reader should note the change in sign to keep $x^1, x^2, \ldots x^n$ in the increasing order (which is not the case with the expression for x_k).

There is a simple geometrical method for obtaining Chebyshev's accuracy points given by Equation (6.2.3). A circle of radius h is drawn with the center at a distance a from the origin O on the x-axis as shown in Fig. 6.2-2a. A regular polygon of $2n$ sides is inscribed within this circle so that two of its sides are normal to the x-axis. The projections of the vertices of this polygon on the x-axis locate the n accuracy points. Figures 6.2-2a and 6.2-2b show this geometric method for $n = 3$ and $n = 4$, respectively.

---

[1] We may recall that $\cos n\theta = 2\cos(n-1)\theta \cos\theta - \cos(n-2)\theta$ for $n > 2$ and $\cos 2\theta = 2\cos^2\theta - 1$, which together suggest that $\cos n\theta$ is a polynomial of nth order in $\cos\theta$.

6.2. EXACT AND APPROXIMATE SYNTHESIS

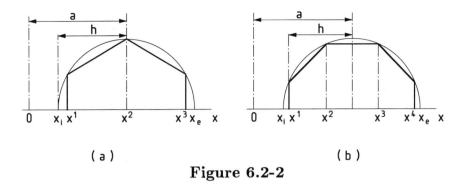

Figure 6.2-2

It should be mentioned that Chebyshev's accuracy points are to be taken only as a first approximation. This is because $\epsilon(x)$, in general, is not proportional to a Chebyshev's polynomial of degree n. The nature of $\epsilon_s(x)$ is entirely governed by the characteristics of $F(x, D_1, D_2, \ldots D_n)$ and $f(x)$. However, $F(x, D_1, D_2, \ldots D_n)$ and $f(x)$ can be approximated by Taylor's series expansions, retaining terms up to x^{n-1} and x^n, respectively. Consequently, $\epsilon_s(x)$ is approximated as proportional to a Chebyshev's polynomial. In a real-life problem of dimensional synthesis, the mechanism is first designed on the basis of Chebyshev's accuracy points. Then $\epsilon_s(x)$, for the mechanism so designed, is plotted (versus x). The spacing of the two accuracy points containing $\mid \epsilon_s(x) \mid_{\max}$ is reduced and the process is repeated. An optimum design minimizing the value of $\mid \epsilon_s(x) \mid_{\max}$ is, thus, obtained only by a number of trials

If a desired functional relationship is to be maintained between the output and input velocities (rather than the displacements), then the maximum value of the first derivative of the structural error, i.e., $\mid d\epsilon_s(x)/dx \mid_{\max}$ is to be minimized. In such a situation, some accuracy points, i.e., where $\epsilon_s(x) = 0$, lie outside the range $x_i \leq x \leq x_e$.[2]

Furthermore, a function generation problem can also be approached by matching the function value, slope and higher order derivatives at a point rather than only matching the function values at a number of precision points.

---

[2] For a detailed discussion of this and other mathematically intricate aspects of Chebyshev's accuracy points the reader may refer to R.S. Hartenberg and J. Denavit - Kinematic Synthesis of Linkages, McGraw Hill Book Co., New York 1964.

236 CHAPTER 6. DIMENSIONAL SYNTHESIS

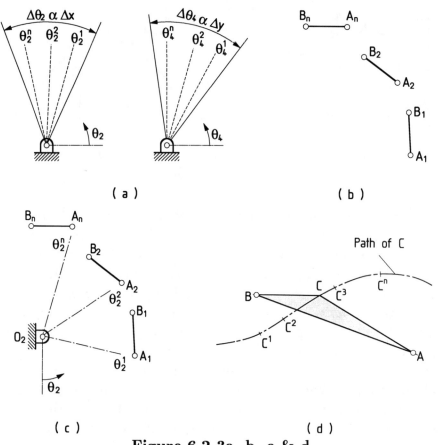

Figure 6.2-3a, b, c & d

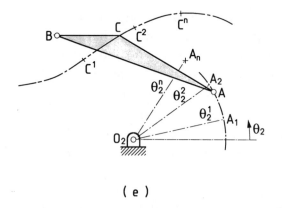

Figure 6.2-3e

6.3. GRAPHICAL METHODS (THREE POSITIONS)

In terms of precision points, different classes of synthesis problems are explained in Figs. 6.2-3(a) to 6.2-3(e), with reference to a 4R linkage. In Fig. 6.2-3a, the problem of function generation is indicated where θ_2 represents the input variable and θ_4 the output variable. The superscripts $1, 2, 3, \ldots n$ refer to the precision points[3]. Figures 6.2-3b and 6.2-3c represent, respectively, the motion generation problem without and with coordination between the output and input movements. In the first situation, the coupler AB has to be guided through the positions $A_1B_1, A_2B_2, \ldots A_nB_n$, whereas in the latter, the input movements corresponding to these positions are also prescribed in terms of $\theta_2^1, \theta_2^2, \ldots \theta_2^n$. The problems of path generation without and with coordination between the output and input movements are indicated in Figs. 6.2-3d and 6.2-3e, respectively. The coupler point C may or may not lie on the line AB.

6.3 Graphical Methods (Three Positions)

Graphical methods occupy a very important place in the synthesis of planar mechanisms. They are quick and also provide a good physical understanding. In this section, we shall discuss various graphical methods for synthesizing slider-crank and 4R mechanisms, which are immensely useful to a mechanism designer. It may be pointed out that these methods are meant only for the precision-point approach. For a clear understanding of the basic approach, we shall restrict our discussions in this section mainly to three precision points (or two coordinated movements). The problems of motion generation, path generation and function generation are solved following the same methodology.

6.3.1 Motion Generation

Let a rigid body be guided through two prescribed positions, labelled as I and II in Fig. 6.3-1. We shall treat the rigid body as the coupler

---
[3]Freudenstein has suggested that to obtain higher accuracy (i) $\Delta \theta$' s $> 120^o$ and (ii) generation of symmetric functions like $y = x^2$ and $-1 \leq x \leq 1$, both should be avoided.

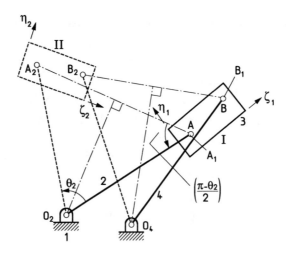

Figure 6.3-1

of a 4R linkage. We can take any two points on the rigid body, say A and B, which occupy positions A_1, B_1 and A_2, B_2, respectively, in configurations I and II. Assuming the moving hinges to be located at A and B, the fixed hinges O_2 and O_4 can be located anywhere on the midnormals of A_1A_2 and B_1B_2, respectively, as shown in Fig. 6.3-1. By moving the 4R linkage O_2ABO_4, the rigid body can be guided through I and II. Further, if the movement of the coupler from I to II has to take place during a prescribed rotation of the input link 2, say θ_2, then O_2 must be located so that $\angle A_1O_2A_2 = \theta_2$, as indicated in Fig. 6.3-1.

If we consider three prescribed positions, I, II and III, of the rigid body, then again we can arbitrarily choose two points, A and B, on the body, as the locations of the moving hinges. The fixed pivot O_2 is uniquely located at the center of the circle passing through A_1, A_2 and A_3. Similarly, O_4 is located at the center of the circle passing through B_1, B_2 and B_3.

If four positions of the rigid body are prescribed, then A and B can no longer be arbitrarily chosen. The locations of the moving hinges must then be two circle-points corresponding to the four prescribed positions (see Section 6.5.2). The fixed hinges O_2 and O_4 will then be the centre-points corresponding to A and B, respectively. For five

6.3. GRAPHICAL METHODS (THREE POSITIONS)

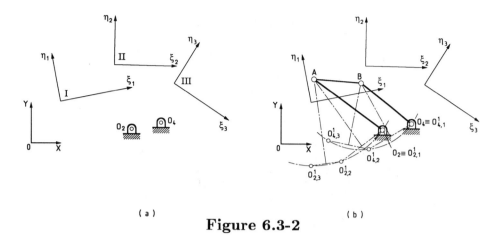

Figure 6.3-2

prescribed positions a solution exists only if Burmester points exist for these positions. The moving hinges must then be chosen as the Burmester points. It may be noted that at this stage, there is no guarantee that all the positions can be taken up by the same branch of the 4R linkage or that the prescribed positions will be taken in the prescribed order. This is true even when we obtain a solution in the manner explained above. For details of this aspect, see Section 6.7.

In most real-life designs, the locations of the ground pivots (i.e., the fixed hinges) are prescribed, since these are to be located conveniently depending on factors like space constraints. In such a situation, the methodology discussed above in this section can be used after considering a kinematic inversion with the coupler (i.e., the body to be guided) as the fixed link. The procedure is explained below, with the help of a three positions problem.

Referring to Fig. 6.3-2a, O_2 and O_4 are the prescribed locations of the fixed pivots in the fixed frame OXY. Three prescribed positions of the coupler I, II and III are indicated by a coordinate frame $\xi - \eta$ attached to the coupler. To locate the moving hinges A, B in the first position of the coupler, we make a kinematic inversion of the required 4R-linkage by fixing the coupler at its first position. The inverted positions of the fixed hinges O_2 and O_4 (signifying the same relative movements) are most conveniently located by using a tracing paper[4].

---

[4]It can be done by simple geometric construction also, but too many lines and

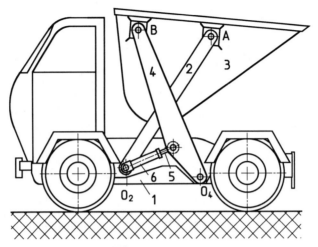

Figure 6.3-3a

For example, locate ξ_2, η_2 axes and the point O_2 on a tracing paper. Place the tracing paper such that ξ_2, η_2 axes coincide, respectively, with ξ_1, η_1 axes. The position of O_2 on the tracing paper is located as $O_{2,2}^1$ (Fig. 6.3-2b). In this manner, all the inverted positions of O_2 and O_4 are located and labelled. The reader should carefully note the labelling. The superscript refers to the position where the inversion has been done and the second subscript to the position being inverted. Since O_2A is of constant length, the moving hinge A is located in its first configuration at the center of the circle passing through $O_{2,1}^1 (\equiv O_2)$, $O_{2,2}^1$ and $O_{2,3}^1$. Similarly, the other moving hinge B is located at the center of the circle passing through $O_{4,1}^1 (\equiv O_4)$, $O_{4,2}^1$ and $O_{4,3}^1$. For better understanding and practice, the reader is advised to obtain the same solution by considering an inversion at the second position.

Problem 6.3-1

Figure 6.3-3a shows the design of a tipper-dumper truck. Three desired configurations of the bin are indicated as I, II and III in Fig. 6.3-3b. The fixed hinges O_2 and O_4 are conveniently located on the body of the truck. Obtain the locations of the hinges A and B on the body of the bin.

---

arcs make the drawing clumsy. Use of a tracing paper keeps the drawing clean and easy to understand.

6.3. GRAPHICAL METHODS (THREE POSITIONS)

Solution

Referring to Fig. 6.3-3b, we choose any two convenient points X and Y on the bin, which are labelled as $(X_1, Y_1), (X_2, Y_2)$ and (X_3, Y_3) in the three desired configurations. On a tracing paper we mark the points X_2, Y_2, O_2 and O_4. We then move the tracing paper so that X_2 coincides with X_1 and Y_2 with Y_1, when O_2 moves to $O^1_{2,2}$ and O_4 moves to $O^1_{4,2}$. Similarly, making (X_3, Y_3) coincide with (X_1, Y_1), O_2 moves to $O^1_{2,3}$ and O_4 moves to $O^1_{4,3}$. The hinge A on the bin in its configuration I, is located at the center of the circle passing through $O^1_{2,1} (\equiv O_2), O^1_{2,2}$ and $O^1_{2,3}$. Similarly, B is located at the center of the circle passing through $O^1_{4,1} (\equiv O_4), O^1_{4,2}$ and $O^1_{4,3}$.

6.3.2 Path Generation

A typical path generation problem, illustrated below, can also be solved by the technique of kinematic inversion discussed above. Referring to Fig. 6.3-4a, let the fixed hinges of a 4R linkage be located at O_2 and O_4. The moving hinge A at the first configuration is located at A_1, with a particular coupler point at C_1. It is desired that this coupler point pass through two more prescribed locations, C_2 and C_3.

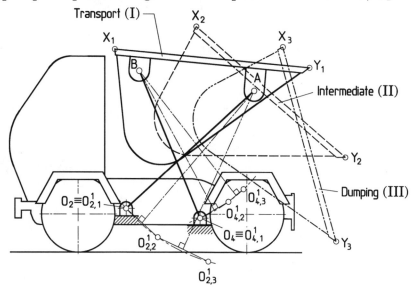

Figure 6.3-3b

242   CHAPTER 6. DIMENSIONAL SYNTHESIS

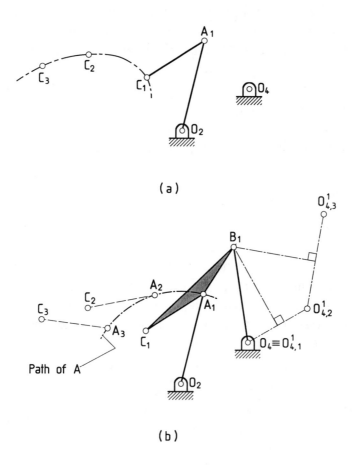

Figure 6.3-4

6.3. GRAPHICAL METHODS (THREE POSITIONS)

To design the complete linkage the other moving hinge B must be located in its first configuration (B_1). Since the length AC is known, first we locate A_2 and A_3 on the path of A, corresponding to the second and third configurations, as shown in Fig. 6.3-4b. Using then $A_1 C_1$ we make a kinematic inversion of the mechanism with the coupler fixed at its first position. With the help of a tracing paper we locate the inverted positions of the fixed hinge O_4 at $O^1_{4,1} (\equiv O_4), O^1_{4,2}$ and $O^1_{4,3}$. The moving hinge B_1 is located at the center of the circle passing through $O^1_{4,1}, O^1_{4,2}$ and $O^1_{4,3}$.

6.3.3 Function Generation

The approach developed in Section 6.3.1 can easily be extended to solve a function generation problem with three precision points. Referring to Fig. 6.3-5a, two pairs of coordinated movements between the input and output links of a 4R linkage are to be generated. Let us denote $\theta_2^{12} = \theta_2^2 - \theta_2^1, \theta_4^{12} = \theta_4^2 - \theta_4^1, \theta_2^{23} = \theta_2^3 - \theta_2^2$ and $\theta_4^{23} = \theta_4^3 - \theta_4^2$.[5] The desired coordinated movements are expressed as $(\theta_2^{12}, \theta_4^{12})$ and $(\theta_2^{23}, \theta_4^{23})$. Referring to Fig. 6.3-5b, with prescribed (or assumed) locations of O_2, O_4 and A_1 (i.e., l_1, l_2 and θ_2^1), the desired linkage can be synthesized as explained below.

First, we make a kinematic inversion with the follower (the output link) fixed at its first configuration. If the follower is held fixed, then the same relative movements from configurations I to II are obtained by rotating the fixed link through $-\theta_4^{12}$. Therefore, the inverted position $O^1_{2,2}$ is obtained at the location of O_2 after rotating $O_4 O_2$ about O_4 through $-\theta_4^{12}$. Similarly, $O^1_{2,3}$ is obtained by rotating $O_4 O_2$ about O_4 through $-\theta_4^{13}$. The inverted position of A_2, i.e., A_2^1 can be obtained with $O^1_{2,2} A_2^1 = O_2 A_2$ and $\angle A_2^1 O^1_{2,2} O_4 = \angle A_2 O_2 O_4 = \theta_2^2$. It is easy to see that A_2^1 can also be easily obtained at the location taken up by A_2, after rotating $O_4 A_2$ about O_4 through $-\theta_4^{12}$. Following a similar procedure, A_3^1 can be obtained as indicated in Fig. 6.3-5b. The moving hinge B_1 is located at the center of the circle passing through $A_1^1 (\equiv A_1), A_2^1$ and A_3^1.

---

[5] In general, $\theta_k^{ij} = \theta_k^j - \theta_k^i$.

CHAPTER 6. DIMENSIONAL SYNTHESIS

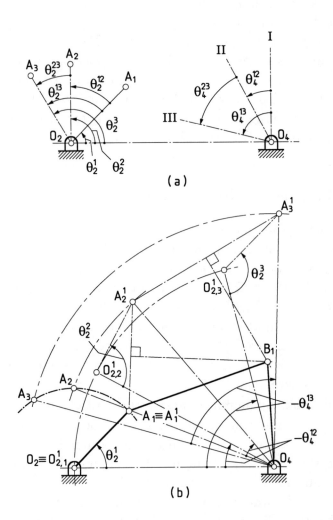

Figure 6.3-5

6.3. GRAPHICAL METHODS (THREE POSITIONS)

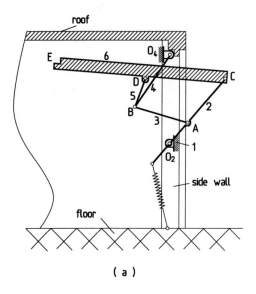

(a)

Figure 6.3-6a

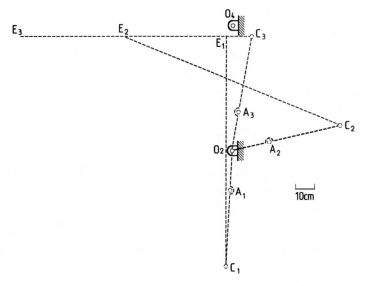

Figure 6.3-6b

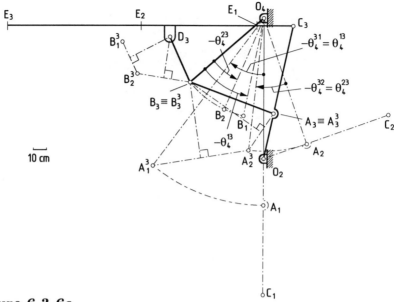

Figure 6.3-6c

Problem 6.3-2

Figure 6.3-6a shows the sketch of an overhead garagedoor mechanism. In this six-link mechanism, link 6 represents the door-board. Three desired positions of the door-board are indicated in Fig. 6.3-6b as C_1E_1 (closed), C_2E_2 (intermediate) and C_3E_3 (open). The rotations of link 4 corresponding to three positions are $\theta_4^{12} = 12°$(CW) and $\theta_4^{23} = 26°$(CW). Obtain the location of the hinge B corresponding to position III (open), and hence, the link-lengths O_4B and AB. Also determine the link-lengths BD, CD and $\angle DCE$.

Solution

Since the mechanism is desired at position III, we invert the 4R mechanism O_2ABO_4 with link 4 fixed at this position (Fig. 6.3-6c). Accordingly, the inverted positions of A_1 and A_2 are located at A_1^3 and A_2^3, respectively, whereas $A_3^3 \equiv A_3$. The point A_1^3 is obtained by rotating O_4A_1 about O_4 through an angle $-\theta_4^{31} = \theta_4^{13} = \theta_4^{12} + \theta_4^{23} = 38°$(CW). Similarly, A_2^3 is obtained by rotating O_4A_2 about O_4 through an angle $-\theta_4^{32} = \theta_4^{23} = 26°$(CW). The point B_3 is now located at the center of the circle passing through $A_3^3(\equiv A_3), A_2^3$ and A_1^3. Once B_3 is obtained, D_3 is located by considering a kinematic inversion with

6.3. GRAPHICAL METHODS (THREE POSITIONS)

link 6 (i.e., CE) fixed at its III position, C_3E_3. Towards this end, first we locate B_2 and B_1 by rotating O_4B_3 about O_4 through angles $-\theta_4^{23}(=26° \text{ CCW})$ and $-\theta_4^{13}(=38° \text{ CCW})$, respectively. The inverted positions of B_1 and B_2 are located at B_1^3 and B_2^3, respectively, whereas $B_3^3 \equiv B_3$. The points B_1^3 and B_2^3 are most conveniently located with the help of a tracing paper. On this tracing paper, the points C_1, E_1 and B_1 are marked. The tracing paper is then laid so that C_1 and E_1 coincide with C_3 and E_3, respectively. The location that B_1 now occupies is marked as B_1^3. Similarly, B_2^3 is obtained by marking C_2, E_2 and B_2 on the tracing paper. The point D_3 is then obtained at the center of the circle passing through B_1^3, B_2^3 and $B_3^3 (\equiv B_3)$. From measurements, we finally get $O_4B = 72.3$ cm, $AB = 66.9$ cm, $BD = 36$ cm, $CD = 92.8$ cm and $\angle DCE = 5°$.

It is important to mention here that if one attempts to design the mechanism using a 4R linkage, it will not be possible to achieve optimum inside space. The available space is determined by the envelope of the straight lines, representing the garage door at various positions.

6.3.4 Relative Poles

In Sections 2.8 and 2.11 we came across the concept of a pole, which is used to describe a finite movement of a lamina (i.e., the coupler) with respect to a fixed plane (i.e., the frame). The same concept can be extended to study the relative motion between two laminae e.g., the crank and the follower of a 4R linkage. In Section 6.3.3 we studied this relative motion assuming the follower to be fixed. In this section the concept and use of *relative poles* are elaborated, assuming the crank to be fixed. It is needless to say that one gets the same answer to a problem whether the method illustrated in Section 6.3.3 is used or the method to be developed in this section with the help of relative poles is used.

However, it should be noted that it is generally not possible to apply the method of inversion (presented in Section 6.3.3) to problems involving more than three precision points (configurations) except for some special choices. On the other hand, the method using the concept of relative poles can be extended to four (or even five) precision points synthesis problems (See Section 6.5) because the results derived

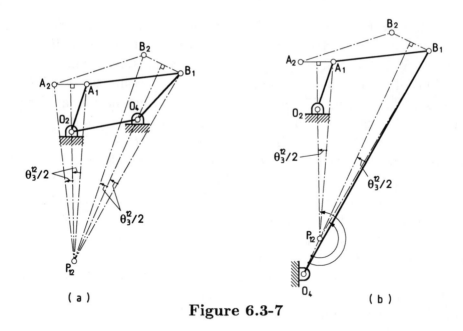

Figure 6.3-7

in Section 2.11 for poles are valid for relative poles, if one considers the motion of the follower relative to the crank.

Referring to Fig. 6.3-7a, if the coupler AB of a 4R linkage must be transferred from position I to II, indicated by A_1B_1 and A_2B_2, respectively, then we know that the fixed hinges O_2 and O_4 can be located anywhere on the midnormals of A_1A_2 and B_1B_2, respectively. The pole P_{12} for this movement is at the point of intersection of these two midnormals. The rotation of the coupler about P_{12} is θ_3^{12}. As indicated in Fig. 6.3-7a, it is easy to see that the coupler (at both positions) and the frame subtend equal angles at the pole. Similarly, the crank and the follower also subtend equal angles at the pole. This is also true for both of the configurations involved. However, the angles subtended at the pole by the coupler and the frame can also differ by π, as shown in Fig. 6.3-7b. In such a situation, the angles subtended at the pole by the crank and the follower also differ by π. It may be mentioned that angles should be measured algebraically, i.e., say CCW as positive. For example, in Fig. 6.3-7b, the angle subtended at P_{12} by the crank, $\angle O_2P_{12}A_1$, is negative, whereas that by the follower, $\angle O_4P_{12}B_1$, is positive. We can now summarize that

6.3. GRAPHICAL METHODS (THREE POSITIONS)

the angles subtended at the pole by the coupler and the frame are either equal or differ by π. The same is true for the angles subtended by the crank and follower. The last two statements are valid for both of the configurations involved.

Let us now study the relative motion between the follower and crank as the linkage moves from configuration I to II (Fig. 6.3-8a), when the rotations of the crank and follower are θ_2^{12} and θ_4^{12}, respectively. Assuming the crank to be fixed at its first position, the inverted position of the follower (with the same relative movement) corresponding to its second configuration is indicated as $O_{4,2}^1 B_2^1$ in Fig. 6.3-8b. As already explained in Fig. 6.3-5, the locations of $O_{4,2}^1$ and B_2^1 are obtained by rotating $O_2 O_4$ and $O_2 B_2$, respectively, about O_2 through an angle $-\theta_2^{12}$. Thus, the relative movement between the follower and crank, depicted in Fig. 6.3-8a, is same as moving the coupler of Fig. 6.3-8b from $O_4 B_1$ to $O_{4,2}^1 B_2^1$, with $O_2 A_1$ as the frame. Following the construction of Fig. 6.3-7, we can now locate the pole for this movement at R_{12} (Fig. 6.3-8c), which is called the relative pole. The rotation of $O_4 B_1$ about R_{12} to reach $O_{4,2}^1 B_2^1$ is obviously $\theta_4^{12} - \theta_2^{12}$. Hence, the angle $\psi = (\theta_4^{12} - \theta_2^{12})/2$ (CCW). From $\triangle O_2 O_4 R_{12}$, it is seen that $\phi = -(\theta_2^{12}/2) - \psi = -\theta_4^{12}/2$.[6] Thus, the relative pole R_{12} is located at the intersection of two lines, one of which is drawn through O_2 at an angle $-\theta_2^{12}/2$ with $O_2 O_4$, and the other through O_4

---

[6]The reader should carefully note the directions of the angles ϕ and ψ.

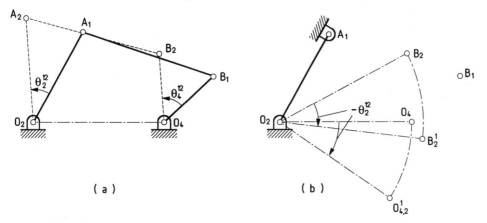

Figure 6.3-8a & b

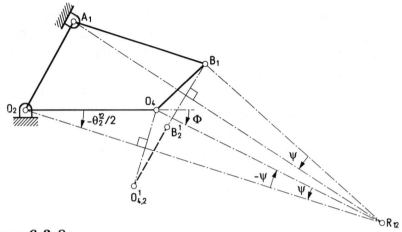

Figure 6.3-8c

at an angle $-\theta_4^{12}/2$ with O_2O_4. Further, from the properties of poles explained in Fig. 6.3-7a, we conclude that O_2A_1 and O_4B_1 subtend equal angles at R_{12} and, likewise, O_2O_4 and A_1B_1 also subtend equal angles at R_{12}.

In the previous chapters we have seen that a 3R-1P slider-crank mechanism is obtained as a limit of a 4R linkage where one R pair is replaced by a P pair. In other words, the R pair at O_4 of the follower can be regarded as being at infinity in a direction perpendicular to the line of movement of the slider. Using this concept we can determine

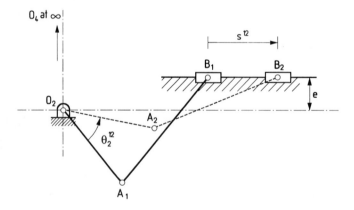

Figure 6.3-9

6.3. GRAPHICAL METHODS (THREE POSITIONS)

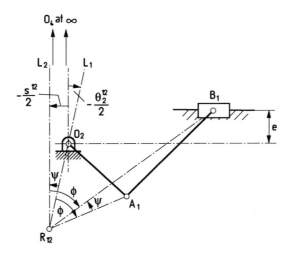

Figure 6.3-10

the relative pole R_{12} corresponding to a movement (θ_2^{12}, s^{12}), where s^{12} is the slider displacement corresponding to a crank rotation θ_2^{12} (Fig. 6.3-9). Referring to Fig. 6.3-10, a line, L_1, is drawn through O_2 making an angle $-\theta_2^{12}$ with O_2O_4. Another line, L_2, is drawn that is perpendicular to the line of movement of the slider and is at a distance $-s^{12}/2$ from O_2. The relative pole R_{12} is located at the intersection of L_1 and L_2. The reader is invited to justify this construction in view of Fig. 6.3-8 for a 4R linkage. Here again, as in the case of a 4R linkage, the connecting rod A_1B_1 and the frame line O_2O_4 subtend equal angles ψ at R_{12} (Fig. 6.3-10). Similarly, the crank O_2A_1 and the line O_4B_1 subtend equal angles ϕ at R_{12}.

Problem 6.3-3

Design a 4R linkage to generate the following pairs of prescribed movements: $(\theta_2^{12} = 26°, \theta_4^{12} = 54°)$ and $(\theta_2^{13} = 52°, \theta_4^{13} = 80°)$. Assume the frame-length $(l_1)O_2O_4 = 5.1$cm, crank-length $(l_2)O_2A_1 = 6.4$ cm and $\theta_2^1 = -40°$.

Solution

With the given data the points O_2, O_4 and A_1 are located first (Fig. 6.3-11). The relative poles R_{12} and R_{13} are located for the prescribed pairs of movements. Let $\angle O_2R_{12}O_4 = \psi_2$ and $\angle O_2R_{13}O_4 = \psi_3$. At R_{12}, a line N_1 is drawn making an angle ψ_2 with A_1R_{12}. The moving hinge

B_1 must lie on N_1 so that the coupler and the frame subtend equal angles at R_{12}. Similarly, at R_{13} a line N_2 is drawn at an angle ψ_3 with $A_1 R_{13}$. Then, B_1 is located at the intersection of N_1 and N_2. From measurements, we get the coupler length $(l_3) A_1 B_1 = 1.9$ cm and the follower length $(l_4) O_4 B_1 = 2.9$ cm.

It may be noted that since we used the point A_1 and the relative poles R_{12} and R_{13} (the common subscript being 1), the desired mechanism is obtained at its first configuration. The reader is advised to use A_2, the relative poles R_{21} (mirror image of R_{12} with $O_2 O_4$ as the mirror)[7] and R_{23} and obtain the same mechanism at its second configuration. Further, the same solution can also be obtained by inversion, with the follower as the frame, as explained in Fig. 6.3-5.[8]

Problem 6.3-4

Figure 6.3-12 shows a slider-crank mechanism where the slider movement is desired to be proportional to the crank rotation. The desired initial position of the slider is at B_i. Given $\triangle s = 1.5$m and $\triangle \theta = 60°$
(CCW), design the mechanism with three Chebyshev's accuracy points.

Solution

First, let us locate three Chebyshev's accuracy points in the interval $B_f B_i$, which are indicated as B_1, B_2 and B_3 in Fig. 6.3-13b.

---

[7]We may note that $R_{12} = R_{12,1}^1$ and $R_{21} = R_{12,1}^2$.
[8]For more discussion on the solution obtained, see Section 6.9.

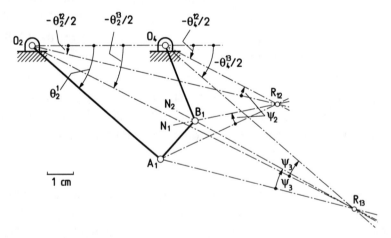

Figure 6.3-11

6.3. GRAPHICAL METHODS (THREE POSITIONS)

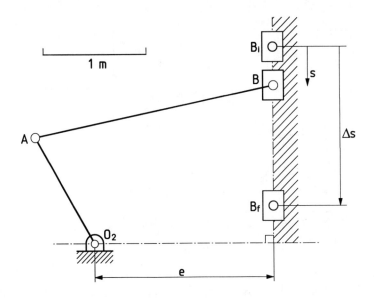

Figure 6.3-12

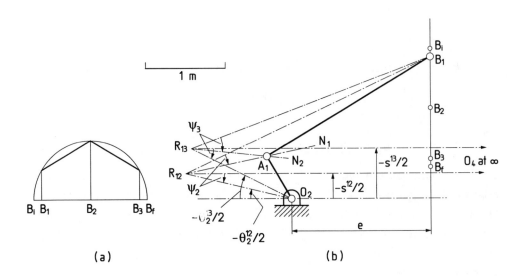

Figure 6.3-13

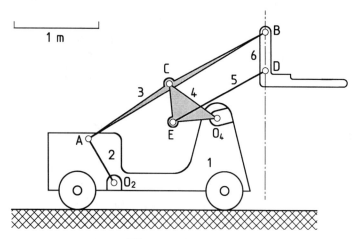

Figure 6.3-14a

This can be done either by using Equation (6.2.3) or graphically, as shown in Fig. 6.3-13a (see Fig. 6.2-2a). Since the slider displacement is desired to be proportional to the crank-rotation, we can write

$$\theta_2^{12}/s^{12} = \Delta\theta/\Delta s$$

or

$$\theta_2^{12} = (s^{12}/\Delta s) \times 60°(\text{CCW}) = (B_1B_2/B_iB_f) \times 60°(\text{CCW})$$

$$= 26°(\text{CCW}).$$

Similarly,

$$\theta_2^{13} = 52°(\text{CCW}) \quad \text{and} \quad s^{13} = B_1B_3.$$

With the known values (θ_2^{12}, s^{12}) and (θ_2^{13}, s^{13}), we locate the relative poles R_{12} and R_{13} (Fig. 6.3-13b). The crank-pin A_1 is located at the intersection of the lines N_1 and N_2, which are drawn so that O_4O_2 and B_1A_1 subtend equal angles at each relative pole. Thus, $O_2A_1B_1$ is the desired slider-crank mechanism at the first accuracy point.

Problem 6.3-5

Figure 6.3-14a shows the kinematic sketch of a forklift without a vertical guide (i.e., a prismatic pair with consequent large friction and wear). It is desired that the point B moves approximately along a vertical straight line and the fork remains approximately horizontal

6.3. GRAPHICAL METHODS (THREE POSITIONS)

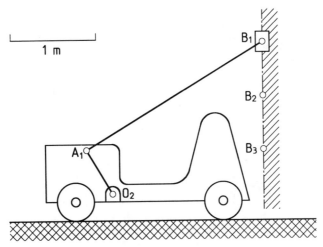

Figure 6.3-14b

during the entire range of movement. The fixed hinges should be conveniently located on the body of the truck. The vertical movement of the fork should be proportional to the rotation of the input link O_2A. Design the linkage with three accuracy points.

Solution

The solution to this design problem is carried out in a number of steps:

Step 1: The fixed hinge O_2 (on the body of the truck) and the desired line and range of movement of the point B are chosen suitably. Let the vertical movement of the point B be 1.5 m, during which the input link 2 rotates through $60°$. Accordingly, we can design with three accuracy points, a slider-crank mechanism O_2AB, maintaining a linear relation between the input rotation and output sliding movement, as explained in Problem 6.3-4. This design is shown in Fig. 6.3-14b at the first accuracy point.

Step 2: The slider at B can be removed by adding a suitable link (4) O_4C to guide B through the three accuracy points B_1, B_2 and B_3 (Fig. 6.3-14c). Of course the point B will no longer

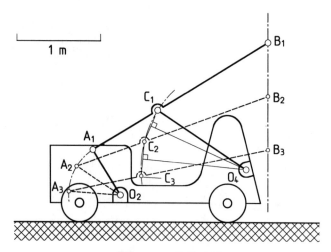

Figure 6.3-14c

move along an exact vertical line. The deviation of the path of B from the vertical line can be determined by obtaining the coupler curve of the point B of the resulting 4R linkage O_2ACO_4. Different approaches can be taken for determining the link O_4C. In one such approach, C is chosen near the midpoint of AB. This choice of C near the line AB renders link 3 as a slender member (with low inertia). Further, if C is chosen near B, then O_4 goes very far off (the path of C near B, obviously, is close to a straight line). On the other hand, if C is chosen near A, then O_4 goes very near to O_2 and the resulting 4R linkage O_2ACO_4 will have poor performance characteristics. Let us take C just to the left of the midpoint of AB. The positions of link AB, corresponding to the three accuracy points, are shown in Fig. 6.3-14c, where C_1, C_2 and C_3 are the locations of the point C. The fixed hinge (on the truck body) O_4 is at the center of the circle passing through C_1, C_2 and C_3. If the location of O_4 becomes unacceptable, then a few trials are necessary for the choice of C.

In another approach, O_4 may be chosen at a convenient location (Fig. 6.3-14d) and the 4R linkage may be inverted with the coupler fixed at its first configuration. Using a tracing paper with the lines A_1B_1, A_2B_2 and A_3B_3, locate the inverted positions of

6.3. GRAPHICAL METHODS (THREE POSITIONS)

O_4, as $O_{4,1}^1 (\equiv O_4), O_{4,2}^1$ and $O_{4,3}^1$ (see also Fig. 6.3-4). Next C_1 is located at the center of the circle passing through $O_{4,1}^1, O_{4,2}^1$ and $O_{4,3}^1$. It may be noted that for a good design, C_1 should come out near the midpoint of AB (in this case). Luckily, this happens in Fig. 6.3-14d and subsequently, we continue with this design.

Step 3: So far we have not made any attempt to keep the fork approximately horizontal, as it is simply hinged at B. To keep the fork horizontal, another conveniently chosen point on it, say D (Fig. 6.3-14a), has to be constrained in a suitable manner. Let D_1, D_2 and D_3 be the positions of D at the three configurations under consideration (Fig. 6.3-14e) when C_1, C_2 and C_3 are the respective locations of the point C. To locate the point E_1 (i.e., E at configuration I), we invert the mechanism with link 4 fixed at its first position. Using the points $O_4, C_1, C_2, C_3, D_1, D_2$ and D_3, with the help of a tracing paper, we locate the inverted positions of D as $D_1^1 (\equiv D_1), D_2^1$ and D_3^1. The point E_1 is at the center of the circle passing through $D_1^1 (\equiv D_1), D_2^1$ and D_3^1. Thus, we finally we get the desired mechanism, as shown in Fig. 6.3-14e.

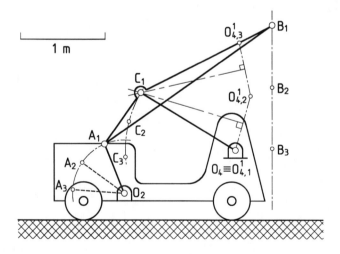

Figure 6.3-14d

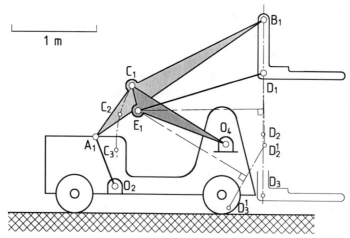

Figure 6.3-14e

6.4 Dead-Center Problems

Let us first consider the problem of synthesizing a slider-crank mechanism with prescribed stroke-length and quick-return ratio. Referring to Fig. 6.4-1, B_1 and B_2 represent the extreme positions of the slider 4 (when the crank 2 and the connecting rod 3 become collinear). During the forward stroke of the slider (from B_1 to B_2), the crank rotates through an angle θ_2^* (CCW) from the outer to the inner dead-center position. The stroke of the slider $S_H = B_1 B_2$. The quick-return ra-

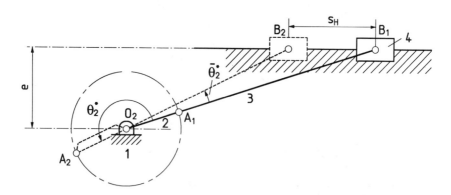

Figure 6.4-1

6.4. DEAD-CENTER PROBLEMS

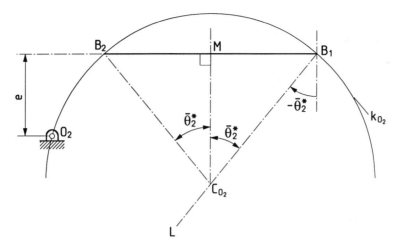

Figure 6.4-2

tio q is given by $q = \theta_2^*/(2\pi - \theta_2^*)$. It should be noted from Fig. 6.4-1 that no unique solution exists for prescribed values of S_H and θ_2^*. This is so because the point O_2 in Fig. 6.4-1 can be anywhere on the circular arc, with B_1B_2 as the chord and with a circumferential angle $\bar{\theta}_2^* = \theta_2^* - \pi$. Thus, for a unique design, besides S_H and θ_2^* (i.e., the quick-return ratio), one more kinematic dimension, viz., l_2 (crank length) or l_3 (connecting rod length) or e (offset), must be either prescribed or assumed. Depending on which one of the three dimensions is given (or assumed), three different geometric methods of determining the other two are discussed below. All constructions are illustrated with $\bar{\theta}_2^*$ positive (CCW). The little variation necessary when $\bar{\theta}_2^*$ is negative (i.e., when $q < 1$) is left as an exercise to the reader.

Case 1. S_H, θ_2^* and e are prescribed, l_2 and l_3 are to be determined.

Draw $B_1B_2 = S_H$ (Fig. 6.4-2). At B_1 draw the line L at an angle $-\bar{\theta}_2^*$ (i.e., CW $\bar{\theta}_2^*$) to the normal on B_1B_2. The line L intersects the midnormal of B_1B_2 at C_{O_2}. With C_{O_2} as center and $C_{O_2}B_1$ as radius, a circle, k_{O_2}, is drawn. The point O_2 is located on k_{O_2} at a distance e below the line B_1B_2. Now $O_2B_1 = l_3 + l_2$ and $O_2B_2 = l_3 - l_2$.

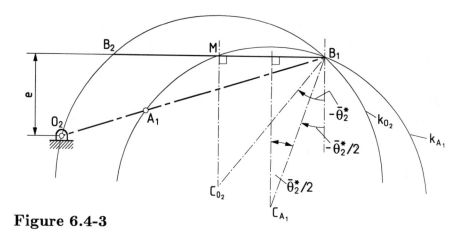

Figure 6.4-3

Therefore,

$$l_3 = (O_2B_1 + O_2B_2)/2 \quad \text{and} \quad l_2 = (O_2B_1 - O_2B_2)/2.$$

The proof of this construction is evident by comparing Figs. 6.4-1 and 6.4-2, and it is left as an exercise.

Case 2: S_H, θ_2^* and l_3 are prescribed, l_2 and e are to be determined.

Draw k_{O_2} (Fig. 6.4-3) as in Fig. 6.4-2. Draw the circle k_{A_1} with center at C_{A_1} (on the midnormal of B_1M) and radius $C_{A_1}B_1$ as ex-

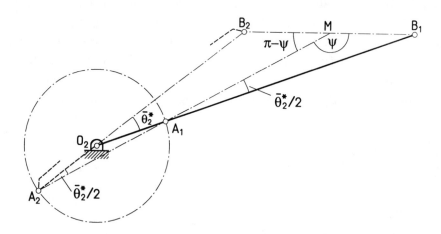

Figure 6.4-4

6.4. DEAD-CENTER PROBLEMS

plained in Fig. 6.4-3. Locate A_1 on k_{A_1}, with $B_1A_1 = l_3$. Extend B_1A_1 to intersect k_{O_2} at O_2, when $l_2(= O_2A_1)$ and e are determined.

It is easy to note that in the above construction, $\angle B_1A_1M = \bar{\theta}_2^*/2$, where M is the midpoint of B_1B_2, and this will be true as long A_1 lies on k_{A_1}. The requirement $\angle B_1A_1M = \bar{\theta}_2^*/2$ can be seen from Fig. 6.4-4, where the mechanism is drawn at both its dead-center configurations and we prove that M is the midpoint of B_1B_2. Let A_1A_2, when extended, meet B_1B_2 at M. From $\triangle A_1BM$ we get $A_1B_1/\sin\psi = B_1M/\sin(\bar{\theta}_2^*/2)$ and from $\triangle A_2B_2M$ we get $A_2B_2/\sin\psi = B_2M/\sin(\bar{\theta}_2^*/2)$ Since $A_1B_1 = A_2B_2 = l_3$, the above two relations imply $B_1M = B_2M$, i.e., M is the midpoint of B_1B_2.

Case 3: S_H, θ_2^* and l_2 are prescribed, l_3 and e are to be determined.

Draw k_{O_2} as in Figs. 6.4-2 and 6.4-3. Draw the circle $k_{\bar{A}_1}$ with center at $C_{\bar{A}_1}$ and radius $C_{\bar{A}_1}B_1$ as explained in Fig. 6.4-5. Locate $\bar{A}_1$ on $k_{\bar{A}_1}$ with $B_1\bar{A}_1 = l_2$. Extend $B_1\bar{A}_1$ to intersect k_{O_2} at O_2, when $l_3(= O_2B_1 - l_2)$ and e are determined. The reader can prove this construction by taking the point $\bar{A}_1$ on the line A_1B_1, in Fig. 6.4-4, with $B_1\bar{A}_1 = l_2$ and showing that $\angle B_1\bar{A}_1M = (\pi + \bar{\theta}_2^*)/2$, as satisfied by the construction shown in Fig. 6.4-5.

In the case of a 4R crank-rocker linkage (Fig. 6.4-6) unequal periods for the forward and return movements imply that the line B_1B_2 does not pass through O_2 (see Section 4.3). The distance of B_1B_2 from O_2 is defined as the offset e. If the rocker length l_4 and the rocker swing-angle θ_4^* are prescribed, then $B_1B_2 = 2l_4\sin(\theta_4^*/2)$ and it can be

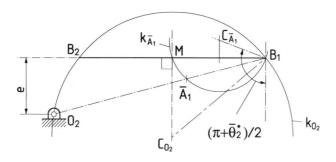

Figure 6.4-5

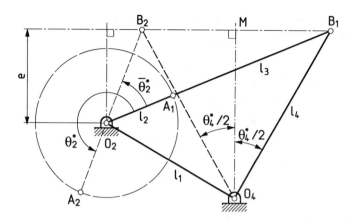

Figure 6.4-6

treated as the stroke length S_H. The constructions described above for slider-crank mechanisms can be used for synthesizing a crank-rocker linkage with prescribed quick-return ratio. It may be noted that besides θ_2^*, θ_4^* and l_4, one more kinematic dimension (l_1, or l_2, or l_3, or e) must be either prescribed or assumed, to yield a unique solution. Furthermore, the reader is advised to note the differences between Figs. 4.3-2 and 6.4-6 due to different left-right relative dispositions of the crank and the rocker. In Fig. 4.3-2, θ_2^* is measured CCW from the inner to the outer dead-center position, whereas in Fig. 6.4-6 it is measured CCW from the outer to the inner dead-center position.

In real life, very often one needs to design a 4R crank-rocker linkage with prescribed l_1 (i.e., a convenient frame length or prescribed locations of the fixed-pivots), θ_2^* (i.e., quick-return ratio) and θ_4^* (rocker swing-angle). In such a situation, the solution is obtained by Alt's construction[9]. Alt's constructions for various situations are presented below. In all cases θ_2^* and θ_4^* are measured CCW from the outer to the inner dead-center configurations.

Case 1: $\theta_2^* > \pi$ (Fig. 6.4-7):

O_2 and O_4 are the prescribed fixed pivots. At O_2, a line is drawn

---

[9]The underlying theory behind this construction, which needs four-position synthesis, is omitted here. Some discussions on this topic will be presented at the end of this section.

6.4. DEAD-CENTER PROBLEMS

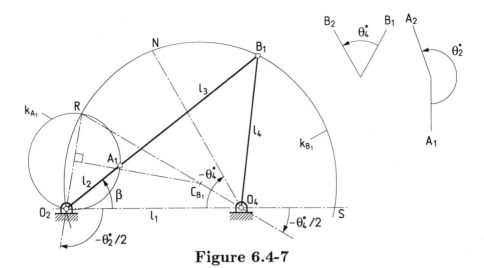

Figure 6.4-7

at an angle $-\theta_4^*/2$ to O_2O_4, and another line is drawn through O_4 at an angle $-\theta_4^*/2$ to O_2O_4. These two lines intersect at R. The circle k_{A_1} is drawn with O_2R as the diameter. The midnormal of O_2R intersects O_4R at C_{B_1}. Another circle, k_{B_1}, is drawn with its center at C_{B_1} and radius as $C_{B_1}R$. The circle k_{B_1} intersects O_2O_4 at S. A line O_4N is drawn through O_4 at an angle $-\theta_4^*$, with O_4O_2 intersecting k_{B_1} at N. The point B_1 can now be taken anywhere on the arc NS. The line O_2B_1 intersects k_{A_1} at A_1 and $O_2A_1B_1O_4$ is the required 4R crank-rocker linkage at its outer dead-center configuration.

Case 2: $\theta_2^* < \pi$ (Fig. 6.4-8):

The only difference between Figs. 6.4-7 and 6.4-8 is that in the latter, the line O_4N intersects k_{B_1} at two points above the line O_2O_4. The second point of intersection is labelled as S. The moving hinge B_1 can now be located anywhere on k_{B_1} within the arc NS.

Case 3: $\theta_2^* = \pi$ (i.e., no quick-return):

In this case the point S becomes coincident with O_4 and the rest of the construction remains same as above.

Case 4: $\theta_2^* = \pi + \theta_4^*$:

In this special situation, the point C_{B_1} goes to infinity (since $\angle O_2RO_4 = \pi/2$) and the circle k_{B_1} degenerates into the straight line O_2R (Fig. 6.4-7). The point N on O_2R is easily seen to be located so that $O_2R = RN$. The point B_1 can now be located anywhere on the

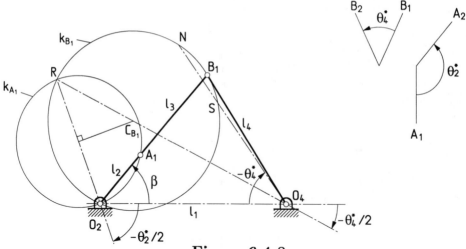

Figure 6.4-8

line O_2R above N and the crank-length ($= O_2R$) remains the same for all solutions.

It should be noted that Alt's construction yields an infinite number of solutions when only l_1, θ_2^* and θ_4^* are prescribed. The designer should normally choose the solution which provides the best transmission quality, i.e., the solution which has the largest magnitude of the minimum transmission angle. To achieve this, Volmer's nomogram (Fig. 6.4-9) is very useful. The nomogram shows the maximum possible value of the minimum transmission angle $(\mu_{\min})_{\max}$ for prescribed values of θ_2^* and θ_4^*. It also indicates the corresponding required values of the angle β (see Figs. 6.4-7 and 6.4-8). The following points regarding Alt's construction and Volmer's nomogram may be noted.

1. If the point B_1 is taken outside the region NS, then the rocker has to cross the frame-line during its movement, which is not possible in a crank-rocker mechanism.

2. For a given value of θ_4^*, the minimum and maximum values of θ_2^* (for a crank-rocker linkage to exist) are given, respectively, by

$$(\theta_2^*)_{\min} = (\pi + \theta_4^*)/2 \quad and \quad (\theta_2^*)_{\max} = (3\pi + \theta_4^*)/2$$

as was seen in Exercise Problem 4.9. In these limiting situations, the line O_4N in Alt's construction becomes tangent to k_{B_1}.

6.4. DEAD-CENTER PROBLEMS

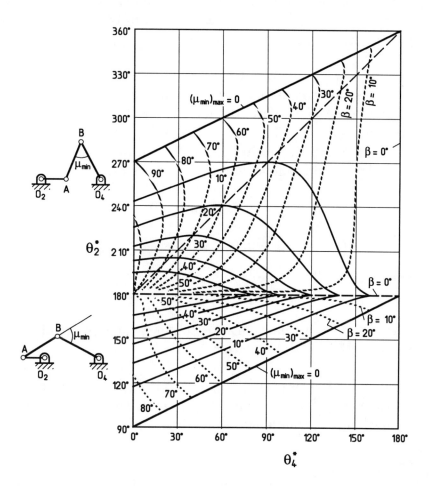

Figure 6.4-9

3. For $\theta_2^* = \pi$, the value of $(\mu_{min})_{max}$ is indicated in the nomogram but the required value of β (for the optimum solution) cannot be obtained. To get the optimum solution in this case, substitute $\mu_{min} = (\mu_{min})_{max}$ in Problem 4.4-1.

4. When $\theta_2^* = \pi + \theta_4^*$, (indicated by the broken straight line in Fig. 6.4-9), as already noted, there exists only one value of β (for all solutions). Therefore, the knowledge of β does not yield the optimum solution. The value of $(\mu_{min})_{max}$ can be read from the nomogram and the optimum link-length ratios are obtained as the solution to the following three equations:

$$l_2 = l_1 \sin(\theta_4^*/2), \tag{6.4.1}$$

$$(l_1 - l_2)^2 = l_3^2 + l_4^2 - 2l_3 l_4 \cos(\mu_{min})_{max}, \tag{6.4.2}$$

and

$$l_4^2 = l_1^2 + l_3^2 - l_2^2. \tag{6.4.3}$$

The above three equations are nonlinear and the reader is advised to derive these equations.

5. Without quick-return $(\mu_{min})_{max} < 30°$ if $\theta_4^* > 120°$. With quick-return a satisfactory crank-rocker, with $(\mu_{min})_{max} > 30°$ has rather limited swing angle of the rocker.

Problem 6.4-1

Figure 6.4-10a shows a symmetrical toggle press. The driving mechanism for the right half, at the working instant, is shown in Fig. 6.4-10b. In this mechanism the working instant corresponds to the inner dead-center configuration of the 4R crank-rocker linkage O_2ABO_4. Also, O_4C and CD are almost collinear at this instant, in order to transmit the required large force to the die. The transmission angle of the 4R linkage O_2ABO_4 should also be high at this instant. One can think of another design where the working instant corresponds to the outer dead-center configuration of O_2ABO_4. For higher productivity, a quick- return ratio of 1.25 is desired and the rocker swing-angle is prescribed as $50°$.

6.4. DEAD-CENTER PROBLEMS

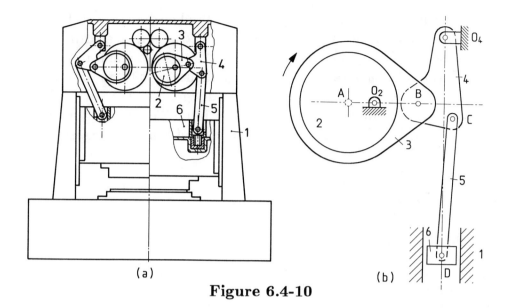

(a)   (b)

Figure 6.4-10

(a) Sketch the possible alternatives of the crank-rocker linkage in order to satisfy the requirements. In each case, indicate the direction of the crank rotation and the working stroke of the follower.

(b) If the follower length $O_4 B (= l_4)$ is prescribed as 100 mm, with transmission angle $\mu = 75°$ at the working instant, design the possible variations of the 4R linkage and determine μ_{min} in each case.

(c) If the linkages are designed for $(\mu_{min})_{max}$ condition, with a frame length of 100 mm, synthesize the possible variations and determine the values of μ at the working instant.

(d) From the solutions obtained in (b) and (c), which alternative, suggested in (a) should be finally chosen?

Solution

(a) Four possible variations of the design are shown in Figs. 6.4-11a to 6.4-11d. These are numbered as design I to design IV. In Figs. 6.4-11a and 6.4-11b the crank rotation is CCW, whereas that in Figs. 6.4-11c and 6.4-11d is CW. The direction of the follower movement during the working stroke is indicated in each case. With a quick-return ratio of 1.25, we get crank rotations during the forward and return movements as $200°$ and $160°$, respectively. It may be noted that for designs I and IV, $\theta_2^* = 200°$, whereas for designs II and III, $\theta_2^* = 160°$. The working instant in designs I and III is at the inner dead-center configuration,

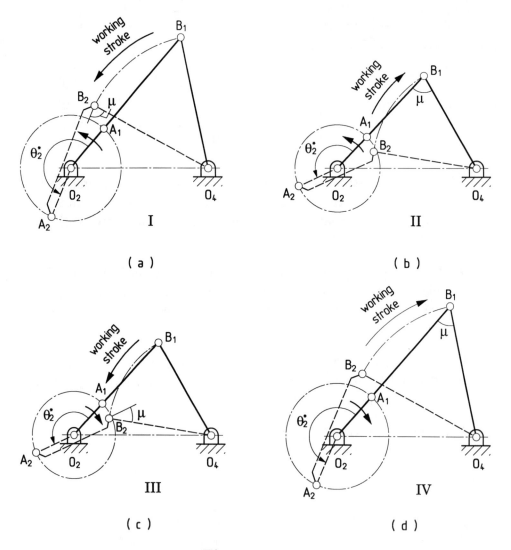

Figure 6.4-11

6.4. DEAD-CENTER PROBLEMS

whereas that in Figs. 6.4-11b and 6.4-11d is at the outer dead-center configuration.

(b) It is obvious from Figs. 6.4-11a to 6.4-11d that the link lengths for designs I and IV are identical if all other requirements are satisfied by these two linkages. Similarly, the link lengths for designs II and III are also identical. In each of these figures the transmission angle at the working instant is indicated by μ. Hence, basically, there are two different designs (for the same direction of crank rotation), one with the working instant at the inner dead-center configuration and the other with the working instant at the outer dead-center configuration. From here forwards we shall consider only CCW rotation of the crank, i.e., designs I and II.

Design I

With $l_4 = 100$ mm and $\theta_4^* = 50°$, we get $B_1B_2 = 2l_4 \sin(\theta_4^*/2) = 84.5$mm. Treating $B_1B_2 = S_H$ and $\theta_2^* = 200°$, we draw k_{O_2} (see Fig. 6.4-6 and 6.4-2) as shown in Fig. 6.4-12a. The fixed hinge O_2 is located on k_{O_2} such that $\angle O_4B_2O_2 = 75°$, as desired. The link lengths are finally obtained as follows:

$$l_1 = O_2O_4 = 113\text{mm}, \quad l_2 = (O_2B_1 - O_2B_2)/2 = 37.5\text{mm},$$

$$l_3 = (O_2B_1 + O_2B_2)/2 = 122\text{mm} \quad \text{and} \quad l_4 = O_4B_1 = 100\text{mm}.$$

The minimum transmission angle for this mechanism is obtained by considering the configuration when the crank (l_2) is in the line of frame with $\theta_2 = \pi$. Thus, from Equation (4.4.1)

$$\mu_{\min} = \cos^{-1}\left(\left\{(l_3^2 + l_4^2) - (l_1 - l_2)^2\right\}/2l_3l_4\right) = 38°.$$

Design II

In this case $\theta_2^* = 160°$ and the construction described above is repeated with $\angle O_4B_1O_2 = 75°$ (Fig. 6.4-12b). The link lengths are obtained as follows:

$$l_1 = 137\text{mm}, \quad l_2 = 40.5\text{mm}, \quad l_3 = 83\text{mm} \quad \text{and} \quad l_4 = 100mm.$$

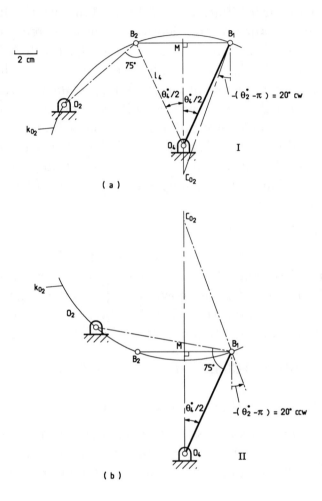

Figure 6.4-12

6.4. DEAD-CENTER PROBLEMS

The minimum transmission angle for this mechanism occurs with $\theta_2 = 0$, when from Equation (4.4.1),

$$\mu_{\min} = \cos^{-1}\left(\left\{(l_3^2 + l_4^2) - (l_1 + l_2)^2\right\}/2l_3l_4\right) = 27°.$$

(c) From the nomogram (Fig. 6.4-9) we get the following values:

For design I $\theta_2^* = 200°, \theta_4^* = 50°, (\mu_{\min})_{\max} = 41°$ and $\beta = 48°$.
For design II $\theta_2^* = 160°, \theta_4^* = 50°, (\mu_{\min})_{\max} = 28°$ and $\beta = 46°$.

Carrying out Alt's construction with the above data (Figs. 6.4-13a and 6.4-13b), the link lengths and the transmission angle at the working instant for the two cases are determined as follows:

Design I (see Fig. 6.4-13a):

$$l_1 = 100\text{mm}, \quad l_2 = 37\text{mm}, \quad l_3 = 84.5\text{mm} \quad \text{and} \quad l_4 = 93mm.$$

Referring to Fig. 6.4-11a, the transmission angle μ at the working instant is given by

$$\mu = \cos^{-1}\left(\left\{l_4^2 + (l_3 - l_2)^2 - l_1^2\right\}/2l_4(l_3 - l_2)\right) = 84°.$$

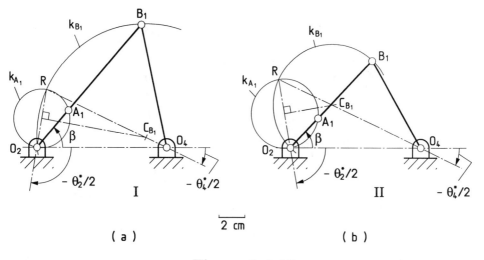

Figure 6.4-13

Design II (see Fig. 6.4-13b):

$$l_1 = 100\text{mm}, \quad l_2 = 30.5\text{mm}, \quad l_3 = 59\text{mm} \quad \text{and} l_4 = 75\text{mm}.$$

Referring to Fig. 6.4-11b, the transmission angle μ at the working instant is given by

$$\mu = \cos^{-1}\left(\left\{l_4^2 + (l_3 + l_2)^2 - l_1^2\right\}/2l_4(l_3 + l_2)\right) = 74°.$$

(d) From the solutions obtained in (b) and (c), we should choose design I. This is because in solution (b), design I, when compared with design II, has a higher $\mu_{\min}$ and the same μ at the working instant. In solution (c), again design I has both $\mu_{\min}$ and μ at the working instant, higher than the respective values obtained in design II.

If the coupler length l_3, over and above θ_2^*, θ_4^* and l_1, is also prescribed, the Alt's construction can be extended as shown in Fig. 6.4-14, which has been drawn for $\theta_2^* > \pi$. In this figure, after drawing k_{A_1} as before, a circle $k_{\bar{B}_1}$ is drawn, with O_2H as the diameter where H is

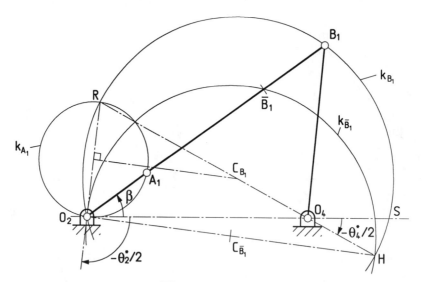

Figure 6.4-14

6.4. DEAD-CENTER PROBLEMS

the second point of intersection of the line O_4R with k_{B_1}. The point $\bar{B}_1$ is located on $k_{\bar{B}_1}$, with $O_2\bar{B}_1 = l_3$. The line $O_2\bar{B}_1$ intersects k_{A_1} at A_1, and when extended, intersects k_{B_1} at B_1. Then, $O_2A_1B_1O_4$ is the required linkage at its outer dead-center configuration. The proof of this construction is left as an exercise for the reader, with the hint that it can be shown that

$$l_3 = A_1B_1 = O_2\bar{B}_1 = 2r_{B_1}\sin[(\theta_2^* - \theta_4^*)/2]\sin(\beta + \theta_2^*/2)$$

where r_{B_1} is the radius of k_{B_1}.

It should be noted that the dead-center problem, solved by Alt's construction, can be seen as a synthesis problem to coordinate four positions of the crank and the rocker, as explained below. Since the velocity of the rocker is zero at both the extreme positions, we can say that at the extreme positions an infinitesimal rotation $d\theta_2$ of the crank produces no movement of the rocker. This is explained in terms of four coordinated positions I to IV in Fig. 6.4-15[10].

Referring to Fig. 6.4-16a, a fifth coordinated position of the crank and the follower, labelled as V, can also be considered. The synthesis of the desired crank-rocker, satisfying this extra requirement, is explained in Fig. 6.4-16b. Draw k_{A_1} and k_{B_1} as in Fig. 6.4-8 ($\theta_2^* < \pi$). Locate the point R' at the intersection of the lines drawn at O_2 and O_4, respectively, at angles $-\theta_2^V/2$ and $-\theta_4^V/2$ with O_2O_4. The line O_4R''

---

[10] For more discussion, see Section 6.5.2.

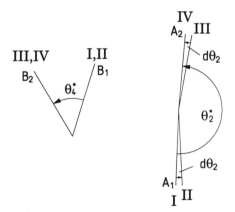

Figure 6.4-15

CHAPTER 6. DIMENSIONAL SYNTHESIS

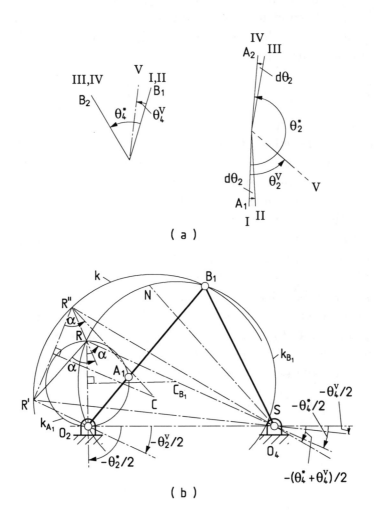

Figure 6.4-16

is drawn at O_4 at an angle $-(\theta_4^* + \theta_4^V)/2$ with O_2O_4. The point R'' is located on this line so that $\angle R'RR'' = \pi - \alpha$, where $\alpha = \angle O_2RO_4$. At R'' a line is drawn at an angle α with $R'R''$, which intersects the midnormal of $R'R''$ at C. With C as the center and CR'' as the radius, the circle k is drawn, which intersects k_{B_1} at B_1. The line O_2B_1 intersects k_{A_1} at A_1 when $O_2A_1B_1O_4$ is the required linkage at its outer dead-center configuration. If k and k_{B_1} do not intersect within NS, then no crank-rocker linkage exists that would satisfy all the requirements.

6.5 Graphical Methods (Four Positions)

While carrying out three-position synthesis in Section 6.3, we had arbitrarily chosen some dimensions in order to obtain a unique solution. For example, in a motion generation problem (Fig. 6.3-2) we had to choose both ground pivots O_2 and O_4 arbitrarily. Similarly in a function generation problem, besides O_2 and O_4, the moving hinge A was arbitrarily chosen at one of the desired configurations (Figs. 6.3-5 and 6.3-11). If we need to synthesize a mechanism for four positions, i.e., three coordinated movements, then obviously we have to leave out one more (than that required for three-position synthesis) design parameter as unknown. Before discussing a general procedure for four-position synthesis, a simple method, known as point-position reduction, based on kinematic inversion,[11] will be presented. This method is illustrated with reference to problems of function generation and path generation.

6.5.1 Point-Position Reduction

Refer to Fig. 6.3-5b. If we add a fourth position A_4, then after inversion let A_4 move to A_4^1. The problem is that with arbitrary location of A (as in Fig. 6.3-5b), the points $A_1^1 (\equiv A_1)$, A_2^1, A_3^1 and A_4^1 do not lie on a circle. However, if A_1 is so chosen that after inversion two of these four points coincide, then again we get only three different points for the four inverted positions of A. So, B_1 can be located

---

[11]The method is due to Kurt Hain. For further details see Hain, K: Applied Kinematics, 2nd Edition, McGraw-Hill, New York, 1967.

CHAPTER 6. DIMENSIONAL SYNTHESIS

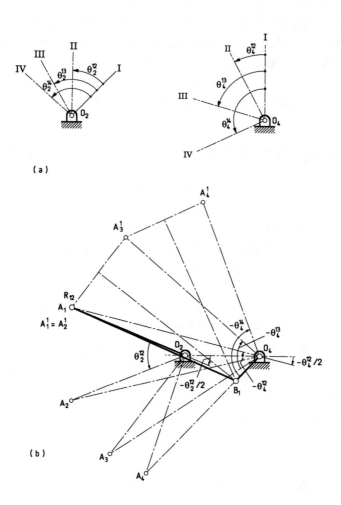

Figure 6.5-1

6.5. GRAPHICAL METHODS (FOUR POSITIONS)

at the center of the circle passing through these three points. In Fig. 6.5-1a we have added a fourth position to the three positions shown in Fig. 6.3-5a. For the three pairs of coordinated movements $(\theta_2^{12}, \theta_4^{12})$, $(\theta_2^{13}, \theta_4^{13})$ and $(\theta_2^{14}, \theta_4^{14})$, if A_1 is chosen at R_{12}, as indicated in Fig. 6.5-1b, then it is obvious, from the method of constructing the inverted positions, that A_2^1 coincides with $A_1 (\equiv A_1^1)$. Thus, with this particular choice of A_1 we can locate B_1 at the center of the circle passing through $A_1^1 (\equiv A_2^1)$, A_3^1 and A_4^1. The relative movement between links 4 and 2, as they move from position I to II, is a pure rotation about R_{12}, which is why A_2^1 coincides with A_1 (because point A_1 is identical with the pole R_{12}). It is needless to mention that either R_{13} or R_{14} could also have been chosen as A_1. In the first case A_3^1 would coincide with A_1, whereas in the latter option A_4^1 would coincide with A_1. Similarly, we could choose A_2 at R_{23} and invert the mechanism at its second configuration, when A_3^2 coincides with A_2 $(\equiv A_2^2)$, and B_2 can be located at the center of the circle passing through A_1^2, A_2^2 ($\equiv A_2^2 \equiv A_3^2$) and A_4^2. It may be noted that whichever pole is chosen as one of the locations for A, link 2 has to cross the line of frame during its movement and hence, this link cannot be the rocker of a crank-rocker linkage.

For the solution of a problem of path generation with four positions, let us consider a fourth point, C_4, in Fig. 6.3-4a, as shown in Fig. 6.5-2a. In Fig. 6.3-4b, besides the location of O_2, the location of O_4 and link lengths $O_2 A$ and AC were also arbitrarily chosen. In that case, however, the four inverted positions of O_4, viz., $O_{4,1}^1 (\equiv O_4)$, $O_{4,2}^1$, $O_{4,3}^1$ and $O_{4,4}^1$, did not lie, in general, on a circle. It will now be shown how, with a proper choice of location of O_4, we can make two of the four inverted positions of O_4 coincide. In Fig. 6.5-2b the method is explained in which $O_{4,1}^1 (\equiv O_4)$ and $O_{4,3}^1$ are made to coincide. We may note that in Fig. 6.5-1b the points A_1 and A_2 are located symmetrically about the line $O_2 O_4$, and similarly in Fig. 6.5-2b, the points A_1 and A_3 are located symmetrically about the line $O_2 O_4$, which is accomplished in the following way:

With a convenient location of O_2 and assumed length of link 2, draw k_A. With an assumed length of AC, locate A_1, A_2, A_3 and A_4 on k_A using the prescribed locations C_1, C_2, C_3 and C_4, respectively. Locate O_4 at the intersection of the midnormals of $A_1 A_3$ and $C_1 C_3$.

Making an inversion with the coupler fixed at its first position A_1C_1, the inverted positions of O_4 are labelled as $O_{4,1}^1$ ($\equiv O_{4,3}^1 \equiv O_4$), $O_{4,2}^1$ and $O_{4,4}^1$. It is obvious that $O_{4,1}^1 \equiv O_{4,3}^1$ because by construction we have made $O_4(\equiv O_{4,1}^1)$ the pole P_{13} for the coupler movement from position I to position III. Finally, B_1 is located at the center of the circle passing through $O_{4,1}^1 (\equiv O_{4,3}^1)$, $O_{4,2}^1$ and $O_{4,4}^1$.

If we want to synthesize the linkage so as to generate a coupler curve through five accuracy points C_1, C_2, C_3, C_4 and C_5 (Fig. 6.5-3a), then the point-position reduction technique can be extended as

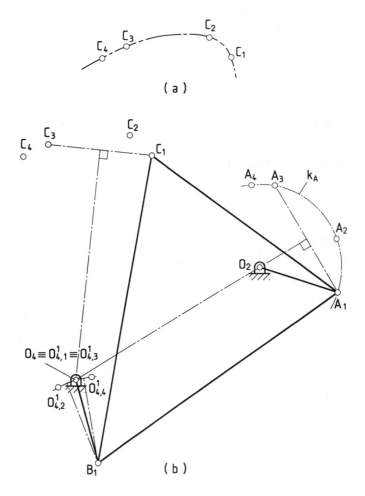

Figure 6.5-2

6.5. GRAPHICAL METHODS (FOUR POSITIONS)

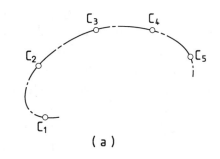

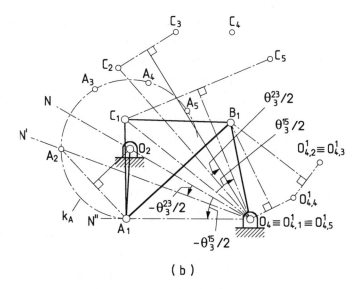

Figure 6.5-3

follows. In this case two pairs of inverted positions of O_4 are made to coincide so that B_1 can again be located at the center of the circle passing through three distinct locations of the inverted positions of O_4. In Fig. 6.5-3b the two pairs $(O_{4,1}^1, O_{4,5}^1)$ and $(O_{4,2}^1, O_{4,3}^1)$ are made to coincide. (Any other two pairs can be chosen). Accordingly, O_4 is located at the point of intersection of the midnormals of C_1C_5 and C_2C_3. Our objective is to make O_4 the location for both P_{23} and P_{15} of the coupler motion, which is achieved in the following way. A line N is drawn conveniently at any orientation through O_4. The lines N' and N'' are drawn through O_4, as indicated in the figure (using the values of θ_3^{23} and θ_3^{15} obtained with the assumption that O_4 is the pole of the coupler movement between the chosen pairs of configurations, i.e., (I, V) and (II, III)). With arbitrarily chosen link-length AC and prescribed locations of C_1 and C_2, the locations of A_1 and A_2 are obtained on the lines N'' and N', respectively. The fixed hinge O_2 is located at the point of intersection of N and the midnormal of A_1A_2.

The points A_3, A_4 and A_5 are now located on k_A with the help of prescribed locations of C_3, C_4 and C_5 and the assumed length AC. Making an inversion, with the coupler fixed at its first configuration A_1C_1, the inverted positions of O_4 are labelled as $O_{4,1}^1$ ($\equiv O_{4,5}^1$), $O_{4,2}^1$ ($\equiv O_{4,3}^1$) and $O_{4,4}^1$. Then B_1 is located at the center of the circle passing through O_4, $O_{4,2}^1$ and $O_{4,4}^1$. The reader is advised to note that (i) several options for the two chosen pairs exist and (ii) the line N can be drawn at any orientation through O_4. Further, the method of construction followed in Fig. 6.5-2b should be reexamined in light of Fig. 6.5-3b, i.e., in the first figure, we could have also used an arbitrary location of O_4 on the midnormal of C_1C_3, and the angle θ_3^{13} could have been obtained with the assumption that $O_4 \equiv P_{13}$.

6.5.2 Application of Burmester Theory

In Section 2.11 we discussed the rudiments of the Burmester theory for four finitely separated positions of a moving lamina. In this section we briefly discuss how this theory can be applied towards solving problems of motion generation and function generation involving four positions. Referring to Fig. 6.5-4a, let the coupler of a 4R linkage need to be guided through four prescribed positions, labelled as I to

6.5. GRAPHICAL METHODS (FOUR POSITIONS)

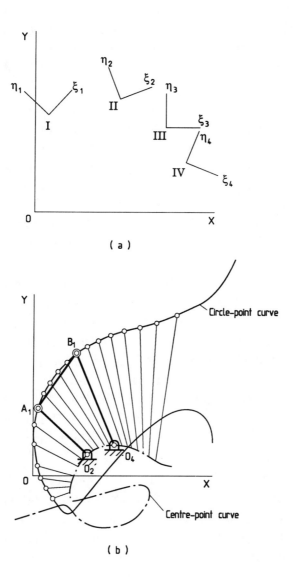

Figure 6.5-4

IV and indicated by the $\xi - \eta$ axes attached to the coupler. The six poles corresponding to this sequence of movements can easily be obtained by considering two points attached to the $\xi - \eta$ axes (refer to Fig. 2.8-3). Once the poles are obtained, the circle-point curve and the center-point curve can be drawn in a manner explained in Figs. 2.11-7a to 2.11-7d.

The circle-point curve (in position I) and the center-point curve, corresponding to Fig. 6.5-4a, are shown in Fig. 6.5-4b. For clarity, the poles and all the construction details have been omitted in Fig. 6.5-4b, whereas for identification, the corresponding circle-points and centre-points are joined by rays. The moving hinges (at their first positions) can be chosen as any two circle-points and the fixed hinges as the corresponding centre-points. Thus $O_2 A_1 B_1 O_4$ is a possible solution that, when moved, takes the coupler AB through the four prescribed positions.[12]

If we have to synthesize a 4R linkage or a slider-crank mechanism for a function generation problem with four precision points, then we use the relative poles (see Section 6.3.4). Thereafter, the center-point and circle-point curves are obtained using these relative poles and the results of Section 2.11. The moving hinge B is located at a circle-point and the moving hinge A is obtained at the center-point corresponding to B. We may recall that all the poles (i.e., the relative poles in the present case) themselves are centre-points. Thus, in Section 6.5.1 the moving hinge A was located at one of the relative poles for four-position function generation.

At this stage we are in a position to justify the Alt's construction presented in Figs. 6.4-7 and 6.4-8. If we draw the relative poles (and their images) corresponding to Fig. 6.4-15, we find that in this special situation the circle-point curve and the center-point curves split into a circle and a straight line (see Problem 2.11-3 and Fig. 2.11-9c). For example, in Fig. 6.4-7, the center-point curve is given by k_{A_1} and the line $O_2 R$ and the circle-point curve is given by k_{B_1} and the line $O_4 R$. The restricted region NS, for the choice of B_1, results from the fact

---

[12] For synthesis with five prescribed positions, the choice of the moving hinges is restricted to only Burmester points. The maximum number of Burmester points being four, by taking two at a time, one can get a maximum of six solutions.

6.6 Analytical Methods (Three Positions)

Analytical methods of synthesis using the accuracy point approach are based on the displacement equations discussed in Section 4.2. In this section, we shall discuss these methods for synthesizing a 4R linkage and a slider-crank mechanism, with reference to the problems of function generation. The discussion will be restricted to only three accuracy points when, as will be seen later, only one needs to handle a set of linear algebraic equations. Further, the reader should note that different design parameters are assumed in different analytical methods, so that a set of linear equations, involving suitably selected unknown design parameters, is obtained. The assumed parameters are also different from those assumed in the graphical method presented in Section 6.3. Finally, the analytical methods, in contrast to the graphical methods, can conveniently handle prescribed instantaneous kinematic conditions involving velocity, acceleration, etc. (i.e., on the derivatives of the desired function).[13] Therefore, in this and in the next sections, by three or four positions synthesis, we shall mean three or four prescribed conditions to be satisfied. Synthesis problems involving four or more precision points invariably result in nonlinear equations.

Two different methods, one due to Bloch and the other due to Freudenstein, are discussed[14]. The first one involves the coupler movement in addition to the movements of the input and the output links. Thus, for a function generation problem involving prescribed correlation between the input and output movements, the associated coupler movements must be arbitrarily chosen. Freudenstein's method, on the

---

[13] For a graphical method of synthesizing 4R function generators involving prescribed conditions on velocity, see Ghosh, A. and Dittrich, G.: A new approach to the synthesis of 4-bar function generators, Mechanism and Machine Theory, Vol.14, pp. 255-265, 1979.

[14] For another method of solving interesting synthesis problems for specified instantaneous kinematic conditions (e.g., prescribed extreme values of ω_4 for a prescribed constant ω_2) see Hirschhorn J.: Kinematics and Dynamics of Plane Mechanisms, McGraw-Hill, New York, 1962.

other hand, does not involve the coupler motion and hence cannot be used for any problem with prescribed condition on the coupler. These points will be made clearer through examples.

6.6.1 Bloch's Method

Bloch's method of synthesis of a 4R linkage is based on the loop-closure equation expressed in complex notation (see Sections 4.2 and 5.4). Referring to Fig. 4.2-1, the loop-closure equation is written as[15]

$$l_1 + l_2 e^{i\theta_2} + l_3 e^{i\theta_3} - l_4 e^{i\theta_4} = 0. \tag{6.6.1}$$

Differentiating Equation (6.6.1), successively with respect to time, we get the following velocity and acceleration equations:

$$\omega_2 l_2 e^{i\theta_2} + \omega_3 l_3 e^{i\theta_3} - \omega_4 l_4 e^{i\theta_4} = 0 \tag{6.6.2}$$

$$(\alpha_2 + i\omega_2^2) l_2 e^{i\theta_2} + (\alpha_3 + i\omega_3^2) l_3 e^{i\theta_3}$$
$$-(\alpha_4 + i\omega_4^2) l_4 e^{i\theta_4} = 0 \tag{6.6.3}$$

For three-position synthesis, the three link-length ratios, l_2/l_1, l_3/l_1 and l_4/l_1 are left as unknowns and all other quantities, if not prescribed, have to be suitably assumed. Finally, we get three linear equations involving the three unknown link-length ratios. As will be seen in the examples to follow, different algebraic manipulations of Equations (6.6.1) to (6.6.3) for different prescribed quantities may lead to computational conveniences. Let us denote the link-length ratios as $l'_2 = l_2/l_1$, $l'_3 = l_3/l_1$, and $l'_4 = l_4/l_1$.

Problem 6.6-1

Synthesize a 4R linkage to generate $y = \log_{10} x$ in the interval $1 \leq x \leq 10$, with three Chebyshev's accuracy points. Assume $\Delta\theta_2 = 60°$(CCW) and $\Delta\theta_4 = 90°$(CCW). The length of the smallest link is 10 cm.

---

[15]The reader should again note the different left-right dispositions of the input and output members as compared to what has been used in general, while discussing graphical methods.

6.6. ANALYTICAL METHODS (THREE POSITIONS)

Solution

For $x_1 = 1$ and $x_e = 10$, let us first determine the accuracy points. Using Equation (6.2.3), with $a = (x_e + x_i)/2$ and $h = (x_e - x)/2$, we get $x^1 = 1.6$, $x^2 = 5.5$ and $x^3 = 9.4$. According to $y = \log_{10} x$, the corresponding values of y are obtained as $y^1 = 0.204$, $y^2 = 0.741$ and $y^3 = 0.974$. Assuming a linear relationship between $\Delta\theta_2$ and Δx, we obtain

$$\theta_2^{12} = (x^2 - x^1)\frac{\Delta\theta_2}{\Delta x} = (5.5 - 1.6) \times \frac{60}{9} = 26°,$$

$$\theta_2^{13} = (x^3 - x^1)\frac{\Delta\theta_2}{\Delta x} = (9.4 - 1.6) \times \frac{60}{9} = 52°.$$

Similarly, with a linear relationship between $\Delta\theta_4$ and Δy, we get

$$\theta_4^{12} = (y^2 - y^1)\frac{\Delta\theta_4}{\Delta y} = (0.741 - 0.204) \times \frac{90}{1} = 48.33°,$$

$$\theta_4^{13} = (y^3 - y^1)\frac{\Delta\theta_4}{\Delta y} = (0.974 - 0.204) \times \frac{90}{1} = 69.3°.$$

Dividing Equation (6.6.1) by l_1 the following relation is obtained:

$$l'_2 e^{i\theta_2} + l'_3 e^{i\theta_3} - l'_4 e^{i\theta_4} = -1 \tag{a}$$

Considering the configuration corresponding to the first accuracy point we now substitute

$$z_2 = l'_2 e^{i\theta_2^1},$$
$$z_3 = l'_3 e^{i\theta_3^1} \quad \text{and}$$
$$z_4 = l'_4 e^{i\theta_4^1}$$

when Equation (a) can be written for the three accuracy points as follows:

$$z_2 + z_3 - z_4 = -1 \tag{b}$$

$$z_2 e^{ip_2} + z_3 e^{iq_2} - z_4 e^{ir_2} = -1 \tag{c}$$

$$z_2 e^{ip_3} + z_3 e^{iq_3} - z_4 e^{ir_3} = -1 \tag{d}$$

where $p_m = \theta_2^{1m}$, $q_m = \theta_3^{1m}$ and $r_m = \theta_4^{1m}$.

Equations (b) to (d) are three linear (complex) equations, in three complex variables z_2, z_3 and z_4, which can be easily solved with known values of p_2, p_3, r_2 and r_3 if we assume suitable values of q_2 and q_3. We should note that the following three sets of choices are forbidden since they give rise to a trivial solution, $z_2 = z_3 = z_4 = 0$, to the set of Equations (b) to (d):

(i) $q_2 = q_3 = 0$, (ii) $q_2 = p_2$, $q_3 = p_3$ and (iii) $q_2 = r_2$, $q_2 = r_3$.

With any of the above choices, the determinant of the coefficient matrix

$$\begin{pmatrix} 1 & 1 & -1 \\ e^{ip_2} & e^{iq_2} & -e^{ir_2} \\ e^{ip_3} & e^{iq_3} & -e^{ir_3} \end{pmatrix}$$

turns out to be zero. Let us choose $q_2 = 30°$ and $q_3 = 60°$. Substituting the numerical values for the angular movements of the three links between the precision points in Equations (b) to (d) and solving we get

$$z_2 = 7.137 e^{i149.44°}, \quad z_3 = 6.274 e^{i320.34°},$$

and

$$z_4 = 0.490 e^{i229.98°}.$$

Thus, the link-length ratios $l'_2 = |z_2| = 7.137$, $l'_3 = |z_3| = 6.274$ and $l'_4 = |z_4| = 0.490$. Hence, the shortest link is the follower (l_4). Taking $l_4 = 10$ cm, the other link-lengths are obtained as $l_1 = 20.408$ cm,

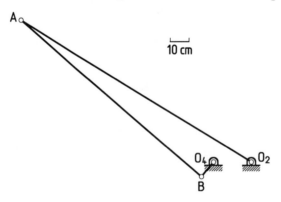

Figure 6.6-1

6.6. ANALYTICAL METHODS (THREE POSITIONS)

$l_2 = 145.652$ cm and $l_3 = 128.040$ cm. The designed linkage is shown in Fig. 6.6-1, at the configuration corresponding to the first accuracy point.

Problem 6.6-2

Synthesize a 4R linkage to generate the following conditions at a particular configuration:

$$\omega_2 = 10 \text{rad/s}, \quad \omega_3 = 4 \text{rad/s}, \quad \omega_4 1 = 20 \text{rad/s},$$

$$\alpha_2 = 40 \text{rad/s}, \quad \alpha_3 = 60 \text{rad/s}, \quad \alpha_4 = 0.$$

The length of the fixed link $l_1 = 10$ cm.

Solution

Dividing Equations (6.6.1) to (6.6.3) throughout by l_1 and substituting $z_k = l'_k e^{i\theta_k}$ and $k = 2, 3$ and 4, we get the following equations:

$$z_2 + z_3 - z_4 = -1 \tag{a}$$

$$\omega_2 z_2 + \omega_3 z_3 - \omega_4 z_4 = 0 \tag{b}$$

$$(\alpha_2 + i\omega_2^2)z_2 + (\alpha_3 + i\omega_3^2)z_3 - (\alpha_4 + i\omega_4^2)z_4 = 0 \tag{c}$$

Substituting the prescribed values of ω_k and α_k and $k = 2, 3, 4$ in Equations (a) to (c) and solving, we obtain

$$z_2 = 1.826 e^{i40.77°}, \quad z_3 = 2.242 e^{i199.42°}$$

and

$$z_4 = 0.522 e^{i59°}.$$

With $l_1 = 10$ cm, the synthesized linkage is shown in Fig. 6.6-2 at the configuration where the prescribed conditions are satisfied.

6.6.2 Freudenstein's Method

Freudenstein's method is based on the direct input-output relationship (displacement equation), which can be obtained by analytic geometry for four-link mechanisms. Again referring to Fig. 4.2-1, for a 4R linkage and rearranging Equation (q) of Problem 4.2-1, we obtain

$$2l_2 l_4 (\cos \theta_4 + \sin \theta_2 \sin \theta_4)$$

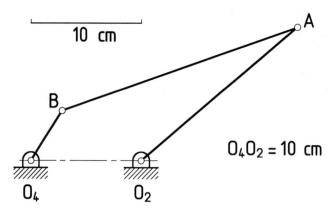

Figure 6.6-2

$$= 2l_1 l_2 \cos\theta_2 - 2l_1 l_4 \cos\theta_4 + l_1^2 + l_2^2 - l_3^2 + l_4^2.$$

Dividing both sides by $2l_2 l_4$,

$$\cos(\theta_2 - \theta_4) = \frac{l_1}{l_4} \cos\theta_2 - \frac{l_1}{l_2} \cos\theta_4 + \frac{l_1^2 + l_2^2 - l_3^2 + l_4^2}{2l_2 l_4}$$

$$= D_1 \cos\theta_2 - D_2 \cos\theta_4 + D_3 \qquad (6.6.4)^{16}$$

where

$$D_1 = \frac{l_1}{l_4}, \quad D_2 = \frac{l_1}{l_2}, \quad D_3 = \frac{l_1^2 + l_2^2 - l_3^2 + l_4^2}{2l_2 l_4} \qquad (6.6.5)$$

Equation 6.6.4, known as *Freudenstein's equation*, is linear in the three design variables D_1, D_2 and D_3 if the values of θ_2 and θ_4 are known. Thus, if the locations of the input and output members are chosen at the three accuracy points as (θ_2^k, θ_4^k), with $k = 1, 2$ and 3, then D_1, D_2 and D_3 can be easily solved.[17] Comparing with Bloch's method we may note that here we assume θ_2^1 and θ_4^1 instead of θ_3^{12}

---

[16] We may note that Equation (6.6.4) is obtained by separating the real and imaginary parts of Equation (6.6.1) and eliminating θ_3 from the resulting two equations.

[17] Assuming that the chosen values of (θ_2^k, θ_4^k), with $k = 1, 2, 3$, do not make the resulting three equations linearly dependent.

6.6. ANALYTICAL METHODS (THREE POSITIONS)

and θ_3^{13}. From the values of D_1, D_2 and D_3 we get the three link-length ratios, l_1/l_2, l_1/l_3 and l_1/l_4, in a manner to be explained in the examples to follow. We should note that there exists a possibility of some link-length ratios coming out as negative quantities. The lengths of the frame and coupler, i.e., l_1 and l_3, are always taken as positive. The negative lengths of the input (l_2) and output (l_4) links should be interpreted in the vector sense as will be explained in Fig. 6.6-4.

Problem 6.6-3

Design a 4R linkage to generate the following pairs of prescribed movements: ($\theta_2^{12} = 54°$, $\theta_4^{12} = 26°$) and ($\theta_2^{13} = 80°$, $\theta_4^{13} = 52°$). Assume the frame-length $l_1 = 5.1$ cm.

Solution

First we choose the values of θ_2^1 and θ_4^1 so that we get (θ_2^k, θ_4^k) for $k = 1, 2$ and 3 corresponding to three accuracy points. Let $\theta_2^1 = -68°$ and $\theta_4^1 = -40°$, when

$$\theta_2^2 = \theta_2^1 + \theta_2^{12} = -14°, \quad \theta_4^2 = \theta_4^1 + \theta_4^{12} = -14°,$$

and

$$\theta_2^3 = \theta_2^1 + \theta_2^{13} = 12°, \quad \theta_4^3 = \theta_4^1 + \theta_4^{13} = 12°.$$

Substituting three pairs of values, (θ_2^1, θ_4^1), (θ_2^2, θ_4^2) and (θ_2^3, θ_4^3), by turn in Equation (6.6.4), we get the following set of three equations:

$$\cos(-28°) = D_1 \cos(-68°) - D_2 \cos(-40°) + D_3 \quad (a)$$

$$\cos(0°) = D_1 \cos(-14°) - D_2 \cos(-14°) + D_3 \quad (b)$$

$$\cos(0°) = D_1 \cos(12°) - D_2 \cos(12°) + D_3 \quad (c)$$

The solution of the above set of equations is as follows:

$$D_1 = D_2 = 0.299, \quad D_3 = 1$$

Using these values in Equation (6.6.5), we finally get[18]

$$l_2 = l_4 = 5.1/0.299 \text{ cm} = 17.06 \text{ cm and } l_1 = l_3 = 5.1 \text{ cm}.$$

The designed linkage is shown at its first configuration in Fig. 6.6-3. The reader should compare this problem with Problem 6.3-3 and

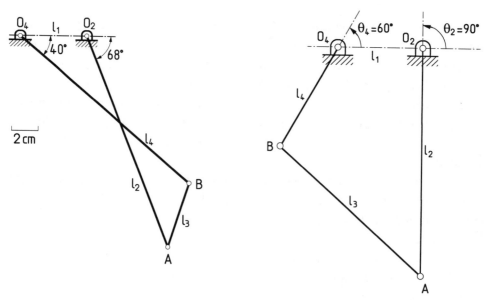

Figure 6.6-3 **Figure 6.6-4**

note the similarity in the problem statement and the differences in the assumptions and final solutions.

Problem 6.6-4

The following three conditions are to be satisfied by a 4R linkage:

$$= \theta_2 = 90°, \quad = \theta_4 = 60°$$

$$\omega_2 = 2\,\text{rad/s}, \quad \omega_4 = 3\,\text{rad/s}$$

$$\alpha_2 = 0, \quad \alpha_4 = -1\,\text{rad/s}$$

Determine the required link-length ratios.

Solution

Equation (6.6.4) and the first and second time derivatives of its both sides can be written as follows:

$$\cos(\theta_2 - \theta_4) = D_1 \cos\theta_2 - D_2 \cos\theta_4 + D_3 \quad (a)$$

$$(\omega_2 - \omega_4)\sin(\theta_2 - \theta_4) = D_1\omega_2 \sin\theta_2 - D_2\omega_4 \sin\theta_4 \quad (b)$$

$$(\alpha_2 - \alpha_4)\sin(\theta_2 - \theta_4) + (\omega_2 - \omega_4)^2 \cos(\theta_2 - \theta_4)$$

---

[18] For more discussion on the solution obtained, see Section 6.9.

6.6. ANALYTICAL METHODS (THREE POSITIONS)

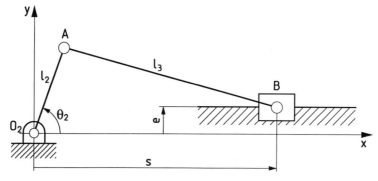

Figure 6.6-5

$$= D_1(\alpha \sin \theta_2 + \omega_2^2 \cos \theta_2) - D_2(\alpha_4 \sin \theta_4 + \omega_4^2 \cos \theta_4) \quad (c)$$

Substituting the desired conditions in Equations (a), (b) and (c) we get the following three linear equations in D_1, D_2 and D_3:

$$0.86 = -0.5D_2 + D_3 \quad (d)$$

$$-0.5 = 2D_1 - 2.598D_2 \quad (e)$$

$$1.366 = -3.634D_2 \quad (f)$$

Solving these three equations we obtain

$$D_1 = -0.738, \quad D_2 = -0.376 \quad \text{and} \quad D_3 = 0.678.$$

From Equation (6.6.5) with the above values of D_1, D_2 and D_3, the link-length ratios are finally obtained as

$$l_1/l_4 = -0.738, \quad l_1/l_2 = -0.376 \quad \text{and} \quad l_1/l_3 = 0.446.$$

The linkage is drawn in Fig. 6.6-4 at the desired configuration. The reader should note the interpretation of negative link-length ratios.

Problem 6.6-5

Synthesize an offset slider-crank mechanism so that the displacement of the slider is proportional to the square of the crank rotation in the interval $45° \leq \theta_2 \leq 135°$. The distance of the slider from the crank-shaft, s, should be 10 cm for $\theta_2 = 45°$ and 3 cm for $\theta_2 = 135°$. Use three Chebyshev's accuracy points.

Solution

Referring to Fig. 6.6-5, first let us derive the displacement equation for this mechanism. The coordinates of the points A and B in the $x - y$ coordinate system are

$$x_A = l_2 \cos \theta_2, \quad y_A = l_2 \sin \theta_2; \quad x_B = s \quad \text{and} \quad y_B = e. \quad (a)$$

Now,

$$AB^2 = l_3^2 = (x_A - x_B)^2 + (y_A - y_B)^2$$
$$= (l_2 \cos \theta_2 - s)^2 + (l_2 \sin \theta_2 - e)^2 \quad (b)$$

or

$$D_1 s \cos \theta_2 + D_2 \sin \theta_2 - D_3 = s^2 \quad (c)$$

where

$$D_1 = 2l_2, \quad D_2 = 2l_2 e \quad \text{and} \quad D_3 = l_2^2 - l_3^2 + e^2. \quad (d)$$

Thus, the displacement equation is given by Equation (c), which is linear in the three design variables, viz. D_1, D_2 and D_3, given by Equation (d). These design variables and subsequently, the link-lengths, can be solved from the three prescribed conditions. As in the case of a 4R linkage, here again l_2 and e can be negative, which are interpreted as vectors. The connecting rod length l_3 is always positive.

For the present problem,

$$s - s^i = C(\theta_2 - \theta_2^i)^2 \quad (e)$$

where C is the constant of proportionality and the superscript i refers to the initial values. If the superscript e is used to indicate the end values, then, substituting the data, we get

$$C = \frac{s^e - s^i}{(\theta_2^e - \theta_2^i)^2} = \frac{7}{90^2} \quad (f)$$

Using Equation (6.2.3) with $n = 3$, the Chebyshev's accuracy points are obtained as $\theta_2^1 = 51.3°$, $\theta_2^2 = 90°$ and $\theta_2^3 = 128.7°$. Substituting these values in Equation (e) and using Equation (f), we get the values of s corresponding to the three accuracy points as $s^1 = 9.97$ cm, $s^2 = 8.26$ cm and $s^3 = 3.95$ cm. Using the values of the related pairs

(51.3°, 9.97 cm), (90°, 8.26 cm) and (128.7°, 3.95 cm) in Equation (c), we get

$$6.230 D_1 + 0.780 D_2 - D_3 = 99.401, \qquad (g)$$
$$D_2 - D_3 = 68.228 \qquad (h)$$

and

$$-2.472 D_1 + 0.780 D_2 - D_3 = 15.602. \qquad (i)$$

The solution to the above set of three equations yields $D_1 = 9.61$, $D_2 = 130.75$ and $D_3 = 62.75$. Using these values in Equation (d), we finally get the link lengths as $l_2 = 4.805$ cm, $l_3 = 12.05$ cm and $e = 13.606$ cm.

6.7 Analytical Methods (Four Positions)

When using analytical methods for synthesizing a four-link mechanism which will satisfy four prescribed conditions, we again start from the loop-closure or the displacement equation (and their time derivatives when instantaneous kinematics conditions are also involved). However, for three-position synthesis, except the three link-length ratios in a 4R linkage (or the three link-lengths in a slider-crank mechanism), all other parameters were chosen arbitrarily. In the case of four-position synthesis no parameters, except the three link-length parameters, can be chosen arbitrarily, as four equations are to be satisfied. Either one more parameter is to be left out as unknown, or a compatibility condition must be satisfied by these chosen parameters. In any case, the resulting equations turn out to be nonlinear[19] and, with arbitrary values of the chosen parameters, there may exist zero, one or two solutions. In the following sections, these aspects are clarified with reference to a 4R linkage, using Bloch's method, and a slider-crank mechanism, using Freudenstein's method. In the latter, the resulting nonlinear equations in four unknowns are solved using the technique of linear superposition.

---

[19] A geared-linkage function generator (having in total 5 links and 6 gears, ultimately giving rise to 7 interconnected rigid bodies) can be designed using only linear equations, even for four-position synthesis. For details, see Sandor, G.N. and Erdman, A.G.: Advanced Mechanism Design: Analysis and Synthesis, Vol. 2, Prentice-Hall, Inc., New Jersey, 1984.

6.7.1 Bloch's Method

Assume that a 4R linkage is to be designed to satisfy three coordinated pairs of movements $(\theta_2^{12}, \theta_4^{12})$, $(\theta_2^{13}, \theta_4^{13})$ and $(\theta_2^{14}, \theta_4^{14})$.[20] Following Equations (b) to (d) of Problem 6.6-1, we can write the loop-closure equations corresponding to four accuracy points as follows:

$$z_2 + z_3 - z_4 = -1 \qquad (6.7.1)$$

$$z_2 e^{ip_2} + z_3 e^{iq_2} - z_4 e^{ir_2} = -1 \qquad (6.7.2)$$

$$z_2 e^{ip_3} + z_3 e^{iq_3} - z_4 e^{ir_3} = -1 \qquad (6.7.3)$$

$$z_2 e^{ip_4} + z_3 e^{iq_4} - z_4 e^{ir_4} = -1 \qquad (6.7.4)$$

where the link-length ratios are given by $l_2/l_1 = |z_2|$, $l_3/l_1 = |z_3|$ and $l_4/l_1 = |z_4|$. Since four equations are to be satisfied we cannot choose all the coupler movements (q_2, q_3 and q_4) arbitrarily. In fact, only one of these three, say q_2, can be chosen arbitrarily when the other two, i.e., q_3 and q_4, have to satisfy a so-called compatibility condition, as shown below.

Subtracting Equations (6.7.2), (6.7.3) and (6.7.4) by turn from Equation (6.7.1), we get

$$z_2(1 - e^{ip_2}) + z_3(1 - e^{iq_2}) - z_4(1 - e^{ir_2}) = 0, \qquad (6.7.5)$$

$$z_2(1 - e^{ip_3}) + z_3(1 - e^{iq_3}) - z_4(1 - e^{ir_3}) = 0 \qquad (6.7.6)$$

and

$$z_2(1 - e^{ip_4}) + z_3(1 - e^{iq_4}) - z_4(1 - e^{ir_4}) = 0. \qquad (6.7.7)$$

For nontrivial solution of Equations (6.7.5) to (6.7.7) we need

$$\begin{pmatrix} (1 - e^{ip_2}) & (1 - e^{iq_2}) & (1 - e^{ir_2}) \\ (1 - e^{ip_3}) & (1 - e^{iq_3}) & (1 - e^{ir_3}) \\ (1 - e^{ip_4}) & (1 - e^{iq_4}) & (1 - e^{ir_4}) \end{pmatrix} = 0 \qquad (6.7.8)$$

---

[20] For other combinations of four prescribed conditions, e.g., three specified coordinated positions (i.e., two pairs of coordinated movements) and specified velocity at one of these positions, etc., see Mclarnan, C.W.: Four Bar Linkage Function Generators, Bulletin 197, Engineering Experiment Station, Ohio State University, Columbia, Ohio, 1966.

6.7. ANALYTICAL METHODS (FOUR POSITIONS)

which is the *compatibility condition*.

We should note that Equation (6.7.8) is a complex equation. Hence, two real unknowns, viz., q_3 and q_4, can be obtained when (p_2, r_2), (p_3, r_3) and (p_4, r_4) are prescribed and q_2 is assumed. Further, Equation (6.7.8) is not linear in q_3 and q_4.

Expanding the left-hand-side of Equation (6.7.8) and carrying out algebraic manipulations[21], we finally get a quadratic equation in $\tan(q_3/2)$ after substituting $\cos q_3 = [1 - \tan^2(q_3/2)]/[1 + \tan^2(q_3/2)]$ and $\sin q_3 = 2\tan(q_3/2)/[1 + \tan^2(q_3/2)]$. No real value of q_3 is obtained if the discriminant of the quadratic equation is negative. One real root for q_3 is obtained if the discriminant is zero, whereas two real roots for q_3 are obtained if the discriminant is positive. With the assumed value of q_2 and the values of q_3 and q_4 satisfying Equation (6.7.8), we can take any three equations from (6.7.1) to (6.7.4) and solve these three linear equations for three unknowns, z_2, z_3 and z_4.

6.7.2 Freudenstein's Method

Assume that the slider-crank mechanism, shown in Fig. 6.6-5, must satisfy three pairs of coordinated movements (θ_2^{1k}, s^{1k}) with $k = 2, 3$ and 4 when Equation (c) of Problem 6.6-5 must be satisfied at all the four accuracy points. Thus, we can write

$$D_1 s^m \cos\theta_2^m + D_2 \sin\theta_2^m - D_3 = (s^m)^2; \quad m = 1, 2, 3, 4 \quad (6.7.9)$$

where

$$\theta_2^m = \theta_2^1 + \theta_2^{1m}, \quad s^m = s^1 + s^{1m}$$

with

$$\theta_2^{11} = 0 \quad \text{and} \quad s^{11} = 0.$$

When the values of (θ_2^{1m}, s^{1m}) are prescribed we can no longer (like three positions synthesis) choose both θ_2^1 and s^1 arbitrarily, because then the set of four equations of (6.7.9) will involve only three unknowns, D_1, D_2 and D_3. Let us, therefore, choose one more quantity, say s^1, as unknown with θ_2^1 assumed and rewrite Equation (6.7.9) as

$$D_1(s^1 + s^{1m})\cos\theta_2^m + D_2 \sin\theta_2^m - D_3 = (s^1 + s^{1m})^2;$$

---

[21]For details see Mclarnan, C.W., op. cit.

$$m = 1, 2, 3, 4. \qquad (6.7.10)$$

We should note that Equation (6.7.10) is a set of four simultaneous nonlinear equations in four unknowns, D_1, D_2, D_3 and s^1. This set of nonlinear equations is solved below using the technique of linear superposition.

First, we expand Equation (6.7.10) and use Equation (d) of Problem 6.6-5 as follows:

$$2l_2(s^1 + s^{1m})\cos\theta_2^{m\cdot} + 2l_2 e \sin\theta_2^m - (l_2^2 - l_3^2 + e^2)$$
$$= (s^1)^2 + (s^{1m})^2 + 2s^1 s^{1m}$$

or

$$R_1 \cos\theta_2^m + R_2 s^{1m}\cos\theta_2^m + R_3 \sin\theta_2^m - R_4$$
$$= R_5 s^{1m} + (s^{1m})^2; m = 1, 2, 3, 4 \qquad (6.7.11)$$

where

$$R_1 = 2l_2 s^1, \quad R_2 = 2l_2, \quad R_3 = 2l_2 e, \quad R_4 = l_2^2 - l_3^2 + e^2 + (s^1)^2, R_5 = 2s^1.$$

It may appear that the system of four equations of (6.7.11), involving five unknowns, R_1 to R_5, is indeterminate. However, this is not so since R_1 to R_5 are not independent. In fact, by definition

$$2R_1 - R_2 R_5 = 0. \qquad (6.7.12)$$

Thus, Equations (6.7.11) and (6.7.12) constitute five equations in five unknowns where those of (6.7.11) are linear and (6.7.12) is nonlinear. This nonlinear Equation (6.7.12) represents the compatibility condition. The set of five equations from (6.7.11) and (6.7.12) can be easily solved, as explained below.

Let us denote[22] $R_5 = \lambda$ and write Equation (6.7.12) as

$$2R_1 - \lambda R_2 = 0. \qquad (6.7.13)$$

To satisfy Equation (6.7.11), we define

$$R_j = P_j + \lambda Q_j; \quad j = 1, 2, 3, 4 \qquad (6.7.14)$$

---

[22] λ is also known as the nonlinearity coefficient.

6.7. ANALYTICAL METHODS (FOUR POSITIONS)

where P_j's and Q_j's satisfy the following two sets of four linear equations:

$$P_1 \cos \theta_2^m + P_2 s^{1m} \cos \theta_2^m + P_3 \sin \theta_2^m - P_4 = (s^{1m})^2 \qquad (6.7.15)$$

$$Q_1 \cos \theta_2^m + Q_2 s^{1m} \cos \theta_2^m + Q_3 \sin \theta_2^m - Q_4 = (s^{1m}) \qquad (6.7.16)$$

$$m = 1, 2, 3, 4$$

To solve for λ we first solve P_j's and Q_j's from Equations (6.7.15) and (6.7.16). Hence, using Equation (6.7.14), we get

$$R_1 = P_1 + \lambda Q_1 \qquad (6.7.17)$$

and

$$R_2 = P_2 + \lambda Q_2. \qquad (6.7.18)$$

Substituting Equations (6.7.17) and (6.7.18) in Equation (6.7.13), we obtain

$$2(P_1 + \lambda Q_1) - \lambda(P_2 + \lambda Q_2) = 0$$

or

$$\lambda^2 Q_2 + \lambda(P_2 - 2Q_1) - 2P_1 = 0$$

or

$$\lambda = [2Q_1 - P_2 \mp \sqrt{(P_2 - 2Q_1)^2 + 8P_1 Q_2}]/2Q_2. \qquad (6.7.19)$$

Once λ is known, the design variables R_1 to R_4 are easily obtained from Equation (6.7.14). Thus, once all the design parameters are known, we can get the link-length parameters l_2, e, l_3 and s^1. We should note from Equation (6.7.19) that, just as in the case of a 4R linkage, there may exist

(i) no solution if $(P_2 - 2Q_1)^2 + 8P_1 Q_2 < 0$

(ii) one solution if $(P_2 - 2Q_1)^2 + 8P_1 Q_2 = 0$, or

(iii) two solutions if $(P_2 - 2Q_1)^2 + 8P_1 Q_2 > 0$.

6.8 Optimization Method

As mentioned in Section 6.2, with the precision point approach of kinematic synthesis the number of precision points are limited by the number of available design parameters in the linkage. In the entire range of movement of the linkage, the desired motion characteristics are achieved at these precision points and there will be inherent structural errors at all other points. In the optimization approach, however, the desired motion characteristic is not attempted to be exactly satisfied at any point in the range of movement. Rather, any number of design positions are chosen and some error quantity, suitably defined using these chosen positions, is minimized. The number of such design positions is not limited by the number of available design parameters. A considerable amount of work has been done to solve function generation and path generation problems using optimization methods.[23] These methods are not limited to four-link mechanisms. Without going into the details of all kinds of optimization approaches for mechanism synthesis, we shall discuss a least-square technique applied to a six-link Stephenson's linkage function generator [see reference (iii) below]. The technique of linear superposition, already explained in Section 6.7.2, is used.

Referring to Fig. 6.8-1a, let us consider a Stephenson's mechanism of type I to be synthesized as a function generator. Let the position parameter θ_2 represent the independent variable x, while θ_6 represents the desired function $y = f(x)$. The link-vector diagram corresponding to Fig. 6.8-1a is shown in Fig. 6.8-1b. The design variables are l_1, l_2, l_3, l_4, l_3', ψ, l_5, l_6, l_1' and ϕ. As already discussed in Section 6.6

---

[23] For the use of Galerkin's function and other references see

(i) Akcali, I.D. and Dittrich, G.: "Function Generation by Galerkin's Method", Mechanism and Machine Theory, Vol. 24, No. 1, pp. 39-43, 1989.

(ii) Akcali, I.D. and Dittrich, G.: "Path Generation by Subdomain Method", Mechanism and Machine Theory, Vol. 24, No. 1, pp. 45-52, 1989.

(iii) Bagci, C.: "Optimum Synthesis of Multi-Loop Planar Mechanisms for Function Generation via the Linear Superposition Technique", Proc. I. Mech. E., Vol. 189, pp. 855-859, 1975.

6.8. OPTIMIZATION METHOD

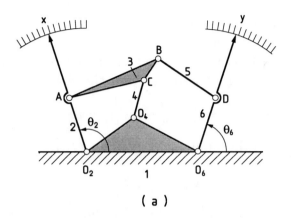

(a)

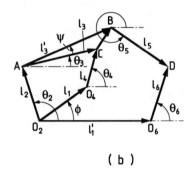

(b)

Figure 6.8-1

the values of θ_2 and θ_6 at N number of design points are evaluated using the prescribed functional relationship, chosen scale factors and the initial values. The value of N (arbitrary) dictates the computational effort and level of accuracy. Out of the 10 listed link design parameters one link-length is arbitrarily chosen to fix the overall size of the mechanism. (Recall that in Section 6.6, for a 4R linkage only three link-length ratios were treated as unknowns). Considering the loop $O_2ABDO_6O_2$ (Fig. 6.8-1b), we find the closure equation

$$l_2 + l'_3 + l_5 - l_6 - l'_1 = 0. \tag{6.8.1}$$

Similarly, for the other independent loop $O_2ACO_4O_2$, the closure equation is

$$l_2 + l_3 - l_4 - l_1 = 0. \tag{6.8.2}$$

Using complex exponential notation for the link-vectors and noting that $\theta'_1 = 0$ and $\theta_1 = \phi$, we can obtain four equations by separating the real and imaginary parts of Equations (6.8.1) and (6.8.2). Eliminating θ_5 and θ_4 from these four equations, we get

$$l'^2_1 + l^2_6 + l'^2_3 - l^2_5 + 2l'_1 l_6 \cos\theta_6 - 2l_2 l'_1 \cos\theta_2$$
$$-2l'_1 l'_3 \cos(\theta_3 + \psi) - 2l_6 l'_3 \cos(\theta_6 - \theta_3 - \psi)$$
$$-2l_6 l_2 \cos(\theta_6 - \theta_2) + 2l_2 l'_3 \cos(\theta_2 - \theta_3 - \psi) = 0 \tag{6.8.3}$$

and

$$l^2_2 + l^2_3 + l^2_1 - l^2_4 + 2l_2 l_3 \cos(\theta_2 - \theta_3)$$
$$-2l_1 l_2 \cos(\theta_2 - \phi) - 2l_1 l_3 \cos(\theta_3 - \phi) = 0. \tag{6.8.4}$$

As was done in Bloch's method, the values of θ_3 at the design points are chosen arbitrarily. At this stage, it must be appreciated that Equations (6.8.3) and (6.8.4) are not satisfied exactly at the design positions since the values of θ_2 and θ_6 have already been coordinated according to the prescribed function. Let the left-hand-sides of Equations (6.8.3) and (6.8.4) turn out, respectively, as ϵ_1 and ϵ_2 when the values of θ_2, θ_3 and θ_6 (for a design position) are substituted. The values of ϵ_1 and ϵ_2 represent the error in the loop-closure equations at a design position. Our objective is to determine the design parameters so as to minimize the overall error (considering all the design

6.8. OPTIMIZATION METHOD

positions) in the least square sense. From here forward, that is what we shall do.

We define design parameters (functions of linkage design parameters, e.g., D_1, D_2 and D_3 in Equation (6.6.4) in such a way that a design parameter appears only in either Equation (6.8.3) or (6.8.4). In other words, Equations (6.8.3) and (6.8.4) get decoupled in terms of unknown design parameters. Some of the linkage design parameters can, of course, be conveniently chosen beforehand. Then the expressions for ϵ_1 and ϵ_2 for the qth design position can generally be written in the form

$$\epsilon_{1q} = \sum_{p=1}^{M_1} D_p \mu_{pq} - \nu_q - \lambda_1 \rho_q - \lambda_2 \delta_q \qquad (6.8.5)$$

and

$$\epsilon_{2q} = \sum_{p=1}^{M_2} D'_p \mu'_{pq} - \nu'_q - \lambda'_1 \rho'_q - \lambda'_2 \delta'_q \qquad (6.8.6)$$

where D_p and D'_p are the unknown design parameters and μ_{pq}, ν_q, ρ_q, δ_q, μ'_{pq}, ν'_q, ρ'_q and δ'_q are functions of θ_2, θ_3, θ_6 and the chosen link parameters. The first loop-closure equation has M_1 and the second loop-closure equation has M_2 design parameters. The nonlinearity coefficients λ_1, λ_2, λ'_1 and λ'_2 determine the compatibility conditions (recall λ in Section 6.7.2) for the system of nonlinear equations in (6.8.5) and (6.8.6). The number of these nonlinearity coefficients depends on the number of linkage design parameters assumed. There may be one or more, or no such coefficient in Equations (6.8.5) and (6.8.6). Ideally,

$$\epsilon_{1q} = \epsilon_{2q} = 0; \quad q = 1, 2, \ldots N. \qquad (6.8.7)$$

Following the linear superposition technique, let D_p and D'_p be expressed as

$$D_p = \alpha_p + \lambda_1 \beta_p + \lambda_2 \gamma_p$$

and

$$D'_p = \alpha'_p + \lambda'_1 \beta'_p + \lambda'_2 \gamma'_p. \qquad (6.8.8)$$

If α's, β's, γ's and α'''s β'''s and γ'''s satisfy the following equations

$$\epsilon_{1\alpha_q} = \sum_{p=1}^{M_1} \alpha_p \mu_{pq} - \nu_q = 0,$$

$$\epsilon_{1\beta_q} = \sum_{p=1}^{M_1} \beta_p \mu_{pq} - \rho_q = 0.$$

$$\epsilon_{1\gamma_q} = \sum_{p=1}^{M_1} \gamma_p \mu_{pq} - \delta_q = 0,$$

$$\epsilon_{2\alpha'_q} = \sum_{p=1}^{M_2} \alpha'_p \mu'_{pq} - \nu'_q = 0,$$

$$\epsilon_{2\beta'_q} = \sum_{p=1}^{M_2} \beta'_p \mu'_{Pq} - \rho'_q = 0$$

and

$$\epsilon_{2\gamma'_q} = \sum_{p=1}^{M_2} \gamma'_q \mu'_{pq} - \delta'_q = 0, \qquad (6.8.9)$$

then, using Equations (6.8.5) and (6.8.6), we find that Equation (6.8.7) is satisfied. Now, let the overall error (in the least-square sense) be expressed as

$$E = \sum_{q=1}^{N} (\epsilon_{1\alpha_q}^2 + \epsilon_{1\beta_q}^2 + \epsilon_{1\gamma_q}^2 + \epsilon_{2\alpha'_q}^2 + \epsilon_{2\beta'_q}^2 + \epsilon_{2\gamma'_q}^2). \qquad (6.8.10)$$

For E to be minimum, each term on the right-hand-side of Equation (6.8.10) must be minimum. To achieve this, the set of quantities α's, β's, γ's and α'''s, β'''s, γ'''s (both of which determine the design parameters D's and D''s) should satisfy the equations

$$\frac{\delta}{\delta \alpha_p}(\sum_{q=1}^{N} \epsilon_{1\alpha_q}^2) = \frac{\delta}{\delta \beta_p}(\sum_{q=1}^{N} \epsilon_{1\beta_q}^2) = \frac{\delta}{\delta \gamma_p}(\sum_{q=1}^{N} \epsilon_{1\gamma_q}^2) = 0;$$

$$p = 1, 2, \ldots, M_1 \qquad (6.8.11a)$$

6.8. OPTIMIZATION METHOD

and

$$\frac{\delta}{\delta \alpha'_p}(\sum_{q=1}^{N} \epsilon^2_{2\alpha'_q}) = \frac{\delta}{\delta \beta'_p}(\sum_{q=1}^{N} \epsilon^2_{2\beta'_q}) = \frac{\delta}{\delta \gamma'_p}(\sum_{q=1}^{N} \epsilon^2_{2\gamma'_q}) = 0;$$
$$p = 1, 2, \ldots, M_2 \tag{6.8.11b}$$

Equations (6.8.11a) and (6.8.11b) provide $3(M_1 + M_2)$, i.e., $3M$ linear equations to determine $3M$ unknowns, namely, M_1 numbers of α's, β's, γ's and M_2 numbers of α''s, β''s and γ''s. In matrix notation, equations obtained from (6.8.11a) can be written as

$$\begin{pmatrix} a_{11} & a_{12} & \cdots & a_{1M_1} \\ a_{21} & a_{22} & \cdots & a_{2M_1} \\ \vdots & \vdots & & \vdots \\ a_{M_11} & a_{M_12} & \cdots & a_{M_1M_1} \end{pmatrix} \begin{pmatrix} \alpha_1 & \beta_1 & \gamma_1 \\ \alpha_2 & \beta_2 & \gamma_2 \\ \vdots & \vdots & \vdots \\ \alpha_{M_1} & \beta_{M_1} & \gamma_{M_1} \end{pmatrix}$$
$$= \begin{pmatrix} b_{\alpha_1} & b_{\beta_1} & b_{\gamma_1} \\ b_{\alpha_2} & b_{\beta_2} & b_{\gamma_2} \\ \vdots & \vdots & \vdots \\ b_{\alpha_{M_1}} & b_{\beta_{M_1}} & b_{\gamma_{M_1}} \end{pmatrix} \tag{6.8.12a}$$

where

$$a_{rs} = \sum_{q=1}^{N} \mu_{rq}\mu_{sq}, \; b_{\alpha_r} = \sum_{q=1}^{N} \mu_{rq}\nu_q, \; b_{\beta_r} = \sum_{q=1}^{N} \mu_{rq}\rho_q,$$

$$b_{\gamma_r} = \sum_{q=1}^{N} \mu_{rq}\delta_q.$$

Similarly, equations obtained from (6.8.11b) can be written as

$$\begin{pmatrix} a'_{11} & a'_{12} & \cdots & a'_{1M_2} \\ a'_{21} & a'_{22} & \cdots & a'_{2M_2} \\ \vdots & \vdots & & \vdots \\ a'_{M_21} & a'_{M_22} & \cdots & a'_{M_2M_2} \end{pmatrix} \begin{pmatrix} \alpha'_1 & \beta'_1 & \gamma'_1 \\ \alpha'_2 & \beta'_2 & \gamma'_2 \\ \vdots & \vdots & \vdots \\ \alpha'_{M_2} & \beta'_{M_2} & \gamma'_{M_2} \end{pmatrix}$$
$$= \begin{pmatrix} b'_{\alpha'_1} & b'_{\beta'_1} & b_{\gamma'_1} \\ b'_{\alpha'_2} & b'_{\beta'_2} & b_{\gamma'_2} \\ \vdots & \vdots & \vdots \\ b'_{\alpha'_{M_2}} & b'_{\beta'_{M_2}} & b_{\gamma'_{M_2}} \end{pmatrix} \tag{6.8.12b}$$

where

$$a'_{rs} = \sum_{q=1}^{N} \mu'_{rq}\mu'_{sq}, \quad b'_{\alpha'_r} = \sum_{q=1}^{N} \mu'_{rq}\nu'_q, \quad b'_{\beta'_r} = \sum_{q=1}^{N} \mu'_{rq}\rho'_q,$$

$$b'_{\gamma'_r} = \sum_{q=1}^{N} \mu'_{rq}\delta'_q.$$

Thereafter, the design parameters D's and D''s are obtained from Equation (6.8.8), once the nonlinearity coefficients are obtained from the compatibility equations, as demonstrated below for the Stephenson's mechanism (Fig. 6.8-1a).

Of the ten possible design parameters let us assume that l_1, l_2, l_3, l_4, l_5 and l_6 are unknown and all other parameters are assumed. Dividing Equation (6.8.3) by $2l_2l_6$ and rearranging, we get

$$\frac{l'_3}{l_2}\cos(\theta_6 - \theta_3 - \psi) - \frac{l'_1}{l_2}\cos\theta_6$$

$$-\frac{{l'_1}^2 + l_6^2 + l_2^2 + {l'_3}^2 - l_5^2}{2l_2l_6} + \frac{l'_1 l'_3}{l_2 l_6}\cos(\theta_3 + \psi)$$

$$+\frac{l'_1}{l_6}\cos\theta_2 - \frac{l'_3}{l_6}\cos(\theta_2 - \theta_3 - \psi) + \cos(\theta_6 - \theta_2) = 0.$$

Thus, for the qth design position

$$D_1[\cos(\theta_6^q - \theta_3^q - \psi) - \frac{l'_1}{l'_3}\cos\theta_6^q]$$

$$+D_2[\cos\theta_2^q - \frac{l'_3}{l'_1}(\cos\theta_2^q - \theta_3^q - \psi)]$$

$$-D_3 + \lambda_1\cos(\theta_3^q + \psi) + \cos(\theta_6^q - \theta_2^q) = 0 \qquad (6.8.13)$$

where

$$D_1 = \frac{l'_3}{l_2}, \quad D_2 = \frac{l'_1}{l_6},$$

$$D_3 = \frac{{l'_1}^2 + l_6^2 + l_2^2 + {l'_3}^2 - l_5^2}{2l_2l_6}$$

6.8. OPTIMIZATION METHOD

and
$$\lambda_1 = \frac{l'_1 l'_3}{l_2 l_6} = D_1 D_2.$$

Comparing Equation (6.8.13) with Equation (6.8.5), we see that $M_1 = 3$ with

$$\mu_{1q} = \cos(\theta_6^q - \theta_3^q - \psi) - \frac{l'_1}{l'_3} \cos \theta_6^q,$$

$$\mu_{2q} = \cos \theta_2^q - \frac{l'_3}{l'_1} \cos(\theta_2^q - \theta_3^q - \psi). \quad (6.8.14)$$

$$\mu_{3q} = -1, \quad \nu_q = -\cos(\theta_6^q - \theta_2^q), \quad \rho_q = -\cos(\theta_3^q + \psi)$$

and
$$\lambda_2 = 0.$$

Similarly, dividing Equation (6.8.4) by $2l_2 l_3$, we finally get (with $M_2 = 3$)

$$D'_1 = \frac{l_1}{l_2}, \quad D'_2 = \frac{l_1}{l_3}.$$

$$D'_3 = \frac{l_2^2 + l_3^2 + l_1^2 - l_4^2}{2l_2 l_3},$$

$$\mu'_{1q} = -\cos(\theta_3^q - \phi), \quad \mu'_{2q} = -\cos(\theta_2^q - \phi).$$

$$\mu'_{3q} = 1, \quad \nu'_q = -\cos(\theta_2^q - \theta_3^q), \quad \lambda'_1 = 0 \quad \text{and} \quad \lambda'_2 = 0. \quad (6.8.15)$$

From the first equation in (6.8.8), we get

$$D_1 = \alpha_1 + \lambda_1 \beta_1 \quad \text{and} \quad D_2 = \alpha_2 + \lambda_1 \beta_2.$$

Using $\lambda_1 = D_1 D_2$, we get the compatibility condition for determining λ_1 as

$$\beta_1 \beta_2 \lambda_1^2 + (\alpha_1 \beta_2 + \alpha_2 \beta_1 - 1)\lambda_1 + \alpha_1 \alpha_2 = 0. \quad (6.8.16)$$

Each real root of this equation yields a possible design.

We should note that a six-link mechanism has four geometric inversions (i.e., four different configurations are possible for given link dimensions and a value of θ_2). Thus, with two real values of λ_1 there are, in total, eight geometric inversions. One of these eight is the optimum solution being sought. Further, the resulting solution generates values of θ_3^q that are different from those assumed in the beginning. These generated values can be used to improve the solution iteratively.

If besides l_1, l_2, l_3, l_4, l_5 and l_6, two more parameters, ϕ and θ_6^1, are also left as unknowns, then we can write $\theta_6^q = \theta_6^i + x^q$, with x^q as unknown quantities. Proceeding in the usual manner, we get $M_1 = M_2 = 4$, with three nonlinearity coefficients λ_1, λ_2 and λ_1', when the compatibility condition in λ_1 turns out to be a cubic equation.[24]

6.9 Branch and Order Defects

In this section we draw the reader's attention to an inherent problem of linkage synthesis using precision points. This problem may occur irrespective of whether the synthesis is carried out using the graphical or analytical method. A synthesized linkage should be checked for two kinds of defects, viz., branch defect and order defect. These aspects are discussed below with reference to the synthesis of 4R linkages.

By branch defect we refer to a situation in which the synthesized 4R linkage must change its assembly configuration [recall that a Grashof linkage has two distinct assembly configurations (Section 4.3)] in order to satisfy all the precision points. Referring to Figs. 4.3-1a to 4.3-1c, we should note that if the coordinated pairs (θ_2^q, θ_4^q) are distributed over both the branches, then theoretically the same linkage (with unique link-lengths) satisfies all the coordinated positions. However, such a solution is obviously of no use to a designer. Therefore, a synthesized linkage (i.e., the solution to a problem) should always be checked against possible branch defect.

For example, let us consider Problem 6.6-3 and its solution. The synthesized linkage turned out to have $l_1 = l_3$ and $l_2 = l_4$. Thus, the linkage can be either a parallelogram or an antiparallelogram. The reader may easily verify that while the first accuracy point is satisfied by the antiparallelogram configuration (Fig. 6.6-3), the second and third accuracy points are satisfied by the parallelogram configuration. The linkage under discussion, being transitional between Grashof and non-Grashof ($l_{\min} + l_{\max} = l' + l''$), need not be dismantled in order to transform from parrallelogram to anti-parallelogram configuration

---

[24]For details see Bagci, C., op. cit.

6.9. BRANCH AND ORDER DEFECTS

and vice-versa. These transformations can take place at the uncertainty configurations when all links become collinear. The plot of θ_2 vs. θ_4 for this linkage has a linear and a curved segment, both passing through (0, 0) and (π, π), and thus a branch defect in the strict sense does not appear. However, the synthesized linkage, requiring such transformations, is not satisfactory for any technical application.

Incidentally, if we examine the solution obtained for Problem 6.3-3, another interesting feature is revealed. Here the synthesized linkage turned out to be non-Grashof (Fig. 6.3-11). In this linkage, as link 2 goes from position I to II (i.e. θ_2 goes from θ_2^1 to θ_2^2) by CCW movement, link 4 also moves from position I to II (i.e., θ_4 changes from θ_4^1 to θ_4^2). But when link 2 goes further from position II to III (i.e., θ_2 changes from θ_2^2 to θ_2^3 by CCW movement), link 4 does not move to position III (i.e., $\theta_4 \neq \theta_4^3$). However, after link 2 reaches its extreme position and while returning, as θ_2 becomes equal to θ_2^3, link 4 takes up position III (i.e., $\theta_4 = \theta_4^3$). In a real-life design, even the break of these monotonic movements of links 2 and 4 through positions I, II and III may not be desirable. Referring to Figs. 4.3-1c and 4.3-1d we note that θ_2 vs. θ_4 plots of double rockers are always closed loops.[25] This type of break in monotonic movements through precision points always exist in case of double rockers, if all the coordinated pairs of input-output angles (θ_2^q, θ_4^q) do not lie on the same segment between the vertical tangents of the $\theta_2 - \theta_4$ plot, which is a closed loop.

By order defect we refer to a situation in which the synthesized linkage (though satisfying all the precision points) fails to pass through the design positions in the specified order. Referring to Fig. 6.9-1a, let the coordinated pairs (θ_2^q, θ_4^q), in which $q = 1, 2, 3, 4,$, be located as P_1, P_2, P_3 and P_4 and it is desired that the linkage should take up these configurations in the sequence 1-2-3-4. We may obtain a solution for the linkage for which the $\theta_2 - \theta_4$ plot passes through all the four positions, as indicated in Fig. 6.9-1b. We should note that the obtained solution fails to satisfy the prescribed order independent of the direction of rotation of the input link. It is obvious that for a

---

[25]The reader may note that the symbols θ_2 and θ_4 in Figs. 6.3-11 and 4.3-1 are different. However, this change is of no consequence in the context of present discussion.

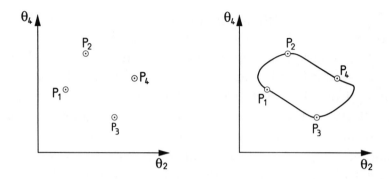

Figure 6.9-1

three position synthesis problem, an order defect is easily rectified by reversing the direction of rotation of the input link.

To take care of branch and order defects, while synthesizing a 4R linkage using the Burmester Theory (graphical method), the choice of circle-point is restricted in some definite regions (or segments) on the circle-point curve (see Fig. 6.5-4)[26].

6.10 Mechanical Error

The generated output of a linkage, synthesized by either precision points or optimization method, deviates from the desired output. The maximum deviation is referred to as the mathematical or structural error. Besides this structural error in the design stage, another source of error in a linkage when it is fabricated is due to inherent inaccura-

---

[26] For details see:

(i) Filemon, E.: "In Addition to Burmester Theory", Proceedings of the Third World Congress - Theory of Machines and Mechanisms D., pp- 63-78 (1971).

(ii) Modler, K.H.: "Reihenfolge der Homologen Punkte," Maschinenbautechnik, Vol. 21, pp. 258-265, (1972).

(iii) Waldron, K.J. and Strong, R.T.: "Improved Solutions of the Branch and Order Problems of Burmester Linkage Synthesis", Mechanism and Machine Theory, Vol. 13, pp. 199-207 (1978).

6.10. MECHANICAL ERROR

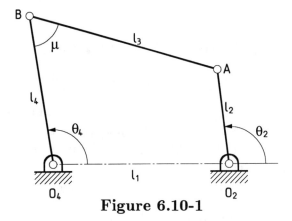

Figure 6.10-1

cies in the dimensions of the links. These inaccuracies are introduced by the tolerances on the link dimensions and clearances in the joints (kinematic pairs). The maximum possible error in the generated output, due to tolerances and clearances, is called the mechanical error. Very often the mechanical error is an order of magnitude higher than the structural error. In this section, we shall analyse this mechanical error in a 4R linkage assuming the links to be rigid, i.e., no deformation or deflection of the links will be considered. This analysis helps the designer in the following two ways:

1. It provides an idea about the maximum possible deviation from the desired output when the tolerances and clearances are known.

2. The designer can allocate the tolerances and clearances suitably, so that the maximum error does not exceed a prescribed limit and the cost of fabrication is minimized.

Two different approaches, viz., deterministic and stochastic, can be adopted while analysing mechanical errors.

6.10.1 Deterministic Approach

Let us consider the 4R linkage shown in Fig. 6.10-1, whose nominal link-lengths are l_1, l_2, l_3 and l_4. The output (θ_4) and input (θ_2) of this linkage are related through Freudenstein's equation, Equation (6.6.4), which can be rewritten as

$$(2l_1l_4 + 2l_2l_4 \cos\theta_2)\cos\theta_4 + (2l_2l_4 \sin\theta_2)\sin\theta_4$$
$$= 2l_1l_2 \cos\theta_2 + l_1^2 + l_2^2 - l_3^2 + l_4^2. \quad (6.10.1)$$

If the mechanical error in the link-lengths are respectively Δl_1, , Δl_2, Δl_3 and Δl_4, then our objective is to determine the error $\Delta\theta_4$, which is obviously a function of l's, Δl's and θ_2. We may note that Δl's may be positive or negative.

First, we rewrite Equation (6.10.1) as follows:

$$C_1 \cos\theta_4 + C_2 \sin\theta_4 = C_3 \quad (6.10.2)$$

where

$$C_1 = 2l_1l_4 + 2l_2l_4 \cos\theta_2, \quad (6.10.3a)$$
$$C_2 = 2l_2l_4 \sin\theta_2 \quad (6.10.3b)$$

and

$$C_3 = 2l_1l_2 \cos\theta_2 + l_1^2 + l_2^2 - l_3^2 + l_4^2. \quad (6.10.3c)$$

Due to the errors in link lengths, for a given θ_2, let the errors in C_1, C_2, C_3 and θ_4 be ΔC_1, ΔC_2, ΔC_3 and $\Delta\theta_4$, respectively. Thus, from Equation (6.10.2), we can write

$$(C_1 + \Delta C_1)\cos(\theta_4 + \Delta\theta_4) + (C_2 + \Delta C_2)\sin(\theta_4 + \Delta\theta_4)$$
$$= C_3 + \Delta C_3. \quad (6.10.4)$$

Assuming all the errors to be small, so that their squares and products are neglected and $\cos(\Delta\theta_4) \approx 1$, $\sin(\Delta\theta_4) \approx \Delta\theta_4$, using Equation (6.10.2) in Equation (6.10.4), we get

$$\Delta\theta_4 \approx (\Delta C_3 - \Delta C_1 \cos\theta_4 - \Delta C_2 \sin\theta_4)/(C_2 \cos\theta_4 - C_1 \sin\theta_4). \quad (6.10.5)$$

With small errors in link lengths, from Equations (6.10.3a), and (6.10.3b), we obtain

$$\Delta C_1 \approx 2l_4 \Delta l_1 + 2l_4 \cos\theta_2 \Delta l_2 + 2(l_1 + l_2 \cos\theta_2)\Delta l_4, \quad (6.10.6a)$$

$$\Delta C_2 \approx 2l_4 \sin\theta_2 \Delta l_2 + 2l_2 \sin\theta_2 \Delta l_4 \quad (6.10.6b)$$

6.10. MECHANICAL ERROR

and
$$\Delta C_3 \approx 2(l_1 + l_2 \cos \theta_2)\Delta l_1 + 2(l_1 \cos \theta_2 + l_2)\Delta l_2$$
$$-2l_3 \Delta l_3 + 2l_4 \Delta l_4. \qquad (6.10.6c)$$

Thus, from Equations (6.10.5) to (6.10.6a, b and c) we are in a position to estimate $\Delta \theta_4$ (note that θ_4 is decided by the l's and θ_2).

In real life, the limits on Δl's are prescribed mostly as say $|\Delta l_1|$, $|\Delta l_2|$, $|\Delta l_3|$ and $|\Delta l_4|$, and we need to estimate $|\Delta \theta_4|_{\max}$. The most conservative estimate can be obtained by assuming $\Delta l_q = |\Delta l_q|$ and $q = 1, 2, 3,$ and 4. A more rational approach can be as follows. Let $(\Delta \theta_4)_1$ be the value of $\Delta \theta_4$, with $\Delta l_1 = |\Delta l_1|$, and all other Δl's equal to zero. Similarly, we evaluate $(\Delta \theta_4)_2$, $(\Delta \theta_4)_3$ and $(\Delta \theta_4)_4$, assuming in each case only one link-length to be erroneous. Then the root-mean-square (r.m.s.) error in the output,

$$(\Delta \theta_4)_{r.m.s} = [(\Delta \theta_4)_1^2 + (\Delta \theta_4)_2^2 + (\Delta \theta_4)_3^2 + (\Delta \theta_4)_4^2]^{1/2}. \qquad (6.10.7)$$

Another interesting feature, mentioned in Section 4.4, is revealed from the denominator of $\Delta \theta_4$ appearing in Equation (6.10.5). Using Equations (4.4.1), (6.10.2) and (6.10.3), the reader can verify (after some algebraic manipulations) that this denominator,

$$C_2 \cos \theta_4 - C_1 \sin \theta_4 = \pm 2l_3 l_4 \sin \mu \qquad (6.10.8)$$

where μ is the transmission angle. Thus, a poor transmission quality (i.e., a small $\sin \mu$) implies high mechanical error (due to high sensitivity of the output variable to link length errors).

6.10.2 Stochastic Approach

In Section 6.10.1 it was assumed that the errors in link lengths, Δl's, were either known or replaced by their maximum values. This approach is too conservative and hence gives rise to higher cost of fabrication. Since both tolerances in the link lengths and clearances in the joints, which govern the link length errors, are normally described in a statistical sense, the link length errors can be treated as random variables. With this consideration, stochastic models have been proposed for handling mechanical errors in both function generating and path generating mechanisms. [27]

---

[27] For details see:

312				CHAPTER 6. DIMENSIONAL SYNTHESIS

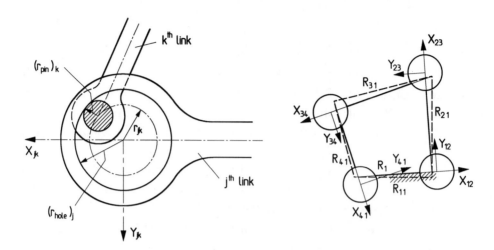

Figure 6.10-2

In this section we shall briefly discuss the stochastic modelling of a 4R linkage function generator.

Let the nominal link length for the jth link be l_j, with ϵ as the tolerance per unit length. Hence, the actual link lengths are

$$R_j = l_j + \epsilon l_j; \quad j = 1, 2, 3, 4. \tag{6.10.9}$$

The clearance between the pin of the kth link and the hole of one of the adjacent links (say jth) is shown in Fig. 6.10-2a. A Cartesian coordinate system (X_{jk}, Y_{jk}) is set-up at the center of the hole in the jth link, with X axis along the axis of the jth link (along which R_j is measured). The center of the pin in the kth link can then lie anywhere within the circle of radius r_{jk}, where

$$r_{jk} = (r_{hole})_j - (r_{pin})_k \tag{6.10.10}$$

Dhande, S.G. and Chakraborty, J.: "Analysis and Synthesis of Mechanical Error in Linkages - A Stochastic Approach", Jr. of Eng. for Industry, Trans. ASME, Ser. B, Vol. 95 (3), pp. 672-676, 1973.

Mallik, A.K. and Dhande, S.G.: "Analysis and Synthesis of Mechanical Error in Path-generating Linkages using a Stochastic Approach," Int. J. of Mechanism and Machine Theory, Vol. 22 (2), pp. 115-123, 1987.

6.10. MECHANICAL ERROR

with $(r_{hole})_j$ and $(r_{pin})_k$ as the radii of the hole in the jth link and pin in the kth link, respectively.

It is assumed that the pin-center can lie anywhere within the circle with equal probability. Thus, the probability density function can be written as

$$f(x_{jk}, y_{jk}) = \begin{cases} \frac{1}{\pi r_{jk}^2} & \text{if } x_{jk}^2 + y_{jk}^2 \leq r_{jk}^2 \\ 0 & \text{if } x_{jk}^2 + y_{jk}^2 > r_{jk}^2. \end{cases} \qquad (6.10.11)$$

where (x_{jk}, y_{jk}) are the coordinates of the pin-centre.

Due to the clearances in the joints, the effective link lengths change further from Equation (6.10.9), as indicated in Fig. 6.10-2b. In this figure the equivalent 4R linkage is shown by dotted lines. The effective link lengths are easily seen to be

$$R_{jk}^2 = (R_j + x_{jk})^2 + y_{jk}^2 \approx (R_j + x_{jk})^2$$

or

$$R_{jk} \approx R_j + x_{jk} \qquad (6.10.12)$$

where

$$k = \begin{cases} j+1 & \text{for } j = 1, 2, 3 \\ 1 & \text{for } j = 4. \end{cases}$$

Thus, the output-input relation of the linkage, e.g., Equation (6.6.4), which depends on these effective link lengths can, finally, be written in terms of eight random variables (V_j's) as follows:

$$\theta_4 = f(\theta_2, V_j) \quad j = 1, 2, \ldots 8 \qquad (6.10.13)$$

where

$$V_1 = R_1, V_2 = R_2, V_3 = R_3, V_4 = R_4,$$

$$V_5 = x_{12}, V_6 = x_{23}, V_7 = x_{34} \text{ and } V_8 = x_{41}.$$

Our objective is to correlate the statistical behaviours of the output, θ_4, with those of V_j's. Towards this goal, we expand the right-hand-side of Equation (6.10.13) in Taylor's series around the mean values of V_j's and neglect the terms having order two and higher. With this linearization and assuming V_j's to be uncorrelated (normally they are), we can write the mean (m) and variance (D) of θ_4 as

$$m[\theta_4] = f[\theta_2, m(V_j)] \qquad (6.10.14)$$

and

$$D[\theta_4] = \sum_{j=1}^{8} (\frac{\delta f}{\delta V_j})\mid_m^2 D(V_j). \qquad (6.10.15)$$

The subscript m indicates that the derivative is evaluated at the mean values of V_j's. From Equation (6.10.11), we get

$$m(V_j) = 0 \quad \text{for} \quad j = 5, 6, 7, 8 \qquad (6.10.16)$$

and

$$D(V_5) = r_{12}^2/4, \quad D(V_6) = r_{23}^2/4,$$
$$D(V_7) = r_{34}^2/4, \quad D(V_8) = r_{41}^2/4. \qquad (6.10.17)$$

Assuming the tolerances to be normally distributed (as observed in practice), we can write

$$m(V_j) = l_j, D(V_j) = (\epsilon l_j/3)^2; \quad j = 1, 2, 3, 4. \qquad (6.10.18)$$

6.11 Exercise Problems

6.1 Figure 6.11-1 shows the sketch of a paper cutting machine. It is desired that a $40°$ CCW rotation of the handle produces a $20°$ CCW rotation of the cutter from the configuration indicated by solid lines. Determine the required offset e.

6.2 Figure 6.11-2 shows a tipper-dumper. The gear sector is driven by a rack, which in turn is driven by a hydraulic cylinder. The tipping position II should be the outer dead-center configuration of the 4R linkage O_2ABO_4, as indicated in the figure. From start to position II, the gear sector, i.e., link 2, rotates through $90°$ CCW and the dumper bin rotates through $45°$ CW. Synthesize the 4R linkage, i.e., determine the dimensions O_2A and AB.

6.11. EXERCISE PROBLEMS

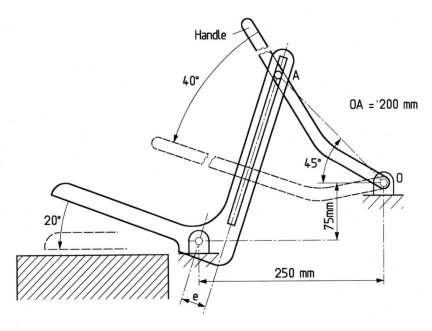

Figure 6.11-1

6.3 A level-luffing crane, used for handling cargo in ports, is shown in Fig. 6.11-3a. The coupler point C has to pass through three points C_1, C_2 and C_3. Determine the required link lengths O_2A and AB. In a real-life design, some trials (regarding the already chosen parameters, e.g., the locations of O_2, O_4, B and the height h) are needed to arrive at a final design like the one shown in Fig. 6.11-3b, where A lies near the midpoint of BC.

6.4 An inverted slider-crank mechanism with offset is shown in Fig. 6.11-4. Given that the crank-length $O_2A = 18$ mm and the eccentricity $e = 12.5$mm, design the linkage so that the output member 4 oscillates through an angle $60°$. Verify whether the input link 2 can rotate completely or not.

6.5 In certain applications a crank-rocker 4R linkage is required to take up a high load at either one or both the dead-center configurations. Consequently, the transmission angle at the dead-center configurations becomes an important factor in the design.

Design a crank-rocker with the following specifications: Frame length = 48.5 mm, $\theta_2^* = 210°$, $\theta_4^* = 30°$ and the transmission angle at both of the dead-centers = 67.5°. What is the minimum transmission angle of the designed mechanism?

6.6 Synthesize a crank-rocker 4R linkage for $(\mu_{min})_{max}$ condition, with the following specifications: Frame length = 48.5 mm, $\theta_2^* = 210°$, $\theta_4^* = 30°$. What are the values of the transmission angle at the dead-center configurations?

6.7 It is required to make the return motion of the slider in the mechanism O_2AB (Fig. 6.11-5) quicker so that forward time/return time = 1.3. This can be achieved by driving the crank O_2A through a 4R double-crank linkage (shown by the broken lines) and using O_6C, instead of O_2A, as the input link with constant angular velocity. Another option is to provide a suitable offset in the slider-crank mechanism O_2AB, to achieve a quick-return ratio of 1.3. Of course, the stroke length in the second option must be less than twice the crank length (O_2A). Design the

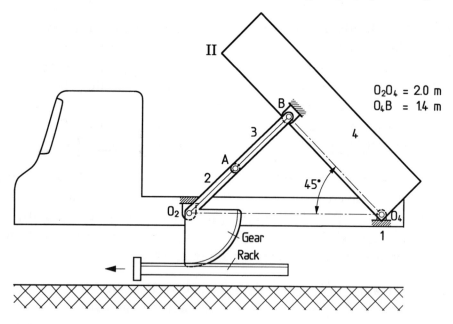

Figure 6.11-2

6.11. EXERCISE PROBLEMS

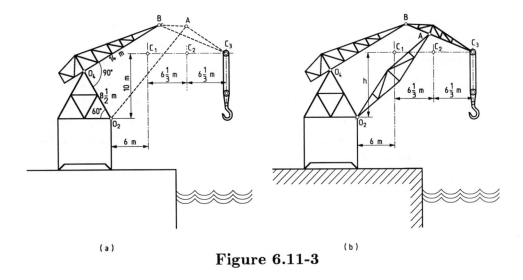

Figure 6.11-3

second option with $O_2A = 30$ cm and the stroke length equal to 50 cm.

6.8 In Problem 6.3-5, the proportional relation between the vertical movement of the fork and the rotation of the input link O_2A was not important. Hartenberg and Denavit[28] discuss in details the synthesis of a pen-recorder mechanism where such a proportional movement is a must. In a strip-chart recorder, if the pen is driven directly by the galvanometer coil, then the pen moves

---

[28] Hartenberg, R.S. and Denavit, J.: Kinematic Synthesis of Linkages, McGraw-Hill Book Co., New York, 1964.

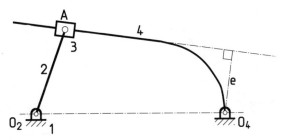

Figure 6.11-4

along a circular arc. The amount of rotation of the coil is proportional to the signal received. The circular movement of the pen necessitates the use of a curvilinear chart paper and distorts the signal record. A 4R linkage, deriving its input from the galvanometer coil, can be designed to drive the pen approximately along a straight line by placing the pen at a properly chosen coupler point. The pen movement must be closely proportional to the input rotation. Synthesize the mechanism assuming 10 cm as the range of pen movement that corresponds to a 90° rotation of the input member. Use both three and four Chebyshev's accuracy points.

6.9 Hunt[29] proposes the design of a wall-mounted reading light following the two-stage synthesis procedure used in Problem 6.3-5 (also required in Problem 6.8). A sketch of the proposed mechanism is shown in Fig. 6.11-6. Design the mechanism so that the lamp moves close to horizontal from S to F. The lamp-axis should also remain essentially vertical during this movement. The swing-hinge on the wall takes the entire mechanism to different planes in order to increase the work-space of the lamp.

6.10 The sketch of a folding-desk mechanism to be used in class rooms is shown in Fig. 6.11-7. Two extreme positions (folded and working) of the desk are indicated. Assuming one more intermediate position (III) locate the fixed pivots O_2 and O_4 within

---

[29] Hunt, K.H.: Kinematic Geometry of Mechanisms, Oxford University Press, Oxford, 1978.

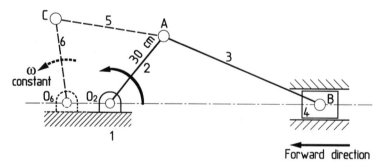

Figure 6.11-5

6.11. EXERCISE PROBLEMS

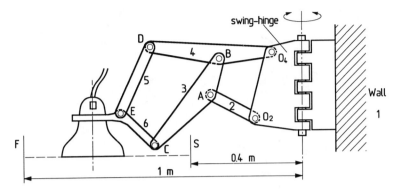

Figure 6.11-6

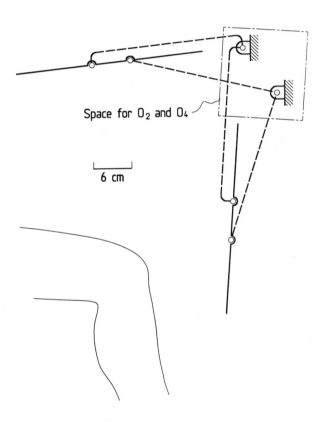

Figure 6.11-7

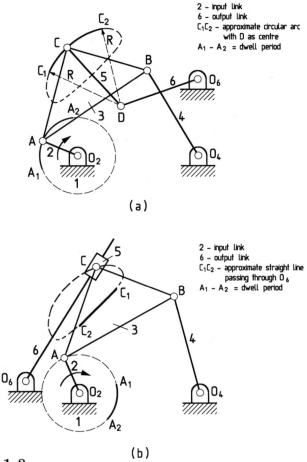

Figure 6.11-8

the space indicated. Compactness of the design is an essential requirement. The mechanism is, obviously, operated with the coupler (desk) as the input link.

6.11 Dwell mechanisms (where the output member remains stationary for a range of continuous input motion) are normally designed using cams. These mechanisms are very commonly used in packaging and other industries. Approximate dwell linkages (having six links and seven lower pairs) can be designed utilizing an approximate circular arc and approximate straight line portions of the coupler curve of a 4R linkage, as explained in

6.11. EXERCISE PROBLEMS

Figs. 6.11-8a and 6.11-8b, respectively. Synthesize these linkages, taking 4 to 5 suitable coupler points.[30]

6.12 Solve Problem 6.3-4 graphically, with four Chebyshev's accuracy points.

6.13 Solve Problem 6.6-1 using Freudenstein's method, assuming $\theta_2^i = 45°$, $\theta_4^i = 135°$ corresponding to $x^i = 1$ and $y^i = 0$.

6.14 Solve Problem 6.6-4 using Bloch's method.

6.15 Synthesize a slider-crank mechanism to satisfy the following three conditions: $\theta_2^1 = 60°$, $s^1 = 20$ cm, $\dot{\theta}_2^1 = 5$ rad/s, $\dot{s}^1 = 30$ cm/s; $\theta_2^2 = 90°$, $s^2 = 25$ cm.

6.16 Synthesize a 4R steering linkage for a vehicle with the basic arrangement and dimensions as shown in Fig. 6.11-9. The minimum radius of curvature of the path traced by the midpoint (G) of the vehicle is 10 m. The steering linkage is symmetrical so as to generate similar left and right turns. Use Freudenstein's method.

---

[30] For such a synthesis with further specified coordination between the output (6) and input (2) link movements see Hartenberg, R.S. and Denavit, J., op. cit.

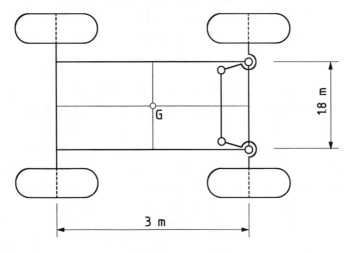

Figure 6.11-9

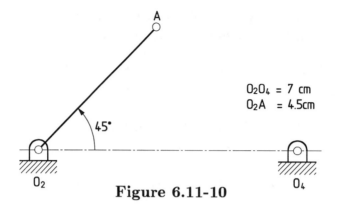

Figure 6.11-10

6.17 Solve Problem 6.3-4 applying Freudenstein's method, using four accuracy points.

6.18 Figure 6.11-10 shows part of a crank-rocker 4R linkage. It is desired that as crank $2(O_2A)$ rotates through $90°$ in the CCW direction from the position indicated in the figure, rocker link $4(O_4B)$ should rotate monotonically in the CCW direction through $45°$. Further, link 4 should be of minimum length. Determine the required coupler length (AB) and follower length (O_4B). Check for the monotonic movements in your solution.

6.19 Evaluate the deviation in y in the solution of Exercise Problem 6.13 at $x = 3.5$, due only to structural error. Also, evaluate the r.m.s. mechanical error at the same configuration, assuming a maximum of $\pm 1\%$ error in all link lengths.

6.20 Figure 6.11-11 shows a schematic arrangement of the paper-feeding mechanism of a printing machine. The main cylinder of radius r_a rotates with a constant angular velocity ω as shown. The arm, driving the smaller cylinder (containing a series of vacuum-operated pick-up holes) of radius r_b, rocks between the extreme positions 1 and 2. The exact locations of these extreme positions 1 and 2 are not prescribed, but 2 should not be much beyond the position where the paper tip gets caught between the two cylinders (i.e., $K = 180°$). Similarly, 1 should not be much beyond the vertical. The relative motion between the two

6.11. EXERCISE PROBLEMS

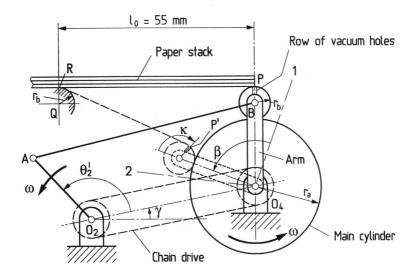

Figure 6.11-11

cylinders is one of pure rolling. The rocking motion of the arm is produced by a crank-rocker 4R linkage O_2ABO_4, where the crank O_2A is driven with the speed ω through a direct chain drive. Determine the required lengths of the links O_2O_4, O_2A and AB with suitable choices for γ and θ_2^i ($l_o = 55$ mm, $r_a = 62.5$ mm, and $r_b = 12.5$ mm) in order to satisfy the following conditions:

(a) The length (l) of the free portion of the paper at any instant ($P'R$) should be as close to its original length, $PR(= l_o)$, as possible.

(b) The velocity of the tips of the vacuum holes, at the instant they touch the paper tip P, should be as close to zero as possible.

After obtaining the solution, plot $(l - l_o)$ versus θ, where θ is the rotation of the main cylinder from the starting position, which corresponds to the vertical position of the arm.
Hint:

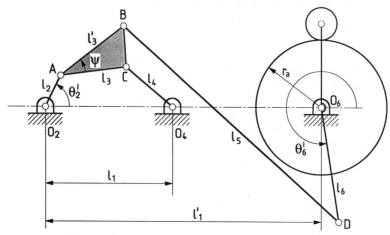

Figure 6.11-12

(1) First derive l

$$= -\beta(r_a + r_b) + r_a\theta + \frac{l_0(1 - \lambda \sin \beta)}{\cos\left[\tan^{-1} \frac{\lambda(1-\cos\beta)}{1-\lambda \sin \beta}\right]}$$

where

$$\lambda = (r_a + r_b)/l_0.$$

(2) Show that condition (ii) is satisfied if $\dot{\beta}/\omega = 0.417$.

(3) Select three suitable values of θ as the accuracy points and evaluate the corresponding values of β so that $l = l_o$, using the expression given in hint 1.

(4) Ensure that the solution yields a crank-rocker type linkage and $\dot{\beta}/\omega$ at $\beta = 0$ is close to 0.417.

(5) If necessry, try with different choices for γ and θ_2^i and different locations of 1 and 2.

6.21 A far superior solution to Problem 6.20, maintaining a much less difference between l and l_o during the entire motion, can be obtained by using a six-link Stephenson's first mechanism, shown in Fig. 6.11-12. Synthesize the mechanism, following the method outlined in Section 6.8, assuming $l_1 = 100$ mm, $l_3' = 35.5$

mm, $\psi = 15°$, $\phi = 0°$, $\theta_2^i = 55°$ and $\theta_6^i = 282.5°$. After obtaining the solution, plot $(l - l_o)$ versus θ.

Chapter 7
COUPLER CURVES

The curve generated on the fixed plane, by a point on the plane of the floating link of a four-link mechanism (e.g., the coupler of a 4R linkage, the connecting rod of a slider-crank mechanism), is referred to as a *coupler-point curve* or a *coupler curve*. In Chapter 6, we saw the use of a coupler curve as a solution to the path generation problems. The desired path may also be used to drive a dwell mechanism or an intermittent motion mechanism. General characteristics of the coupler curves of a 4R linkage have been studied in great detail for more than a century. Many results of immense use to a mechanism designer have been obtained. In this chapter we shall briefly discuss some of the frequently useful results with reference to coupler curves of a 4R linkage. Before going into the details, we may note that the coupler curves of a double-crank linkage are of little practical use, because

(i) the coupler curves are more or less oval-shaped without any striking contrast in curvature, and [1]

(ii) a serious constructional difficulty is encountered if the cranks are to rotate completely.

The most useful information for a designer, regarding the coupler curves of crank-rocker linkages, is contained in the Atlas of Hrones

---

[1] See Hunt, K.H.: "Kinematic Geometry of Mechanisms", Oxford University Press, Oxford, 1978.

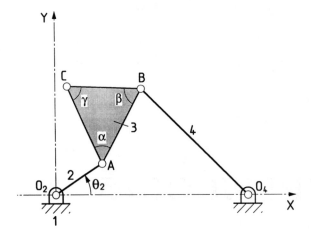

Figure 7.1-1

and Nelson[2]. This atlas contains approximately 7000 coupler curves of various shapes. The extent of the coupler point movement, corresponding to each 10° of crank rotation, is indicated in each curve.

---

[2]Hrones, J.A. and Nelson, G.L.: " Analysis of the Four Bar Linkage" - The Technology Press, MIT, Cambridge, Massachusetts, 1951.

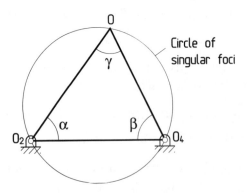

Figure 7.1-2

7.1 Crunodes, Cusps, Symmetry

Referring to Fig. 7.1-1, the coordinates (x, y) of the coupler point C are a function of the link parameters and input-link configuration parameter θ_2. By eliminating θ_2 from the expressions of x and y, one gets the equation of the coupler curve. It can be shown that this curve turns out to be a sixth order curve, implying that a straight line can intersect a coupler curve at no more than six points[3].

Multiple points such as crunodes and cusps may occur in a coupler curve. Since the existence of such points can be utilized by a mechanism designer, a brief discussion of such points in a coupler curve is in order. At a multiple point, the tangent to the curve is not uniquely defined. By crunode we mean a double-point where the curve crosses itself (like a figure eight). As shown below, a coupler curve has a double point (if it exists) at its point of intersection with a circle, called its *circle of singular foci*. Referring to Fig. 7.1-2, let O be such a point on the fixed plane that the triangle OO_2O_4 and the coupler triangle CAB (See Fig. 7.1-1) are similar. The circle then passing through O, O_2 and O_4 is called the circle of singular foci, for the coupler curve of the point C. The points O, O_2 and O_4 are referred to as the real singular foci of the coupler curve corresponding to C.

If the coupler curve for the point C has to cross itself, then there must be two distinct configurations of the linkage when the coupler point locations (C_1, C_2) remain the same, as shown in Fig. 7.1-3. It is easy to see that

$$\angle O_2C_1A_1 = \angle O_2C_1A_2 = \mu \quad \text{(say)}$$

---

[3]If we take a parallelogram linkage, then it is very easy to show that the paths of all coupler points turn out to be circles (a second order curve). In fact, for such linkages passing through a transitional position (all links collinear), one can say that there exist two connected branches of a coupler curve. Each branch is associated with one mode. For example, if a parallelogram linkage transforms into an antiparallelogram configuration at the transition point, then the resulting coupler curve is a fourth-order curve, so that the sum of the two branches is still of the order six. Similarly, for a rhombus-linkage, having two transitional positions, the coupler curve is degenerated into three circles. (For two of three possible modes of a rhombus linkage, the coupler behaves like a simple crank!)

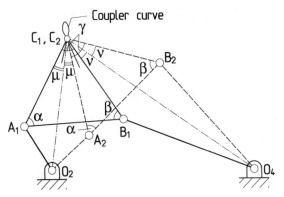

Figure 7.1-3

since $O_2A_1 = O_2A_2$, $A_1C_1 = A_2C_2 (\equiv C_1)$.

Similarly, $\angle O_4C_1B_1 = \angle O_4C_1B_2 = \nu$ (say).
Now, $\angle A_2C_1B_1 = \angle A_1C_1B_1 - 2\mu = \angle A_2C_2B_2 - 2\nu$

Since $\angle A_1C_1B_1 = \angle A_2C_2B_2 = \gamma$, from the above equation, we get

$$\gamma - 2\mu = \gamma - 2\nu$$

or, $\mu = \nu$.

We can write

$$\angle O_2C_1O_4 = \mu + \angle A_2C_1B_1 + \nu = \mu + \gamma - 2\mu + \nu = \gamma.$$

Thus, O_2O_4 subtends an angle γ (the vertex angle of the coupler triangle at C) at the double point. So, referring to Fig. 7.1-4, it is obvious that the double point ($C_1 = C_2$) lies on the circle of singular foci.

A cusp is a point on a curve where the slope changes suddenly, as indicated in Fig. 7.1.5a. The motion of the coupler plane, as we already know, can be represented by rolling (without slipping) its moving centrode C_m over the fixed centrode C_f. If we consider a

7.1. CRUNODES, CUSPS, SYMMETRY

coupler point on C_m, then during the motion of the coupler, this point has zero velocity at the instant it comes in contact with the fixed centrode. With continued motion of the coupler, this point then starts moving with zero velocity in a direction different from that during the approach and a cusp is formed in its path. Thus, the coupler curves, generated by the points on C_m, have cusp (s) whenever they come in contact with C_f. This point can be easily understood if we consider the cycloids generated by the points on the periphery of a rolling wheel (See Fig. 2.8-1). We have reproduced, in Fig. 7.1.5b, C_f and C_m obtained in Fig. 2.8-5. In Fig. 7.1-5b, portions of the coupler curves generated by two coupler points C_1 and C_2, lying on C_m, are seen to have a cusp each when they intersect C_f.

The cusps in coupler curves are used in different mechanisms in order to generate a reversal without impact and shock[4]. A crank-rocker mechanism can always be designed to generate two cusps in a coupler curve, at two prescribed locations, with prescribed amount of crank-rotation as the coupler point moves from one cusp to another. The design of such a linkage is based on the following result.

For every crank-rocker linkage, we can determine a coupler point which has two cusps. Such a coupler point is located at the instantaneous center corresponding to the configurations when the crank-pin

---

[4]For a few such applications, see Hain K.: "Applied Kinematics", 2nd Ed., McGraw-Hill Book Co., New York, 1967.

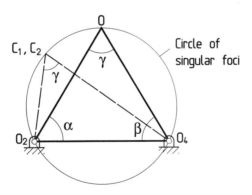

Figure 7.1-4

CHAPTER 7. COUPLER CURVES

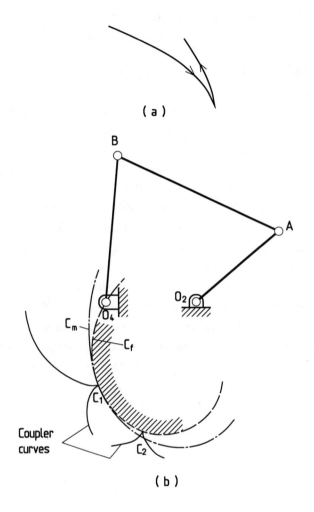

Figure 7.1-5

7.1. CRUNODES, CUSPS, SYMMETRY

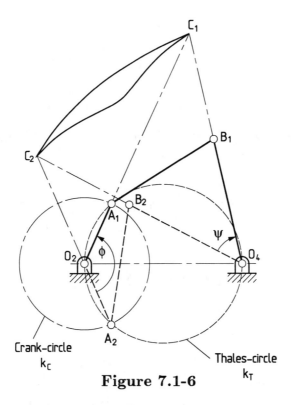

Figure 7.1-6

A lies on the circle, with the frame O_2O_4 as the diameter (this circle is known as the *Thales circle*).

Referring to Fig. 7.1-6, let A_1 and A_2 be the intersections of the crank circle (k_c) and Thales circle (k_T) of a 4R linkage. C_1 and C_2 are the instantaneous centers of the coupler (where cusps of a coupler curve exist) corresponding to configurations $O_2A_1B_1O_4$ and $O_2A_2B_2O_4$, respectively. The result mentioned above is proved if we can show that the coupler triangles $A_1B_1C_1 \equiv A_2B_2C_2$. The proof is as follows:

Consider right angled triangles $O_2A_1O_4$ and $O_2A_2O_4$, where

$$O_2A_1 = O_2A_2, \angle O_2A_1O_4 = \angle O_2A_2O_4 = \pi/2 \text{ and } O_2O_4 \text{ is common.}$$

Thus, $\triangle O_2A_1O_4 \equiv \triangle O_2A_2O_4$. Hence, $O_4A_1 = O_4A_2$.

Now consider Δ's $O_4A_1B_1$ and $O_4A_2B_2$, where

$$O_4A_1 = O_4A_2, \ A_1B_1 = A_2B_2 \text{ and } O_4B_1 = O_4B_2.$$

Therefore, $\Delta O_4A_1B_1 \equiv \Delta O_4A_2B_2$,

which yields $\angle B_1O_4A_1 = \angle B_2O_4A_2$ or $\angle C_1O_4A_1 = \angle C_2O_4A_2$. Next, let us consider triangles $O_4A_1C_1$ and $O_4A_2C_2$, where

$$O_4A_1 = O_4A_2, \ \angle C_1O_4A_1 = \angle C_2O_4A_2, \ \angle C_1A_1O_4 = \angle C_2A_2O_4 = \pi/2.$$

Therefore,

$$\Delta O_4A_1C_1 \equiv \Delta O_4A_2C_2$$

which yields

$$\begin{aligned} A_1C_1 &= A_2C_2 \\ \text{and } O_4C_1 &= O_4C_2 \\ \text{or, } O_4C_1 - O_4B_1 &= O_4C_2 - O_4B_2, \end{aligned}$$

i.e., $B_1C_1 = B_2C_2$. Since by construction $A_1B_1 = A_2B_2$, we finally get

$$\begin{aligned} A_1B_1 &= A_2B_2, \\ A_1C_1 &= A_2C_2 \text{ and} \\ B_1C_1 &= B_2C_2 \\ \text{i.e., } \Delta A_1B_1C_1 &\equiv \Delta A_2B_2C_2. \end{aligned}$$

One can further prove that $\phi + \psi = \pi$, where ϕ and ψ represent the amount of rotation of the crank and the rocker, respectively, as the coupler point moves from one cusp to another.[5] Note the directions of rotation of the crank and the rocker (indicated in Fig. 7.1-6) and also

---

[5] $\Delta O_4A_1B_1 = \Delta O_4A_2B_2$, so by rotating the $\Delta O_4A_1B_1$ about O_4 we get $\Delta O_4A_2B_2$. Therefore, the amount of rotation $\angle B_1O_4B_2 = \angle A_1O_4A_2 = \psi$. Since O_2, A_1, O_4 and A_2 lie on a circle, $\phi + \angle A_1O_4A_2 = \pi$.

7.1. CRUNODES, CUSPS, SYMMETRY

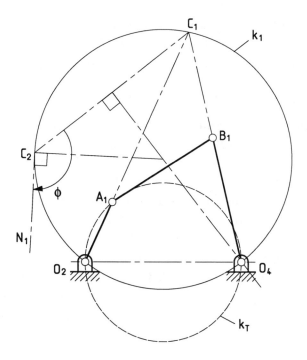

Figure 7.1-7

the fact that ψ does not represent the total swing angle of the follower (since O_4B_1 and O_4B_2 are not the extreme positions of the rocker). Since $\phi + \psi = \pi$, $\angle C_2O_2C_1 = \pi - \phi = \psi = \angle C_1O_4C_2$. Therefore, C_1, C_2, O_2 and O_4 also lie on a circle.

Now we are in a position to synthesize a crank-rocker linkage such that a coupler curve has two cusps at two prescribed locations, say at C_1 and C_2 (Fig. 7.1-7) and the crank rotates in the CCW direction through an angle ϕ as the coupler point moves from C_2 to C_1. The synthesis is carried out in a manner described below.

Referring to Fig. 7.1-7, draw a line N_1 through C_2 making an angle ϕ (in the CW direction) from C_2C_1. Draw the circle k_1 passing through C_1 and C_2, and having N_1 as its tangent. Locate O_4 at the intersection of k_1 with the midnormal of C_1C_2. Choose O_2 anywhere on k_1, when A_1 is located at the intersection of Thales circle (k_T) with the line O_2C_1. The point B_1 is chosen anywhere on the line O_4C_1.

While choosing O_2 and B_1 we must ensure that the resulting linkage $O_2A_1B_1(C_1)O_4$ is a Grashof-linkage, with O_2A_1 as the shortest link. The proof of this synthesis procedure is left as an exercise for the reader to derive, keeping in mind the results mentioned in connection with Fig. 7.1-6.

Another important geometric characteristic of a coupler curve is symmetry, which is attained under special conditions. We shall now show that the sufficient (not necessary) condition for a coupler curve to be symmetric is that the coupler base $AB(l_3)$ is equal to the follower length $O_4B(l_4)$ and that the coupler point lies on a circle centered at B, with radius equal to $l_3(=l_4)$. Under this situation, the symmetric points on the coupler curve correspond to symmetric positions of the crank, with respect to the frame-line O_2O_4, and the axis of symmetry of the coupler curve passes through O_4. Figure 7.1-8 shows a symmetrical coupler curve generated under the conditions mentioned above. We may note that in this figure, $\angle AO_4C = \beta/2$, a result that will be used shortly. It will be further shown that the axis of symmetry of the coupler curve makes an angle $\beta/2$ with O_2O_4 as indicated in this figure.

Let us now consider two configurations of the 4R linkage shown

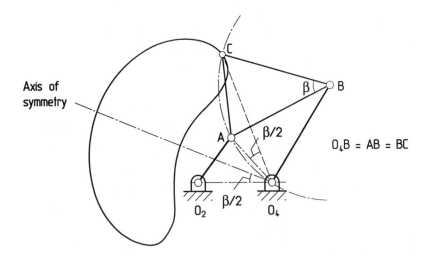

Figure 7.1-8

7.1. CRUNODES, CUSPS, SYMMETRY

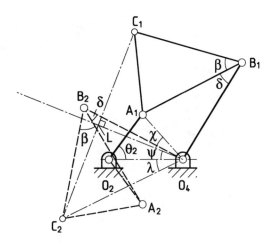

Figure 7.1-9

in Fig. 7.1-8, in which the crank O_2A occupies two positions, O_2A_1 and O_2A_2, which are symmetrical with respect to O_2O_4 (Fig. 7.1-9). Since $O_2A_1 = O_2A_2$ and $\angle A_1O_2O_4 = \angle A_2O_2O_4$ we get

$$\Delta O_2A_1O_4 = \Delta O_2A_2O_4.$$

Therefore, $O_4A_1 = O_4A_2$. Now, since $O_4A_1 = O_4A_2$, $O_4B_1 = O_4B_2$ and $A_1B_1 = A_2B_2$, we obtain $\Delta O_4A_1B_1 = \Delta O_4A_2B_2$. Therefore, $\angle O_4B_1A_1 = \angle O_4B_2A_2 = \delta$ (say). Considering the triangles $O_4B_1C_1$ and $O_4B_2C_2$, we can now write $O_4B_1 = O_4B_2$, $B_1C_1 = B_2C_2$ and $\angle O_4B_1C_1 = \beta + \delta = \angle O_4B_2C_2$. Hence, $\Delta O_4B_1C_1 = \Delta O_4B_2C_2$, which yields $O_4C_1 = O_4C_2$. Thus, the midnormal of the side C_1C_2 of the isosceles triangle $O_4C_1C_2$ passes through O_4 (for any two symmetric locations A_1 and A_2). We shall now prove that this midnormal (O_4L) makes a constant
angle (ψ) with O_2O_4, independent of θ_2 (Fig. 7.1-9), which makes the coupler curve of C symmetric, with O_4L as the axis of symmetry.

Since the triangle $O_4C_1C_2$ is isosceles, with O_4L as the perpendicular bisector of C_1C_2, we get $\angle C_1O_4L = \angle C_2O_4L$, or,

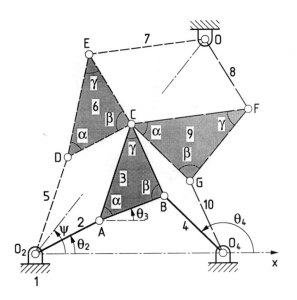

Figure 7.2-1

$$\chi + \frac{\beta}{2} = \lambda + \psi, \qquad (7.1.1)$$

since $\angle A_1 O_4 C_1 = \beta/2$ as shown in Fig. 7.1-8. Again, because A_1 and A_2 are located symmetrically with respect to $O_2 O_4$, we get

$$\angle A_1 O_4 O_2 = \angle A_2 O_4 O_2,$$

or,

$$\chi + \psi = \lambda + \frac{\beta}{2} \qquad (7.1.2)$$

since $\angle A_2 O_4 C_2 = \beta/2$.
From Equations (7.1.1) and (7.1.2), it is easy to see that $\psi = \beta/2$, i.e. constant.

7.2 Cognate Linkages

The most important result regarding the coupler curves of a 4R linkage is probably what is known as Roberts-Chebyshev Theorem, which

7.2. COGNATE LINKAGES

states that three different planar 4R linkages trace identical coupler curves. These three linkages are referred to as coupler curve cognate linkages.

We shall prove Roberts-Chebyshev Theorem by expressing link-vectors through complex exponential notation. To start with, from any 4R linkage O_2ABO_4 with a coupler point C, we construct a ten-link mechanism with only R-pairs as shown in Fig. 7.2-1. In this figure, O_2ACD, $OECF$ and O_4GCB are all parallelograms and triangles ABC, DCE and CGF are all similar. We should note that angles α, β and γ appear in the same cyclic order in all these triangles and also at the point C.

In the ten-link mechanism so obtained, if we apply Equations (3.1.3) and (3.1.4) (with $n = 10$, $h = 0$, $j_2 = 4$, one each at O_2, O_4, O and C, $j_1 = 6$), we get $F = -1$. Thus, apparently this ten-link assembly is a statically indeterminate structure. However, as we shall see shortly, because of special dimensions imposed (through parallelograms and similar triangles), the assembly turns out to be a constrained mechanism with $F = 1$. For the time being, if we asssume $F = 1$, then Roberts-Chebyshev Theorem is proved because three 4R linkages, namely, $1 - 2 - 3 - 4$, $1 - 5 - 6 - 7$ and $1 - 8 - 9 - 10$ obviously generate the same coupler-curve (the path of C in the ten-link mechanism) for the coupler point C, even when they are disconnected from one another at C. Each individual 4R linkage and the ten-link assembly all provide the same constraints on the movement of C.

To prove that the ten-link assembly is a constrained mechanism we shall show that there is a redundant constraint that was overlooked while obtaining $F = -1$ in the preceeding paragraph. Let us consider the ten-link assembly shown in Fig. 7.2-2, which is exactly same as the one shown in Fig. 7.2-1, except that in Fig. 7.2-1 the revolute pair at O (between links 7 and 8) is also connected to the fixed link. Thus, the revolute pair at O in Fig. 7.2-2 is a simple hinge, whereas that in Fig. 7.2-1 is a higher order hinge. Hence if we apply Equations (3.1.3) and (3.1.4) for the mechanism of Fig. 7.2-2, we get $F = 1$, with $n = 10$, $h = 0$, $j_1 = 7$ and $j_2 = 3$. Consequently, this mechanism is a constrained mechanism. In what follows, we shall show that when this constrained mechanism moves, the point O remains stationary. Therefore, the revolute pair with the fixed link at O, in Fig. 7.2-1, is redundant and the two mechanisms shown in Figs. 7.2-1 and 7.2-2 are identical, i.e., the ten-link mechanism shown in Fig. 7.2-1 is constrained, as assumed.

To see that the point O in Fig. 7.2-2 does not move, we use the complex exponential notation for link-vectors and express the vector

$$\begin{aligned}
\boldsymbol{O_2O} \quad &= \quad \boldsymbol{O_2D} + \boldsymbol{DE} + \boldsymbol{EO} \\
&= \quad \boldsymbol{AC} + \boldsymbol{DC}\tfrac{DE}{DC}e^{i\alpha} + \boldsymbol{CF} \\
&= \quad \boldsymbol{AB}\tfrac{AC}{AB}e^{i\alpha} + \boldsymbol{O_2A}\tfrac{DE}{DC}e^{i\alpha} + \boldsymbol{CG}\tfrac{CF}{CG}e^{i\alpha} \\
&= \quad (\boldsymbol{AB} + \boldsymbol{O_2A} + \boldsymbol{BO_4})\tfrac{AC}{AB}e^{i\alpha}
\end{aligned}$$

since $[\tfrac{AC}{AB} = \tfrac{DE}{DF} = \tfrac{CF}{CG}, CG = BO_4]$. Finally, we get,

$$\boldsymbol{O_2O} \quad = \quad \boldsymbol{O_2O_4}\tfrac{AC}{AB}e^{i\alpha}. \qquad (7.2.1)$$

We note from the right-hand-side of Equation (7.2.1) that the vector $\boldsymbol{O_2O}$ does not contain any variable parameter (like θ_2, θ_3, etc) and remains invariant as the mechanism moves. Thus, the point O is stationary.

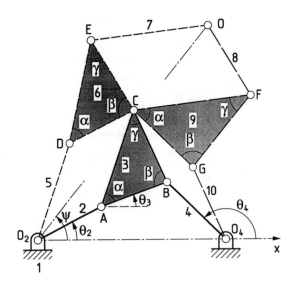

Figure 7.2-2

7.2. COGNATE LINKAGES

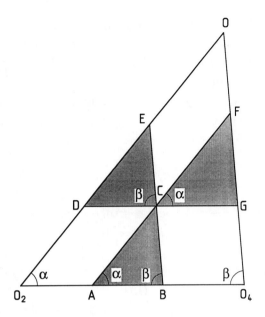

Figure 7.2-3

From Equation (7.2.1) it is obvious that the angle ψ (Fig. 7.2-2) is equal to α and

$$\frac{O_2 O}{O_2 O_4} = \frac{AC}{AB}.$$

Hence, the traingles OO_2O_4 and ABC are similar, i.e., O is the third singular foci of the coupler curve generated by C of the 4R linkage O_2ABO_4.[6]

The 4R linkage $1 - 8 - 9 - 10$ is referred to as the right-hand cognate of the 4R linkage $1 - 2 - 3 - 4$. Similarly, the 4R linkage $1 - 5 - 6 - 7$ is called the left-hand cognate. The construction (known as Cayley's construction), shown in Fig. 7.2-3, may be conveniently used to determine the dimensions of the cognate linkages from those of the original one. In this construction, O_2A, AB and BO_4 are drawn as collinear and thereafter one needs to draw only parallel lines, as explained in Fig. 7.2.3.

---

[6]The unique relationship of the three real singular foci with the coupler curve led Roberts to think that there must be two other 4R linkages, one with O_4 and O and the other with O_2 and O as their fixed hinges, which generate the same coupler curve with C as the coupler point.

Returning to Fig. 7.2-1, it should be noted at this stage that the similarity between cognate linkages is only in respect to the identical coupler curve they produce. In fact, in all other respects they may differ considerably. For example, let us consider the following obvious relationships (emerging from parallelogram configurations in the ten-link mechanism) between the angular motions of various links of different cognate linkages:

Original linkage	*Left − hand cognate*	*Right − hand cognate*
Angular motion of link 2 ≡	Angular motion of link 6 ≡	Angular motion of link 8
Angular motion of link 3 ≡	Angular motion of link 5 ≡	Angular motion of link 10
Angular motion of link 4 ≡	Angular motion of link 7 ≡	Angular motion of link 9

Therefore, if the original linkage is a crank-rocker with say 2, as the crank (when 3 and 4 must oscillate), then the right-hand cognate is also a crank-rocker, with 8 as the crank; however, the left-hand cognate then must be a double-rocker (with the coupler 6 undergoing full rotation).[7] Cognate linkages offer alternative solutions to the designer to generate the same coupler curve. A cognate linkage having different work-space, different locations of fixed pivots or less sensitivity to mechanical error, may be preferred to the others.

In Sections 6.3.2 and 6.5 we discussed the problem of path generation without any coordination with the input motion. The fact that the angular motion of link 3 (in the original linkage) is identical to that of link 5 (in the left-hand cognate) or link 10 (in the right-hand

---

[7]The solution to the design of a 4R linkage, with five prescribed positions of a coupler point coordinated with prescribed input link rotations, is discussed in Hartenberg, R.S. and Denavit, J. : "Kinematic Synthesis of Linkages", McGraw-Hill Book Co., New York, 1964. It is shown that there may exist zero, two or twelve solutions. In the second situation, the two solutions are cognate linkages and in the third situation, the solutions are six-pairs of cognate linkages. The third cognate, in each case, is ruled out because of the imposed coordination with the input-link rotation.

7.2. COGNATE LINKAGES

cognate) can be easily exploited to synthesize a 4R linkage, in order to generate a path coordinated with the input motion. This is illustrated in the following example.

Problem 7.2-1

Referring to Fig. 7.2-4a, let us assume that the coupler point of a 4R linkage must pass through C_1, C_2 and C_3. Furthermore, as the coupler point moves from C_1 to C_2 and C_2 to C_3, the input link should rotate by 20° (CW) and 18° (CW), respectively. Synthesize the required 4R linkage.

Solution

Referring to Fig. 7.2-4b, the crank pin location A_1 is chosen arbitrarily. Through C_2 and C_3 we draw C_2A_2 and C_3A_3, respectively, so that $C_2A_2 = C_1A_1 = C_3A_3$ and C_2A_2 is rotated 20° (CW) from C_1A_1 and C_3A_3 is rotated 18° (CW) from C_2A_2. The fixed hinge O_2 is located at the center of the circle passing through A_1, A_2 and A_3.[8] Next, the fixed hinge O_4 is chosen arbitrarily. Inverting on the coupler AC, fixed at its first position A_1C_1, the inverted positions of $O^1_{4,1}(\equiv O_4)$ are obtained at $O^1_{4,2}$ and $O^1_{4,3}$ (see also Fig. 6.3-4). The moving hinge B_1 is located at the center of the circle passing through $O^1_{4,1}$, $O^1_{4,2}$ and $O^1_{4,3}$. Thus, we get a 4R linkage $O_2A(C)BO_4$ where the coupler point C passes through C_1, C_2 and C_3, and during the movement from C_1 to C_2, the coupler rotation (θ_3^{12}) is 20° (CW) and during the movement from C_2 to C_3, the coupler rotation (θ_3^{23}) is 18° (CW).

Following the construction of Fig. 7.2-1, the cognate linkages $O_2D(C)EO$ and $OF(C)GO_4$ are obtained as shown in Fig. 7.2-4c. The angular motions of links O_2D and O_4G, in the first and second cognate linkages, respectively, are the same as that of link 3. Hence, the 4R linkages shown in Figs. 7.2-4d and 7.2-4e, with input links as O_2D and O_4G, respectively, are the required linkages. The reader should note that the angular motion of link 3 in Fig. 7.2-4b is identical to those of cranks O_2D in Fig. 7.2-4d and O_4G in Fig. 7.2-4e.

---

[8]Some trials regarding the choice of location of A_1 may be necessary if the location of the fixed hinge O_2 must be within a specified region.

344 CHAPTER 7. COUPLER CURVES

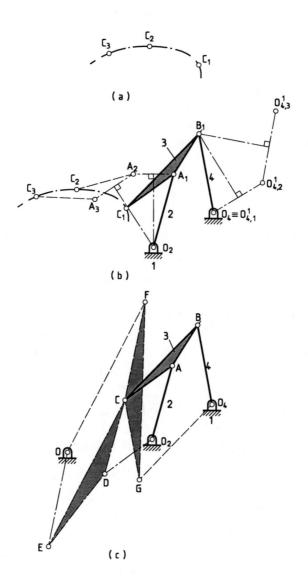

Figure 7.2-4a, b & c

7.2. COGNATE LINKAGES

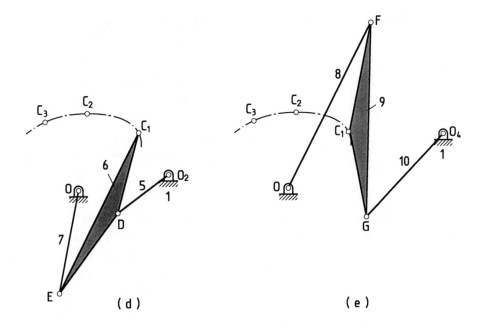

Figure 7.2-4d & e

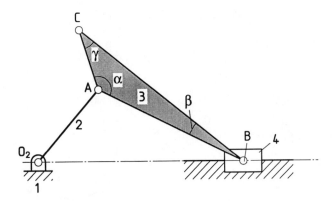

Figure 7.3-1

7.3 Extension of Cognate Linkages

The idea of coupler curve cognate linkages, explained with reference to a 4R linkage in Section 7.2, can be extended to various situations. For example, let us consider a slider-crank mechanism $O_2A(C)B$, shown in Fig. 7.3-1. We can determine, as explained below, another slider-crank mechanism that generates the same curve at the point C as generated by $O_2A(C)B$. To obtain this cognate slider-crank mechanism, let us first consider the six-link mechanism shown in Fig. 7.3-2. This six-link mechanism is obtained, through the parallelogram (O_2ACD) and the triangle DCE (similar to $\triangle ABC$), from $O_2A(C)B$, as explained in the figure. The degree-of-freedom of the six-link mechanism so obtained is one [as evident from Equations (3.1.3) and (3.1.4), with $j_2 = 1$ at O_2, $j_1 = 5$, $n = 6$, and $h = 0$]. Therefore, the path of C remains unchanged by the addition of links 5 and 6. We shall now show that when this six-link mechanism moves, the point E moves along a straight line (see also Exercise Problem 4.7-1). Towards this goal, we write

$$\begin{aligned} O_2E &= O_2D + DE \\ &= AC + \tfrac{DE}{DC} e^{i\alpha} DC \\ &= \tfrac{AC}{AB} e^{i\alpha} AB + \tfrac{DE}{DC} e^{i\alpha} O_2A \end{aligned}$$

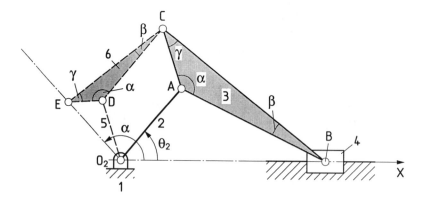

Figure 7.3-2

7.3. EXTENSION OF COGNATE LINKAGES

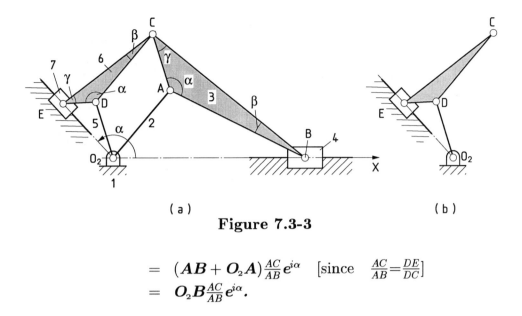

Figure 7.3-3

$$= (\boldsymbol{AB} + \boldsymbol{O_2A})\tfrac{AC}{AB}e^{i\alpha} \quad [\text{since } \tfrac{AC}{AB}=\tfrac{DE}{DC}]$$
$$= \boldsymbol{O_2B}\tfrac{AC}{AB}e^{i\alpha}.$$

As the vector $\boldsymbol{O_2B}$ is always along the X-axis, the vector $\boldsymbol{O_2E}$ has a fixed orientation (at an angle α from the X-axis). Thus, E moves along O_2E. Consequently, if we hinge a slider (7) to link 6 at E, with a prismatic pair (between 1 and 7) having its axis of sliding along O_2E (Fig. 7.3-3a), no extra constraint is imposed and the resulting overclosed seven-link mechanism is a constrained mechanism. The path of C in this seven-link mechanism also remains the same as that generated by $O_2A(C)B$. Hence, the path of C in the (cognate) slider-crank mechanism $O_2D(C)E$, shown in Fig. 7.3-3b, is identical to that generated by the (original) slider-crank mechanism $O_2A(C)B$. The reader is advised to note the angular motion relationships that exist between various links of the original and the cognate slider-crank mechanisms.

Using parallelograms and similar triangles, we can also construct six-link cognate linkages for a 4R linkage. For a given 4R linkage and its coupler point, the two possible cognate six-link mechanisms are obtained, as explained by dashed lines in Figs. 7.3-4a and 7.3-4b. In Fig. 7.3-4a, O_2ACD, O_2O_4ED and O_4BCF are parallelograms when $\triangle ACB \equiv \triangle EO_4F$. In Fig. 7.3-4b, O_2ACE, O_4BCD and O_2O_4DF are parallelograms when $\triangle ACB \equiv \triangle EO_2F$. The reader may note

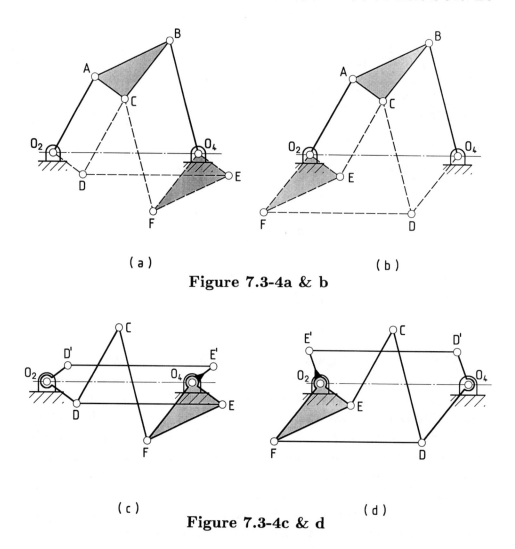

Figure 7.3-4a & b

Figure 7.3-4c & d

that in the six-link cognate linkages the point C is an R pair. Moreover, it is necessary to maintain the parallelograms, O_2O_4ED in Fig. 7.3-4a and O_2O_4DF in Fig. 7.3-4b, at the transitional configurations. This can be easily done, as explained in Figs. 7.3-4c and 7.3-4d, for Figs. 7.3-4a and 7.3-4b, respectively (recall Fig. 3.3-2).

The proof that the six-link mechanisms obtained in the aforementioned manner generate, at the joint C the same curve as the coupler-curve of the original 4R linkage $O_2A(C)BO_4$, can be carried out using the approach explained in this and the last sections. One can show

7.3. EXTENSION OF COGNATE LINKAGES

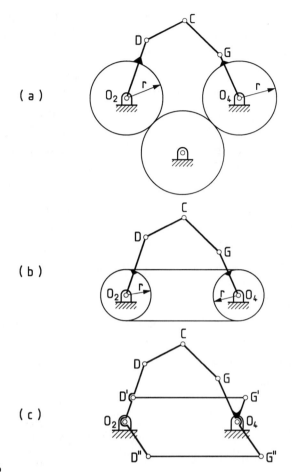

Figure 7.3-5

that the overall nine-link mechanism (original + cognate together) is, again, an overclosed constrained mechanism. The presence of the ternary link CD with R pairs at its ends does not impose any extra constraint. The details of this proof are left as an exercise for the reader.

Returning again to Fig. 7.2-1 and the relationships between the angular motion of various links of the ten-link mechanism shown in this figure, we can decipher some geared five-link cognate mechanisms. Two such mechanisms are discussed below.

It is obvious that if we consider the five-link mechanism O_2DCGO_4 (1-5-6-9-10) and connect 5 and 10 in such a manner that they have the same angular motion, then the path of C generated by this mechanism is same as the coupler curve of C produced by the original 4R linkage $O_2A(C)BO_4$. The same angular (synchronous) motion of O_2D and O_4G can be achieved either through gearing, by using a chain and sprockets or by means of a parallelogram linkage, as shown in Figs. 7.3-5a, 7.3-5b and 7.3-5c, respectively. Similarly, we could have considered the five-link mechanism $O_2ACFO(1-2-3-9-8)$ (Fig. 7.2-1) with synchronous motion of links 2 and 8. In this mechanism, not only the path of C, but the motion of the entire coupler plane remains same as that produced by the original 4R linkage. The five link mechanism O_2ACFO is, therefore, referred to as the coupler cognate (rather than just the coupler curve cognate) of $O_2A(C)BO_4$.[9] We may further note that, O_2A being the input link of both the coupler cognate and original linkages, any specified motion coordination between the coupler and input links remains invariant.

Problem 7.3-1

The motion of the entire coupler-plane of the 4R linkage O_2ABO_4, shown in Fig. 7.3-6a, must be reproduced by a geared five-link mechanism, with O_2 and O as the locations of the gear-shafts. Obtain this five-link mechanism.

Solution

Referring to Fig. 7.3-6b, the point C is located by making $\triangle CAB$ similar to $\triangle OO_2O_4$. The parallelogram O_2ACD is then completed and $\triangle EDC$ is drawn similar to $\triangle CAB$. Finally, the parallelogram $OECF$ is constructed. The required five-link mechanism is thus obtained as O_2ACFO, where the synchronous motion of links O_2A and OF must be maintained by connecting two equal sized gears (one each attached to O_2A and OF) through an idler, as explained in Fig. 7.3-5.

7.4 Parallel Motion Generator

The theory of the coupler curve cognate linkages, developed in the last two sections, can be applied to design six-link (7R) mechanisms where

---

[9]Referring to the Kempe-Burmester focal mechanism, (Fig. 3.3-4), one can observe six-link coupler cognates of a 4R linkage. For example, in Fig. 3.3-4, the six-link mechanism, $O_2SAPTR(1-2-3-5-6-7)$, is a coupler cognate of O_2ABO_4.

7.4. PARALLEL MOTION GENERATOR

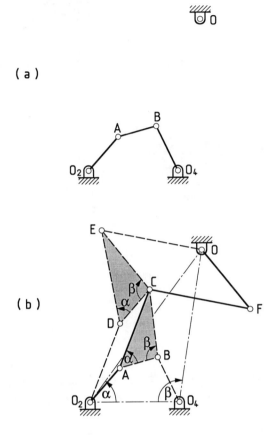

Figure 7.3-6

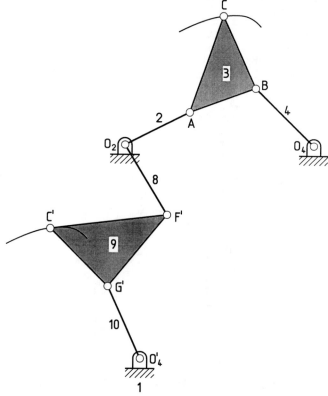

Figure 7.4-1

one link always moves parallel to itself (i.e., undergoes curvilinear translation). Such a mechanism, called a parallel motion generator, is discussed in this section.

In Fig. 7.4-1, the 4R linkage $O_2A(C)BO_4$ of Fig. 7.2-1 is reproduced. Let us consider the right-hand cognate linkage $OF(C)GO_4$, shown in Fig. 7.2-1. In Fig. 7.4-1, the configuration $OF(C)GO_4$ of Fig. 7.2-1 has been translated without rotation, as a rigid body, so that O coincides with O_2, and the shifted configuration is identified with primed symbols. We note that the motions of links 2 and 8 are synchronous and the paths of C and C' are parallel but otherwise similar curves. Hence, links 2 and 8 can be rigidly attached, whereas C and C' (their distance always remaining the same) can be connected by a rigid link, as shown in Fig. 7.4-2. The links are renumbered

7.4. PARALLEL MOTION GENERATOR

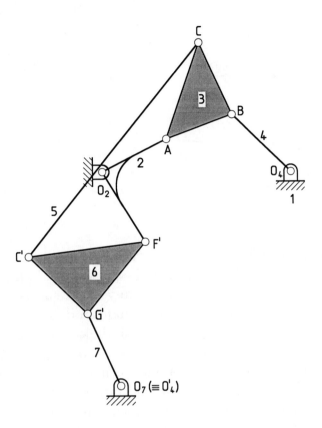

Figure 7.4-2

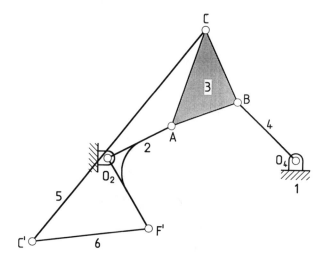

Figure 7.4-3

(continuously) in Fig. 7.4-2. This seven-link mechanism is an overclosed constrained mechanism, since the presence of link $CC'(5)$, with R pairs at its ends, does not impose any extra constraint. We can reproduce the exact motion of this mechanism by removing link 7, as shown in Fig. 7.4-3. Thus, we get a six-link (7R) mechanism with constrained motion. In both the mechanisms of Figs. 7.4-2 and 7.4-3, the link $CC'(5)$ always moves parallel to itself because the points C and C' move identically on two parallel but otherwise similar curves. It may be emphasized that the motions of link 3 and $O'G'$ (i.e., 7 in Fig. 7.4-2) being synchronous, the seven-link (overclosed) mechanism of Fig. 7.4-2 can be easily designed where the movement of C can be coordinated with the input motion by using link 7 as the input member (basically the same principle was used in Problem 7.2-1). Another set of such six-link and seven-link parallel motion generators can be designed by using the left-hand cognate linkage $O_2D(C)EO$ of Fig. 7.2-1. Here, of course, the cognate linkage should be translated without rotation, as a rigid body, so that O coincides with O_4 and links 4 and 7 having, synchronous motions, must be rigidly attached. The resulting seven- and six-link parallel motion generators are shown in Figs. 7.4-4 and 7.4-5, respectively.

7.4. PARALLEL MOTION GENERATOR

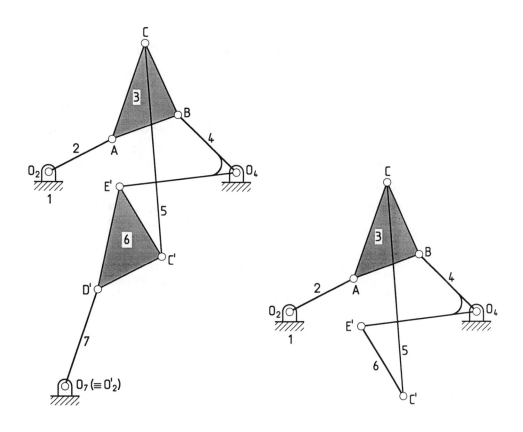

Figure 7.4-4

Figure 7.4-5

Starting from a 4R linkage, the development of a six-link parallel motion generator can also be explained through the principle of stretch-rotation operation.[10] This principle states the rather obvious geometric fact that the relative angular motions between various links of a constrained mechanism remain invariant if the kinematic diagram is drawn to a different scale (stretched or shrinked) and orientation (rotated as a rigid body). For example, let us consider an instantaneous configuration of a 4R linkage, O_2ABO_4 (Fig. 7.4-6). A second 4R linkage configuration, $O'_2A'B'O'_4$, is obtained by first drawing O_2ABO_4 to a different scale $(O'_2A''B''O''_4)$ and then rotating the scale drawing as a rigid body (through any angle α). It is quite obvious, then, that the relative angular motions between various links in the two mechanisms at these configurations remain the same. The linkage $O'_2A'B'O'_4$ is obtained from the original linkage O_2ABO_4 by a stretch-rotation operation.

Let us now consider the original 4R linkage, $O_2A(C)BO_4$, of Fig. 7.2-1 which, is reproduced in Fig. 7.4-7. We shall apply a stretch-

---

[10]For a discussion of the entire theory of cognate linkages through the principle of stretch-rotation and some applications of parallel motion generators, See Dijksman, E.A.: " Motion Geometry of Mechanisms", Cambridge Univ. Press, London, 1976.

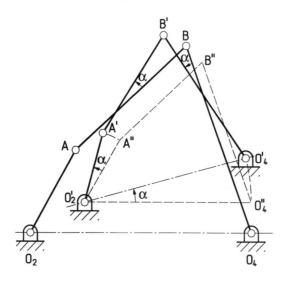

Figure 7.4-6

7.4. PARALLEL MOTION GENERATOR

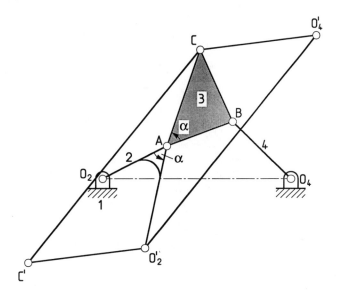

Figure 7.4-7

rotation (about the point A) operation on this linkage so that the second 4R linkage has its moving hinges at A and C. In other words, we use a scale factor AC/AB and rotate the first linkage about A (as a rigid body) through an angle α (CCW) to get the second 4R linkage as $O'_2ACO'_4$, as shown. The relative angular motion between O_2O_4 and O_2A in the first linkage is same as that between $O'_2O'_4$ and O'_2A in the second linkage. Therefore, if O_2A and O'_2A are rigidly attached (i.e., have same angular motion), as indicated in the figure, then $O'_2O'_4$ (having same angular motion as O_2O_4, i.e., zero) does not rotate. So, if we complete the parallelogram $O'_2O'_4CC'$, then the link CC' also does not rotate, i.e., it moves parallel to itself. From this eight-link constrained mechanism (Fig. 7.4-7), if the links $O'_2O'_4$ and O'_4C are removed, we get the six-link parallel motion generator shown in Fig. 7.4-8, which is identical to Fig. 7.4-3. The mechanism shown in Fig. 7.4-5 can be similarly obtained by applying the stretch-rotation operation about the point B. This is left as an exercise for the reader.

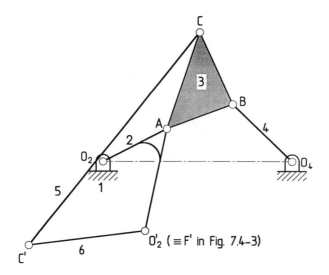

Figure 7.4-8

7.5 Approximate Straight-Line Linkages

In Section 7.1 we noted that a straight line can intersect the coupler curve of a 4R linkage at no more than six points. However, with proper choices of the link-length ratios and the coupler point, 4R linkages have been devised that can generate straight lines with reasonably good approximation as a part of their coupler curves.[11] Some of the well known such linkages are named after their inventors, e.g., Watt's, Roberts', Evans' and Chebyshev's. These four linkages and the type of coupler curves (with an approximate straight line portion) they produce are illustrated in Figs. 7.5-1a to 7.5-1d. A designer can also judiciously use his knowledge of the theory of planar kinematics of a rigid body (Chapter 2) to generate the approximate straight line

---

[11]For a historical account of the need and development of straight-line mechanisms, see Hartenberg, R.S. and Denavit, J.:" Kinematic Synthesis of Linkages"., McGraw-Hill Book Co., New York, 1964. Before the planing machine was invented, making a long guide to provide a prismatic pair was difficult. This provided the incentive to design linkages with only revolute pairs that can generate approximate or exact straight-lines.

7.5. APPROXIMATE STRAIGHT-LINE LINKAGES

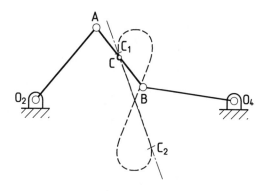

$O_2A = O_4B$, $AC = BC$
$O_2O_4 : O_2A : AB = 10.77 : 5 : 4$

(a) Watt's

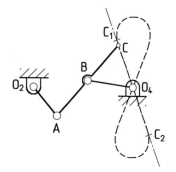

$AB = O_4B = BC$
$O_2A : AB : O_2O_4 = 2 : 2.5 : 5.39$

(a) Evans'

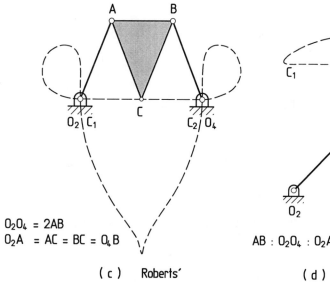

$O_2O_4 = 2AB$
$O_2A = AC = BC = O_4B$

(c) Roberts'

$AB : O_2O_4 : O_2A : O_4B = 1 : 2 : 2.5 : 2.5$

(d) Chebyshev's

Figure 7.5-1

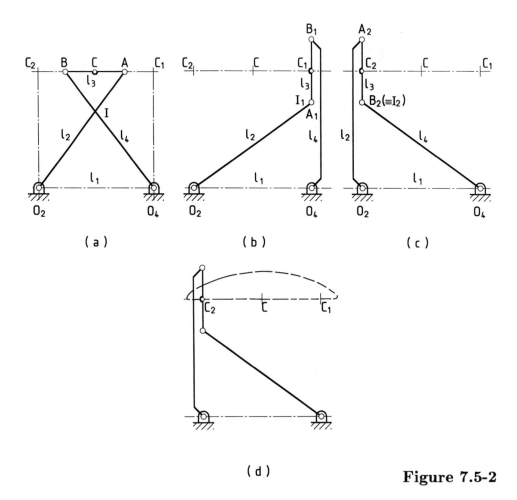

Figure 7.5-2

portion of a coupler curve, as illustrated below through some examples.

Problem 7.5-1

Figure 7.5-2a shows a symmetrical double-rocker 4R linkage with $l_2 = l_4$. It is desired that the midpoint, C, of the coupler AB should have the same tangent (parallel to O_2O_4) at three positions, C, C_1 and C_2. Determine the required values of the link-length ratios l_1/l_3 and l_1/l_2. Also draw the complete coupler curve and observe the expected approximate straight line portion (parallel to O_2O_4) between C_1 and C_2.

7.5. APPROXIMATE STRAIGHT-LINE LINKAGES

Solution

A tangent to the coupler-curve parallel to O_2O_4 implies that the coupler point velocity is parallel to O_2O_4, i.e., the line joining the coupler point and the instantaneous center (I) of the coupler should be perpendicular to O_2O_4. The instantaneous center of the coupler is at the point of intersection of the input and output links. Thus, the symmetrical linkage, in order to satisfy the requirements, must take up the configurations shown in Figs. 7.5-2a, 7.5-2b and 7.5-2c.

From Fig. 7.5-2b (or Fig. 7.5-2c), with $l_2 = l_4$ we get

$$l_2^2 = l_1^2 + (l_2 - l_3)^2$$

or, $$(\frac{l_1}{l_3})^2 + 1 = 2\frac{l_2}{l_3}. \qquad (a)$$

Similarly, from Fig. 7.5-2a, we can write

$$l_2^2 = (\frac{l_1 + l_3}{2})^2 + (l_2 - \frac{l_3}{2})^2$$

or

$$\frac{1}{2}(\frac{l_1}{l_3})^2 + \frac{l_1}{l_3} + 1 = 2\frac{l_2}{l_3}. \qquad (b)$$

Solving Equations (a) and (b), we obtain

$$\frac{l_3}{l_1} = \frac{1}{2}, \quad \text{i.e.,} \frac{l_1}{l_3} = 2 \text{ and } \frac{l_2}{l_3} = \frac{5}{2} \text{ i.e., } \frac{l_1}{l_2} = \frac{4}{5}.$$

The reader may note that the designed linkage is the well known Chebyshev's straight-line mechanism.

The coupler curve of C is shown separately in Fig. 7.5-2d. Note the symmetry and the approximate straight line portion of the coupler curve.

362   CHAPTER 7. COUPLER CURVES

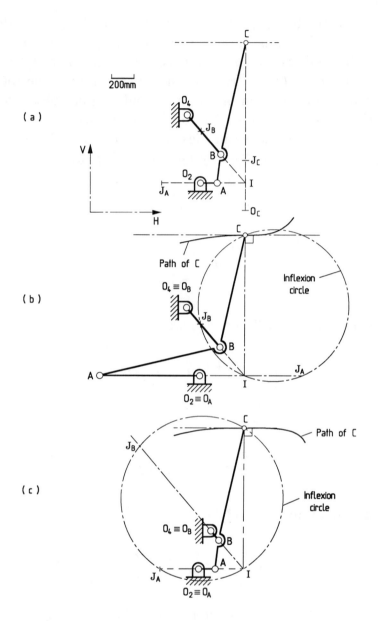

Figure 7.5-3

7.5. APPROXIMATE STRAIGHT-LINE LINKAGES 363

Problem 7.5-2

(i) Figure 7.5-3a shows the kinematic diagram of a part of a textile machinery, where the point C is supposed to carry a shuttle approximately along a horizontal straight line around the given configuration. Determine the instantaneous radius of path-curvature for the point C.

(ii) Modify the given mechanism with a new location for A only, with all other locations given as in (i), such that $\rho_C \to \infty$, i.e., a better approximation of the desired horizontal straight line motion of C is expected.

(iii) Modify the mechanism given in (i) with a new location for O_4 only, with all other locations unchanged, such that again $\rho_C \to \infty$.

(iv) Draw portions of the coupler curves around C, for the designs obtained in (ii), and (iii) and comment on the choice between these two designs.

Solution

1. Following the method of solution for Problem 2.10-1, using the Euler-Savary equation, we get $O_c C = \rho_C = 1480$ mm, as explained in Fig. 7.5-3a.

2. Since the path of C is desired to be horizontal at C, the instantaneous center I should lie on the vertical line through C (Fig. 7.5-3b). Thus, I is located at the intersection of $O_B B$ and the vertical line through C. As $\rho_C \to \infty$, the point C must lie on the inflexion circle. Using Equation (2.10.6) for the point B, we locate J_B. Then the inflexion circle passing through I, J_B and C is obtained. The point J_A is located at the intersection of $O_2 I$ and the inflexion circle. The location of the point A on the line $O_2 I$ is determined by using either Equation (2.10.7) or Equation (2.10.10). The link-length $O_2 A$ is found to be (from measurement) 890 mm.

3. Referring to Fig. 7.5-3c, first I is located at the intersection of O_2A and the vertical line through C. The point J_B is located on O_2A using Equation (2.10.6). The inflexion circle passing through I, J_A and C is drawn. The point J_B is located at the intersection of the line IB and the inflexion circle. Finally, $O_4 (\equiv O_B)$ is determined on the line IB using Equation (2.10.6). The link-length O_4B is found to be (from measurement) 110 mm.

4. In Figs. 7.5-3b and 7.5-3c a portion of the coupler curves of the point C for the corresponding designs are drawn. The design obtained in (iii) is better than that obtained in (ii) for the following two reasons:

    (a) The linkage is more compact.

    (b) The coupler curve around the given location remains approximately straight over a longer length.

We should note that the use of a point on the inflexion circle for generating an approximate straight path may or may not give satisfactory results.

Problem 7.5-3

The sketch of a pair of parallel-jaw pliers[12] (providing a better grip than a pair of simple hinged pliers) is shown in Fig. 7.5-4a. Consider the motion of all the links relative to the lower jaw (numbered as 1). It is desired that the upper jaw 5 always moves parallel to itself, with all the points on it moving approximately along a straight line perpendicular to the jaw-faces. Design the pliers for a jaw-movement of 25 mm.

Solution

Recalling the discussion of Section 7.4 we note that we need to design a parallel motion generator where the point C moves along an approximate straight line perpendicular to the jaw-faces. We solve the problem in two stages. In the first stage, we design a 4R linkage, $O_2A(C)BO_4$, such that C moves approximately along a straight line

---

[12]See Hain, K: "Getriebebeispiel - Atlas", Eine Zusammenstellung ungleichförmig übersetzender Getriebe für den konstrunkteur, VDI-Verlag, Düsseldorf, 1973.

7.5. APPROXIMATE STRAIGHT-LINE LINKAGES

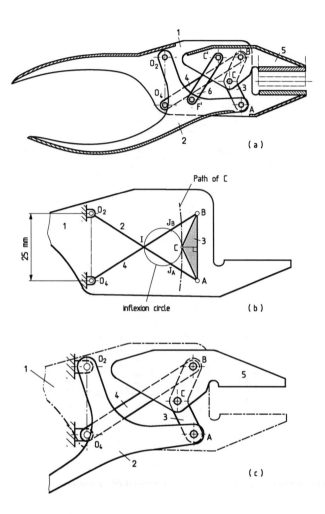

Figure 7.5-4a, b & c

CHAPTER 7. COUPLER CURVES

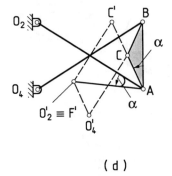

(d)

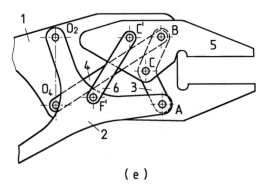

(e)

Figure 7.5-4d & e

parallel to O_2O_4 (which is perpendicular to the jaw-faces). Towards this end, we choose O_2O_4 (= 25 mm) perpendicular to the jaw-faces. A symmetrical 4R linkage with $O_2A = O_4B$ (i.e., $l_2 = l_4$) = 47 mm (assumed arbitrarily) and $O_2O_4 = AB$ is tried (Fig. 7.5-4b). Only a portion of the total range of movement of this (transitional) double-crank linkage will be used. To produce an approximate straight-line path parallel to O_2O_4 and symmetric around this configuration, the coupler point C should be chosen at the point of intersection of the midnormal of AB and the inflexion circle of the coupler at this instant. Thus, the 4R linkage $O_2A(C)BO_4$ can be obtained as explained in Fig. 7.5-4b. The instantaneous center I is located at the intersection of O_2A and O_4B. Using Equation (2.10.6), the points J_A and J_B are located on the lines O_2A and O_4B, respectively. The inflection circle passing through I, J_A and J_B is drawn. The point C is located at the intersection of the midnormal of AB and the inflexion circle. The path of C, shown in Fig. 7.5-4b, is found to be reasonably straight (and parallel to O_2O_4) around the considered configuration for a total travel of 25 mm.

The actual construction of the design obtained so far is shown in Fig. 7.5-4c. Using the theory of developing a parallel motion generator (from a 4R linkage), elaborated in Section 7.4, we obtain the points C' and F', as shown in Fig. 7.5-4d. The method explained in Fig. 7.4-7 has been used. The link CC' (i.e., the upper jaw) moves parallel to itself. Thus, the final design is obtained, which is shown in Fig. 7.5-4e.

7.6 Exact Straight-Line Linkages

Exact straight-line linkages consisting of only R-pairs must involve more than four links. In fact, six- and eight-link mechanisms have been invented in which some point of a link generates an exact straight line when the mechanism moves. The design of these mechanisms is based on a geometric principle known as inversion.[13] If two points P

---

[13]Not to be confused with kinematic inversion (i.e., the process of fixing different links of the same kinematic chain) or geometric inversion (which refers to different modes of assembly of a linkage with given link-lengths).

and Q move in such a fashion that the line PQ always passes through a fixed point O, maintaining the relation $OP \times OQ = $ constant, then the curves traced out by P and Q are called the *inverse* of each other. It can be shown that if P traces out a circle not passing through O, then Q also traces a circle. If the circle traced by P passes through O, then Q moves along a straight line (a circle of infinite radius) perpendicular to OO_P where O_P is the center of the circle on which P moves. Exact straight-line mechanisms designed on this principle are called *inversors*. We shall start with the proof that the inverse of a circle is also a circle.

Referring to Fig. 7.6-1, let O be a fixed point and the points P and Q move in a manner so that $OP \times OQ = $ constant (say λ), with O, P and Q collinear. Further, the path of P is the circle k_P (not passing through O) with its center at O_P and radius r_P. Let OO_P intersect k_P at S and S' and let OP (extended) intersect k_P again at P'. Since the triangles OPS and $OP'S'$ are similar, we get

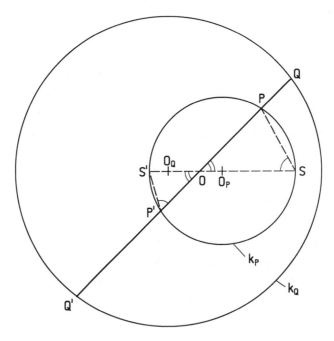

Figure 7.6-1

7.6. EXACT STRAIGHT-LINE LINKAGES

$$\frac{OP}{OS'} = \frac{OS}{OP'}$$

or, $OP \times OP' = OS \times OS' = (OO_P + O_P S)(OO_P - S'O_P)^{14}$

$$= OO_P^2 - r_P^2 = constant. \qquad (7.6.1)$$

It is given that

$$OP \times OQ = \lambda \quad (constant). \qquad (7.6.2)$$

Dividing Equation (7.6.2) by Equation (7.6.1), we obtain

$$\frac{OQ}{OP'} = \frac{\lambda}{OO_P^2 - r_P^2} = constant. \qquad (7.6.3)$$

Thus, Q moves on a curve similar to that of P' (same as that of P), i.e., k_P, with the center of similitude at O. Therefore, Q moves on a circle (say, k_Q in Fig. 7.6-1). Let the line OPQ intersect k_Q again at Q' and the circle k_Q have its center at O_Q and radius r_Q. From similitude we can show that O_Q lies on OO_P and the following relations are satisfied:

$$\frac{OO_Q}{OO_P} = \frac{\lambda}{OO_P^2 - r_P^2}. \qquad (7.6.4)$$

$$\frac{r_Q}{r_P} = \left|\frac{\lambda}{OO_P^2 - r_P^2}\right|. \qquad (7.6.5)$$

We should note that in Fig. 7.6-1, $OO_P < r_P$. Therefore, from Equation (7.6.4) we find that $\frac{OO_Q}{OO_P}$ is negative, i.e., O_Q and O_P lie on opposite sides of O. If k_P passes through O, then $OO_P = r_P$, which from Equation (7.6.5) yields $r_Q \to \infty$. In other words, the path of Q is a straight line (perpendicular to OO_P). This last result (i.e., the straight line path of Q) can also be directly obtained, as shown below.

Referring to Fig. 7.6-2, let k_P pass through O and $OP \times OQ = \lambda$. From the point Q we draw a line QT perpendicular to OS (the

---
[14]Note that the distances are measured in algebraic sense.

diameter of k_P through O). Since the triangles OQT and OPS are similar, we get

$$\frac{OT}{OP} = \frac{OQ}{OS}$$

or,

$$OT = \frac{OP \times OQ}{OS} = \frac{\lambda}{2r_P} = \text{constant}.$$

Since the distance OT is constant (independent of θ), Q moves along the line QT, i.e., along a straight line perpendicular to OO_P.

The Peucellier mechanism was the first inversor designed on the aforementioned principle. This mechanism, consisting of eight links, is constructed as shown in Fig. 7.6-3. In this figure, $APBQ$ is a rhombus (of side s) that is connected to the fixed point O by two equal links OA and OB of length l. It is easily seen from symmetry that when this assembly is deformed, the points O, P and Q always lie on a straight line. Further,

$$OP \times OQ = (OC - PC)(OC + QC)$$

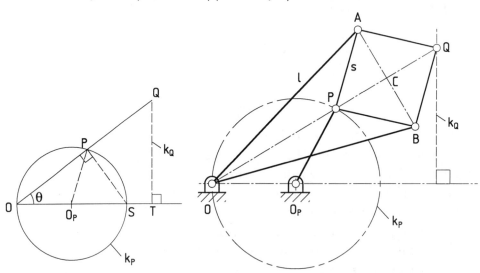

Figure 7.6-2   Figure 7.6-3

7.7. EXERCISE PROBLEMS

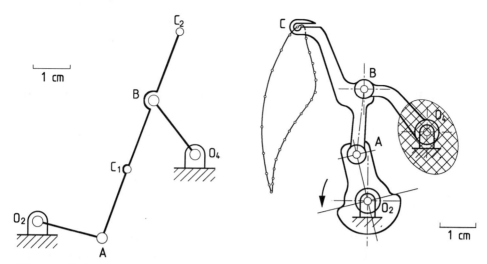

Figure 7.7-1

Figure 7.7-2

$$\begin{aligned}
&= OC^2 - PC^2 \\
&= (OA^2 - AC^2) - (AP^2 - AC^2) \\
&= OA^2 - AP^2
\end{aligned}$$

(since diagonals of a rhombus are perpendicular bisectors of each other).

$$= l^2 - s^2 = \text{constant}. \qquad (7.6.6)$$

Thus, the condition of inversion is satisfied. Next, the link O_2P of length OO_P is added so that k_P becomes a circle passing through O. Hence, Q (in this constrained eight-link mechanism) moves along a straight line perpendicular to OO_P. We may note that by making OO_P different but very close to O_PP, one can generate a circular arc of very large radius by the path of Q.

7.7 Exercise Problems

7.1 Show that for the ten-link mechanism of Fig. 7.2-1, the instantaneous centers 13, 16, 19 and the point C always remain collinear.

7.2 Figure 7.7-1 shows a 4R linkage with two coupler points C_1 and

C_2 lying on the line AB, where $AC_1 = BC_1$ and $AC_2 = 3BC_2$. Obtain the coupler curve cognate 4R linkages for the points C_1 and C_2.

7.3 Figure 7.3-3b shows a slider-crank cognate linkage for the point C of the original slider-crank mechanism $O_2A(C)B$, shown in Fig. 7.3-3a. Obtain the third cognate four-link mechanism for the point C.

7.4 Determine all the six-link cognate linkages for the points C_1 and C_4 of Fig. 7.7-1.

7.5 Figure 7.7-2 shows a 4R linkage designed to be used in a sewing machine. The coupler point C, carrying the thread, must move along a prescribed path at a given rate as the crank O_2A rotates at a constant speed. The mechanism shown in the figure is found to satisfy all the requirements. However, the follower link O_4B is found to foul with some other members (not shown) present in the machine around O_4 in the region shown cross-hatched. Synthesize a geared five-link mechanism satisfying all the requirements and keeping the cross-hatched region free.

7.6 Refer to the problem and solution of Problem 7.2-1. Synthesize a geared five-link mechanism with O_2 and O_4 as the gear-shaft axes so that the point C moves from C_1 to C_2 and C_2 to C_3 as the input link rotates by 20° (CW) and 18 (CW), respectively. Note that this solution gives easier control over the choices of the fixed-pivot locations as compared to that in the 4R linkages obtained in Problem 7.2-1.

7.7 Figure 7.7-3 shows another inversor, known as Hart's straight-line mechanism. This mechanism consists of only six links, as compared to eight for the Peucellier mechanism. In Hart's mechanism $AB = CD$ and $BC = AD$, i.e., $ABCD$ is an antiparallelogram. If the points O, P, Q lie on a line parallel to AC with $O_P P = OO_P$, prove that Q traces an exact straight line perpendicular to OO_P.

7.7. EXERCISE PROBLEMS

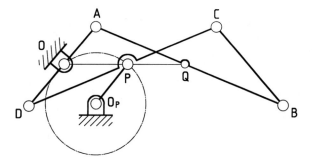

Figure 7.7-3

Chapter 8

SPHERICAL AND SPATIAL LINKAGES

We have already defined spherical and spatial linkages in Section 1.5. A number of such linkages are commonly used for coupling nonaligned shafts. In recent times, spatial linkages have been found to have other applications as well[1].

All the methods of kinematic analysis and synthesis discussed in the previous chapters were in the context of planar linkages, and graphical methods were found to be very useful. However, projection on to a single plane does not reveal the true motion of all the points of a nonplanar linkage. Consequently, graphical methods are not convenient for studying the kinematics of spatial linkages and special analytical tools have been developed towards this end.

The degrees-of-freedom of a free (unconnected) link in a three dimensional space is six, three of which are translational and the other three rotational. A free link undergoing spheric motion has only three rotational degrees-of-freedom. We may recall that a free link undergoing planar movement has also three degrees-of-freedom, two of which are translational and one rotational. In fact, using spherical trigonometry, the counterparts of some of the well known features of planar linkages (e.g., Grashof criterion, Euler-Savary equation, dead center

---

[1]See Sandor, G.N. and Erdman, A.G.: "Advanced Mechanism Design: Analysis and Synthesis," Vol. 2., Prentice Hall., Inc., New Jersey, 1984.

376   CHAPTER 8. SPHERICAL AND SPATIAL LINKAGES

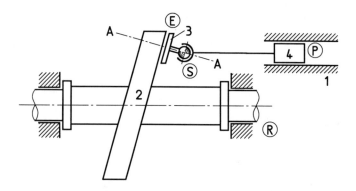

Figure 8.1-1

problems) have been obtained for spherical linkages[2]. The methods used for studying the kinematics of three dimensional mechanisms are also useful for robot manipulators, which are open loop spatial chains. In this chapter, we shall mainly discuss the methods for obtaining the displacement equation of three dimensional linkages. A simple spherical linkage consists of four links, with four revolute pairs having intersecting axes. A spatial linkage, on the otherhand, can have more numbers of links and all six types of lower pairs discussed in Section 1.2. We shall, however, limit our discussions mainly to four-link mechanisms.

8.1 Degrees of Freedom

The degrees-of-freedom of a spatial linkage can be obtained (except in special cases with particular kinematic dimensions) by extending the Kutzbach equation discussed in Section 3.1. The degrees-of-freedom of a rigid body undergoing spatial motion is six, three of which describe translations along three mutually perpendicular axes and the other three describe rotational movements. The number of the degrees-of-freedom curtailed at each kinematic pair depends on

---
[2]For a comprehensive treatment of spherical linkages see Chiang, C.H.: "Kinematics of Spherical Mechanisms", Cambridge University Press, Cambridge, 1988.

8.1. DEGREES OF FREEDOM

the nature of the kinematic pair (see Section 1.2). Thus, with n number of total links (including the frame), the degrees-of-freedom of a spatial linkage are given by

$$F = 6(n-1) - 5(R + P + H) - 4C - 3(S + E) \qquad (8.1.1)$$

where R, P, H, C, S and E are respectively, the numbers of revolute, prismatic, screw, cylindric, spheric and planar pairs. While using Equation (8.1.1), one must be, as usual, careful to check for some redundant degrees-of-freedom. For example, let us consider the four-link spatial linkage shown in Fig. 8.1-1, which is used in a hydraulic pump (or motor) and known as a swash plate drive. From Equation (8.1.1) we get the degrees-of-freedom $F = 2$, since $n = 4$ and $R = S = E = P = 1$. However, the effective degree-of-freedom is one (i.e., the mechanism is a constrained one having unique output with a single input), since three is a redundant degree-of-freedom for link 3. This link can rotate freely about the axis $A - A$ (which is normal to the surface of the planar pair (E) and passes through the center of the spheric pair) without causing any movement of any other link. This redundant degree-of-freedom does not affect the output-input relationship in anyway.

A simple spatial chain (with one fixed link) consisting of only binary links has $F = 1$ if the total degrees-of-freedom of all the kinematic pairs is seven. The proof of this statement is self evident if we recall that six out of these seven degrees-of-freedom are curtailed by the fixed link, which could have six degrees-of-freedom if left free. This statement is referred to in this text as *the rule of seven* (see Problem 8.8-1). However, this rule of seven or Equation (8.1.1) has exceptions for special kinematic dimensions[3]. One such exception is the Bennett's (4R) linkage, shown in Fig. 1.5-3. The special kinematic dimensions that make $F = 1$ for this linkage are discussed in Problem 8.3-1. Another example of an exception to Equation (8.1.1)

---

[3] For a systematic development of all possible constrained spatial linkages consisting of only binary links and some examples of exceptions to Equation (8.1.1) see VDI - Richtlinien - VDI 2156 "Einfache Raumliche Kurbelgetriebe - Sysmatik und Begriffsbestimmungen," Dusseldorf, VDI - Verlag, 1975.

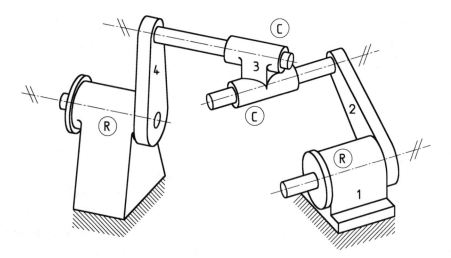

Figure 8.1-2

(or the rule of seven, for a constrained simple chain) is shown as an R-C-C-R linkage in Fig. 8.1-2. This linkage has $F = 1$ if the axes of each adjacent R-C pairs are parallel (consequently, the angle between the axes of the R pairs is same as that between the axes of the C pairs), as indicated.

8.2 Displacement Equation (Analytic Geometry)

As discussed in Chapter 4, the displacement analysis of (spatial) linkages involves the solution to the following problem: "Given the kinematic dimensions and input movement(s) of a (spatial) linkage, determine the movements of all other links."

Before going into the details of the generalized matrix method (analogous to loop-closure, complex exponential equations used for planar linkages) of displacement analysis, we shall discuss how the output-input relationship (i.e., only the displacement equation) of some simple spatial linkages can be obtained through analytic geometry. This approach is similar to Freudenstein's approach for obtaining displacement equations (e.g., Equation (6.6.4)) of planar linkages.

8.2. DISPLACEMENT EQUATION (ANALYTIC GEOMETRY)

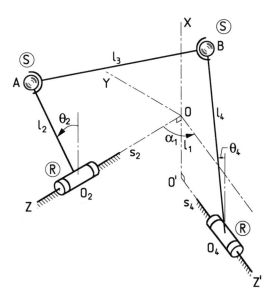

Figure 8.2-1

Let us consider an R-S-S-R linkage in which the fixed link has R pairs at both its ends, as shown in Fig. 8.2-1. The effective degree-of-freedom of this linkage is one (discounting the redundant degree-of-freedom of rotation of link 3 about AB). Our objective is to correlate the rotations of the output and input links (i.e., those connected to the fixed link) in terms of the link parameters. Towards this end, first we identify the link parameters after choosing a coordinate system for measurement of the link rotations. As usual, the input link is numbered as 2, with the output link numbered as 4 and the coupler as 3.

The revolute axis between links 1 and 2 is defined as the Z axis, with its positive direction chosen arbitrarily. Similarly, the revolute axis between links 1 and 4 is defined as the Z' axis, with its positive direction chosen arbitrarily. The common perpendicular to the Z and Z' axes, $O'O$, oriented always from Z' to Z, is defined as the X axis. A right-handed orthogonal coordinate system is completed with XYZ as its axes. The centers of the spherical joints are indicated as A and B. The lines AO_2 and BO_4 are perpendiculars from A to the Z axis and from B to the Z' axis, respectively. Hence, the following link parameters completely define the linkage:

(i) The link lengths $l_1 = O'O$, $l_2 = O_2A$, $l_3 = AB$ and $l_4 = O_4B$, and

(ii) The offsets $s_2 = OO_2$ and $s_4 = O'O_4$, and

(iii) The skew-angle α_1, through which the Z axis must rotate about the X axis[4] in order to be parallel to the Z' axis. The positions of the input and output links are expressed by the angles θ_2 and θ_4, respectively. The angles θ_2 and θ_4 indicate, respectively, the rotations of O_2A about the Z axis and of O_4B about the Z' axis. Both of these angles are measured from lines parallel to X axis, as indicated in the figure. The angles θ_2 and θ_4 can be related through the seven link parameters as explained below.

The coordinates of the points A and B in the XYZ system are given by

$$\begin{aligned} x_A &= l_2 \cos\theta_2, & y_A &= l_2 \sin\theta_2, & z_A &= s_2, \\ x_B &= l_4 \cos\theta_4 - l_1, & y_B &= l_4 \sin\theta_4 \cos\alpha_1 - s_4 \sin\alpha_1 \\ z_B &= s_4 \cos\alpha_1 + l_4 \sin\theta_4 \sin\alpha_1. \end{aligned} \quad (8.2.1)$$

The displacement equation, correlating θ_4 and θ_2, is obtained from

$$l_3^2 = (x_A - x_B)^2 + (y_A - y_B)^2 + (z_A - z_B)^2.$$

Using Equation (8.2.1) in the above equation and simplifying we get

$$\begin{aligned} 2l_1l_2C\theta_2 &- 2l_1l_4C\theta_4 + 2l_2s_4S\alpha_1S\theta_2 - 2l_2l_4 \\ &\quad (C\theta_2 C\theta_4 + S\theta_2 S\theta_4 C\alpha_1) \\ &- 2s_2l_4S\alpha_1S\theta_4 + (l_1^2 + l_2^2 - l_3^2 + l_4^2 \\ &+ s_2^2 + s_4^2 - 2s_2s_4C\alpha_1) = 0 \end{aligned} \quad (8.2.2)$$

where $S \equiv \sin$ and $C \equiv \cos$.

Equation (8.2.2) is the displacement equation of the R-S-S-R linkage. We shall now consider some limiting cases of Equation (8.2.2)

---

[4]The right-hand screw convention for rotation is followed consistently.

8.2. DISPLACEMENT EQUATION (ANALYTIC GEOMETRY)

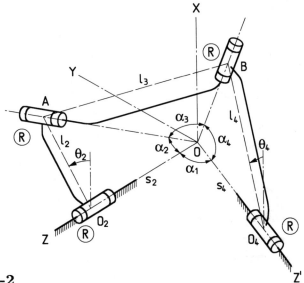

Figure 8.2-2

and obtain the displacement equations of several planar, spherical and spatial linkages.

For example, if $s_2 = s_4 = 0$ and $\alpha_1 = 0$, then the R-S-S-R linkage is effectively converted to a planar 4R linkage moving in the XY plane, when Equation (8.2.2) is reduced to

$$2l_1l_2C\theta_2 - 2l_1l_4C\theta_4 - 2l_2l_4C(\theta_2 - \theta_4)$$
$$+ l_1^2 + l_2^2 - l_3^2 + l_4^2 = 0 \qquad (8.2.3)$$

which is identical to Equation (6.6.4). We may note that with $\alpha_1 = 0$, O_2A and O_4B move in parallel planes. Consequently, the spheric pairs at A and B effectively work as revolute pairs with their axes parallel to those at O_2 and O_4.

Referring to Fig. 8.2-2 for a spherical 4R linkage, and comparing it with Fig. 8.2-1, it is easy to see that the displacement equation of a 4R spherical linkage can be obtained from Equation (8.2.2) by substituting $l_1 = O$. Furthermore, the link parameters of a 4R spherical linkage (Fig. 8.2-3) are the angles between the successive joint axes ($\alpha_1, \alpha_2, \alpha_3$ and α_4). Therefore, to obtain the displacement equation

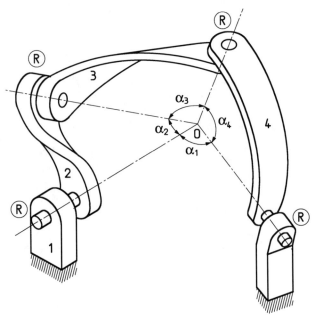

Figure 8.2-3

in terms of link parameters, we make the following substitutions in Equation (8.2.2) (see Fig. 8.2-2):

$$l_2/s_2 = \tan\alpha_2, \quad l_4/s_4 = \tan\alpha_4$$

and

$$\begin{aligned} l_3^2 &= OA^2 + OB^2 - 2OA.OB\cos\alpha_3 \quad \text{(consider the } \triangle OAB) \\ &= s_2^2 + l_2^2 + s_4^2 + l_4^2 - 2\sqrt{s_2^2 + l_2^2}\sqrt{s_4^2 + l_4^2}\cos\alpha_3 \end{aligned} \quad (8.2.4)$$

along with $l_1 = 0$.

Substituting Equation (8.2.4) in Equation (8.2.2) and simplifying we get the displacement equation of a 4R spherical linkage as

$$T\alpha_2 T\alpha_4 C\theta_2 C\theta_4 (T\alpha_4 S\alpha_1 + T\alpha_2 T\alpha_4 C\alpha_1 S\theta_2) S\theta_4 \\ (C\alpha_1 - S\alpha_1 T\alpha_2 S\theta_2 - \frac{C\alpha_3}{C\alpha_2 C\alpha_4}) = 0 \quad (8.2.5)$$

where $T \equiv \tan$.

8.2. DISPLACEMENT EQUATION (ANALYTIC GEOMETRY)

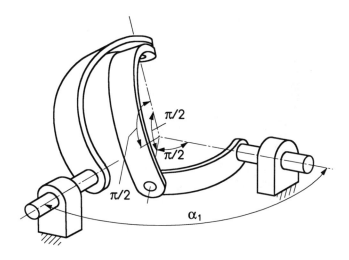

Figure 8.2-4

If we consider a Hooke's joint (Fig. 8.2-4), which is a 4R spherical linkage with $\alpha_2 = \alpha_3 = \alpha_4 = \pi/2$, then its displacement equation is obtained from Equation (8.2.5), after dividing throughout by $T\alpha_2 T\alpha_4$ and then substituting $\alpha_2 = \alpha_3 = \alpha_4 = \pi/2$, as

$$\tan\theta_2 \tan\theta_4 = -\sec\alpha_1. \tag{8.2.6}$$

Let us now consider an R-S-S-P linkage (Fig. 8.2-5a) where the axes of the R and P pairs have an offset l_1 and a skew angle α_1. It is effectively a spatial slider-crank mechanism. We may note that this linkage has a redundant degree-of-freedom for link 3. The link parameters are l_1, l_2, l_3, s_2 and α_1, and our objective is to correlate the output movement, $s_4 \,(= O'B)$, with the input movement, θ_2, in terms of these link parameters. The displacement equation (s_4 vs θ_2) can be easily obtained from Equation (8.2.2), after substituting $l_4 = 0$, as

$$\begin{aligned} s_4^2 &+ 2(l_2 S\alpha_1 S\theta_2 - s_2 C\alpha_1)s_4 \\ &+ (l_1^2 + l_2^2 + s_2^2 - l_3^2 + 2l_1 l_2 C\theta_2) = 0. \end{aligned} \tag{8.2.7}$$

As expected, Equation (8.2.7) is quadratic in s_4, implying two

384 CHAPTER 8. SPHERICAL AND SPATIAL LINKAGES

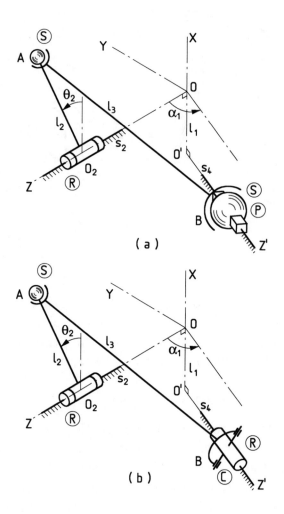

Figure 8.2-5

8.2. DISPLACEMENT EQUATION (ANALYTIC GEOMETRY)

possible modes of assembly for a given value of θ_2 with prescribed link parameters.

It may be pointed out that for the R-S-(RC) linkage, shown in Fig. 8.2-5b, where the axes of the R and C pairs of (RC) are intersecting, the displacement equation (s_4 vs θ_2) is given by Equation (8.2.7). We may note that in Fig. 8.2-5b, unlike in Fig. 8.2-5a, there is no redundant degree-of-freedom. Of course, we have no information as yet, so far as the rotational output at the cylindric pair is concerned.

If the axes of input rotation and output translation in the mechanisms shown in Figs. 8.2-5a and 8.2-5b are intersecting, i.e., $l_1 = 0$, then the displacement equation for such mechanisms is obtained from Equation (8.2.7) as

$$s_4^2 + 2(l_2 S\alpha_1 S\theta_2 - s_2 C\alpha_1)s_4 + (l_2^2 + s_2^2 - l_3^2) = 0. \qquad (8.2.8)$$

From Equation (8.2.8), we may note that if the link parameters are so chosen that $l_2^2 + s_2^2 = l_3^2$, then the displacement equation reduces to

$$s_4^2 + 2(l_2 S\alpha_1 S\theta_2 - s_2 C\alpha_1)s_4 = 0$$

i.e.,

$$\text{either} \quad s_4 = 0, \qquad (8.2.9)$$
$$\text{or} \quad s_4 = 2(s_2 C\alpha_1 - l_2 S\alpha_1 S\theta_2). \qquad (8.2.10)$$

The above two equations imply that there exist two possibilities, viz., either the slider remains stationary at the intersection of the Z and Z' axes, with the input link rotating and link 3 generating a conical surface, or the output translation is a harmonic function of the input rotation, with an amplitude $2l_2 S\alpha_1$.[5]

---

[5] For further details, see Hunt, K.H.: "Kinematic Geometry of Mechanisms", Oxford University Press, Oxford, 1978.

386   CHAPTER 8. SPHERICAL AND SPATIAL LINKAGES

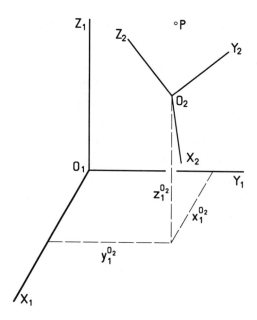

Figure 8.3-1

8.3 Matrix Method

In Section 8.2 we saw that for some simple, (four-link) spatial linkages, the displacement equation can be obtained by using analytic geometry. No information regarding the movement(s) of the intermediate link(s) can be obtained by this method. For complete displacement analysis of spatial linkages, the most commonly used method is known as *Denavit-Hartenberg's matrix method*. In this method, a Cartesian coordinate system is attached to every link of the mechanism, following certain conventions. The coordinates of a point (fixed in space) expressed in two such link-coordinate systems are related through a 4 x 4 matrix, known as a *homogeneous transformation matrix*. The displacement analysis is carried out by following the loop closure equation through these transformation matrices. The elements of the transformation matrices are functions of link parameters and motion variables, as will be seen in the sections to follow.

8.3.1 Coordinate Transformation

Referring to Fig. 8.3-1, let $(XYZ)_1$ and $(XYZ)_2$ be two Cartesian coordinate systems with their origins at O_1 and O_2, respectively. The coordinates of a point P in these two systems are (x_1^P, y_1^P, z_1^P) and (x_2^P, y_2^P, z_2^P), respectively, and let the coordinates of origin O_2 in $(XYZ)_1$ be $(x_1^{O_2}, y_1^{O_2}, z_1^{O_2})$. The coordinates of the point P, expressed in two systems are related as follows:

$$x_1^P = x_1^{O_2} + x_2^P \cos(X_2, X_1) + y_2^P \cos(Y_2, X_1) + z_2^P \cos(Z_2, X_1)$$
$$y_1^P = y_1^{O_2} + x_2^P \cos(X_2, Y_1) + y_2^P \cos(Y_2, Y_1) + z_2^P \cos(Z_2, Y_1)$$
$$z_1^P = z_1^{O_2} + x_2^P \cos(X_2, Z_1) + y_2^P \cos(Y_2, Z_1) + z_2^P \cos(Z_2, Z_1)$$

where $\cos(,)$ are the direction cosines of the second set of axes, e.g., $\cos(X_2, X_1)$ is the projection along X_1 of a unit vector along X_2, and so on. Adding an identity $(1 = 1)$ to the above set of three equations, we can correlate the coordinates of a point, expressed in two systems, through the following matrix equation:

$$\begin{Bmatrix} x_1 \\ y_1 \\ z_1 \\ 1 \end{Bmatrix} = \begin{pmatrix} C(X_2, X_1) & C(Y_2, X_1) & C(Z_2, X_1) & x_1^{O_2} \\ C(X_2, Y_1) & C(Y_2, Y_1) & C(Z_2, Y_1) & y_1^{O_2} \\ C(X_2, Z_1) & C(Y_2, Z_1) & C(Z_2, Z_1) & z_1^{O_2} \\ 0 & 0 & 0 & 1 \end{pmatrix} \begin{Bmatrix} x_2 \\ y_2 \\ z_2 \\ 1 \end{Bmatrix}$$

$$= [A_1] \begin{Bmatrix} x_2 \\ y_2 \\ z_2 \\ 1 \end{Bmatrix} \qquad (8.3.1)$$

The 4×4 matrix

$$[A_1] = \begin{pmatrix} C(X_2, X_1) & C(Y_2, X_1) & C(Z_2, X_1) & x_1^{O_2} \\ C(X_2, Y_1) & C(Y_2, Y_1) & C(Z_2, Y_1) & y_1^{O_2} \\ C(X_2, Z_1) & C(Y_2, Z_1) & C(Z_2, Z_1) & z_1^{O_2} \\ 0 & 0 & 0 & 1 \end{pmatrix} \quad (8.3.2)$$

is called the homogeneous transformation matrix (also commonly known as A matrix). The homogeneous transformation matrix has 16 elements, out of which only six are independent. Three independent elements are $x_1^{O_2}$, $y_1^{O_2}$ and $z_1^{O_2}$. Since X_2, Y_2 and Z_2 are orthogonal and the squares of the direction cosines of a line must add up to 1, only three of the nine direction cosine elements can be independent. The inverse of $[A_1]$ matrix, which obviously correlates the two sets of coordinates in the manner

$$\begin{Bmatrix} x_2 \\ y_2 \\ z_2 \\ 1 \end{Bmatrix} = [A_1]^{-1} \begin{Bmatrix} x_1 \\ y_1 \\ z_1 \\ 1 \end{Bmatrix} \quad (8.3.3)$$

is given (in terms of elements of $[A_1]$) by

$$[A_1]^{-1} = \begin{pmatrix} C(X_2, X_1) & C(X_2, Y_1) & C(X_2, Z_1) & -[arr1] \\ C(Y_2, X_1) & C(Y_2, Y_1) & C(Y_2, Z_1) & -[arr2] \\ C(Z_2, X_1) & C(Z_2, Y_1) & C(Z_2, Z_1) & -[arr3] \\ 0 & 0 & 0 & 1 \end{pmatrix} \quad (8.3.4)$$

with

$$\begin{aligned} arr1 &= x_1^{O_2} C(X_2, X_1) + y_1^{O_2} C(X_2, Y_1) + z_1^{O_2} C(X_2, Z_1) \\ arr2 &= x_1^{O_2} C(Y_2, X_1) + y_1^{O_2} C(Y_2, Y_1) + z_1^{O_2} C(Y_2, Z_1) \\ arr3 &= x_1^{O_2} C(Z_2, X_1) + y_1^{O_2} C(Z_2, Y_1) + z_1^{O_2} C(Z_2, Z_1) \end{aligned}$$

8.3. MATRIX METHOD

Since the direction cosines are the components of a unit vector, we can write Equation (8.3.2) in the following form:

$$[A_1] = \begin{pmatrix} e_{1x} & e_{2x} & e_{3x} & (O_1O_2)_x \\ e_{1y} & e_{2y} & e_{3y} & (O_1O_2)_y \\ e_{1z} & e_{2z} & e_{3z} & (O_1O_2)_z \\ 0 & 0 & 0 & 1 \end{pmatrix} \qquad (8.3.5)$$

where e_1, e_2 and e_3 are the unit vectors along X_2, Y_2 and Z_2, respectively and O_1O_2 is the position vector of O_2 in the first coordinate system, with subscripts x, y, z referring to the components along X_1, Y_1 and Z_1, respectively. Now, Equation (8.3.4) can be rewritten as

$$[A_1]^{-1} = \begin{pmatrix} e_{1x} & e_{1y} & e_{1z} & -(O_1O_2 \cdot e_1) \\ e_{2x} & e_{2y} & e_{2z} & -(O_1O_2 \cdot e_2) \\ e_{3x} & e_{3y} & e_{3z} & -(O_1O_2 \cdot e_3) \\ 0 & 0 & 0 & 1 \end{pmatrix} \qquad (8.3.6)$$

From Equations (8.3.5) and (8.3.6), we note that the elements of the first three rows and columns of $[A_1]^{-1}$ are obtained by transposing (interchanging row and column) the corresponding part of $[A_1]$.

8.3.2 Link Coordinate System

As already stated, to use the matrix method for displacement analysis of a spatial linkage, we need to attach a coordinate system to each link. These coordinate systems are established in a systematic manner, as explained below.

Let us consider a cylindric pair between two successive links (Fig. 8.3.2). The pair is numbered as 1 when the links connected by it are numbered as 1 and 2. For an n-link mechanism (closed chain, i.e., link n is again connected to link 1 if the linkage consists of a simple chain), assuming a simple chain, we shall number the pair between links n

390   CHAPTER 8. SPHERICAL AND SPATIAL LINKAGES

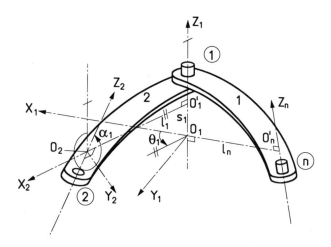

Figure 8.3-2

and 1 as n. The axis of each pair is taken as the Z axis (Fig. 8.3-2), with its positive direction chosen arbitrarily. The axis X_1 is along the common perpendicular between Z_n and Z_1 and is directed from Z_n towards Z_1. Similarly, X_2 is along the common perpendicular between Z_1 and Z_2 and is directed from Z_1 towards Z_2. At the intersection of the Z_i and X_i axes we establish the origins O_i and the Y_i axes to complete right-handed orthogonal coordinate systems ($i = 1, 2, ..n$).

With the coordinate systems so established, the link parameters and the pair variables (link movements) are defined as follows:

(i) Link-length l_1 is the common perpendicular distance from Z_1 to Z_2, measured along X_2 (O'_1O_2 in Fig. 8.3-2) and hence is always positive.

(ii) Link-twist α_i is the angle through which Z_1 should rotate about X_2 in order to be parallel to Z_2.

(iii) Pair-variable θ_1 (the rotational movement at pair 1, i.e., the rotation of link 2 with respect to link 1 about Z_1 axis) is the angle through which X_1 should rotate about Z_1 in order to be parallel to X_2.

8.3. MATRIX METHOD

(iv) Pair-variable s_1 (the translational movement at pair 1, i.e., the translation of link 2 with respect to link 1 along Z_1 axis) is the distance from X_1 to X_2, measured along Z_1 ($O_1 O_1'$ in Fig. 8.3-2). This can be both positive or negative. In Fig. 8.3-2 it is positive.

The reader may note that according to the numbering system followed in this convention, the length of the fixed link comes out to be l_n (rather than l_1 as used in planar linkages).

Though the link-coordinate system, link parameters and pair variables are discussed above with reference to a cylindric pair, we can easily extend these for any kinematic pair with a single degree-of-freedom. For example, for a revolute pair, the parameter s_1 remains constant and ceases to be a pair-variable, while everything else remains the same are for a cylindric pair. Similarly, for a prismatic pair the only pair-variable is s_1, while the parameter θ_1 remains constant. It may be noted that for a P-pair, only the direction of the Z axis is uniquely determined. It can be conveniently located anywhere. For a screw pair, the pair-variables θ_1 and s_1 are related through the lead (L) of the screw pair in the form $\Delta\theta_1/2\pi = \Delta s_1/L$. Since all other lower pairs can be thought of as superpositions of R and P pairs (see Section 1.2), the same methodology can be used to handle all kinds of lower pairs.

Problem 8.3-1

Figure 1.5-3 shows a 4R spatial linkage, known as Bennett's linkage, which, surprisingly, is a constrained mechanism. The constrained movement is resulted from the following special kinematic dimensions:

$$l_1 = l_3, \quad \alpha_1 = \alpha_3, \quad l_2 = l_4, \alpha_2 = \alpha_4,$$

$$s_1 = s_2 = s_3 = s_4 = 0 \quad \text{and} \quad \frac{l_1}{\sin \alpha_1} = \pm \frac{l_2}{\sin \alpha_2}$$

Figure 8.3-3a shows the locations of the four revolute pairs of a Bennett's linkage. Establish all the X and Z axes. Also identify the twist angles and the pair-variables. Note that all the successive X axes are intersecting to yield $s_1 = s_2 = s_3 = s_4 = 0$.

392   CHAPTER 8. SPHERICAL AND SPATIAL LINKAGES

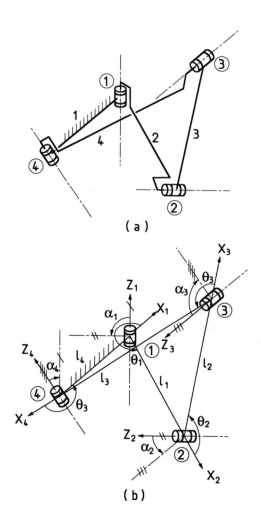

Figure 8.3-3

8.3. MATRIX METHOD

Solution

Figure 8.3-3b shows the coordinate axes X_1-X_4 and Z_1-Z_4 where the positive directions of all the Z axes are chosen arbitrarily. The twist angles (α's) and pair variables (θ's) are indicated according to the convention explained in Fig. 8.3-2.

8.3.3 Derivation of Homogeneous Transformation Matrix

With the link-coordinate system established in a manner explained in the previous section, we are now in a position to obtain the homogeneous transformation matrix $[A_i]$, which correlates the coordinates of a point expressed in $(XYZ)_i$ and $(XYZ)_{i+1}$ according to the following relation:

$$\begin{Bmatrix} x \\ y \\ z \\ 1 \end{Bmatrix}_i = [A_i] \begin{Bmatrix} x \\ y \\ z \\ 1 \end{Bmatrix}_{i+1} . \quad (8.3.7)$$

Referring to Fig. 8.3-2, we derive the matrix $[A_1]$ using Equation (8.3.2) as detailed below. The coordinates of O_2 in $(XYZ)_1$ are seen to be

$$x_1^{O_2} = l_1 \cos\theta_1, \quad y_1^{O_2} = l_1 \sin\theta_1 \quad \text{and} \quad z_1^{O_2} = s_1. \quad (8.3.8)$$

To obtain the direction cosines, first we consider a unit vector along X_2. The components of this vector along the axes of $(XYZ)_1$ are easily obtained as

$$C(X_2, X_1) = \cos\theta_1, \quad C(X_2, Y_1) = \sin\theta_1 \quad \text{and} \quad C(X_2, Z_1) = 0. \quad (8.3.9)$$

Considering a unit vector along Y_2, first we take its two orthogonal projections along Z_1 and in the $X_1 - Y_1$ plane (Fig. 8.3-4a), which are given, respectively, by $\cos(2\pi - \alpha_1 + \pi/2) = \sin\alpha_1$ and $\cos(2\pi - \alpha_1) = \cos\alpha_1$. The components of $\cos\alpha_1$ along X_1 and Y_1 are given, respectively, by $-\cos\alpha_1 \sin\theta_1$ and $\cos\alpha_1 \cos\theta_1$. Thus, we get

CHAPTER 8. SPHERICAL AND SPATIAL LINKAGES

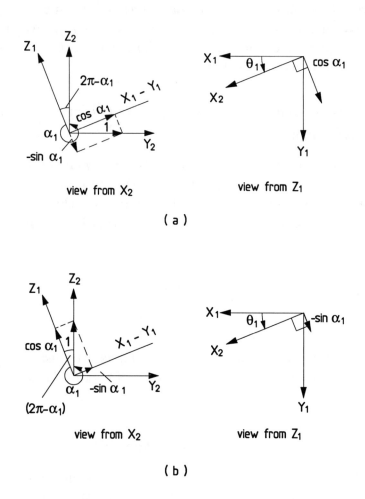

Figure 8.3-4

8.3. MATRIX METHOD

$$C(Y_2, X_1) = -\cos\alpha\sin\theta_1,$$
$$C(Y_2, Y_1) = \cos\alpha_1\cos\theta_1 \text{ and } C(Y_2, Z_1) = \sin\alpha_1. \quad (8.3.10)$$

Finally, we consider a unit vector along Z_2 and again take its two orthogonal projections along Z_1 and in the X_1–Y_1 plane (Fig. 8.3-4b), which are given, respectively, by $\cos(2\pi - \alpha_1) = \cos\alpha_1$ and $\sin(2\pi - \alpha_1) = -\sin\alpha_1$. The components of $-\sin\alpha_1$ along X_1 and Y_1 are given, respectively, by $\sin\alpha_1\sin\theta_1$ and $-\sin\alpha_1\cos\theta_1$. Thus, we get

$$C(Z_2, X_1) = \sin\alpha_1\sin\theta_1,$$
$$C(Z_2, Y_1) = -\sin\alpha_1\cos\theta_1, \text{ and } C(Z_2, Z_1) = \cos\alpha_1. \quad (8.3.11)$$

Using Equations (8.3.8) to (8.3.11) in Equation (8.3.2), we obtain

$$[A_1] = \begin{pmatrix} \cos\theta_1 & -\cos\alpha_1\sin\alpha_1 & \sin\alpha_1\sin\theta_1 & l_1\cos\theta_1 \\ \sin\theta_1 & -\cos\alpha_1\cos\theta_1 & -\sin\alpha_1\cos\theta_1 & l_1\sin\theta_1 \\ 0 & \sin\alpha_1 & \cos\alpha_1 & s_1 \\ 0 & 0 & 0 & 1 \end{pmatrix} \quad (8.3.12)$$

Thus, in general for the ith pair we can write

$$[A_i] = \begin{pmatrix} \cos\theta_i & -\cos\alpha_i\sin\alpha_i & \sin\alpha_i\sin\theta_i & l_i\cos\theta_i \\ \sin\theta_i & -\cos\alpha_i\cos\theta_i & -\sin\alpha_i\cos\theta_i & l_i\sin\theta_i \\ 0 & \sin\alpha_i & \cos\alpha_i & s_i \\ 0 & 0 & 0 & 1 \end{pmatrix} \quad (8.3.13)$$

Another way of deriving $[A_i]$ matrix is through the use of translation and rotation operators. This method may be especially convenient for pairs with more than two degrees-of-freedom, since such pairs, as already stated, permit a series of translation and rotation between two links. In what follows we shall discuss this method of deriving $[A_i]$.

Referring to Fig. 8.3-5, let a coordinate system $(XYZ)_1$ move to $(XYZ)_2$ through translations (without any rotation) ΔX, ΔY and

CHAPTER 8. SPHERICAL AND SPATIAL LINKAGES

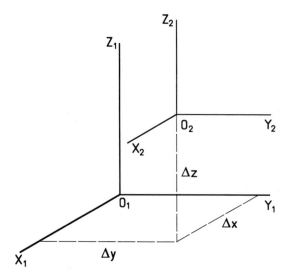

Figure 8.3-5

ΔZ along X_1, Y_1 and Z_1 axes, respectively. The coordinates of any point in these two coordinate systems are related as

$$x_1 = x_2 + \Delta x, \quad y_1 = y_2 + \Delta y \quad \text{and} \quad z_1 = z_2 + \Delta z.$$

Including the identity $1 = 1$, the two sets of coordinates can be related in matrix notation as

$$\begin{Bmatrix} x \\ y \\ z \\ 1 \end{Bmatrix}_1 = \begin{pmatrix} 1 & 0 & 0 & \Delta x \\ 0 & 1 & 0 & \Delta y \\ 0 & 0 & 1 & \Delta z \\ 0 & 0 & 0 & 1 \end{pmatrix} \begin{Bmatrix} x \\ y \\ z \\ 1 \end{Bmatrix}_2$$

$$= [T(\Delta x, \Delta y, \Delta z)] \begin{Bmatrix} x \\ y \\ z \\ 1 \end{Bmatrix}_2 \qquad (8.3.14)$$

where $[T(\Delta x, \Delta y, \Delta z)]$ is the translation operator defined as

8.3. MATRIX METHOD

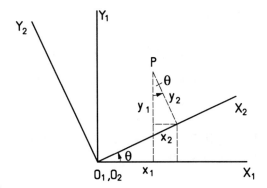

Figure 8.3-6

$$[T(\Delta x, \Delta y, \Delta z)] = \begin{pmatrix} 1 & 0 & 0 & \Delta x \\ 0 & 1 & 0 & \Delta y \\ 0 & 0 & 1 & \Delta z \\ 0 & 0 & 0 & 1 \end{pmatrix} \qquad (8.3.15)$$

Next, referring to Fig. 8.3-6, let a coordinate system $(XYZ)_1$ move to $(XYZ)_2$ by rotating about Z_1 axes through an angle θ, when X_1 goes to X_2, Y_1 to Y_2 and Z_2 remains same as Z_1. The coordinates of any point (P) expressed in these two coordinate systems are related as (Fig. 8.3-6).

$$\begin{aligned} x_1 &= x_2 \cos\theta - y_2 \sin\theta, \\ y_1 &= x_2 \sin\theta + y_2 \cos\theta \\ \text{and} \quad z_1 &= z_2. \end{aligned}$$

As usual, the above relationships are written in matrix notation as

$$\begin{Bmatrix} x \\ y \\ z \\ 1 \end{Bmatrix}_1 = \begin{pmatrix} \cos\theta & -\sin\theta & 0 & 0 \\ \sin\theta & \cos\theta & 0 & 0 \\ 0 & 0 & 1 & 0 \\ 0 & 0 & 0 & 1 \end{pmatrix} \begin{Bmatrix} x \\ y \\ z \\ 1 \end{Bmatrix}_2$$

$$= [R(\theta, z)] \begin{Bmatrix} x \\ y \\ z \\ 1 \end{Bmatrix}_2 \qquad (8.3.16)$$

where $[R(\theta, z)]$ is a rotation operator defined as

$$[R(\theta, z)] = \begin{pmatrix} \cos\theta & -\sin\theta & 0 & 0 \\ \sin\theta & \cos\theta & 0 & 0 \\ 0 & 0 & 1 & 0 \\ 0 & 0 & 0 & 1 \end{pmatrix} \quad (8.3.17)$$

We can define rotation operators signifying rotations about Y_1 and X_1 axes in an identical manner and write these as

$$[R(\theta, y)] = \begin{pmatrix} \cos\theta & 0 & \sin\theta & 0 \\ 0 & 1 & 0 & 0 \\ -\sin\theta & 0 & \cos\theta & 0 \\ 0 & 0 & 0 & 1 \end{pmatrix} \quad (8.3.18)$$

and

$$[R(\theta, x)] = \begin{pmatrix} 1 & 0 & 0 & 0 \\ 0 & \cos\theta & -\sin\theta & 0 \\ 0 & \sin\theta & \cos\theta & 0 \\ 0 & 0 & 0 & 1 \end{pmatrix} \quad (8.3.19)$$

We should note that the movements signified by these operators are expressed in coordinate system 1(old), and the coordinates in this old system (1) are obtained by premultiplying those in the new system (2) by these operators. Furthermore, the rotation operators do not commute (their ordering is important), i.e., $[R_i][R_j] \neq [R_j][R_i]$ where $[R_i]$ and $[R_j]$ are two different rotation operators.

Let us now derive the $[A_1]$ matrix given by Equation (8.3.12) through these $[T]$ and $[R]$ operators. Towards this goal, we consider the $(XYZ)_1$ and $(XYZ)_2$ coordinate systems, shown in Fig. 8.3-2 and see that the following sequence of movements takes $(XYZ)_1$ to $(XYZ)_2$ (Fig. 8.3-7):

(i) $T\,(l_1 \cos\theta_1,\, l_1 \sin\theta_1,\, s_1)$ takes $(XYZ)_1$ to $(XYZ)'_1$

(ii) $R\,(\theta_1, z')$ takes $(XYZ)'_1$ to $(XYZ)''_1$

(iii) $R\,(\alpha_1, x'')$ takes $(XYZ)''_1$ to $(XYZ)_2$

8.3. MATRIX METHOD

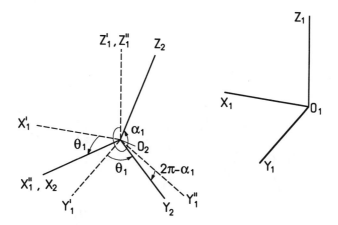

Figure 8.3-7

From Equations (8.3.15), (8.3.17) and (8.3.19) we get

$$[T(l_1 \cos\theta_1, l_1 \sin\theta_1, s_1)] = \begin{pmatrix} 1 & 0 & 0 & l_1 \cos\theta_1 \\ 0 & 1 & 0 & l_1 \sin\theta_1 \\ 0 & 0 & 1 & s_1 \\ 0 & 0 & 0 & 1 \end{pmatrix} \quad (8.3.20)$$

$$[R(\theta_1, z')] = \begin{pmatrix} \cos\theta_1 & -\sin\theta_1 & 0 & 0 \\ \sin\theta_1 & \cos\theta_1 & 0 & 0 \\ 0 & 0 & 1 & 0 \\ 0 & 0 & 0 & 1 \end{pmatrix} \quad (8.3.21)$$

$$[R(\alpha_1, x'')] = \begin{pmatrix} 1 & 0 & 0 & 0 \\ 0 & \cos\alpha_1 & -\sin\alpha_1 & 0 \\ 0 & \sin\alpha_1 & \cos\alpha_1 & 0 \\ 0 & 0 & 0 & 1 \end{pmatrix} \quad (8.3.22)$$

The reader can easily verify that $[A_1]$, given by Equation (8.3.12), is obtained from Equations (8.3.20) to (8.3.22) from the relation

$$[A_1] = [T(l_1 \cos\theta_1, l_1 \sin\theta_1, s_1)][R(\theta_1, z')][R(\alpha_1, x'')]. \quad (8.3.23)$$

CHAPTER 8. SPHERICAL AND SPATIAL LINKAGES

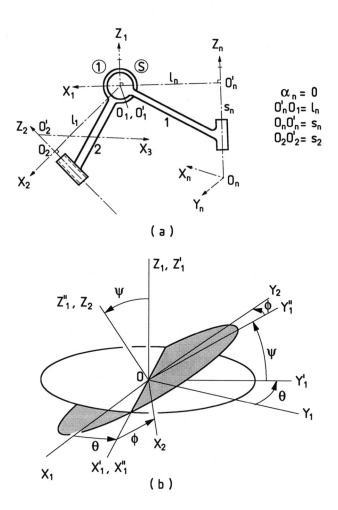

Figure 8.3-8

8.3. MATRIX METHOD

Problem 8.3-2

Derive the $[A_1]$ matrix for a spheric pair (Fig. 8.3-8a) using three Euler angles (θ, ψ, ϕ) as the pair variables.

Solution

Referring to Fig. 8.3-8b, the Euler angles (θ, ψ, ϕ) are defined as follows:

We go from $(XYZ)_1$ to $(XYZ)_2$ (with the same origin O) through the following sequence of rotations:

(i) $R(\theta, z_1)$ takes $(XYZ)_1$ to $(XYZ)'_1$

(ii) $R(\psi, x'_1)$ takes $(XYZ)'_1$ to $(XYZ)''_1$

(iii) $R(\phi, z''_1)$ takes $(XYZ)''_1$ to $(XYZ)_2$

Referring now to Fig. 8.3-8a, first we set up the link-coordinate systems in a manner explained in Fig. 8.3-2. with a difference in the choices of X_1 and X_2 axes. In the present case Z_1 is chosen to be parallel to Z_n and passing through the center (O_1) of the spheric pair. The axis X_1 is taken along the perpendicular from O_1 to Z_n and directed from Z_n to Z_1. Similarly, X_2 is taken along the perpendicular, dropped from O_1 on to Z_2 and directed from Z_1 to Z_2. Thus comparing with Fig. 8.3-2, we get $O_1 \equiv O'_1$, i.e., $s_1 = O_1O'_1 = 0$, $\alpha_n = 0$, $O_1O_2 = O'_1O_2 = l_1$. We obtain the coordinate system $(XYZ)_2$ from $(XYZ)_1$ by first having three rotations expressed by the Euler angles described above, and followed by a translation of l_1 along X_2. Thus,

$$[A_1] = R(\theta, z_1)R(\psi, x'_1)R(\phi, z''_1)T(l_1, 0, 0) \qquad (a)$$

Using Equations (8.3.15), (8.3.17) and (8.3.19) in Equation (a) we obtain

$$[A_1] = \begin{pmatrix} C\theta & -S\theta & 0 & 0 \\ S\theta & C\theta & 0 & 0 \\ 0 & 0 & 1 & 0 \\ 0 & 0 & 0 & 1 \end{pmatrix} \begin{pmatrix} 1 & 0 & 0 & 0 \\ 0 & C\psi & -S\psi & 0 \\ 0 & S\psi & C\psi & 0 \\ 0 & 0 & 0 & 1 \end{pmatrix}$$
$$\begin{pmatrix} C\phi & -S\phi & 0 & 0 \\ S\phi & C\phi & 0 & 0 \\ 0 & 0 & 1 & 0 \\ 0 & 0 & 0 & 1 \end{pmatrix} \begin{pmatrix} 1 & 0 & 0 & l_1 \\ 0 & 1 & 0 & 0 \\ 0 & 0 & 1 & 0 \\ 0 & 0 & 0 & 1 \end{pmatrix}$$

$$= \begin{pmatrix} a_{11} & a_{12} & a_{13} & a_{14} \\ a_{21} & a_{22} & a_{23} & a_{24} \\ a_{31} & a_{32} & a_{33} & a_{34} \\ 0 & 0 & 0 & 1 \end{pmatrix}$$

where

$$\begin{aligned}
a_{11} &= C\theta C\phi - S\theta C\psi S\phi \\
a_{12} &= -C\theta S\phi - S\theta C\psi C\phi \\
a_{13} &= S\theta S\psi \\
a_{14} &= l_1(C\theta C\phi - S\theta C\psi S\phi) \\
a_{21} &= S\theta C\phi + C\theta C\psi S\phi \\
a_{22} &= -S\theta S\phi + C\theta C\psi C\phi \\
a_{23} &= -C\theta S\psi \\
a_{24} &= l_1(S\theta C\phi + C\theta C\psi S\phi) \\
a_{31} &= S\psi S\phi \\
a_{32} &= S\psi C\phi \\
a_{33} &= C\psi \\
a_{34} &= l_1 S\psi S\phi
\end{aligned}$$

8.3.4 Displacement Analysis

The displacement analysis of a spatial linkage using the matrix method is based on following the closed loops existing in the linkage. For a simple mechanism consisting of only binary links there exists only one closed loop. Thus, if we start from link 1 and follow the loop through links 2,3,... n and return to 1, we can write the loop-closure equation as

$$[A_1][A_2][A_3]\ldots[A_n] = [I] \qquad (8.3.24)$$

where $[I]$ is a 4×4 identity matrix given by

$$[I] = \begin{pmatrix} 1 & 0 & 0 & 0 \\ 0 & 1 & 0 & 0 \\ 0 & 0 & 1 & 0 \\ 0 & 0 & 0 & 1 \end{pmatrix} \qquad (8.3.25)$$

8.3. MATRIX METHOD

From Equation (8.3.24), equating element by element of two sides we get the relationships between the motion variables in terms of link parameters. Since the fourth row of each $[A]$ matrix is (0 0 0 1), by equating elements of the fourth row of Equation (8.3.24) we get identities $0 \equiv 0$ or $1 \equiv 1$, i.e., no useful information is yielded. We have already noted that the elements of $[A]$ matrix are not all independent. Likewise, equating elements of the matrix Equation (8.3.24) does not yield independent equations. Furthermore, algebraic manipulation of Equation (8.3.24) may be necessary if we are interested in solving for a particular motion (pair) variable in terms of the input motion and link parameters. The nature of this manipulation depends on the number and type of the kinematic pairs. Very often an explicit solution of a pair variable in terms of the input variables and link parameters is not possible, since the equations generated from Equation (8.3.24) are nonlinear. Consequently, one must use numerical methods for displacement analysis with prescribed link parameters for a specific linkage.[6] All these aspects are illustrated below through some examples.

Problem 8.3-3

Consider a 4R spherical linkage (Fig. 8.3-9a). Using matrix method, derive the output-input relationship for this linkage.

Solution

The Z axes are placed along the axes of the R-pairs, with their positive directions as indicated (chosen arbitrarily). Since all the Z axes are intersecting, all the link-length parameters, l_i's ($i= 1,2,3,4$), are zero, and the positive directions of X axes are also taken arbitrarily. Since all the X axes are also intersecting, all the link-parameters, s_i's, are zero. With X_i and Z_i axes so defined, we get the link-twist angles, α_i's, as indicated in Fig. 8.3-9a. The input variable θ_1 is the rotation about Z_1 of X_1, in order to make X_1 coincide with X_2. Similarly, the output variable θ_4 is the rotation about Z_4 of X_4, in order to make X_4 coincide with X_1 (Fig. 8.3-9b).

Using $l_i = s_i = 0$ in Equation (8.3.12), we can write

---

[6]Uicker (Jr.) J.J., Denavit, J., Hartenberg, R.S.: An iterative method for the displacement analysis of spatial mechanisms, Journal of Applied Mechanics (Trans. ASME), Vol. 31, pp. 309-314 (1964).

CHAPTER 8. SPHERICAL AND SPATIAL LINKAGES

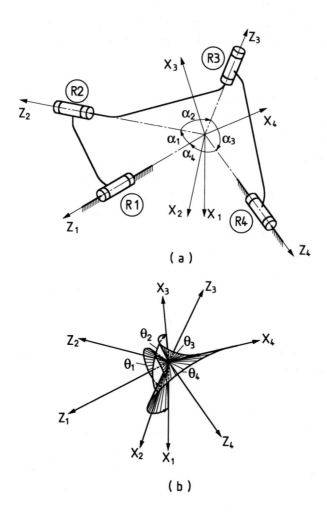

Figure 8.3-9

8.3. MATRIX METHOD

$$[A_i] = \begin{pmatrix} C\theta_i & -C\alpha_i S\theta_i & S\alpha_i S\theta_i & 0 \\ S\theta_i & C\alpha_i C\theta_i & -S\alpha_i C\theta_i & 0 \\ 0 & S\alpha_i & C\alpha_i & 0 \\ 0 & 0 & 0 & 1 \end{pmatrix}. \quad (a)$$

From Equation (8.3.24) we get

$$[A_1][A_2][A_3][A_4] = [I].$$

Premultiplying the above equation by $[A_2]^{-1}[A_1]^{-1}$ we obtain

$$[A_3][A_4] = [A_2]^{-1}[A_1]^{-1}. \quad (b)$$

Using Equations (8.3.5) and (8.3.6) for (a)[7]

$$[A_i]^{-1} = \begin{pmatrix} C\theta_i & S\theta_i & 0 & 0 \\ -C\alpha_i S\theta_i & C\alpha_i C\theta_i & S\alpha_i & 0 \\ S\alpha_i S\theta_i & -S\alpha_i C\theta_i & C\alpha_i & 0 \\ 0 & 0 & 0 & 1 \end{pmatrix}. \quad (c)$$

Using Equations (a) and (c) in Equation (b)

$$\begin{pmatrix} a_{11} & a_{12} & a_{13} & 0 \\ a_{21} & a_{22} & a_{23} & 0 \\ a_{31} & a_{32} & a_{33} & 0 \\ 0 & 0 & 0 & 1 \end{pmatrix}$$

$$= \begin{pmatrix} b_{11} & b_{12} & b_{13} & 0 \\ b_{21} & b_{22} & b_{23} & 0 \\ b_{31} & b_{32} & b_{33} & 0 \\ 0 & 0 & 0 & 1 \end{pmatrix} \quad (d)$$

where

$$a_{11} = C\theta_3 C\theta_4 - C\alpha_3 S\theta_3 S\theta_4$$
$$a_{12} = -C\alpha_4 C\theta_3 S\theta_4 - C\alpha_3 C\alpha_4 S\theta_3 C\theta_4 + S\alpha_3 S\alpha_4 S\theta_3$$
$$a_{13} = S\alpha_4 C\theta_3 S\theta_4 + C\alpha_3 S\alpha_4 S\theta_3 C\theta_4 + S\alpha_3 C\alpha_4 S\theta_3$$

---

[7]Since $l_i = s_i = 0$ for $i = 1,2,3,4$, the reader may note that in this problem, $[A_i]^{-1} = [A_i]^T$, where the superscript T denotes transposed (interchanging of rows of columns).

$$a_{21} = S\theta_3 C\theta_4 + C\alpha_3 C\theta_3 S\theta_4$$
$$a_{22} = -C\alpha_4 S\theta_3 S\theta_4 + C\alpha_3 C\alpha_4 C\theta_3 C\theta_4 - S\alpha_3 S\alpha_4 C\theta_3$$
$$a_{23} = S\alpha_4 S\theta_3 S\theta_4 - C\alpha_3 S\alpha_4 C\theta_3 C\theta_4 - S\alpha_3 C\alpha_4 C\theta_3$$
$$a_{31} = S\alpha_3 S\theta_4$$
$$a_{32} = S\alpha_3 C\alpha_4 C\theta_4 + C\alpha_3 S\alpha_4$$
$$a_{33} = -S\alpha_3 S\theta_4 C\theta_4 + C\alpha_3 C\alpha_4$$
$$b_{11} = C\theta_2 C\theta_1 - C\alpha_1 S\theta_2 S\theta_1$$
$$b_{12} = C\theta_2 S\theta_1 + C\alpha_1 S\theta_2 C\theta_1$$
$$b_{13} = S\alpha_1 S\theta_2$$
$$b_{21} = -C\alpha_2 S\theta_2 C\theta_1 - C\alpha_1 C\alpha_2 C\theta_2 S\theta_1 + S\alpha_1 S\alpha_2 S\theta_1$$
$$b_{22} = -C\alpha_2 S\theta_2 S\theta_1 + C\alpha_1 C\alpha_2 C\theta_2 C\theta_1 - S\alpha_1 S\alpha_2 C\theta_1$$
$$b_{23} = S\alpha_1 C\alpha_2 C\theta_2 + C\alpha_1 S\alpha_2$$
$$b_{31} = S\alpha_2 S\theta_2 C\theta_1 + C\alpha_1 S\alpha_2 C\theta_2 S\theta_1 + S\alpha_1 C\alpha_2 S\theta_1$$
$$b_{32} = S\alpha_1 S\theta_2 S\theta_1 - C\alpha_1 S\alpha_2 C\theta_2 C\theta_1 - S\alpha_1 C\alpha_2 C\theta_1$$
$$b_{33} = -S\alpha_1 S\alpha_2 C\theta_2 + C\alpha_1 C\alpha_2$$

Equating the corresponding elements of the last row and last column from both sides of Equation (d), only identities are obtained. By equating the other corresponding elements from both sides of Equation (d) we get in all nine equations. However, there are only three unknown pair variables, viz., θ_2, θ_3 and θ_4, to be determined for prescribed link parameters (α_i's) and the input variable θ_1. As already mentioned, only three out of the nine resulting equations are independent. To correlate the output (θ_4) and input (θ_1) variables, we shall use, below, the elements of the third row, since θ_3 is absent in these elements and we need to eliminate only θ_2.[8] Equating elements of the third row of matrix equation (d), we get

$$S\alpha_3 S\theta_4 = S\alpha_2 S\theta_2 C\theta_1 + C\alpha_1 S\alpha_2 C\theta_2 S\theta_1 + S\alpha_1 C\alpha_2 S\theta_1, \qquad (e)$$

---

[8]It may not always be possible to perform such an elimination. In that case one has to go through a numerical solution to a set of nonlinear simultaneous equations. For example, see Uicker (Jr.) J.J., Denavit, J., Hartenberg, R.S., op. cit.

8.3. MATRIX METHOD

$$S\alpha_3 C\alpha_4 C\theta_4 + C\alpha_3 S\alpha_4 = S\alpha_2 S\theta_2 S\theta_1 - C\alpha_1 S\alpha_2 C\theta_2 C\theta_1 - S\alpha_1 C\alpha_2 C\theta_1 \tag{f}$$

and $\quad C\alpha_3 C\alpha_4 - S\alpha_3 S\alpha_4 C\theta_4 = C\alpha_1 C\alpha_2 - S\alpha_1 S\alpha_2 C\theta_2.\tag{g}$

From Equation (g), we obtain

$$S\alpha_2 C\theta_2 = (C\alpha_1 C\alpha_2 - C\alpha_3 C\alpha_4 + S\alpha_3 S\alpha_4 C\theta_4)/S\alpha_1. \tag{h}$$

Substituting Equation (h) in Equation (e)

$$S\alpha_2 S\theta_2 = \frac{1}{C\theta_1}[S\alpha_3 S\theta_4 - S\alpha_1 C\alpha_2 S\theta_1 - C\alpha_1 S\theta_1 \\ (\frac{C\alpha_1 C\alpha_2 - C\alpha_3 C\alpha_4 + S\alpha_3 S\alpha_4 C\theta_4}{S\alpha_1})] \tag{i}$$

Substituting Equations (h) and (i) in Equation (f) and simplifying, we finally get the displacement equation as

$$C\alpha_4 C\theta_4 C\theta_1 - S\theta_4 S\theta_1 = \cot\alpha_1 \cot\alpha_3 C\alpha_4 - \frac{C\alpha_2}{S\alpha_1 S\alpha_3} \\ - \cot\alpha_1 S\alpha_4 C\theta_4 - \cot\alpha_3 S\alpha_4 C\theta_1. \tag{j}$$

We may recall that in Section 8.2 we obtained the displacement equation of a 4R spherical linkage, which is given by Equation (8.2.5). Comparing Figs. 8.2-2 and 8.3-9 we see the following correspondence in symbols:

$$\begin{pmatrix} Fig.\ 8.3\text{-}9 & & Fig.\ 8.2\text{-}2 \\ \alpha_4 & = & \alpha_1 \\ \alpha_1 & = & \alpha_2 \\ \alpha_2 & = & \alpha_3 \\ \alpha_3 & = & \alpha_4 \end{pmatrix} \tag{k}$$

Moreover, the angles θ_2 and θ_4 in Fig. 8.2-2 are related, respectively, to the angles θ_1 and θ_4 in Fig. 8.3-9 as follows:

$$\theta_1 = \theta_2 + (3\pi/2)$$

408 CHAPTER 8. SPHERICAL AND SPATIAL LINKAGES

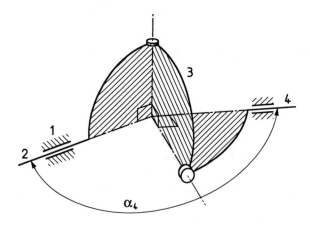

Figure 8.3-10

$$\text{and} \quad \theta_4(\text{Fig.8.3} - 9) = \frac{3\pi}{2} - \theta_4(\text{Fig.8.2} - 2). \tag{l}$$

It is left to the reader to verify [by using Equations (k) and (l)] that Equations (j) and (8.2.5) are identical.

Problem 8.3-4

Using the results of Problem 8.3-3, obtain all the pair variables (i.e., θ_2, θ_3 and θ_4) in terms of the input variable θ_1 and the shaft angle α_4 (i.e., the angle between the input and output shaft) for a Hooke's joint, shown in Fig. 8.3-10.

Solution

A Hooke's joint is a special case of a 4R spherical linkage with $\alpha_1 = \alpha_2 = \alpha_3$ (Fig. 8.3-9) $= \pi/2$. Substituting these values in Equation (d) of Problem 8.3-3, we get

$$\begin{pmatrix} C\theta_3 C\theta_4 & -C\alpha_4 C\theta_3 S\theta_4 + S\alpha_4 S\theta_3 & S\alpha_4 C\theta_3 S\theta_4 + C\alpha_4 S\theta_3 & 0 \\ S\theta_3 S\theta_4 & -C\alpha_4 S\theta_3 S\theta_4 - S\alpha_4 C\theta_3 & S\alpha_4 S\theta_3 S\theta_4 - C\alpha_4 C\theta_3 & 0 \\ S\theta_4 & C\alpha_4 C\theta_4 & -S\alpha_4 C\theta_4 & 0 \\ 0 & 0 & 0 & 1 \end{pmatrix}$$

$$= \begin{pmatrix} C\theta_2 C\theta_1 & C\theta_2 S\theta_1 & S\theta_2 & 0 \\ S\theta_1 & -C\theta_1 & 0 & 0 \\ S\theta_2 C\theta_1 & S\theta_2 S\theta_1 & -C\theta_2 & 0 \\ 0 & 0 & 0 & 1 \end{pmatrix} \tag{a}$$

8.3. MATRIX METHOD

Equating elements (3,1) and (3,2) from both sides of Equation (a), we obtain

$$S\theta_4 = S\theta_2 C\theta_1 \tag{b}$$

and $\quad C\alpha_4 C\theta_4 = S\theta_2 S\theta_1. \tag{c}$

Dividing Equation (b) by Equation (c)

$$\tan\theta_4 = C\alpha_4 \cot\theta_1. \tag{d}$$

Equating elements (3,3) from both sides of Equation (a),

$$S\alpha_4 C\theta_4 = C\theta_2. \tag{e}$$

Dividing Equation (c) by Equation (e),

$$\cot\alpha_4 = \tan\theta_2 S\theta_1$$

or, $\quad \tan\theta_2 = \dfrac{1}{\tan\alpha_4 S\theta_1}. \tag{f}$

Equating elements (1,1) from both sides of Equation (a),

$$C\theta_3 C\theta_4 = C\theta_2 C\theta_1. \tag{g}$$

Dividing Equation (g) by Equation (c),

$$\frac{C\theta_3}{C\alpha_4} = \frac{1}{\tan\theta_2 \tan\theta_1}. \tag{h}$$

Using Equation (f) in Equation (h),

$$C\theta_3 = C\alpha_4 \tan\alpha_4 \frac{S\theta_1}{\tan\theta_1} = S\alpha_4 C\theta_1. \tag{i}$$

By using Equations (d), (f) and (i), the reader may verify that Equation (a) is satisfied for all elements.

8.4 Velocity and Acceleration Analysis

As already mentioned with reference to planar linkages (see Section 5.4), the velocity and acceleration analysis of a spatial linkage can be carried out after the displacement analysis of the same linkage. These analyses are generally simple when both the input and output links are directly connected to the frame. The velocity and acceleration analysis involves solution of linear equations that are obtained from successive differentiation (with respect to time) of various displacement equations. A simple example that demonstrates the procedure is given below.

Problem 8.4-1

The shaft angle of a Hooke's joint is α_4 (Fig. 8.3-10). The input shaft (link 2) rotates at a constant angular velocity $\dot{\theta}_1 = \omega_2$. Determine the angular velocity ($\dot{\theta}_4$) and angular acceleration ($\ddot{\theta}_4$) of the output shaft as functions of θ_1, ω_2 and α_4.

Solution

The output-input displacement equation is given by Equation (d) in Problem 8.3-4 as

$$\tan \theta_4 = C\alpha_4 \cot \theta_1. \tag{a}$$

Differentiating both sides of Equation (a) once, with respect to time,

$$\sec^2 \theta_4 \dot{\theta}_4 = -C\alpha_4 \csc^2 \theta_1 \dot{\theta}_1 \tag{b}$$

Using Equation (a) in Equation (b),

$$\dot{\theta}_4 = \frac{C\alpha_4 \csc^2 \theta_1}{1 + C^2\alpha_4 \cot^2 \theta_1} \dot{\theta}_1 = -\frac{C\alpha_4}{S^2\theta_1 + C^2\alpha_4 C^2\theta_1} \omega_2. \tag{c}$$

Differentiating both sides of Equation (c) with respect to time, we get (with $\dot{\omega}_2 = 0$)

$$\ddot{\theta}_4 = \frac{C\alpha_4(1 - C^2\alpha_4)S2\theta_1}{(S^2\theta_1 + C^2\alpha_4 C^2\theta_1)^2} \omega_2^2 \tag{d}$$

8.5 Kinematic Synthesis

So far as the kinematic synthesis of spatial mechanisms is concerned, we shall restrict our discussion to problems of function generation.[9] Starting from the relevant displacement equation, Freundenstein's approach (explained in Sections 6.6.2 and 6.7.2 in the context of planar linkages) will be used. In Section 6.2, we noted that a 4R planar mechanism is not suitable for generating symmetric functions such as $y = x^2$, $-1 \leq x \leq 1$. It will be shown that with proper choices of link-parameters, the output-input (or the input-output) relationship of several spatial mechanisms can be rendered symmetric. Consequently, such mechanisms are very suitable for generating symmetric functions just mentioned. It is already evident that a four-link spatial linkage (e.g., R-S-S-R) has many more design variables as compared to a four-link planar linkage. Consequently, a function generation problem can be handled with a large number of accuracy points. All these features are illustrated below through examples. The use of both linear and nonlinear (with compatibility condition) equations is demonstrated.

Problem 8.5-1

Formulate a set of linear equations for synthesizing an R-S-S-R linkage (Fig. 8.2-1) as a function generator with six accuracy points. Identify the design variables and outline the procedure for solving them.

Solution

The displacement equation for an R-S-S-R linkage is given by Equation (8.2.2) as

$$2l_1 l_2 C\theta_2 - 2l_1 l_4 C\theta_4 + 2l_2 s_4 S\alpha_1 S\theta_2 - 2l_2 l_4 (C\theta_2 C\theta_4 + S\theta_2 S\theta_4 C\alpha_1)$$

$$-2s_2 l_4 S\alpha_1 S\theta_4 + (l_1^2 + l_2^2 - l_3^2 + l_4^2 + s_2^2 + s_4^2$$

$$-2s_2 s_4 C\alpha_1) = 0. \qquad (a)$$

---

[9] For synthesis of a spatial mechanism for motion generation, see Sandor, G.N. and Erdman, A.G.: "Advanced Mechanism Design - Analysis and Synthesis", Vol. 2, Prentice-Hall, Inc., New Jersey, 1984.

From Equation (a) we note that the design variables for this linkage are l_1, l_2, l_4, l_3, s_2, s_4, and α_1. Out of these seven variables, l_1 and α_1 define the relative position and orientation of the input and output shafts. For a given application, these two variables (l_1, α_1) are normally given to the designer. Therefore, we are left with five variables to be determined. To satisfy six accuracy points, we may leave either θ_2^1 or θ_4^1 (the values of θ_2 and θ_4, respectively, corresponding to the first accuracy point) as one more design variable. It is found (shown below) that using θ_2^1 (and not θ_4^1) as the sixth design variable (besides l_2, l_3, l_4, s_2 and s_4) we can generate six linear equations, in suitably defined six design parameters, corresponding to six accuracy points. Thus, θ_4^1 is also assumed along with l_1 and α_1. After solving for the six design parameters, we are able to determine the six design variables.

First, we write the values of θ_2 and θ_4, corresponding to accuracy points, as

$$\theta_2^m = \theta_2^1 + \theta_2^{1m} \quad m = 1, 2, 3, \ldots 6 \qquad (b)$$

$$\text{and} \quad \theta_4^m = \theta_4^1 + \theta_4^{1m}$$

with $\theta_2^{11} = \theta_4^{11} = 0$. As explained in the context of planar linkages, with an assumed value of θ_4^1, chosen scale factors (for generating a prescribed function within a range) and accuracy points, we know at this stage θ_2^{1m} and θ_4^m for $m = 1, 2, 3, \ldots 6$. Equation (a) is satisfied at all the accuracy points. Therefore, substituting $\theta_2^m = \theta_2^1 + \theta_2^{1m}$ in Equation (a) and dividing throughout by $2l_2 l_4 \cos\theta_2^1$, we get

$$D_1 \cos\theta_2^{1m} + D_2 \sin\theta_2^{1m} + D_3 \cos\theta_4^m + D_4 \sin\theta_4^m + D_5(\sin\theta_2^{1m}\cos\theta_4^m$$

$$- \cos\alpha_1 \cos\theta_2^{1m}\sin\theta_4^m) + D_6 = \cos\theta_2^{1m}\cos\theta_4^m + \cos\alpha_1 \sin\theta_2^{1m}\sin\theta_4^m \quad (c)$$

$$m = 1, 2, 3 \ldots \ldots 6$$

where the unknown design parameters are given by

8.5. KINEMATIC SYNTHESIS

$$D_1 = \frac{l_1 + s_4 \sin \alpha_1 \tan \theta_2^1}{l_4},$$

$$D_2 = \frac{s_4 \sin \alpha_1 - l_1 \tan \theta_2^1}{l_4},$$

$$D_3 = \frac{-l_1}{l_2 \cos \theta_2^1},$$

$$D_4 = \frac{-s_2 \cos \alpha_1}{l_2 \cos \theta_2^1},$$

$$D_5 = \tan \theta_2^1 \qquad\qquad (d)$$

and

$$D_6 = \frac{l_1^2 + l_2^2 - l_3^2 + l_4^2 + s_2^2 + s_4^2 - 2s_2 s_4 \cos \alpha_1}{2 l_2 l_4 \cos \theta_2^1}$$

The system of Equation (c) is linear in the design parameters $D_1, D_2, \ldots D_6$, which can be easily solved. The design variables $\theta_2^1, l_2, s_2, l_4, s_4$ and l_3 can then be determined from Equation (d). The sequence of determining these is as follows:

$(i)\theta_2^1$ from D_5, $(ii) l_2$ from D_3, $(iii) s_2$ from D_4, $(iv) s_4$ and l_4 from D_1 and D_2 and $(v) l_3$ from D_6.

As in the case of a planar linkage, possible negative values of l_2, l_4, s_2 and s_4 are to be treated in vector sense.

Problem 8.5-2

Synthesize an R-S-S-R mechanism to generate $y = x^2$, $-1 \leq x \leq 1$.

Solution

First of all, let us see how, with proper choices of some design variables, the output-input (or the input-output) relationship of this linkage demonstrates an intrinsic symmetry. In this situation, a symmetric function can be generated rather accurately since we can take all the accuracy points in one half (say $0 < x \leq 1$), and the other half (i.e., $-1 \leq x < 0$) will be automatically satisfied.

Let $s_2 = 0$ and $\alpha_1 = \pi/2$, when Equation (8.2.2) is modified as

$$(l_1^2 + l_2^2 - l_3^2 + l_4^2 + s_4^2) + 2l_1 l_2 \cos \theta_2 + 2l_2 s_4 \sin \theta_2 - \cos \theta_4$$

$$(2l_1l_4 + 2l_2l_4 \cos\theta_2) = 0. \tag{a}$$

From Equation (a), it is obvious that if θ_4 is replaced by $-\theta_4$, the value of θ_2 remains unaltered. Thus, if x is represented by θ_4 and y by θ_2, then the generated $y(\equiv \theta_2)$ is a symmetric function of $x(\equiv \theta_4)$ around the zero value.

Following the methodology explained in Problem 8.5-1, we can assume l_1 say, equal to 1. The value of l_1 actually determines the size of the mechanism, and the design variables are l_2, l_3, l_4, s_4 and θ_2^1. Hence, we can use five accuracy points and obtain linear equations in the design parameters. With $s_2 = 0$ and $\alpha_1 = \pi/2$, from Equations (c) and (d) of Problem 8.5-1, we get

$$D_1 \cos\theta_2^{1m} + D_2 \sin\theta_2^{1m} + D_3 \cos\theta_4^m + D_5 \sin\theta_2^{1m} \cos\theta_4^m + D_6$$
$$= \cos\theta_2^{1m} \cos\theta_4^m; \quad m = 1,2,3,4,5 \tag{b}$$

where

$$D_1 = \frac{l_1 + s_4 \tan\theta_2^1}{l_4},$$

$$D_2 = \frac{s_4 - l_1 \tan\theta_2^1}{l_4},$$

$$D_3 = \frac{-l_1}{l_2 \cos\theta_2^1},$$

$$D_5 = \tan\theta_2^1$$

$$D_6 = \frac{l_1^2 + l_2^2 - l_3^2 + l_4^2 + s_4^2}{2l_2l_4 \cos\theta_2^1}. \tag{c}$$

Of the five possible accuracy points, let us take one at the axis of symmetry (i.e., $x = 0$), and the remaining four equally spaced in one half of the interval $0 < x \le 1$. Thus, the accuracy points are obtained as :

$$x^1 = 0, x^2 = 0.25, x^3 = 0.5, x^4 = 0.75, x^5 = 1$$

8.5. KINEMATIC SYNTHESIS

$$y^1 = 0, y^2 = 0.0625, y^3 = 0.25, y^4 = 0.5625, y^5 = 1 \qquad (d)$$

With these accuracy points, we actually have, in total, nine accuracy points in the entire interval $-1 \leq x \leq 1$, since due to intrinsic symmetry about $x(\equiv \theta_4) = 0$, $x = -0.25, -0.5, -0.75$ and -1 will also be accuracy points.

Let $-1 \leq x \leq 1$ be represented by $-90° \leq \theta_4 \leq 90°$ and $0 \leq y \leq 1$ be represented by a total rotation of $90°$ for the variable θ_2. Hence, we get the following values.

$$\theta_4^1 = 0, \theta_4^2 = 22.5°, \theta_4^3 = 45°, \theta_4^4 = 67.5°, \theta_4^5 = 90°, \qquad (e)$$

$$\theta_2^{11} = 0, \theta_2^{12} = 5.625°, \theta_2^{13} = 22.5°, \theta_2^{14} = 50.625° \text{ and } \theta_2^{15} = 90°.$$

Substituting Equation (e) in Equation (b) and solving, we obtain

$$D_1 = 2.712, D_2 = -2.094, D_3 = -3.806,$$

$$D_5 = -1.679 \text{ and } D_6 = 2.094 \qquad (f)$$

Using the above values in Equation (c), the following design is obtained: $\theta_2^1 = -59.22°, l_2 = 0.513, s_4 = 8.270, l_4 = -4.751, l_3 = 9.872$ along with assumed $l_1 = 1, s_2 = 0$ and $\alpha_1 = \pi/2$.

The designed mechanism is shown in Fig. 8.5-1, at the configuration corresponding to the first accuracy point $(x = 0, y = 0)$, i.e., the point of symmetry.

Problem 8.5-3

Figure 8.5-2 shows a special R-S-S-P mechanism (compare this figure with Fig. 8.2-5a) in which the axis of translation (in the $Y - Z$ plane) of the P-pair through the point B intersects the axis of the R-pair at O_2. The angle α_1 is given to be $30°$. Synthesize this mechanism to generate $y = x^2 (-1 \leq x \leq 1)$, with five accuracy points, $x^1 = 0, x^2 = 1/2, x^3 = 1, x^4 = -1/2, x^5 = -1$, where s_4 represents y and θ_2 represents x. Assume $\Delta\theta_2 = 180°$ as x goes from -1 to $+1$ and $\Delta s_4 = 7.5$ cm as y goes from 0 to 1.

Solution

Comparing Fig. 8.5-2 with Fig. 8.2-5a, the displacement equation of the given mechanism is obtained by substituting $l_1 = s_2 = 0$ and $\alpha_1 = 30°$ in Equation (8.2.7) as

CHAPTER 8. SPHERICAL AND SPATIAL LINKAGES

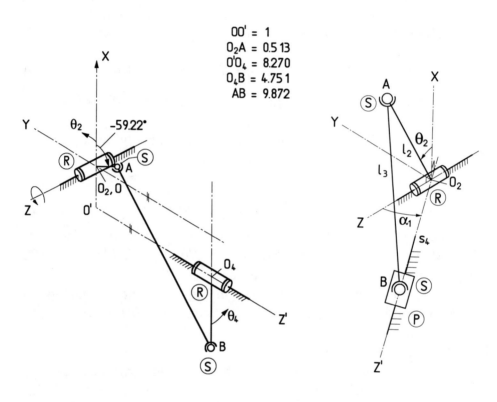

$OO' = 1$
$O_2A = 0.513$
$O'O_4 = 8.270$
$O_4B = 4.751$
$AB = 9.872$

Figure 8.5-1

Figure 8.5-2

8.5. KINEMATIC SYNTHESIS

$$s_4^2 + 2(0.5l_2 \sin \theta_2)s_4 + (l_2^2 - l_3^2) = 0$$

or, $\quad s_4^2 + l_2 s_4 \sin \theta_2 + l_2^2 - l_3^2 = 0.$ \hfill (a)

From Equation (a) it is obvious that s_4 is symmetric about $\theta_2 = 90°$. So if $x^1 = 0$ is represented by $\theta_2^1 = 90°$, then y will be symmetric about $x = 0$. In other words, if we satisfy three accuracy points ($x^1 = 0, y^1 = 0$), ($x^2 = 1/2, y^2 = 1/4$) and ($x^3 = 1, y^3 = 1$), the other two accuracy points will then be automatically satisfied. Thus, we need three design variables. Let us choose three design variables as l_2, l_3 and s_4^1 (i.e., the value of s_4 corresponding to the first accuracy point). Using the prescribed ranges of movements we can now write values of s_4 and θ_2 corresponding to the accuracy points as $(90°, s_4^1)$, $(135°, s_4^1 + 1.875)$ and $(180°, s_4^1 + 7.5)$.

First, we rewrite Equation (a) for the mth accuracy point as

$$(s_4^1 + s_4^{1m}) + l_2 \sin \theta_2^m (s_4^1 + s_4^{1m}) + l_2^2 - l_3^2 = 0$$

or, $\quad l_2 s_4^{1m} \sin \theta_2^m + l_2 s_4^1 \sin \theta_2^m + [l_2^2 + (s_4^1)^2 - l_3^2]$

$$= -(s_4^{1m})^2 - 2s_4^1 s_4^{1m} \quad m = 1, 2, 3 \quad (b)$$

Next, we use the method explained in Section 6.7.2. First define design parameters

$$D_1 = l_2^2 + (s_4^1)^2 - l_3^2, D_2 = l_2 s_4^1, D_3 = l_2 \text{ and } D_4 = 2s_4^1. \quad (c)$$

Thus, Equation (b) can be written as

$$D_1 + D_2 \sin \theta_2^m + D_3 s_4^{1m} \sin \theta_2^m = -(s_4^{1m})^2 - D_4 s_4^{1m}, \; m = 1, 2, 3 \quad (d)$$

where

$$2D_2 = D_3 D_4 \quad (e)$$

is the compatibility condition.

Let us denote $D_4 = \lambda$ and write (e) as

$$2D_2 = \lambda D_3. \quad (f)$$

To satisfy Equation (d), we define

$$D_j = P_j + \lambda Q_j; \; j = 1, 2, 3 \quad (g)$$

where P_j's and Q_j's satisfy the following two sets of linear equations:

$$P_1 + P_2 \sin\theta_2^m + P_3 s_4^{1m} \sin\theta_2^m = -(s_4^{1m})^2 \quad (h)$$

$$m = 1, 2, 3$$

$$Q_1 + Q_2 \sin\theta_2^m + Q_3 s_4^{1m} \sin\theta_2^m = -(s_4^{1m})^2 \quad (i)$$

Now, substituting $\theta_2^1 = 90°$, $\theta_2^2 = 135°$, $\theta_2^3 = 180°$, $s_4^{11} = 0$, $s_4^{12} = 1.875$ and $s_4^{13} = 7.5$ in Equations (h) and (i), and solving, we get

$$P_1 = -56.25, P_2 = 56.25, P_3 = 9.781, \quad (k)$$

$$Q_1 = -7.5, Q_2 = 7.5 \text{ and } Q_3 = 0.243.$$

Using Equation (g) in Equation (f) we obtain

$$2(P_2 + \lambda Q_2) = \lambda(P_3 + \lambda Q_3)$$

or, $\lambda^2 Q_3 + \lambda(P_3 - 2Q_2) - 2P_2 = 0$

$$\text{or, } \lambda_{1,2} = \frac{2Q_2 - P_3 \mp \sqrt{(P_3 - 2Q_2)^2 + 8P_2 Q_3}}{2Q_3} \quad (l)$$

Using Equation (k) in Equation (l)

$$\lambda_{1,2} = -13.309, 34.786.$$

Thus, two possible designs are obtained. Using two values of λ in Equation (g) along with Equation (k), we get

$$D_1 = 43.568, D_2 = -43.568, D_3 = 6.547, D_4 = -13.309$$

and

$$D_1 = -317.145, D_2 = 317.145, D_3 = 18.234, D_4 = 34.786.$$

Using the above values in Equation (c), the design variables for these two sets are finally obtained as

(i) $l_2 = 6.547$ cm, $s_4^1 = -6.655$ cm , $l_3 = 6.602$ cm.

(ii) $l_2 = 18.234$ cm, $s_4^1 = 17.393$ cm, $l_3 = 30.857$ cm.

8.6 Mobility Analysis

In Section 3.4 we discussed the Grashof criterion for four-link planar linkages. This criterion, expressed in terms of simple inequalities involving link parameters, is extremely useful for classifying four-link planar mechanisms according to the possible ranges of movements of various links (i.e., for type identification). In particular, the condition necessary for the rotatability of the input link is of special importance while designing most of the mechanisms. Attempts have been made to obtain similar criteria for four-link spherical and spatial mechanisms. In this section we shall present some of these results, which are obtained analytically starting from the displacement equation. In particular, necessary conditions for the existence of a crank (capable of complete rotation) will be discussed using the limit position analysis (see Problem 4.3-1 for a similar approach in the case of 4R planar linkage).

8.6.1 Type Identification of an R-S-S-R Linkage

In Section 8.2 we noted that a number of four-link spatial mechanisms can be analysed as special cases of an R-S-S-R linkage. Therefore, we start our discussion again with an R-S-S-R linkage and try to obtain

CHAPTER 8. SPHERICAL AND SPATIAL LINKAGES

simple criteria (in terms of link parameters) for classifying such a linkage as double-crank, crank-rocker or double-rocker. Towards this goal, we refer back to Fig. 8.2-1 and reproduce the displacement equation given by Equation (8.2.2) as

$$2l_1l_2C\theta_2 - 2l_1l_4C\theta_4 + 2l_2s_4S\alpha_1S\theta_2 - 2l_2l_4(C\theta_2C\theta_4 + S\theta_2S\theta_4C\alpha_1) - 2s_2l_4S\alpha_1S\theta_4 + (l_1^2 + l_2^2 - l_3^2 + l_4^2 + s_2^2 + s_4^2 - 2s_2s_4C\alpha_1) = 0. \tag{8.6.1}$$

For the input link O_2A to reach an extreme position (when its angular velocity becomes zero and after which it has to reverse its direction of motion), the condition necessary is

$$\frac{d\theta_2}{d\theta_4} = 0. \tag{8.6.2}$$

Differentiating Equation (8.6.1) with respect to θ_4 and simplifying, we get

$$\frac{d\theta_2}{d\theta_4} = \frac{l_4(s_2S\alpha_1C\theta_4 - l_1S\theta_4 + l_2C\alpha_1S\theta_2C\theta_4 - l_2C\theta_2S\theta_4)}{l_2(s_4S\alpha_1C\theta_2 - l_1S\theta_2 - l_4C\alpha_1S\theta_4C\theta_2 + l_4C\theta_4S\theta_2)} \tag{8.6.3}$$

From Equations (8.6.2) and (8.6.3), the condition for O_2A to reach an extreme position can be written as

$$s_2S\alpha_1C\theta_4 - l_1S\theta_4 + l_2C\alpha_1S\theta_2C\theta_4 - l_2C\theta_2S\theta_4 = 0$$

$$\text{or, } s_2S\alpha_1 + l_2C\alpha_1S\theta_2 - (l_1 + l_2C\theta_2)\tan\theta_4 = 0. \tag{8.6.4}$$

Substituting $S\theta_2 = \frac{2\tan(\theta_2/2)}{1+\tan^2(\theta_2/2)}$ and $C\theta_2 = \frac{1-\tan^2(\theta_2/2)}{1+\tan^2(\theta_2/2)}$ in Equation (8.6.4), we obtain

$$[s_2S\alpha_1 + (l_2 - l_1)\tan\theta_4]\tan^2(\theta_2/2) + 2l_2C\alpha_1\tan(\theta_2/2) +$$

$$[s_2S\alpha_1 - (l_1 + l_2)\tan\theta_4] = 0.$$

8.6. MOBILITY ANALYSIS

For θ_2 (i.e., $\tan(\theta_2/2)$) in the above equation to be real, the necessary and sufficient condition is

$$l_2^2 C^2 \alpha_1 - [s_2^2 S^2 \alpha_1 + (l_1^2 - l_2^2)\tan^2 \theta_4 - 2l_1 s_2 S\alpha_1 \tan \theta_4] \geq 0$$

or, $(l_2^2 - l_1^2)\tan^2 \theta_4 + 2l_1 s_2 S\alpha_1 \tan \theta_4 + (l_2^2 C^2 \alpha_1 - s_2^2 S^2 \alpha_1) \geq 0.$

Sufficient conditions for the above relation to be true for any value of θ_4 are

$$l_2^2 - l_1^2 \geq 0 \tag{8.6.5}$$

and

$$(l_1 s_2 S\alpha_1)^2 - (l_2^2 C^2 \alpha_1 - s_2^2 S^2 \alpha_1)(l_2^2 - l_1^2) \leq 0 \tag{8.6.6}$$

After simplification, Condition (8.6.6) reduces to

$$l_2^2 \geq l_1^2 + s_2^2 \tan^2 \alpha_1. \tag{8.6.7}$$

It should be noted that special cases like $\cos \theta_4 = 0$ or $\cos \alpha_1 = 0$ have not been covered so far, since divisions by $\cos \theta_4$ and $\cos \alpha_1^1$ were carried out. However, one can show that for $\cos \theta_4 = 0$, the required condition reduces to Condition (8.6.5). Similarly, for $\cos \alpha_1 = 0$ and $s_4 = 0$, the required condition again reduces to Condition (8.6.5). For $\cos \alpha_1 = 0$ and $s_4 \neq 0$, the required condition becomes impossible to satisfy as also suggested by Condition (8.6.7).

Since Condition (8.6.7) automatically ensures Condition (8.6.5), the sufficient condition for the existence of at least one real value of θ_2 satisfying Equation (8.6.4) for any value of θ_4, is given by Condition (8.6.7).

If $O_2 A$ is to be the crank of a crank-rocker mechanism, then Equation (8.6.1) must yield a single continuous branch (for one mode of assembly of this bimodal mechanism) for the θ_4 vs θ_2 curve, as indicated in Fig. 8.6-1 (also see Fig. 4.3-1a). This curve covers the entire range $O \leq \theta_2 \leq 2\pi$ while θ_4 is limited within $(\theta_4)_{min} \leq \theta_4 \leq (\theta_4)_{max}$. If Condition (8.6.7) is satisfied, then θ_4 vs θ_2 plot satisfying Equation

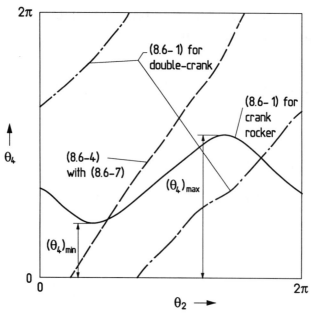

Figure 8.6-1

(8.6.4) covers the entire range $0 \le \theta_4 \le 2\pi$ as indicated in Fig. 8.6-1, since for any value of θ_4 there exists a real root (or roots) for θ_2. Thus, if Condition (8.6.7) is satisfied, then the θ_4 vs θ_2 plot satisfying Equations (8.6.1) and (8.6.4) must intersect. This intersection implies the existence of a configuration (given by the point of intersection) when O_2A attains its extreme position. In other words, Condition (8.6.7) ensures that O_2A cannot be the crank of a crank-rocker mechanism. It may be noted however, that if the mechanism is a double-crank, then the θ_4 vs θ_2 plot of Equation (8.6.1) has two branches as indicated in Fig. 8.6-1 (also see Fig. 4.3-1b). In that case the plots of Equations (8.6.1) and (8.6.4) need not intersect even if Condition (8.6.7) holds good. Consequently, Condition (8.6.7) does not debar O_2A to be a crank of a double crank mechanism. Thus, we finally conclude that Condition (8.6.7) only ensures that O_2A cannot be the crank of a crank rocker mechanism.

For the extreme position of link O_4B, $\frac{d\theta_4}{d\theta_2} = 0$ which implies from Equation (8.6.3) that

8.6. MOBILITY ANALYSIS

$$s_4 S\alpha_1 C\theta_2 - l_1 S\theta_2 - l_4 C\alpha_1 S\theta_4 C\theta_2 + l_4 C\theta_4 S\theta_2 = 0$$

or, $s_4 S\alpha_1 - l_4 C\alpha_1 S\theta_4 - (l_1 - l_4 C\theta_4)\tan\theta_2 = 0.$ (8.6.8)

It is readily seen that Equation (8.6.8) is obtained from Equation (8.6.4) by interchanging s_2 with s_4, l_2 with $-l_4$ and θ_2 with θ_4. Thus, proceeding in a manner explained in preceding paragraphs, the sufficient condition for $O_4 B$ not to be the crank of a crank-rocker mechanism can be written, by interchanging l_2 with $-l_4$ and s_2 with s_4, from Condition (8.6.7), as

$$l_4^2 \geq l_1^2 + s_4^2 \tan^2 \alpha_1. \qquad (8.6.9)$$

Hence, if both Conditions (8.6.7) and (8.6.9) are satisfied, then the mechanism cannot be of the crank-rocker type. If a crank exists fulfilling these conditions, then the mechanism must be a double-crank type. Starting from a 4R planar linkage and continuity argument, one can prove that violation of Conditions (8.6.7) and (8.6.9), respectively, implies that $O_2 A$ and $O_4 B$ can be crank (if at all) of only crank-rocker type linkages.[10] Accordingly, various possible R-S-S-R mechanisms and the associated required conditions can be listed as given in Table. 8.6-1.

Since for each set of conditions, listed in Table 8.6-1, there exists more than one possible type of mechanism, the type identification can be done uniquely

---

[10] Freudenstein, F. and Primrose, E.J.: On the criteria for the rotatability of the cranks of a skew four-bar linkage, J. of Engg. for Industry, Trans. ASME Ser. B., Vol. 98., pp. 1285-1288(1976).

Table 8.6-1[11]

Conditions for various possible R-S-S-R mechanisms (Fig. 8.2-1)

Conditions	Possible Mechanism Types
$l_2^2 \geq l_1^2 + s_2^2 \tan^2 \alpha_1$ $l_4^2 \geq l_1^2 + s_4^2 \tan^2 \alpha_1$	Double-crank, double-rocker
$l_2^2 < l_1^2 + s_2^2 \tan^2 \alpha_1$ $l_4^2 \geq l_1^2 + s_4^2 \tan^2 \alpha_1$	Crank-rocker with $O_2 A$ as crank, double-rocker
$l_2^2 \geq l_1^2 + s_2^2 \tan^2 \alpha_1$ $l_4^2 < l_1^2 + s_4^2 \tan^2 \alpha_1$	Crank-rocker with $O_4 B$ as crank, double-rocker
$l_2^2 < l_1^2 + s_2^2 \tan^2 \alpha_1$ $l_4^2 < l_1^2 + s_4^2 \tan^2 \alpha_1$	Crank-rocker with either $O_2 A$ or $O_4 B$ as crank, double-rocker

only after checking the input link rotatability condition, obtained through limit position analysis. This limit position analysis is illustrated in the next section. It is obvious that for the first three sets of conditions listed in Table 8.6-1, we need to check the rotatability of only one of the links connected to the frame. For the last set of conditions, if one of the links connected to the frame turns out to be a crank, only then the rotatability of the other link connected to the frame needs to be verified as well.

8.6.2 Limit Position Analysis

In Problem 4.3-1 we saw that for a four-link, biomodal planar mechanism, the rotatability condition for the input link can be obtained

---

[11] For further details, see Freudenstein, F. and Primrose, E.J. op. cit.

8.6. MOBILITY ANALYSIS

using the fact that there should not exist any real value of the input variable for which the output variable has a single value. This approach, referred to as the limit position analysis, can also be used for four-link, biomodal spatial linkages. This analysis, of course, assumes that the link parameters are such that the mechanism can be assembled. In the limit position analysis, we derive a discriminant in the form of a polynomial; the nonexistence of any real root of this discriminant polynomial implies rotatability of the input link. The linkage can be assembled only when the discriminant polynomial is positive for some value of the input variable. For spherical and some spatial four-link mechanisms, this polynomial comes out to be of second order and closed form simple criteria for input link rotatability are obtained in terms of link parameters.[12] For an R-S-S-R linkage, this polynomial turns out to be of fourth order and the input link rotatibilty conditions become cumbersome. All these aspects are illustrated below, through some typical four-link mechanisms.

4R Spherical Mechanism

Referring to Fig. 8.2-2, the displacement equation of this 4R spherical mechanism is given by Equation (8.2.5) as

$$(\tan \alpha_2 \tan \alpha_4 C\theta_2)C\theta_4 + (\tan \alpha_4 S\alpha_1 + \tan \alpha_2 \tan \alpha_4 C\alpha_1 S\theta_2)S\theta_4$$

$$+ (C\alpha_1 - S\alpha_1 \tan \alpha_2 S\theta_2 - \frac{C\alpha_3}{C\alpha_2 C\alpha_4}) = 0. \tag{8.6.10}$$

Substituting $S\theta_4 = \frac{2\tan(\theta_4/2)}{1+\tan^2(\theta_4/2)}$ and $C\theta_4 = \frac{1-\tan^2(\theta_4/2)}{1+\tan^2(\theta_4/2)}$ in Equation (8.6.10), we get a quadratic equation in $\tan(\theta_4/2)$. The roots of this quadratic equation refer to two values of θ_4 for a given value of θ_2. The two values of θ_4 so obtained, correspond to two modes of assembly of the given linkage. If the input link has an extreme position, then the two roots of θ_4 coincide. At this configuration, the discriminant of the quadratic equation mentioned above becomes zero. After

---

[12]Mallik, A.K.: Mobility and type identification of four-link mechanisms, Journal of Mechanical Design (Trans ASME), Vol. 115, 1993.

trigonometric manipulation and equating the discriminant to zero, we obtain

$$(S^2\alpha_1 S^2\alpha_2)S^2\theta_2 - 2[S\alpha_1 S\alpha_2(C\alpha_1 C\alpha_2) - C\alpha_3 C\alpha_4)]S\theta_2$$

$$+ (C^2\alpha_1 C^2\alpha_2 + C^2\alpha_3 - S^2\alpha_4 - 2C\alpha_1 C\alpha_2 C\alpha_3 C\alpha_4) = 0. \quad (8.6.11)$$

If no real value of θ_2 satisfies Equation (8.6.11), then it implies that the input link does not reach its extreme position, i.e., the input link can rotate completely.[13] Thus, for a 4R spherical mechanism the discriminant polynomial (i.e., the left hand side of Equation (8.6.11)) has been obtained as a quadratic in $\sin\theta_2$. Solving Equation (8.6.11), we get

$$S\theta_2 = [(C\alpha_1)C\alpha_2 - C\alpha_3 C\alpha_4) \mp S\alpha_3 S\alpha_4]/(S\alpha_1 S\alpha_2). \quad (8.6.12)$$

For θ_2 in Equation (8.6.12), to be nonreal, the right hand side of Equation (8.6.12) must be outside the range -1 to $+1$. Thus, the conditions for complete rotatability of O_2A (i.e., the input link in Fig. 8.2-2) can be written as follows:

$$\text{Either} \quad \cos(\alpha_3 - \alpha_4) < \cos(\alpha_1 + \alpha_2) \quad (8.6.13a)$$

$$\text{or} \quad \cos(\alpha_3 - \alpha_4) > \cos(\alpha_1 - \alpha_2) \quad (8.6.13b)$$

and

$$\text{either} \quad \cos(\alpha_3 + \alpha_4) < \cos(\alpha_1 + \alpha_2) \quad (8.6.13c)$$

$$\text{or} \quad \cos(\alpha_3 + \alpha_4) > \cos(\alpha_1 - \alpha_2) \quad (8.6.13d)$$

It may be noted from Equation (8.6.10) that if $\tan\alpha_2$ and $-\tan\alpha_4$ are interchanged, the roles of the input and output are reversed.

---

[13] If a mechanism cannot be assembled, the discriminant polynomial is negative for all values of the input variables when again, the roots of the discriminant polynomial will be non-real. Such a situation will be encountered later when we discuss an R-S-S-P mechanism.

8.6. MOBILITY ANALYSIS

Hence, the conditions for complete rotatability of the link O_4B are obtained by substituting $\alpha_4 = \pi - \alpha_2$ in Conditions (8.6.13a) to (8.6.13d), as

$$\text{either} \quad \cos(\alpha_2 + \alpha_3) > \cos(\alpha_1 - \alpha_4) \tag{8.6.14a}$$

$$\text{or} \quad \cos(\alpha_2 + \alpha_3) < \cos(\alpha_1 + \alpha_4) \tag{8.6.14b}$$

and

$$\text{either} \quad \cos(\alpha_2 - \alpha_3) > \cos(\alpha_1 - \alpha_4) \tag{8.6.14c}$$

$$\text{or} \quad \cos(\alpha_2 - \alpha_3) < \cos(\alpha_1 + \alpha_4) \tag{8.6.14d}$$

Given the linkage parameters $\alpha_1, \alpha_2, \alpha_3$ and α_4, the type identification of a 4R spherical mechanism can now be completed using Conditions (8.6.13) and (8.6.14). Furthermore, from Equation (8.2.4) and Conditions (8.6.7) and (8.6.9), we can conclude that if

$$\tan^2 \alpha_2 \geq \tan^2 \alpha_1 \tag{8.6.15a}$$

then O_2A cannot be the crank of a crank-rocker mechanism, and if

$$\tan^2 \alpha_4 \geq \tan^2 \alpha_1 \tag{8.6.15b}$$

then O_4B cannot be the crank of a crank-rocker mechanism. Using Conditions (8.6.13), (8.6.14) and (8.6.15), type identification of a 4R spherical mechanism can be summarized as given in Table 8.6-2.

It may be mentioned that Table 8.6-2 is not applicable to special degenerate situations. For example, if the shaft angle is $\pi/2$ in a Hooke's joint, i.e., $\alpha_1 = \alpha_2 = \alpha_3 = \alpha_4 = \pi/2$, then Table 8.6-2 indicates a double-rocker mechanism. However, the reader may verify that with $\alpha_1 = \alpha_2 = \alpha_3 = \alpha_4 = \pi/2$, no motion transmission is possible from the input to the output shaft.[14]

---

[14] Refer to Equation (c) of Problem (8.4-1) for shaft angle equal to $\pi/2$.

Table 8.6-2

Type Identification of a 4R Spherical Mechanism (Fig. 8.2-2)

Conditions to be simultaneously satisfied			Type of Mechanism
$\cos(\alpha_3 - \alpha_4)$ $< \cos(\alpha_1 + \alpha_2)$ or $\cos(\alpha_3 - \alpha_4)$ $> \cos(\alpha_1 - \alpha_2)$	$\cos(\alpha_3 + \alpha_4)$ $< \cos(\alpha_1 + \alpha_2)$ or $\cos(\alpha_3 + \alpha_4)$ $> \cos(\alpha_1 - \alpha_2)$	$\tan^2 \alpha_2 \geq \tan^2 \alpha_1$	Double-crank
$\cos(\alpha_3 - \alpha_4)$ $< \cos(\alpha_1 + \alpha_2)$ or $\cos(\alpha_3 - \alpha_4)$ $> \cos(\alpha_1 - \alpha_2)$	$\cos(\alpha_3 + \alpha_4)$ $< \cos(\alpha_1 + \alpha_2)$ or $\cos(\alpha_3 + \alpha_4)$ $> \cos(\alpha_1 - \alpha_2)$	$\tan^2 \alpha_2 < \tan^2 \alpha_1$	Crank-rocker with O_2A as crank
$\cos(\alpha_2 - \alpha_3)$ $< \cos(\alpha_1 + \alpha_4)$ or $\cos(\alpha_2 - \alpha_3)$ $> \cos(\alpha_1 - \alpha_4)$	$\cos(\alpha_2 + \alpha_3)$ $< \cos(\alpha_1 + \alpha_4)$ or $\cos(\alpha_2 + \alpha_3)$ $> \cos(\alpha_1 - \alpha_4)$	$\tan^2 \alpha_4 < \tan^2 \alpha_1$	Crank-rocker with O_4B as crank
If all the above three combinations fail			Double-rocker

If the link parameters α_i's are such that the sum of no two angles exceeds π, then the type identification of a 4R spherical mechanism can be performed using by Grashof's criteria, with α_i's as the link lengths. In other words, we can write that if

$$\alpha_{min} + \alpha_{max} < \alpha' + \alpha'' \qquad (8.6.16)$$

where α_{min} is the smallest, with α_{max} as the largest and α', α'' as the remaining angles, then we get (Fig. 8.2-2)

(i) a double-crank with $\alpha_1 = \alpha_{min}$,

(ii) a double-rocker with $\alpha_3 = \alpha_{min}$ and

8.6. MOBILITY ANALYSIS

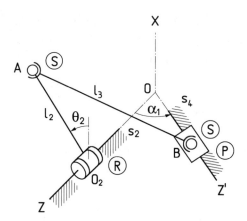

Figure 8.6-2

(iii) two crank rockers with $\alpha_2 = \alpha_{min}$ and $\alpha_4 = \alpha_{min}$.

The reader is advised to verify the above statement in view of Table 8.6-2.[15]

R-S-S-P Mechanism

Let us now consider an R-S-S-P mechanism (Fig. 8.6-2) with intersecting axes of the R and P pairs. Our objective is to derive the criteria for complete rotatability of the input link O_2A from the limit position analysis. The displacement equation of this linkage, given by Equation (8.2.8), is

$$s_4^2 + 2(l_2 S\alpha_1 S\theta_2 - s_2 C\alpha_1)s_4 + (l_2^2 + s_2^2 - l_3^2) = 0. \qquad (8.6.17)$$

For limit positions of the input link O_2A, s_4 should be single-valued and from Equation (8.6.17) this condition is satisfied when

$$(l_2 S\alpha_1 S\theta_2 - s_2 C\alpha_1)^2 - (l_2^2 + s_2^2 - l_3^2) = 0. \qquad (8.6.18)$$

---

[15] For further details, See Duditza, Fl. and Dittrich, G.: Die Bedingungen Für die Umlauffähighkeit sphärisher viergliederiger kurbelgetriebe - Industrie Anzeiger, Vol. 91, No. 71, pp. 1687-1690 (1969).

Here again, as in the case of a 4R spherical mechanism, the discriminant polynomial Equation (8.6.18) has been obtained as a quadratic in $\sin\theta_2$. Solving Equation (8.6.18) we obtain

$$\sin\theta_2 = \frac{s_2 C\alpha_1 \pm \sqrt{l_2^2 + s_2^2 - l_3^2}}{l_2 S\alpha_1} \qquad (8.6.19)$$

For θ_2 in Equation (8.6.19) to be nonreal, one condition is

$$l_3^2 > l_2^2 + s_2^2 \qquad (8.6.20)$$

which renders $\sin\theta_2$ as a complex number. When $l_3^2 < l_2^2 + s_2^2$, for θ_2 to be nonreal, both values of $\sin\theta_2$ given by Equation (8.6.19) must lie outside the range -1 to $+1$. The necessary condition for this to happen is

$$l_3^2 < l_2^2 + s_2^2 - [|s_2\cos\alpha_1| + |l_2\sin\alpha_1|]^2. \qquad (8.6.21)$$

However, Condition (8.6.21), yielding complex roots of θ_2, does not imply full rotation of the input link. Instead, it signifies that the linkage cannot be assembled. Under this condition, the discriminant polynomial (i.e., the left hand side of Equation (8.6.18)) is negative for all values of θ_2. The geometric significance of Condition (8.6.21) is evident. From Fig. 8.6-2, it is readily seen that for the point B to reach the line of reciprocation (of the slider) OZ', the necessary condition is

$$l_3^2 > l_2^2 + s_2^2 - [|s_2\cos\alpha_1| + |l_2\sin\alpha_1|]^2 \qquad (8.6.22)$$

which is contradictory to Condition (8.6.21).

If $l_3^2 = l_2^2 + s_2^2$, then Equation (8.6.17) degenerates to

$$s_4^2 + 2(l_2 S\alpha_1 S\theta_2 - s_2 C\alpha_1)s_4 = 0$$

which always has the following two roots (i.e., s_4 can never be single valued)[16]

---

[16]The first root implies that the slider can remain stationary at the origin while the link $O_2 A$ rotates completely.

8.6. MOBILITY ANALYSIS

$$s_4 = 0$$
$$\text{and} \quad s_4 = 2(s_2 C\alpha_1 - l_2 S\alpha_1 S\theta_2).$$

Thus the condition for complete rotatability of link O_2A can finally be written as

$$l_3^2 \geq l_2^2 + s_2^2 \tag{8.6.23}$$

R-S-S-R Mechanism

Referring to Fig. 8.2-1, if we want to derive the condition for complete rotatability of the input link O_2A, then as usual, we should start from the displacement equation (8.2.2). Substituting for $S\theta_4$ and $C\theta_4$ (in Equation (8.2.2)) in terms of $\tan(\theta_4/2)$, we get a quadratic equation in $(\tan \theta_4/2)$. The condition for θ_4 to be single-valued is obtained by setting the discriminant of this quadratic equation to zero. However, this discriminant polynomial no longer comes out as a second order one in either $\sin\theta_2$ or $\cos\theta_2$. Expressing $S\theta_2$ and $C\theta_2$ in terms of $\tan(\theta_2/2)$, this discriminant polynomial comes out as a fourth order one in $\tan(\theta_2/2)$, in the form $A_4 t^4 + A_3 t^3 + A_2 t^2 + A_1 t + A_0$, where $t = \tan(\theta_2/2)$. The coefficients A_0 to A_4 are complicated functions of the linkage parameters. The conditions for input link rotatability are the same as those for the nonexistence of any real root for t of the quartic equation

$$A_4 t^4 + A_3 t^3 + A_2 t^2 + A_1 t + A_0 = 0 \tag{8.6.24}$$

This in turn implies the nonexistence of any real value of θ_2 corresponding to the extreme position of the input link.

Since the conditions for nonexistence of any real root of a quartic equation are rather cumbersome and the coefficients A_0 to A_4 are also complicated functions of the linkage parameters, no simple criteria for input link rotatability can be written in closed form.[17]

---

[17] For further details, see Williams, R.L.(II) and Reinholtz, F.C.: Mechanism link rotatability and limit position analysis using polynomial discriminants, J. of Mechanism, Transmission and Automation in Design (Trans ASME), Vol. 109, pp. 178-181 (1987).

8.7 Principle of Transference

The concept of dual numbers has been used profitably for displacement analysis of spatial mechanisms. In this section, after discussing some mathematical preliminaries regarding dual numbers and dual angles, we shall complete the displacement analysis of an R-C-C-C spatial linkage by adding the dual symbol to the results already obtained (in Problem 8.3-3) for a 4R spherical linkage. The generalization of this procedure, i.e., passing from spherical to spatial geometry by introducing the dual symbol, is known as the principle of transference. This principle, in fact, confirms the existence of a spherical mechanism corresponding to a given spatial mechanism and vice versa.[18]

8.7.1 Dual Number

A dual number $\hat{a}$ is defined as

$$\hat{a} = a + \epsilon a_o \tag{8.7.1}$$

where a and a_o are real numbers and

$$\epsilon^2 = \epsilon^3 = \text{ all higher powers of } \epsilon = 0. \tag{8.7.2}$$

---

[18] For proofs and details of dual number applications in kinematics of spatial linkage see:

1. Denavit, J.: Displacement analysis of mechanisms based on matrices of dual numbers, VDI Berichte, Vol.29, pp. 81-89 (1958).

2. Keller, M.L.: Analyse und Synthese der Raunm-Kurbelgetriebe mittles Rawnlichen geometrie und dualer Groben. Diss. München (1958). Auszug in Forschung auf d. Gebieted. Ingenieur Wesens, Vol. 25, (1959).

3. Yang, A.T. and Freudenstein, F.: Application of dual-number quaternion algebra to the analysis of spatial mechanisms, J. of Applied Mechanics (Trans ASME - Ser. E) Vol. 31, pp. 300-308 (1964).

4. Keler, M.L.: Analyse und Synthese räumlicher, spharischer und ebener Getriebe in dual-komplexer Darstellung, Feinwerktechnik 74, (1970).

8.7. PRINCIPLE OF TRANSFERENCE

The symbol ^ is used to denote a dual number. The number a is called the *primary part* of $\hat{a}$ and the number a_o is called the *dual part* of $\hat{a}$. Two dual numbers are equal if the respective primary and dual parts are same. The basic arithmetical operations on dual numbers are defined as follows:

If $\hat{a} = a + \epsilon a_o$ and $\hat{b} = b + \epsilon b_o$, then

$$\hat{a} + \hat{b} = (a + b) + \epsilon(a_o + b_o), \tag{8.7.3}$$

$$\hat{a} - \hat{b} = (a - b) + \epsilon(a_o - b_o), \tag{8.7.4}$$

$$\hat{a}.\hat{b} = ab + \epsilon(a_o b + a b_o), \ (\text{see Equation}(8.7.2)) \tag{8.7.5}$$

(i) if $b \neq 0$, then

$$\frac{\hat{a}}{\hat{b}} = \frac{a + \epsilon a_o}{b + \epsilon b_o} = \frac{a + \epsilon a_o}{b + \epsilon b_o} \frac{b - \epsilon b_o}{b - \epsilon b_o}$$

$$= \frac{ab + \epsilon(a_o b - a b_o)}{b^2}$$

$$= \frac{a}{b} + \epsilon \frac{(a_o b - a b_o)}{b^2} \tag{8.7.6a}$$

(ii) if $a = b = 0$ and $b_o \neq 0$, then

$$\frac{\hat{a}}{\hat{b}} = \frac{0 + \epsilon a_o}{0 + \epsilon b_o} = \frac{a_o}{b_o} + \epsilon c_o \tag{8.7.6b}$$

with c_o as an arbitrarily chosen real number.

Functions of a dual number(s) are defined using Taylor's series and Equation (8.7.2) in a manner given below:

$$f(\hat{a}) = f(a + \epsilon a_o) = f(a) + \epsilon a_o \frac{df(a)}{da} \tag{8.7.7}$$

$$f(\hat{a}, \hat{b}, \hat{c}) = f(a + \epsilon a_o, b + \epsilon b_o, c + \epsilon c_o)$$

$$= f(a, b, c) + \epsilon (a_o \frac{\delta}{\delta a} + b_o \frac{\delta}{\delta b} + c_o \frac{\delta}{\delta c}) f(a, b, c) \tag{8.7.8}$$

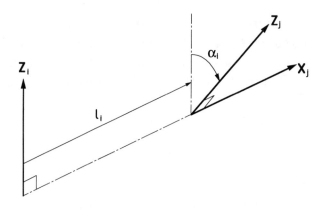

Figure 8.7-1

8.7.2 Dual Angle

The dual angle is used to define the relative positions of two directed skew lines. Referring to Fig. 8.7-1, let Z_i and Z_j be two directed skew lines whose common perpendicular is given by the line X_j, directed from Z_i to Z_j. The relative position of the line Z_j with respect to Z_i is given by two quantities, namely, l_i (the distance of Z_j from Z_i measured along X_j) and α_i (the angle through which Z_i has to rotate about X_j in order to coincide with the direction of Z_j). These two quantities are combined using the dual symbol, and the relative position of Z_j with respect to Z_i is expressed through the dual angle

$$\hat{\alpha}_i = \alpha_i + \epsilon l_i. \tag{8.7.9}$$

Referring to Fig. 8.3-2, we should note that the relative position of the coordinate system $(XYZ)_2$ (attached to the second kinematic pair), with respect to $(XYZ)_1$, can be described in terms of two dual angles, viz.,

$$\hat{\theta}_1 = \theta_1 + \epsilon s_1$$

and

$$\hat{\alpha}_1 = \alpha_1 + \epsilon l_1. \tag{8.7.10}$$

8.7. PRINCIPLE OF TRANSFERENCE

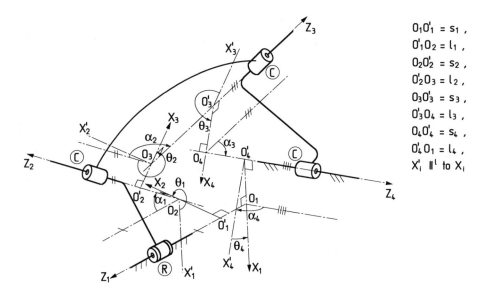

$O_1 O'_1 = s_1$,
$O'_1 O_2 = l_1$,
$O_2 O'_2 = s_2$,
$O'_2 O_3 = l_2$,
$O_3 O'_3 = s_3$,
$O'_3 O_4 = l_3$,
$O_4 O'_4 = s_4$,
$O'_4 O_1 = l_4$,
$X'_i \parallel$ to X_i

Figure 8.7-2

Trigonometric functions of dual angles can be written, using Equations (8.7.7) and (8.7.8), as given below:

$$\sin(\hat{\alpha}_i) = \sin \alpha_i + \epsilon l_i \cos \alpha_i \qquad (8.7.11)$$

$$\cos(\hat{\alpha}_i) = \cos \alpha_i - \epsilon l_i \sin \alpha_i \qquad (8.7.12)$$

It is left for the reader to verify that all trigonometric identities for real angles hold good for dual angles.

8.7.3 Displacement Equation of an R-C-C-C Linkage

An R-C-C-C mechanism is shown in Fig. 8.7-2. First, the link-coordinate systems are established according to Denavit-Hartenberg's convention, explained in Section 8.3.2. Our objective is to obtain the output pair variables s_4 and θ_4 in terms of the input variable θ_1 and the link parameters $\alpha_1, \alpha_2, \alpha_3, \alpha_4$ and s_1. Comparing Figs. 8.7-2 and

8.3-9, we observe that the R-C-C-C linkage is obtained from the 4R spherical linkage by adding dual symbols to θ_i's and α_i's in Fig. 8.3-9, as given below:

$$\hat{\theta}_i = \theta_i + \epsilon s_i$$
$$\hat{\alpha}_i = \alpha_i + \epsilon l_i. \qquad (8.7.13)$$

From the principle of transference, the output-input relationship of the R-C-C-C linkage can be obtained from Equation (j) of Problem 8.3-3 by substituting for each angle its corresponding dual angle. Thus, the displacement equation of the R-C-C-C linkage is obtained as

$$C\hat{\alpha}_4 C\hat{\theta}_4 C\hat{\theta}_1 - S\hat{\theta}_4 S\hat{\theta}_1$$
$$= \cot\hat{\alpha}_1 \cot\hat{\alpha}_3 C\hat{\alpha}_4 - \frac{C\hat{\alpha}_2}{S\hat{\alpha}_1 S\hat{\alpha}_3}$$
$$- \cot\hat{\alpha}_1 S\hat{\alpha}_4 C\hat{\theta}_4 - \cot\hat{\alpha}_3 S\hat{\alpha}_4 C\hat{\theta}_1$$

or,

$$S\hat{\alpha}_1 S\hat{\alpha}_3 C\hat{\alpha}_4 C\hat{\theta}_4 C\hat{\theta}_1 - S\hat{\alpha}_1 S\hat{\alpha}_3 S\hat{\theta}_4 S\hat{\theta}_1$$
$$= C\hat{\alpha}_1 C\hat{\alpha}_3 C\hat{\alpha}_4 C\hat{\alpha}_2 - C\hat{\alpha}_1 S\hat{\alpha}_3 S\hat{\alpha}_4 C\hat{\theta}_4$$
$$- S\hat{\alpha}_1 C\hat{\alpha}_3 S\hat{\alpha}_4 C\hat{\theta}_1. \qquad (8.7.14)$$

We now substitute Equation (8.7.13) in Equation (8.7.14) and use Equations (8.7.11) and (8.7.12); thereafter equating primary and dual parts from both sides, we get the displacement equations for the rotational (θ_4) and translational (s_4) outputs as

$$A(\theta_1)\sin\theta_4 + B(\theta_1)\cos\theta_4 = C(\theta_1) \qquad (8.7.15)$$

and

$$s_4[A(\theta_1)\cos\theta_4 - B(\theta_1)\sin\theta_4] = A_o(\theta_1)\sin\theta_4 + B_o(\theta_1)\cos\theta_4 + C_o(\theta_1)$$
$$(8.7.16)$$

where

$$A(\theta_1) = S\alpha_1 S\alpha_3 S\theta_1,$$
$$B(\theta_1) = -S\alpha_3(S\alpha_4 C\alpha_1 + C\alpha_4 S\alpha_1 C\theta_1),$$
$$C(\theta_1) = C\alpha_2 - C\alpha_3(C\alpha_4 C\alpha_1 - S\alpha_4 S\alpha_1 C\theta_1),$$
$$A_o(\theta_1) = -(l_1 C\alpha_1 S\alpha_3 + l_3 S\alpha_1 C\alpha_3)S\theta_1 - s_1 S\alpha_1 S\alpha_3 C\theta_1,$$
$$B_o(\theta_1) = l_4 S\alpha_3(C\alpha_4 C\alpha_1 - S\alpha_4 S\alpha_1 C\theta_1) - l_1 S\alpha_3(S\alpha_4 S\alpha_1 - C\alpha_4 C\alpha_1 C\theta_1) + l_3 C\alpha_3(S\alpha_4 C\alpha_1 + C\alpha_4 S\alpha_1 C\theta_1)$$
$$- s_1 S\alpha_3 C\alpha_4 S\alpha_1 S\theta_1 \text{ and}$$
$$C_o(\theta_1) = l_4 C\alpha_3(S\alpha_4 C\alpha_1 + C\alpha_4 S\alpha_1 C\theta_1) + l_1 C\alpha_3(C\alpha_4 S\alpha_1 + S\alpha_4 C\alpha_1 C\theta_1) - l_2 S\alpha_2 + l_3 S\alpha_3(C\alpha_4 C\alpha_1 - S\alpha_4 S\alpha_1 C\theta_1)$$
$$- s_1 C\alpha_3 S\alpha_4 S\alpha_1 S\theta_1$$

It is obvious that Equation (8.7.15) is identical to Equation (j) of Problem 8.3-3, which in turn is the same as Equation (8.2.5), with necessary changes of symbols. Therefore, from Equations (8.7.15) and (8.7.16) we can say that Table 8.6-2 is applicable to an R-C-C-C linkage also, with proper identification of symbols for various link parameters (see Exercise Problem 8.9).

8.8 Exercise Problems

8.1 Using "the rule of seven" mentioned in Section 8.2, list all possible spatial mechanisms (disregarding kinematic inversions of the same chain) consisting of only binary links that can be generated using different combinations of R, C and S pairs.

8.2 Establish the Denavit-Hartenberg link-coordinate system for the R-C-R-C-R linkage shown in Fig. 8.8-1. Identify the link parameters and indicate the input and output variables.

8.3 For a Hooke's joint with shaft angle α_4, if the input shaft 2 rotates at a constant speed, determine the ratio of the maximum to minimum speed of the output shaft 4.

CHAPTER 8. SPHERICAL AND SPATIAL LINKAGES

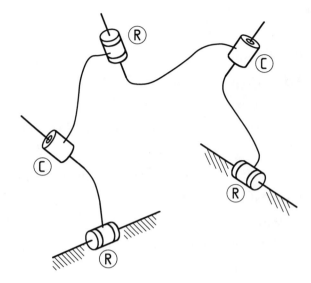

Figure 8.8-1

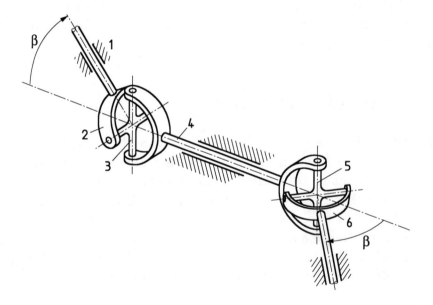

Figure 8.8-2

8.8. EXERCISE PROBLEMS

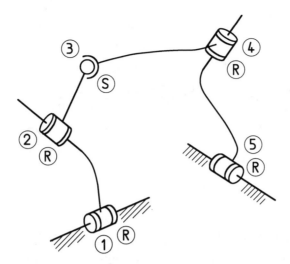

Figure 8.8-3

8.4 Two Hooke's joints are arranged in series (back to back), as shown in Fig. 8.8-2. Show that the resulting linkage transmits constant unit angular velocity ratio between shafts 2 and 6 when the angles between shafts 2 and 4 and 4 and 6 are equal, as indicated in the figure.

8.5 A generalized Clemens coupling in the form of an R-R-S-R-R spatial linkage is shown in Fig. 8.8-3. Establish the link-coordinate systems and write all the [A] matrices for this linkage after identifying the link parameters and pair variables.

8.6 The Clemens coupling (Fig. 8.8-4) is a special case of the R-R-S-R-R linkage shown in Fig. 8.8-3 with the following parameters:
$l_2 = l_3$, $\alpha_1 = \alpha_4 = \pi/2$, $s_2 = s_4 = 0$, $l_5 = 0$, $l_1 = l_4 = 0$, $s_1 = s_5$

Show that this coupling transmits a constant unit angular velocity ratio.

8.7 Starting from the displacement equation of a 4R spherical mechanism, given by Equation (j) of problem 8.3-3, synthesize this mechanism to generate the function

$$y = \sin x \quad 0 \leq x \leq \pi/2$$

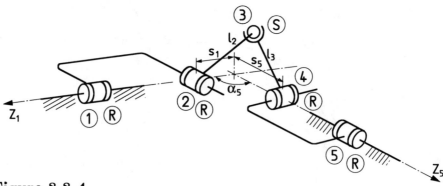

Figure 8.8-4

with three Chebyshev's accuracy points. Choose design parameters in order to obtain linear equations in them. Assume shaft angle $\alpha_4 = 90°$, $\Delta\theta_1 = 60°$, $\Delta\theta_4 = 90°$, and initial values of θ_1 and θ_4 are $\theta_1^i = 45°$ and $\theta_4^i = -45°$.

8.8 Synthesize a 4R spherical mechanism in order to generate a symmetric function

$$y = x^2, \quad -1 \le x \le 1$$

with proper choices of parameters. Obtain solutions for both three and five accuracy points. Use the data of Problem 8.5-2.

8.9 Obtain the mobility criteria for type identification of the R-C-C-C linkage shown in Fig. 8.7-2.

8.10 An R-R-R-(R_3P) spatial linkage is shown in Fig. 8.8-5. Obtain the [A] matrices and convince yourself that only a numerical solution is possible for the displacement analysis, i.e., to obtain the output movement θ_4 in terms of input movement θ_1 and the link parameters.

8.8. EXERCISE PROBLEMS

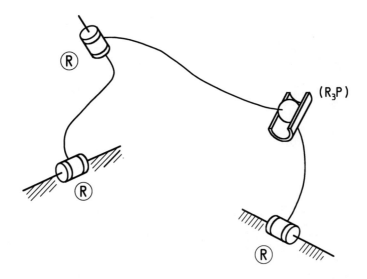

Figure 8.8-5

Chapter 9
CAM MECHANISMS

Cam mechanisms play an extremely important role in modern machinery and are extensively used in mechanical systems. The storage of information through the shape of a cam made it possible not only to generate complex, coordinated movements but also to introduce automation before the concept of numerical control came into existence. Use of linkages to generate a particular desired motion is not always possible, specially when a given complex motion has to be generated very accurately. Besides, cam mechanisms are relatively compact and easy to design. In this chapter we shall discuss the methods required for the analysis and synthesis of cam mechanisms.

9.1 Types and Fundamentals of Cam Mechanisms

The basic principle of all cam mechanisms is to generate a desired relative motion between two links through a higher pair contact. As in the case of linkages, cam mechanisms can be also classified into two primary groups as follows:

(i) Plane cam mechanisms: In such systems all the points of the system move on parallel planes.

(ii) Space cam mechanisms: The points on such mechanisms do not move in parallel planes.

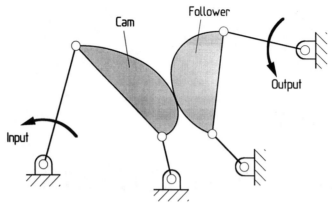

Figure 9.1-1

First, the basic principles involved in the analysis and synthesis of cam mechanisms are explained, with reference to plane cam mechanisms. The discussion on space cam mechanisms is presented later in this chapter.

In general, a desired motion can be generated using a higher pair contact (maintained by a sustained external effort) between the coupler links of two 4R linkages, as indicated in Fig. 9.1-1. Any complex relationship between the output and input motions can be achieved by suitably synthesizing the profiles of the cam and the follower. In its most fundamental (and commonly used) form, a cam mechanism consists of three links connected by two lower pairs and one higher pair. The higher pair contact between the cam (whose shape contains the information regarding the output-input relationship) and the follower link can be

(i) point/line contact (knife-edge follower) of sliding type,

(ii) point/line contact (flat- or curved-face follower) of sliding type, and

(iii) line contact (roller follower) of rolling type.

Roller followers are used to reduce friction and increase the cam life. However, an extra member (the roller) with a redundant degree of freedom is introduced. Figure 9.1-2 shows the above three

9.1. TYPES AND FUNDAMENTALS OF CAM MECHANISMS

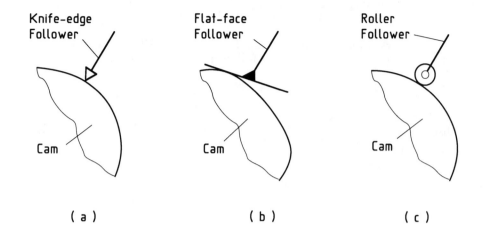

Figure 9.1-2

Figure 9.1-3

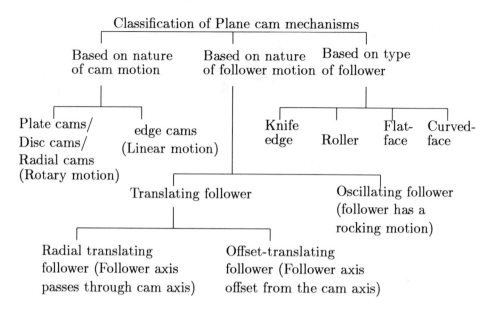

types of followers. The various possible combinations of kinematic pairs, inversions and nature of motion for a three-link (disregarding the roller with its redundant rotary movement) plane cam mechanism, are indicated in Fig. 9.1-3 although not all of these forms are equally popular, the complete range of possibilities become evident from this systematic presentation.

9.1.1 Types of Plane Cam Mechanisms

Figure 9.1-3 shows all the possible combinations of features in plane cam mechanisms. However, in most cases a continuously rotating cam is used. Classification of the plane cam mechanism can be based on different points of view. The cams that are used in plane cam mechanisms are usually in the form of plates (or discs) with a prescribed contour geometry. These cams are called *plate cams* (or *disc cams* or *radial cams*), when mounted on a rotating cam shaft. When cams slide instead of rotating they are normally referred to as *wedge cams*. The tree diagram given above shows the structure of classification of cam-follower systems. Different types of plane cam mechanisms com-

9.1. TYPES AND FUNDAMENTALS OF CAM MECHANISMS

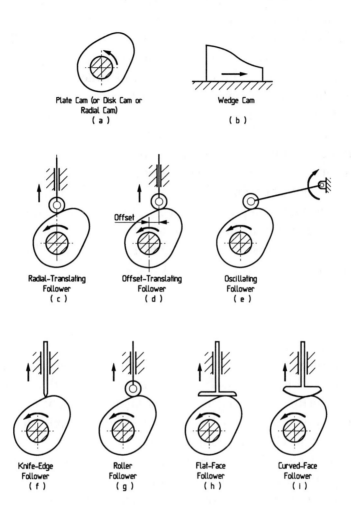

Figure 9.1-4

monly used are indicated in Fig. 9.1-4. In practice, plate cams are used most frequently and we will limit our discussions primarily to this type. The knife-edge follower, though simple from the point of view of analysis, is rarely used because the wear rate is high. The flat-face followers give rise to less side thrust at the follower guide (and consequently, result in a lower friction force and have less chance of jamming) than that for knife-edge and roller followers. The tendency to jam at the follower guide can be also controlled by suitably offsetting the follower from the axis of rotation of the cam. The sliding wear, in case of a flat-face translating follower, is further reduced by offsetting the follower in a direction perpendicular to the plane of cam rotation, so that the follower rotates about its axis of translation. The flat-face (and curved-face) followers are used in situations where space is a problem and compactness is essential. The use of roller followers in such situations is restricted by the minimum size of the pin to be used to connect the roller with the follower. Roller followers are useful when large forces are to be transmitted.

In cam-follower mechanisms the contact between the cam and the follower has to be ensured. In simple situations and with low speeds of operation, this can be done using either the gravity or springs, as indicated in Figs. 9.1-5a to 9.1-5c. If the roller is constrained to move inside a groove with a specific geometry, as shown in Fig. 9.1-5d, then the follower is also constrained to move when the cam shaft rotates. Mechanical constraint can also be achieved by using *dual* or *conjugate cams*, shown in Fig. 9.1-5e. In this arrangement, each cam has its own roller but both the rollers are mounted on the same follower link. *Constant-breadth cams* can also provide mechanical constraint because of two contact points between the cam and the follower, as indicated in Fig. 9.1-5f. However, with such an arrangement the nature of follower motion cannot be chosen at will, only the amount of displacement can be prescribed. In Figs. 9.1-5a to 9.1-5c, the higher pair is force-closed whereas, in Figs. 9.1-5d to 9.1-5f, the higher pair is form-closed.

9.1. TYPES AND FUNDAMENTALS OF CAMMECHANISMS 449

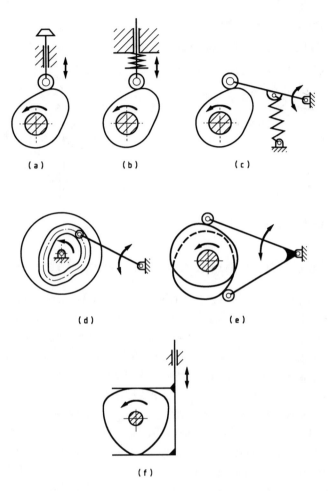

Figure 9.1-5

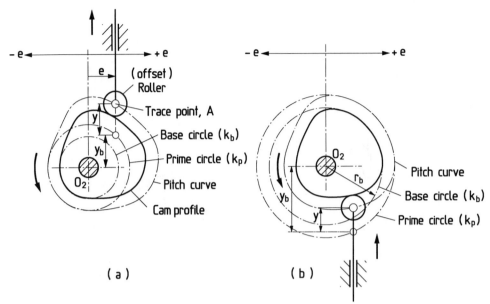

Figure 9.1-6

9.1.2 Basic Features and Nomenclature

Figure 9.1-6a shows a plate (or disc or radial) cam driving an offset translating roller follower. The smallest circle that can be drawn with the cam center O_2 as the center and touching the cam profile, is called the *base circle* k_b. The point on the follower, whose motion is used to describe the follower motion, is called the *trace point*. In the case of a knife-edge follower, the point at the edge is a convenient trace point.[1] For a roller follower, the trace point is at the roller center (as indicated in Fig. 9.1-6a), and for a flat-face follower it is at the point of contact between the follower and the cam surface when the contact is along the base circle of the cam. Obviously, the trace point is not necessarily the point of contact for all other positions of the cam. If we apply the principle of inversion, i.e., if we hold the cam fixed and rotate the follower in a direction opposite to that of the cam, then the curve generated by the locus of the trace point is called the *pitch curve*. Obviously, for a knife-edge follower, the pitch curve

---

[1] As only plane motion is being considered, the projection of the cam-follower system on the plane of motion is sufficient for complete description.

9.1. TYPES AND FUNDAMENTALS OF CAM MECHANISMS

and the cam profile are identical. For a roller follower, the pitch curve is parallel to the cam profile at a distance equal to the roller radius. The smallest circle that can be drawn with the cam center O_2 as the center touching the pitch curve, is called the *prime circle* k_p. Hence, for a roller follower,

$$r_p = r_b + r_r \qquad (9.1.1)$$

where r_p = prime circle radius, r_b = base circle radius and r_r = radius of the roller. The distance of the axis of the translating follower from the cam center O_2 is called the *offset e*. The positive and negative senses of e, for a given direction of cam rotation, are indicated in Fig. 9.1-6a.

The prime circle k_p is a circle of radius r_p from which the motion of the trace point A starts (in case of knife-edge or flat-face follower it is the base circle k_b), and this circle determines the datumn position of the follower for a given amount of offset. In the case shown in Fig. 9.1-6a, the point A moves away from the cam center O_2 during the forward movement of the follower. This type of cam mechanism is called *F-cam mechanisms*.[2] On the other hand, in the arrangement shown in Fig. 9.1-6b, the follower moves toward the cam center O_2,

---

[2]The letters "F" and "P" imply centrifugal and centripetal movements, respectively.

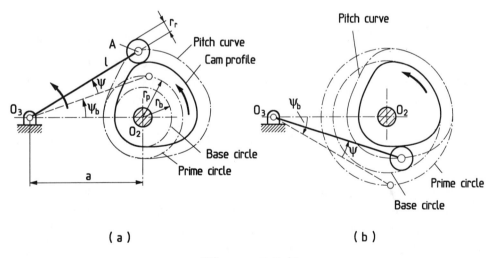

(a)   (b)

Figure 9.1-7

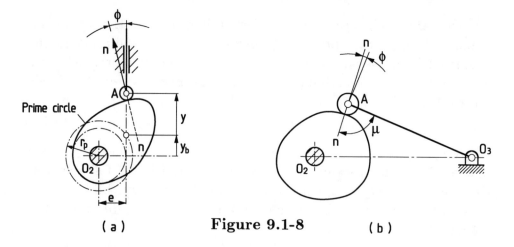

Figure 9.1-8

corresponding to any movement from the datumn. The base circle, in such cases, is obviously the largest circle which can be drawn with O_2 as the center and touching the cam profile. Such mechanisms are called *P-cam mechanisms*. For an oscillating roller follower the F- and P-cam mechanisms are shown in Figs. 9.1-7a and 9.1-7b, respectively. The datumn position angle from the line of frame O_2O_3, ψ_b, is given by

$$\psi_b = \Gamma \cos^{-1}[(a^2 + l^2 - r_p^2)/2al] \qquad (9.1.2)$$

where $\Gamma = +1$ for F-cam mechanisms and -1 for P-cam mechanisms.

The dimensions a, l and r_b are indicated in Fig. 9.1-7a. For translating roller followers the datumn position y_b can be written as

$$y_b = \Gamma\sqrt{r_p^2 - e^2} \qquad (9.1.3)$$

where $\Gamma = +1$ for F-cam mechanisms and -1 for P-cam mechanisms. The prime circle radius (or the base circle radius) specifies the size of a cam. It will be seen later that r_p has an important role to play in determining the transmission characteristics of a cam-follower mechanism.

9.1.3 Pressure Angle

The pressure angle is one of he most important parameters for designing cams with roller followers.[3] However, it does not have any direct influence on the follower displacement-cam rotation relationship. But it is an indication of the force transmission characteristics. Figures 9.1-8a and 9.1-8b show F-cam follower mechanisms with translating and oscillating roller followers, respectively. The *pressure angle* is defined as the angle ϕ between the direction of force applied on the follower and the direction of motion of the trace point A. The direction of force can be taken to be normal to the cam surface at the point of contact. The direction of movement of the roller center (trace point) A is along the axis of translation, in the case of translating followers, and perpendicular to the follower O_3A, in the case of oscillating followers, as shown in Fig. 9.1-8. The reader should note that the pressure angle ϕ is, by definition, equal to $\pi/2$ - transmission angle (μ), defined in Section 4.4.

The pressure angle should be as small as possible. In case of slow cam mechanisms with oscillating followers, it is recommended that

$$\phi_{\max} \leq 45^o.$$

In case of translating followers at higher speeds of operation (i.e. cam r.p.m. $>$ 50), the recommended value is

$$\phi_{\max} \leq 30^o.$$

By increasing the size of the cam (i.e., the base circle), the pressure angle can be reduced, however, other problems, such as space limitations and unbalance at high speeds, become critical. The pressure angle can also be controlled by adjusting the offset in case of translating roller followers. For a given cam size, the pressure angle, as we shall see later, depends on the nature of the follower motion.

---

[3]In case of flat-face followers, the distance of the contact point from the axis of translation is a measure of transmission quality the larger the distance, the higher the tendency of jamming of the follower in the guides.

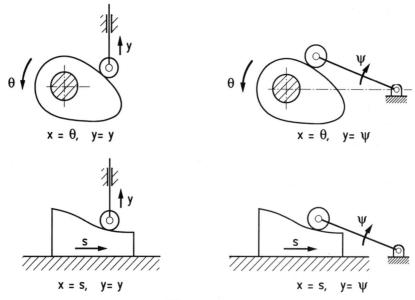

Figure 9.2-1

9.2 Follower Movement, Displacement Diagram

A cam-follower mechanism can be considered to be a function generator in which, for a given input motion (cam movement) a desired output motion (follower movement) is obtained. If x is the input then the output y is a function of x, $y = y(x)$. Commonly, four different combinations are possible. The input can be an angular rotation θ of the cam (mounted on a cam shaft) and the output can be either a linear displacement y of the translating follower or an angular displacement ψ of the oscillating follower. Again, in the case of wedge cams, the input can be a linear displacement s of the cam and the output can be either a linear displacement y of the translating follower or an angular displacement ψ of an oscillating follower. These are illustrated in Fig. 9.2-1. However, the most common situation is that in which a cam, mounted on a shaft, rotates at a constant speed. There can be two different types of follower displacement. In one, the complete relationship of the follower movement and the

9.2. FOLLOWER MOVEMENT, DISPLACEMENT DIAGRAM

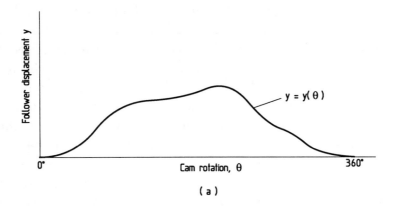

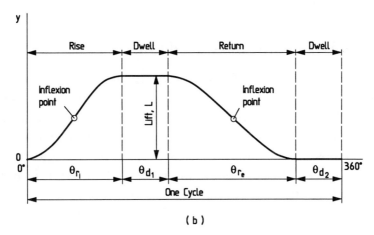

Figure 9.2-2

cam rotation is prescribed (i.e., $y = y(x)$ for the complete range of motion is fully prescribed). In the other case, only the amounts of displacements and their temporal relationships with the input motion are prescribed. Figure 9.2-2a shows the first situation, in which y, as a function of θ, is completely specified. In Fig. 9.2-2b, the follower displacement, during a complete rotation of the cam, shows the amount of total displacement (lift) and its location in the cycle. The second type of situation is more frequent in practice, though in some cases (viz., automatic screw cutting machines) the complete function $y = y(\theta), 0 \leq \theta \leq 360°$, is specified.

The plot showing y vs. θ is called the *displacement diagram*.

Figure 9.2-2b shows a typical displacement diagram in which the maximum follower displacement is referred to as *lift*, L, of the follower. It is seen that, in general, the displacement diagram consists of four parts, namely, (i) the *rise* (which is the movement of the follower away from the cam center)[4], (ii) the *dwell* (when there is no movement of the follower), (iii) the *return* (when the follower comes towards the cam center) and (iv) the *dwell*. Unless otherwise specified, the cam mechanisms will be of F-type and it is always assumed that there is a dwell before and after the rise. The inflexion points (corresponding to maximum and minimum velocities of the follower) separate the accelerating and decelerating portions in the corresponding parts of the displacement diagram.

9.2.1 Displacement Functions

Most often the natures of follower movement during the rise and return of the follower are not prescribed and are to be decided by the designer. The decision is based on minimum cost of production while still maintaining the performance quality. The smoothness of performance depends on the choice of $y(\theta)$ during the rise and return.

It is convenient to handle the displacement function in nondimensionalized form. If we consider the rise portion of the displacement diagram, then the displacement function in non-dimensional form can be expressed as follows:

$$\frac{y}{L} = \bar{y}(\frac{\theta}{\theta_{ri}}) = \bar{y}(z); \quad 0 \leq z \leq 1 \tag{9.2.1}$$

where L is the lift, $z = \theta/\theta_{ri}$ and θ_{ri} is the cam rotation for the rise. The function $\bar{y}(z)$ is such that

$$\bar{y}(0) = 0 \quad \text{and} \quad \bar{y}(1) = 1. \tag{9.2.2}$$

If AB is the portion of the displacement diagram (Fig. 9.2-3) during rise and P is the inflexion point (i.e., the curve has zero curvature

---

[4]In case of P-cam mechanism it will be towards the cam center. But since F-cam mechanisms are much more common, from here forward we will discuss primarily F-cam mechanisms without mentioning it every time.

9.2. FOLLOWER MOVEMENT, DISPLACEMENT DIAGRAM

at P), then the follower accelerates during the part AP and then decelerates (ultimately the velocity becomes again zero at B) during PB. If the overall magnitude of acceleration (or deceleration) during the rise has to be minimum, then the accelerating and decelerating portions must be symmetrical. Such a situation is obtained when the nondimensionalized displacement function $\bar{y}(z)$ satisfies the following relation:

$$\bar{y}(z) = 1 - \bar{y}(1-z) \qquad (9.2.3)$$

The above equation can be easily derived from Fig. 9.2-3. The condition necessary for symmetry of the portions AP and BP is $SU = VT$, which directly yields Equation (9.2.3). If θ_{rs} is the cam rotation during the return phase of the displacement diagram, then the displacement function for the return phase can be written as follows:

$$\frac{y}{L} = 1 - \bar{y}(z) \qquad (9.2.4)$$

where $z = (\theta - \theta_{d_1} - \theta_{ri})/\theta_{re}$, with θ_{d_1} as the cam rotation for the first dwell period.

When the nondimensionalized displacement function $\bar{y}(z)$ is prescribed, with L and the constant rotational speed of the cam given, all

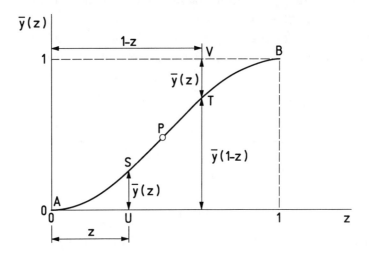

Figure 9.2-3

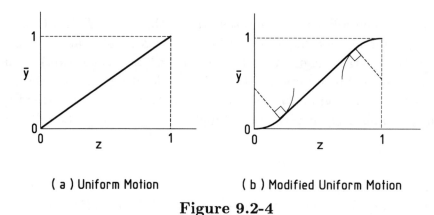

(a) Uniform Motion   (b) Modified Uniform Motion

Figure 9.2-4

the time derivatives of follower motion can be determined as follows:

$$
\begin{array}{ll}
\text{Rise} & \text{Return} \\
y = L\bar{y}(z) & y = L[1 - \bar{y}(z)] \\
\dot{y} = \frac{L}{\theta_{ri}}\omega\bar{y}'(z) & \dot{y} = \frac{L}{\theta_{re}}\omega\bar{y}'(z) \\
\ddot{y} = \frac{L}{\theta_{ri}^2}\omega^2\bar{y}''(z) & \ddot{y} = \frac{L}{\theta_{re}^2}\omega^2\bar{y}''(z)
\end{array}
\qquad (9.2.5)
$$

where dot ($\cdot$) denotes differentiation with respect to time, and prime ($'$) denotes differentiation with respect to the nondimensional input variable z. The cam rotates with a constant angular velocity ω. In case of oscillating followers, the various derivatives are as follows:

$$
\begin{array}{ll}
\text{Rise} & \text{Return} \\
\psi = \psi_L\bar{y}(z) & \psi = \psi_L[1 - \bar{y}(z)] \\
\dot{\psi} = \frac{\psi_L}{\theta_{ri}}\omega\bar{y}'(z) & \dot{\psi} = \frac{\psi_L}{\theta_{re}}\omega\bar{y}'(z) \\
\ddot{\psi} = \frac{\psi_L}{\theta_{ri}^2}\omega^2\bar{y}''(z) & \ddot{\psi} = \frac{\psi_L}{\theta_{re}^2}\omega^2\bar{y}''(z)
\end{array}
\qquad (9.2.6)
$$

where ψ_L is the lift of the oscillating follower.

While deciding on the displacement functions during rise and return the designer should try to keep the jerks (third derivative) and the accelerations as low as possible. In the following sections some standard types of displacement functions are discussed.

9.2.2 Uniform Motion

By uniform motion (Fig. 9.2-4a) we mean that the velocity of the follower is constant. The displacement function is given by

$$\bar{y}(z) = z \text{(for rise) and} \bar{y}(z) = 1 - z \text{(for return)}. \tag{9.2.7}$$

It is obvious that though this is the simplest type of motion, it finds very little use in practice because of the instantaneous change in velocity at the beginning and end of the period under consideration. To avoid these sudden changes, the straight line displacement diagram is connected tangentially to the dwell at both ends, by means of circular arcs of a convenient radius, as shown in Fig. 9.2-4b. The bulk of the displacement takes place at uniform velocity.

9.2.3 Simple Harmonic Motion

Very large higher order derivatives can be avoided by using simple harmonic displacement functions. The graphical determination of the displacement diagram with simple harmonic motion can be obtained as explained in Fig. 9.2-5. The line segment representing $0 \leq z \leq 1$ is divided into a convenient number of equal parts. A semicircle is drawn on the ordinate, with the segment representing $0 \leq \bar{y}(z) \leq 1$ as diameter, and this semicircle is divided into the same number of equal arcs. Horizontal lines are drawn from the points so obtained on the semicircle to meet the corresponding vertical lines through the points on the abscissa. Analytical expression of the displacement function is

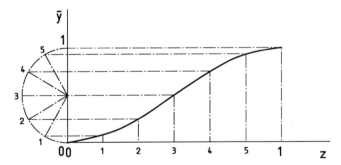

Figure 9.2-5

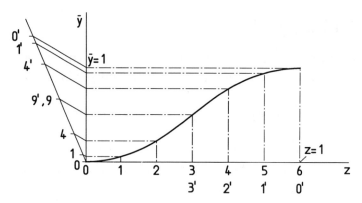

Figure 9.2-6

as follows:

$$\bar{y}(z) = \frac{1}{2}[1 - \cos(\pi z)] \qquad (9.2.8)$$

9.2.4 Parabolic or Piecewise Uniform Acceleration Motion

With dwell at both the ends of the rise-period, if the lift of the follower has to take place in a given time, it is easy to show that the maximum acceleration is the least if the first half of the rise takes place at a constant acceleration and the remaining displacement is at a constant deceleration (of the same magnitude). This fact makes parabolic motion very suitable for high-speed cams, as it minimizes the maximum inertia force. However, it should be noted that it produces an infinite jerk (rate of change of acceleration) at the inflexion point. The graphical construction of the displacement diagram is explained in Fig. 9.2-6. An even number of equal divisions, n (in the figure it is six), are marked on the segment representing $0 \leq z \leq 1$ (i.e., either the rise or the return). For locating the corresponding vertical divisions, we make use of the fact that, with constant acceleration, the displacement is proportional to the square of time (i.e., it is proportional to the square of the cam rotation since the cam rotates at a constant speed) for the accelerating first half. This is also true for the decelerating second half, if the origin is shifted to the end and the displacement direction is reversed (Fig. 9.2-6). Analytical expression

9.2. FOLLOWER MOVEMENT, DISPLACEMENT DIAGRAM

for y(z) is as follows:

Accelerating Part $\bar{y}(z) = 2z^2$; $\quad 0 \leq z \leq 0.5$

Decelerating Part $\bar{y}(z) = 1 - 2(1-z)^2$; $\quad 0.5 \leq z \leq 1$ $\quad$ (9.2.9)

In some cases (viz., cams operating valves of internal combustion engines), the modified uniform acceleration motion is used for the follower. For example, in case of engines, it is desired that the valves open very quickly at the beginning of its opening motion and close very quickly at the end of its closing motion. In this modified parabolic motion, the acceleration during the first part (the accelerating part) of the rise is more than the decelaration in the second part (the decelerating part). Similarly, for the return phase, the acceleration during the first part is less than the deceleration in the second part. If f_1 and f_2 are the magnitudes of the acceleration and deceleration, respectively, during the first and second parts of the rise, and

$$f_1 = Kf_2, \quad (9.2.10)$$

then it is easy to prove that

$$\theta_{ri_2} = K\theta_{ri_1}, \quad (9.2.11)$$

where θ_{ri_1} and θ_{ri_2} are the angles of cam rotation during the first and second parts of the rise, respectively. The lift L is also given by

$$L = L_1 + KL_1, \quad (9.2.12)$$

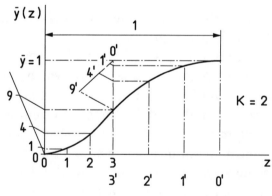

Figure 9.2-7

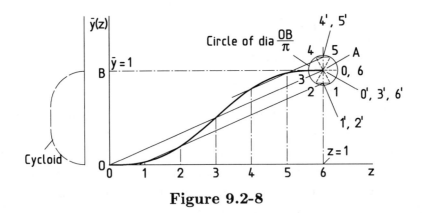

Figure 9.2-8

where L_1 is the amount of rise during the first part. The graphical construction of the displacement diagram is similar to that explained in Fig. 9.2-6 and illustrated in Fig. 9.2-7, using $K = 2$. Analytically, the expressions are as follows:

Accelerating Part: $\bar{y}(z) = (1+K)z^2; \quad 0 \leq z \leq \dfrac{1}{1+K}$

Decelerating Part: $\bar{y}(z) = 1 - \dfrac{1+K}{K}(1-z)^2; \quad \dfrac{1}{1+K} \leq z \leq 1$

(9.2.13)

9.2.5 Cycloidal Motion

Cycloidal motion is obtained by rolling a circle of diameter $1/\pi$ on the ordinate of the displacement diagram (i.e., the $\bar{y}$ axis). A point P on the circle describes a cycloid. Cycloidal displacement can be considered to be a modification of a sinusoid, which gives zero second derivatives at the beginning and at the end. A convenient graphical construction of a cycloidal displacement diagram is shown in Fig. 9.2-8. A circle of diameter $1/\pi$ is drawn with its center at the end A of the displacement diagram as shown. This circle is divided into the same number (say six) as the abscissa of the diagram showing z. The points, so obtained on the circle, are numbered $0, 1, 2, \ldots 6$. The projections of these points on the vertical diameter are numbered as

$0', 1', 2', \ldots 6'$. Lines parallel to OA are drawn from these points as shown. The displacement diagram is obtained from the intersection of the vertical lines through the points on the abscissa and the corresponding lines parallel to OA. The analytical expression for the displacement function is given by

$$\bar{y}(z) = z - \frac{1}{2\pi}\sin(2\pi z). \qquad (9.2.14)$$

9.2.6 Polynomial Functions

In many situations the basic follower displacement functions discussed so far are inadequate for smooth operation. This is particularly so for high speed operation, as the higher derivatives are very large. Sometimes an acceptable solution is found by combining more than one basic form to generate the required cam profile for rise. One way to design a cam profile to satisfy a number of conditions is to use polynomial functions. Such polynomial curves, representing the follower motion, are sometimes called *advanced cam curves*. The nondimensionalized displacement function, represented by a polynomial of order n, can be written in the following form:

$$\bar{y}(z) = \sum_{i=1}^{n} \lambda_i z^i \qquad (9.2.15)$$

The initial condition, $\bar{y}(0) = 0$, is automatically satisfied by Equation (9.2.15). The other condition, $\bar{y}(1) = 1$, and the prescribed values of up to the mth order derivative at each end (i.e., at $z = 0$ and $z = 1$), can be satisfied by choosing an nth order polynomial where $n = 2m + 1$. The resulting equations and the solution are as follows:[5]

---

[5] For details see Schönherr, J.: Synthese der Ubertragungsfunktion von Kurvengetrieben durch stückweise Polynom - Interpolation, Tagung Maschinendynamik und Getriebetechnik, Dresden, 1985.

$$\lambda_1 = \bar{y}'(0)$$
$$2\lambda_2 = \bar{y}''(0)$$
$$\ldots \quad \ldots$$
$$1.2.3\ldots m\lambda_m = \bar{y}^{(m)}(0)$$
$$\lambda_1 + \lambda_2 + \ldots + \lambda_m + \lambda_{m+1} + \ldots + \lambda_{2m+1} = \bar{y}(1) = 1$$
$$\lambda_1 + 2\lambda_2 + \ldots + m\lambda_m + (m+1)\lambda_{m+1} + \ldots + (2m+1)\lambda_{2m+1} = \bar{y}'(1)$$
$$2\lambda_2 + \ldots + (m-1)m\lambda_m + m.(m+1)\lambda_{m+1} + \ldots + 2m.(2m+1)\lambda_{2m+1} = \bar{y}''(1)$$
$$\ldots \quad \ldots$$
$$1.2.3\ldots(m+1)\lambda_{m+1} + \ldots + (m+1)(m+2)\ldots(2m+1)\lambda_{2m+1} = \bar{y}^{(m)}(1)$$

The solution to the above set of equations will be in the following form:

$$\lambda_i = \frac{1}{i!}\bar{y}^{(i)}(0); i = 1, 2, \ldots, m \quad (9.2.16a)$$

$$\lambda_{m+i} = -\sum_{j=1}^{m}[\frac{1}{j!}\sum_{l=1}^{j+1}\{(-1)^{i+1j}C_{l-1}$$

$$\sum_{k=\max(i,l)}^{m+1}(^{k-1}C_{i-1} \cdot ^{m+k-l}C_{k-l})\}\bar{y}^{(j)}(0)$$

$$+\sum_{j=1}^{m+1}\{\frac{1}{(j-1)!}\sum_{k=\max(i,j)}^{m+1}(^{k-1}C_{i-1} \cdot ^{m+k-l}C_{k-j})\}\bar{y}^{(j-1)}(1)] \quad (9.2.16b)$$

For example, if $\bar{y}(1)$, $\bar{y}'(0)$, $\bar{y}''(0)$, $\bar{y}'(1)$ and $\bar{y}''(1)$ are prescribed, then a 5th order polynomial,

$$\bar{y}(z) = \lambda_1 z + \lambda_2 z^2 + \lambda_3 z^3 + \lambda_4 z^4 + \lambda_5 z^5 \quad (9.2.17)$$

can be chosen. From Equations (9.2.16a) and (9.2.16b),

$$\lambda_1 = \bar{y}'(0),$$

9.2. FOLLOWER MOVEMENT, DISPLACEMENT DIAGRAM

$$\lambda_2 = \frac{1}{2}\bar{y}''(0),$$

$$\lambda_3 = 10 - 6\bar{y}'(0) - 4\bar{y}'(1) - \frac{3}{2}\bar{y}''(0) + \frac{1}{2}\bar{y}''(1), \qquad (9.2.18)$$

$$\lambda_4 = -15 + 8\bar{y}'(0) + 7\bar{y}'(1) + \frac{3}{2}\bar{y}''(0) - \bar{y}''(1)$$

and

$$\lambda_5 = 6 - 3[\bar{y}'(0) + \bar{y}(1)] - \frac{1}{2}[\bar{y}''(0) - \bar{y}''(1)].$$

When $\bar{y}(1) = 1$ and $\bar{y}'(0) = \bar{y}''(0) = \bar{y}'(1) = \bar{y}''(1) = 0$ Equation (9.2.17) with Equation (9.2.18) yield the well known 3-4-5 polynomial, as given below:

$$\bar{y}(z) = 10z^3 - 15z^4 + 6z^5$$

If up to 3rd order derivatives are prescribed at the ends, then a 7th order polynomial,

$$\bar{y}(z) = \lambda_1 z + \lambda_2 z^2 + \lambda_3 z^3 + \lambda_4 z^4 + \lambda_5 z^5 + \lambda_6 z^6 + \lambda_7 z^7 \qquad (9.2.19)$$

is necessary to describe the displacement function. The coefficients are as follows:

$$\lambda_1 = \bar{y}'(0)$$

$$\lambda_2 = \frac{1}{2}\bar{y}''(0)$$

$$\lambda_3 = \frac{1}{6}\bar{y}'''(0)$$

$$\lambda_4 = 35 - 20\bar{y}'(0) - 15\bar{y}'(1) - 5\bar{y}''(0) + \frac{5}{2}\bar{y}''(1) - \frac{2}{3}\bar{y}'''(0) + \frac{1}{6}\bar{y}'''(1)$$

$$\lambda_5 = -84 + 45\bar{y}'(0) + 39\bar{y}'(1) + 10\bar{y}''(0) - 7\bar{y}''(1) + \bar{y}'''(0) + \frac{1}{2}\bar{y}'''(1)$$

$$\lambda_6 = 70 - 36\bar{y}'(0) - 34\bar{y}(1) - \frac{15}{2}\bar{y}''(0) + \frac{13}{2}\bar{y}''(1) - \frac{2}{3}\bar{y}'''(0) - \frac{1}{2}\bar{y}'''(1)$$

$$\lambda_7 = -20 + 10[\bar{y}'(0) + \bar{y}'(1)] + 2[\bar{y}''(0) - \bar{y}''(1)] + \frac{1}{6}[\bar{y}'''(0) + \bar{y}'''(1)]$$

$$(9.2.20)$$

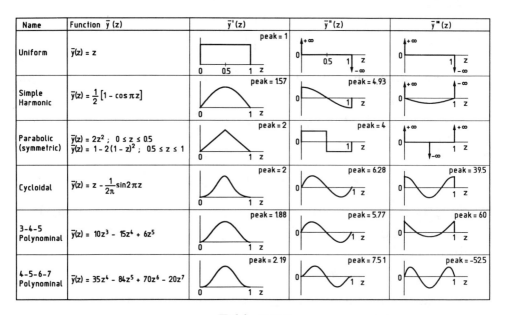

Table 9.2-1

When $\bar{y}(1) = 1$ and $\bar{y}'(0) = \bar{y}''(0) = \bar{y}'''(0) = \bar{y}'(1) = \bar{y}''(1) = \bar{y}'''(1) = 0$, we get the following 4-5-6-7 polynomial from Equations (9.2.19) and (9.2.20):

$$\bar{y}(z) = 35z^4 - 84z^5 + 70z^6 - 20z^7$$

Table 9.2-1 presents important information on the various typical displacement functions.[6]

Sometimes optimization procedures are adopted to achieve minimum possible peak values of the various derivatives.[7] Such displacement functions are useful in developing high precision high-speed cam follower mechanisms.

9.3 Determination of Basic Dimensions

Before the task of determining the cam profile to generate a prescribed displacement function is begun, it is necessary to obtain a few

---

[6] For a more comprehensive treatment of the topic see - Volmer, J.: Getriebetechnik: Kurvengetriebe, VEB Verlag Technik, Berlin, 1989.

[7] See Volmer, J., op. cit.

9.3. DETERMINATION OF BASIC DIMENSIONS

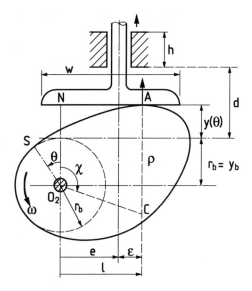

Figure 9.3-1

important basic dimensions. This is so because such dimensions are essential for synthesizing a cam profile. The designer may not be provided with these dimensions and must determine them on the basis of other information. The most important point to be noted by the designer is that a minimum quality of motion transmission during the whole cycle is ensured. An effective way to do that is to select the values of the basic dimensions in such a way that the pressure angle is maintained within certain limit. Thus, the maximum permissible pressure angle, ϕ_{max} (or the minimum permissible transmission angle, μ_{min}), is the starting parameter. In this section, methods to determine the basic dimensions from the prescribed limiting value(s) of the pressure (or transmission) angle, minimum allowed radius of curvature of cam profile (for flat-face followers) and maximum eccentricity of the driving force (for flat-face followers) will be discussed.

9.3.1 Translating Flat-Face Follower

Figure 9.3-1 shows a cam mechanism with an offset translating flat-face follower. The base circle radius is r_b. The follower starts lifting when it touches the point S on the cam profile. At the instant shown (when the cam has rotated through an angle θ from the position the follower has started lifting), the point of contact is at A and the center of curvature for the cam profile corresponding to the point A is at C. The offset is given by e and the eccentricity of the driving effort is ϵ, as indicated in Fig. 9.3-1. As explained in Section 1.9, the velocity and acceleration of the follower, at the instant, will not be altered if the actual cam is replaced by an equivalent one that is a circle of radius $\rho(=CA)$ with its center at C. Let us consider the cam to be replaced by such an equivalent cam. For such a cam, both ρ and the position of C, with respect to the cam remain unaltered. We should also note that as both the lines O_2S and O_2C are fixed to the cam,

$$d\theta = -d\chi \quad \text{or,} \quad \dot{\theta} = -\dot{\chi}. \qquad (9.3.1)$$

Now from the figure,

$$\rho = r_b + y(\theta) - O_2C \cos \chi. \qquad (9.3.2)$$

Differentiating both sides of Equation (9.3.2) with respect to time, and noting that both ρ and O_2C (for the equivalent circular cam) are constants, (up to the second derivatives) we get[8]

$$y'(\theta).\dot{\theta} = -O_2C \sin \chi.\dot{\chi}.$$

Using Equation (9.3.1) in the above equation

$$y'(\theta) = O_2C \sin \chi. \qquad (9.3.3)$$

Differentiating Equation (9.3.3) once more with respect to time, and using Equation (9.3.1) again,

$$y''(\theta) = -O_2C \cos \chi. \qquad (9.3.4)$$

---

[8]Prime(') indicates differentiation with respect to θ.

9.3. DETERMINATION OF BASIC DIMENSIONS

From Equations (9.3.2) and (9.3.4), the expression for ρ becomes as follows:

$$\rho = r_b + y(\theta) + y''(\theta) \tag{9.3.5}$$

From the contact stress point of view, a restriction on the minimum permissible value of ρ can be imposed and the minimum ρ allowed can be $\rho_{\min}$. Then the minimum value of the base circle radius (which should be used to develop the cam profile if no other constraints are imposed),

$$r_{b_{\min}} = \rho_{\min} - [y(\theta) + y''(\theta)]_{\min} \tag{9.3.6}$$

where the minimum value of $[y(\theta) + y''(\theta)]$ can be determined from the prescribed displacement function for the whole cycle.

The eccentricity of the driving effort (i.e., ϵ in Fig. 9.3-1) also cannot be permitted to go beyond a certain limit that would lead to the jamming of the follower in its guides. The limiting value of the magnitude of ϵ can be estimated if the bearing dimension h, its distance d (Fig. 9.3-1) and the coefficient of friction are known. From Fig. 9.3-1 we find

$$O_2 C \sin \chi = l = e + \epsilon.$$

Using this relation in Equation (9.3.3), we get

$$l = e + \epsilon = y'(\theta). \tag{9.3.7}$$

If $\epsilon_{\max}$ is the maximum permissible eccentricity of the driving effort, then the least required offset

$$e_{\min} = [y'(\theta)]_{\max} - \epsilon_{\max} \quad \text{for} \quad y'(\theta) > 0. \tag{9.3.8a}$$

During return the follower is driven by a spring and so the distance of contact point from the translational axis is not so important.

From Equation (9.3.7) we can write an expression for the minimum required width of the follower face, as follows:

$$w_{\min} = [y'(\theta)]_{\max} + |\,[y'(\theta)]_{\min}\,| + Allowance \tag{9.3.8b}$$

9.3.2 Translating Roller Follower

The important basic dimensions of a cam mechanism with a translating roller follower are the prime circle radius and the amount of required offset. In this case, the constraints are imposed by the maximum allowable pressure angles during rise and return[9] and the minimum permissible radius of curvature of the cam profile, due to limitations on contact stress. Both the conditions are to be checked.

Figure 9.3-2 shows a cam mechanism with an offset translating roller follower. The point of intersection, P_0, of the translating axis and the prime circle denotes the lowest position of the trace point (the roller center). If O_2N is the normal on the radius of curvature, ρ_P, of the pitch curve at P, then from Fig. 9.3-2,

$$O_2N = O_2A \sin\phi + AP \cos\phi. \tag{9.3.9}$$

Note that the line CP (C is the center of curvature), being the common normal to the roller and cam at their contact point Q,

---

[9]Since return motion is due to a spring force, the pressure angle during return is less important and can sometimes be even ignored.

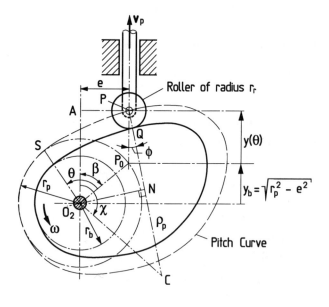

Figure 9.3-2

9.3. DETERMINATION OF BASIC DIMENSIONS

$\angle NPP_0$ = pressure angle = ϕ. The above equation can be also written as

$$O_2N = [y_b + y(\theta)] \sin\phi + e\cos\phi. \tag{9.3.10}$$

Now the velocity of point N (on the cam) must be equal to the component of v_P along NP, which is $v_P \cos\phi$. Again,

$$v_P = \frac{d}{dt}[y_b + y(\theta)] = y'(\theta).\omega. \tag{9.3.11}$$

Therefore,

$$\omega.O_2N = y'(\theta).\omega \cos\phi. \tag{9.3.12}$$

Substituting O_2N in Equation (9.3.12) from Equation (9.3.10) we get

$$[y_b + y(\theta)] \sin\phi + e\cos\phi = y'(\theta)\cos\phi.$$

Rearranging, we get an equation for the transmission angle as follows:

$$\tan\phi = \frac{y'(\theta) - e}{y_b + y(\theta)}$$

or

$$\tan\phi = \frac{y'(\theta) - e}{\sqrt{r_p^2 - e^2} + y(\theta)}. \tag{9.3.13}$$

The above equation can be represented graphically as shown in Fig. 9.3-3a. A plot of $y'(\theta)$ vs $y(\theta)$ is made for the complete cycle, using the prescribed displacement function. If P represents a particular instantaneous configuration of the mechanism and O_2 is a point at a distance r_p from the origin of the plot, P_o, and at a distance e from the $y(\theta)$-axis, then from the figure,[10]

$$\tan \angle VO_2P = \frac{PV}{O_2V} = \frac{PT - VT}{O_2W + P_0T} = \frac{y'(\theta) - e}{\sqrt{r_p^2 - e^2} + y(\theta)}$$

$$= \tan\phi.$$

---

[10] Care should be taken while plotting $y'(\theta)$ vs $y(\theta)$ so that the scales for both axes are the same. Scales for e and r_p will also be the same one used for $y(\theta)$ and $y'(\theta)$.

CHAPTER 9. CAM MECHANISMS

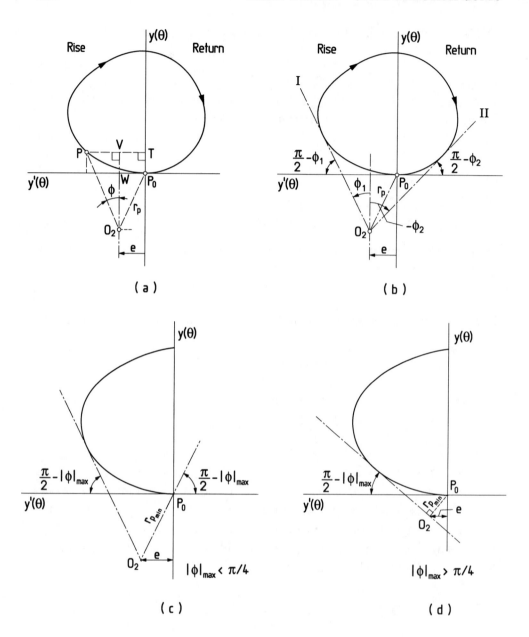

Figure 9.3-3

9.3. DETERMINATION OF BASIC DIMENSIONS

Hence, the inclination of the line O_2P from the $y(\theta)$ axis represents the corresponding pressure angle. The convention of taking positive e towards the left of the $y(\theta)$ axis when the cam rotates in the CCW direction should be carefully noted. Hence, if the translating axis lies to the left of the cam center the corresponding offset is treated as negative (i.e., to the left of point O_2 in Fig. 9.3-2). So far as the pressure angle is concerned, it is treated as positive when measured in the CCW direction (i.e., the direction of ω) from the $y(\theta)$ axis and vice versa. Since $y'(\theta)$ is negative during return, the pressure angle ϕ is also negative (when $e \geq 0$), as seen from Fig. 9.3-3a. If the limiting values of the pressure angle during the rise and return are prescribed as ϕ_1 and $-\phi_2$, respectively, the required values of r_p and e can be determined from the plot of $y'(\theta)$ vs $y(\theta)$. After the plot $y'(\theta)$ vs. $y(\theta)$ for the complete cycle is made, two lines, I and II, are drawn so that they are tangential to the plot and are inclined to the $y(\theta)$ axis at angles ϕ_1 and $-\phi_2$, respectively (Fig. 9.3-3b). These lines intersect at O_2. Hence, the distance of O_2 from the $y(\theta)$ axis represents the required offset and P_0O_2 gives the minimum necessary value of r_p in the same scale used to plot $y(\theta)$ and $y'(\theta)$.

If the maximum allowable magnitude of the pressure angle during rise (or return) is prescribed, the minimum allowable base circle (or prime circle) radius can be determined, as explained in Figs. 9.3-3c and 9.3-3d. It is seen that the method depends on the magnitude of $|\phi|_{max}$. Similar construction is possible if the limiting value of the pressure angle magnitude is prescribed during return.

Problem 9.3-1

The displacement function of a cam mechanism with a translating flat-face follower is given by

$$y(\theta) = 50(1 - \cos\theta) \text{ mm}; \quad 0 \leq \theta \leq 2\pi.$$

The maximum allowable eccentricity of the driving effort from the axis of translation is 30 mm (during rise) and, to avoid excessive stress at the contact point, the radius of curvature of the cam profile should not be less than 80 mm. Find out the minimum required offset and the smallest possible base circle radius.

Solution

The displacement function is given as

$$y(\theta) = 50(1 - \cos\theta) \text{ mm}; \quad 0 \le \theta \le 2\pi$$

Hence,
$$y'(\theta) = 50\sin\theta \text{ mm}; \quad 0 \le \theta \le 2\pi$$

and
$$y''(\theta) = 50\cos\theta \text{ mm}; \quad 0 \le \theta \le 2\pi.$$

Thus,
$$y(\theta) + y''(\theta) = 50 \text{ mm}; \quad 0 \le \theta \le 2\pi$$

which implies, from Equation (9.3.5), that the radius of curvature of the cam profile is constant, or in other words, the cam is circular in shape. From Equation (9.3.6),

$$r_{b_{\min}} = (80 - 50) \text{ mm} = 30 \text{ mm}.$$

It is further mentioned that $\epsilon_{\max}$ during rise can be 30 mm. Now,

$$y'(\theta) = 50\sin\theta \text{ mm}; \quad 0 \le \theta \le 2\pi.$$

Hence,
$$[y'(\theta)]_{\max} = 50 \text{ mm}.$$

From Equation (9.3.8),

$$e_{\min} = (50 - 30) \text{ mm} = 20 \text{ mm}.$$

Since e is positive it will be towards right of the cam center.

Problem 9.3-2

A cam mechanism with translating roller follower has the following displacement function:

$$y(\theta) = 30(1 - \cos\theta) \text{ mm}; \quad 0 \le \theta \le 2\pi.$$

The roller radius is 8 mm. The pressure angle during rise should not go beyond $25°$, whereas during return, it should not be lower than $-45°$. Find out the minimum possible base circle radius and the corresponding offset. The cam rotates in the CCW direction.

9.3. DETERMINATION OF BASIC DIMENSIONS

Solution

For the displacement function $y(\theta) = 30(1 - \cos\theta)$ mm, $0 \leq \theta \leq 2\pi$, $y'(\theta) = 30\sin\theta$ mm, $0 \leq \theta \leq 2\pi$. Thus,

$$[y(\theta) - 30]^2 + [y'(\theta)]^2 = 30^2 \text{ mm}^2; \quad 0 \leq \theta \leq 2\pi.$$

Therefore, when $y'(\theta)$ is plotted against $y(\theta)$, we get a circle of radius 30 mm with its center on the $y(\theta)$ axis at a distance of 30 mm, as shown in Fig. 9.3-4. A line I is drawn at an angle of $25°$, with the $y(\theta)$ axis touching the rise portion of the $y'(\theta)$ - $y(\theta)$ plot. Similarly, line II is drawn touching the return portion of the $y'(\theta)$-$y(\theta)$ plot and making an angle of $-45°$ with the vertical. The point of intersection is located as O_2, when from measurements, $O_2P_0 = r_{p_{min}} = 23$ mm and the corresponding offset $e = 8.7$ mm (positive). As the roller radius is 8 mm, the minimum base circle radius $r_{b_{min}} = r_{p_{min}} - r_r = (23 - 8)$ mm $= 15$ mm.

If the conditions prescribed imply the magnitudes of the pressure angles not to exceed $25°$ and $45°$ during the rise and return, respectively, then the above procedure would give the correct answer only if the inclination of O_2P_o from the $y(\theta)$-axis does not exceed $25°$. Otherwise, point O_2 must be obtained as the intersection of the tangent to the rise portion of the $y(\theta)$-$y'(\theta)$ plot, making an angle $25°$ with the $y(\theta)$-axis (as shown) and a line drawn through P_o, making an angle of $-25°$ with the $y(\theta)$-axis.

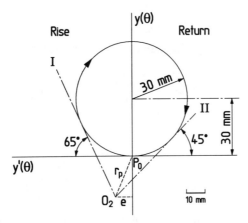

Figure 9.3-4

In the case of roller followers also, contact stress is an important constraint and the radius of curvature of the cam profile cannot be less than a critical value. Referring to Fig. 9.3-2,

$$\rho_P \cos\phi = y(\theta) + y_b - O_2C \cos\chi \qquad (9.3.14)$$

and

$$\rho_P \sin\phi = O_2C \sin\chi - e \qquad (9.3.15)$$

where

$$y_b = \sqrt{r_p^2 - e^2}.$$

Differentiating Equation (9.3.15) with respect to θ, we get[11]

$$\rho_P \cos\phi \frac{d\phi}{d\theta} = O_2C \cos\chi \frac{d\chi}{d\theta}.$$

Since from Fig. 9.3-2, $d\theta = -d\chi$, the above equation becomes

$$\rho_P \cos\phi \cdot \phi' = -O_2C \cos\chi. \qquad (9.3.16)$$

Substituting $O_2C \cos\chi$ from Equation (9.3.16) in Equation (9.3.14),

$$\rho_P \cos\phi = y_b + y(\theta) + \rho_P \cos\phi \cdot \phi'.$$

Rearranging, we get

$$\rho_P = \frac{y_b + y(\theta)}{(1 - \phi') \cos\phi}. \qquad (9.3.17)$$

Next, we differentiate both sides of Equation (9.3.13) with respect to θ, and, after some manipulations, the following relation is obtained:

$$\phi' = \frac{y''(\theta) - y' \tan\phi}{y_b + y(\theta)} \cos^2\phi. \qquad (9.3.18)$$

By using the identity

$$\sec^2\phi = 1 + \tan^2\phi = 1 + \left[\frac{y'(\theta) - e}{y_b + y(\theta)}\right]^2,$$

---

[11] We first replace the actual cam by an equivalent circular cam, with center at C and radius equal to ρ_P, as was done in Section 9.3.1.

9.3. DETERMINATION OF BASIC DIMENSIONS

Equations (9.3.17) and (9.3.18) yield

$$\rho_P = \frac{[\{y_b + y(\theta)\}^2 + \{e - y'(\theta)\}^2]^{3/2}}{[y_b + y(\theta)]^2 + [e - y'(\theta)][2\{e - y'(\theta)\} - e] - \{y_b + y(\theta)\}y''(\theta)} \quad (9.3.19)$$

The radius of curvature of the cam profile

$$\rho = \rho_P - r_r. \quad (9.3.20)$$

After e is evaluated, following the procedure discussed earlier in this section, ρ can be evaluated. To avoid formation of a cusp in the cam profile, the minimum value of ρ is zero when $\rho_P = r_r$. However, in order to avoid a high value of contact stress, it should be checked that the minimum value of ρ is greater than the minimum permissible value, $\rho_{\min}$. Or,

$$\rho \geq \rho_{\min}.$$

If this condition is violated, r_b should be increased (keeping e constant) and the process of checking this condition repeated until the desired result is obtained.

Problem 9.3-3

The displacement function of a cam mechanism with a translating roller follower during rise is

$$y(\theta) = \begin{cases} 20(\theta/\pi)^2 \text{ cm}; & 0 \leq \theta \leq \frac{\pi}{2} \text{ and} \\ 10 - 20(1 - \frac{\theta}{\pi})^2 \text{ cm}; & \frac{\pi}{2} \leq \theta \leq \pi. \end{cases}$$

The radius of the roller is 7.5 mm and the magnitude of the pressure angle during rise is not to exceed $30°$. Find out the minimum base circle radius and the necessary offset. With this offset what will be the minimum radius of curvature of the cam profile during rise? The cam rotates in the CCW direction.

Solution

During rise,

$$y(\theta) = \begin{cases} 20(\theta/\pi)^2 \text{ cm}; & 0 \leq \theta \leq \frac{\pi}{2} \text{ and} \\ 10 - 20(1 - \frac{\theta}{\pi})^2 \text{ cm}; & \frac{\pi}{2} \leq \theta \leq \pi. \end{cases} \quad (a)$$

Differentiating both sides of Equation (a) with respect to time we get

$$y'(\theta) = \begin{cases} (40/\pi^2)\theta \text{ cm}; & 0 \leq \theta \leq \frac{\pi}{2} \text{ and} \\ \frac{40}{\pi}(1 - \frac{\theta}{\pi}) \text{ cm}; & \frac{\pi}{2} \leq \theta \leq \pi. \end{cases} \quad (b)$$

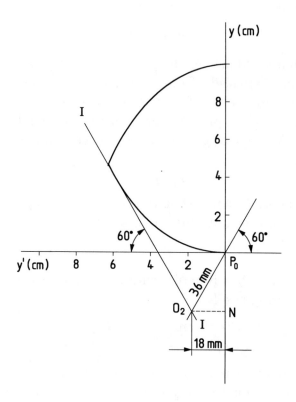

Figure 9.3-5

9.3. DETERMINATION OF BASIC DIMENSIONS

Figure 9.3-5 shows the $y'(\theta) - y(\theta)$ plot during rise. A line I is drawn touching this curve and making an angle of $30°$ (in the CCW direction) with the $y(\theta)$ axis. Next, the line $P_0 O_2$ is drawn from P_0, making an angle of $30°$ with the $y(\theta)$ axis in the CW direction (as shown), which intersects line I at O_2[12]. Then the minimum possible prime circle radius is given by

$$r_{p_{\min}} = P_0 O_2 = 36 \text{ mm}.$$

Hence, the minimum base circle radius $r_{b_{\min}} = r_{p_{\min}} - r_r = 28.5$ mm. The corresponding offset is equal to $e = O_2 N = 18$ mm (to the right of cam center).

Using these values of r_p and e we get $y_b = 31.2$ mm. Substituting the values of y_b and e and using the expressions for $y(\theta)$ and $y'(\theta)$ from Equations (a) and (b) in Equation (9.3.19), we get the expression for ρ_P during rise as follows:

$$\rho_P = \frac{[B^2 + C^2]^{3/2}}{B^2 + C[2C - 18] - B.400/\pi^2} \text{ mm} \quad 0 \leq \theta \leq \pi/2$$

where

$$B = 31.2 + 200(\theta\pi)^2 \quad \text{and} \quad C = 18 - (400.\theta/\pi^2)$$

and

$$\rho_P = \frac{[B'^2 + C'^2]^{3/2}}{B'^2 + C'[2C' - 18] - B'.(-400)/\pi^2} \text{ mm} \quad \pi/2 \leq \theta \leq \pi \quad (c)$$

where

$$B' = 131.2 - 200(1 - \theta\pi)^2 \quad \text{and} \quad C' = 18 - \frac{400}{\pi}(1 - \theta/\pi).$$

From Equation (c), the minimum radius of curvature of the pitch curve during rise is found to be 50.71 mm. Hence, using Equation (9.3.20), the minimum radius of curvature of the rise part of the cam profile,

$$\rho_{\min} = 50.71 \text{ mm} - 7.5 \text{ mm} = 43.21 \text{ mm}.$$

---

[12] If a normal on line I is drawn from P_0, the point of intersection will be above O_2, resulting in a pressure angle whose magnitude will exceed $30°$ during the initial portion of the rise.

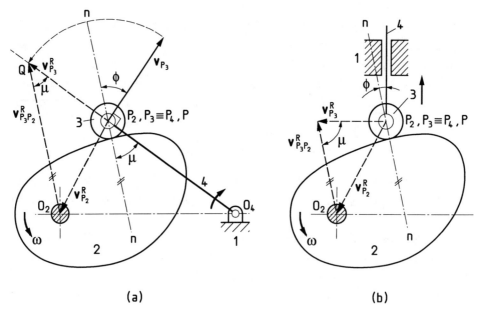

Figure 9.3-6

9.3.3 Flocke's Method

Flocke[13] developed a method for the determination of the basic dimensions of a cam mechanism with a roller follower when the required pressure angle characteristics are prescribed. This method is applicable to both oscillating (Fig. 9.3-6a) and translating (Fig. 9.3-6b) follower mechanisms.

To understand the basic principle of this method let us consider the cam mechanism with an oscillating roller follower shown in Fig. 9.3-6a. The line nn represents the normal to the cam surface at the point of contact. Angles ϕ and $\mu (= 90° - \phi)$ are the pressure and transmission angles, respectively. v_{P_2} is the velocity of a point P_2 on the cam (i.e., link 2) at the roller center ($P_3 \equiv P_4$), drawn to a scale so that $v_{P_2} = O_2 P_2$. If v_{P_2} is rotated in the sense of ω, (i.e. CCW direction) by 90°, we get the rotated velocity vector $v_{P_2}^R$ which coincides with the segment $P_2 O_2$, as indicated. The velocity difference $v_{P_3 P_2}$ must be along the common tangent to the cam and roller surfaces

---

[13]Flocke, Karl Alexander: Zur Konstruktion von Kurvengetrieben bei Verarbeitungsmaschinen, VDI, Forschungsheft 345, Berlin, 1931.

9.3. DETERMINATION OF BASIC DIMENSIONS 481

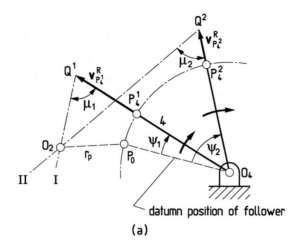

(a)

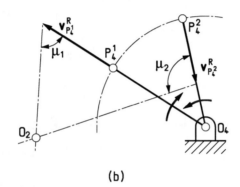

(b)

Figure 9.3-7

at the point of contact and hence, $v^R_{P_3P_2}$ ($v_{P_3P_2}$ rotated by 90° in the CCW direction as per the convention assumed) will be parallel to nn.

Again, since v_{P_3} ($\equiv v_{P_4}$) is in the direction perpendicular to O_4P_4, $v^R_{P_3}$ (v_{P_3} rotated by 90° in the CCW direction) will be along O_4P_4, as shown (away from O_4 if link 4 is rotating in the CW direction at the instant, or towards O_4 if it is rotating in the CCW direction). Thus, O_2P_2Q (in Fig. 9.3-6a) represents the rotated velocity triangle (velocity diagram rotated rigidly by 90° in the CCW direction) and $\angle P_2QO_2 = \mu = 90° - \phi$. This can be used to determine the location of cam center O_2 when the displacement function is prescribed and the required pressure angle at two different positions ($\psi = \psi_1$ and $\psi = \psi_2$) are given as ϕ_1 and ϕ_2. First, we draw the follower link 4 at the two prescribed positions as $O_4P_4^1$ and $O_4P_4^2$, selecting a suitable datumn position of the follower O_4P_0 and the length of the follower to a particular scale (Fig. 9.3-7a).

From the procedure of drawing the rotated velocity triangle explained above, we now know that the length of $v_{P_2}^R$ was chosen to be equal to O_2P_2. Hence,

$$v_{P_2}.S_v = O_2P_2.S_m \qquad (9.3.21)$$

where S_v is the scale factor for drawing the velocity triangle and S_m is the scale factor for drawing the cam mechanism. The relation between these two scale factors from becomes as follows:

$$S_v = S_m \frac{O_2P_2}{v_{P_2}}$$

Again,

$$v_{P_2} = \omega.O_2P_2$$

where ω is the angular velocity of the cam. Thus finally,

$$S_v = S_m/\omega. \qquad (9.3.22)$$

Since the follower (with an assumed length) has been drawn at the two prescribed positions according to a chosen scale S_m, the two rotated velocity vectors $v_{P_4^1}^R$ and $v_{P_4^2}^R$ are now drawn[14] at P_4^1 and P_4^2 to a scale, S_m/ω, where ω is the angular velocity of the cam (prescribed). Thus, the locations of the points Q^1 and Q^2 are determined as indicated in Fig. 9.3-7a. The location of the cam center will then be on the line I, drawn through Q^1 at an angle μ_1 (= 90° - prescribed pressure angle for position 1) with P^1Q^1, as shown. Similarly, the cam center O_2 must be on line II, which is drawn through Q^2, making an angle μ_2 with $P_4^2Q^2$. The intersection of lines I and II determines the location of O_2 with respect to the follower. Thus, if P_0 is the

---

[14]Complete information about v_{P_4} is provided in the displacement function and the angular velocity of the cam. Thus, for $\psi = \psi_1$ and $\psi = \psi_2$, $v_{P_4^1}$ and $v_{P_4^2}$ can be determined.

9.3. DETERMINATION OF BASIC DIMENSIONS

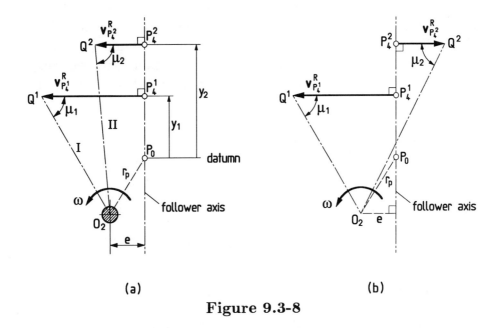

Figure 9.3-8

datumn position of the roller center, obviously O_2P_0 is equal to the prime circle radius r_p. The base circle radius can be determined to be $r_b = r_p$ - roller radius (assumed or prescribed). The two positions of the follower need not be on the same part of the displacement function. Figure 9.3-7b shows the procedure for determining the location of O_2 when the first position of the follower with prescribed pressure angle is during rise whereas the second position is while the follower returns.

Referring to Fig. 9.3-6b, the procedure can be easily extended to cam mechanisms with translating roller followers. If the required pressure angles at two positions of the follower are prescribed, the prime circle radius r_p and the offset can be determined, as illustrated in Fig. 9.3-8, which is self explanatory.

Problem 9.3-4

A cam mechanism has an oscillating roller follower that moves according to the following displacement function during rise:

$$\psi(\theta) = \frac{\pi}{8}(1 - \cos\theta); \quad 0 \le \theta \le 2\pi.$$

The required pressure angle at $\psi(\theta) = 30°$ is $25°$, during the forward motion (rise). During return, when $\psi(\theta) = 15°$, ϕ is $30°$. If the follower length is 150 mm and the roller radius is 15 mm, find out the required base circle radius and the follower position (with respect to the line of frame) at the start of motion. Cam rotates at 120 rpm in the CCW direction.

Solution

The angular speed of the cam

$$\omega = 2\pi \times 120/60 \text{ rad/s} = 12.56 \text{ rad/s}.$$

The displacement function is

$$\psi = \frac{\pi}{8}(1 - \cos\theta) \text{ rad}.$$

Differentiating with respect to time

$$\dot{\psi} = \frac{\pi}{8}.\omega\sin\theta = 4.93\sin\theta \text{ rad/s}.$$

At position 1, $\psi = 30°$, which corresponds to $\theta = 109.5°$, for which the above expression yields $\dot{\psi} = 4.67$ rad/s. With a follower length of 150 mm, the velocity of the roller center at position 1 is 700.3 mm/s. Similarly, at position 2, $\psi = 15°$ and the corresponding cam rotation is $289.5°$ (during return). The velocity of the roller center at this position is 700.3 mm/s (in the opposite direction).

If we make $S_m = 1/3$, then $S_v = 1/(3 \times 12.56) = 1/37.68$. With these scales the construction is shown in Fig. 9.3-9. From the construction (and using the scale factor S_m), $r_p = 93$ mm. Since the roller radius is 15 mm, the required base circle radius

$$r_b = r_p - r_r = (93 - 15) \text{ mm} = 78 \text{ mm}.$$

The corresponding value of the follower inclination at the start,

$$\psi_b = 22°.$$

Quite often the limiting values of ϕ (or μ) are not prescribed for specific positions of the follower but for the whole ranges of motion. For example, if the limiting value of μ, $\mu_{\min}$, is prescribed for a cam

9.3. DETERMINATION OF BASIC DIMENSIONS

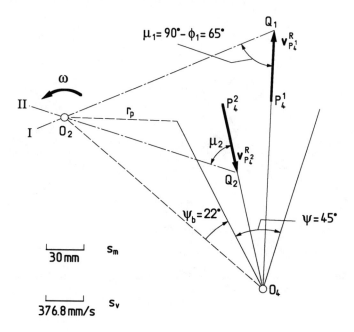

Figure 9.3-9

486 CHAPTER 9. CAM MECHANISMS

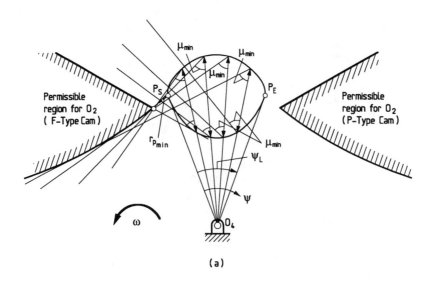

(a)

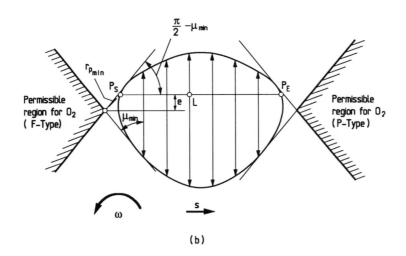

(b)

Figure 9.3-10

9.3. DETERMINATION OF BASIC DIMENSIONS

mechanism, then the regions for the permissible locations of the cam center O_2 can be determined, as shown in Figs. 9.3-10a and b for oscillating and translating followers, respectively. In the case of an oscillating follower the follower hinge O_4 and the follower length O_4P are chosen. The follower at the start and end positions (of the rise period) are drawn as O_4P_S and O_4P_E at an angle of ψ_L apart. The follower positions and the corresponding rotated velocity vectors are drawn (both during rise and return) at a number of intermediate locations, as shown in Fig. 9.3-10a. From the tip of each rotated velocity vector a line is then drawn making an angle $\mu_{\min}$ (measured in the $+\psi$ direction during rise and in the $-\psi$ direction during return), as indicated. The region enclosed by the envelope generated by these lines is the region that can contain the cam hinge O_2 without resulting a value of μ less than $\mu_{\min}$ during the whole cycle. The minimum possible prime circle radius is also indicated. Construction lines for P-type cam have not been shown in the figure. In the case of a translating follower the construction has been illustrated through the self explanatory Fig. 9.3-10b.

It is needless to mention that the above procedure is somewhat time consuming, as the determination of the permissible region for O_2 requires a large number of lines to be drawn. Flocke proposed the following aproximate method to determine the permissible regions for locating O_2 (with respect to the follower). The principle on which the method is based is to satisfy the $\mu_{\min}$ condition when the follower speed is maximum during rise and return. Let the maximum speed during rise (for an oscillating follower) be $\dot\psi$, so that for a given (or chosen) length of the follower the maximum trace point speed is $v_{P_{max_1}}$. The corresponding location is given by $\psi = \psi_1$. Similarly, while returning, the maximum speed is $v_{P_{\max_2}}$, and occurs when $\psi = \psi_2$ (i.e., at P_2). The minimum permissible value of the transmission angle is $\mu_{\min}(= \frac{\pi}{2} - \phi_{\max})$. The approximate determination of the permissible region for O_2 is then quite straightforward, as illustrated in Fig. 9.3-11a. Figure 9.3-11b shows similar construction for a translating follower.

Problem 9.3-5

The oscillating roller follower of a cam mechanism swings through an angle of $30°(\equiv \pi/6)$. During rise, the cam rotates through $180°(\equiv \pi)$ and the follower moves according to the following displacement

function:
$$\psi(\theta) = \frac{\pi}{6}[10(\frac{\theta}{\pi})^3 - 15(\frac{\theta}{\pi})^4 + 6(\frac{\theta}{\pi})^5].$$

During the next 60°, rotation of the cam the follower remains stationary and then returns with a simple harmonic motion as the cam rotates through the remaining 120° of the cycle. Assuming the follower length to be 150 mm, determine the minimum possible prime circle radius, the center distance O_2O_4 and ψ_b, if the pressure angle is not to exceed 20°. Assume F-type cam and use Flocke's approximate method. The cam rotates at 60 rpm.

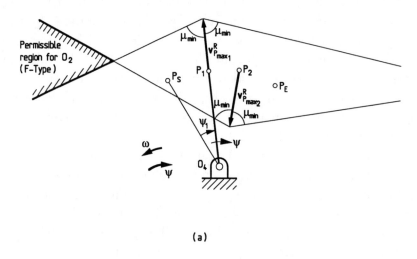

(a)

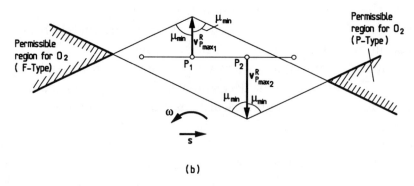

(b)

Figure 9.3-11

9.3. DETERMINATION OF BASIC DIMENSIONS

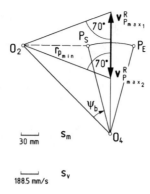

Figure 9.3-12

Solution

The expressions for velocity and acceleration during rise are obtained as follows:

$$\dot{\psi}(\theta) = \frac{\pi}{6}[30(\frac{\theta}{\pi})^2 - 60(\frac{\theta}{\pi})^3 + 30(\frac{\theta}{\pi})^4]\frac{\omega}{\pi}$$

and

$$\ddot{\psi}(\theta) = \frac{\pi}{3\pi}[60(\frac{\theta}{\pi}) - 180(\frac{\theta}{\pi})^2 + 120(\frac{\theta}{\pi})^3]\frac{\omega^2}{\pi^2}$$

where $\omega = 2\pi$ rad/s. Hence, $\dot{\psi}(\theta)$ is maximum when $\theta = \pi/2$ and the magnitude of this peak velocity is $5\pi/8$ rad/s. With a follower length of 150 mm, the maximum trace point velocity is 294.5 mm/s, and occurs when $\psi = 15°$.

During return, $\psi(\theta) = \frac{\pi}{12}[1 + \cos\frac{3(\theta - 4\pi/3)}{2}]$; $\frac{4\pi}{3} \leq \theta \leq 2\pi$.

Hence, the peak return speed is 2.47 rad/s and it occurs when $\psi = \pi/24 (\equiv 15°)$. The corresponding trace point speed is 369.75 mm/s. If we take $S_m = 1/3$ from Equation (9.3.22), $S_v = S_m/\omega = (1/3)/(2\pi) = 1/(6\pi)$. The construction with the above values and $\mu_{min} = 90° - \phi_{max} = 90° - 20° = 70°$, is shown in Fig. 9.3-12, for F-type cam. From the construction and the scale used, the minimum prime circle radius $r_{p_{min}} = 109.5$ mm and the center distance $O_2O_4 = 209.5$ mm. The value of ψ_b is $31°$.

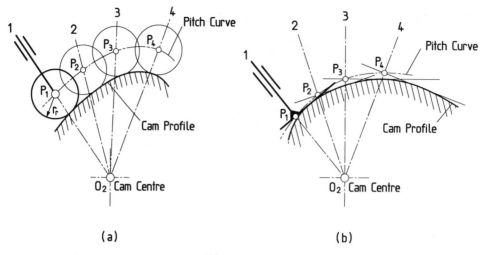

Figure 9.4-1

9.4 Cam Synthesis (Graphical Approach)

Graphical approach to laying out the cam profile, once the information regarding the desired follower motion and the basic dimensions are known, is simple and straightforward. However, as in the case of all graphical approaches, the accuracy of the results is limited and such an approach can be used for synthesising low speed cams only. The method involves the principle of inversion. The cam is considered to be held stationary and the follower is rotated in a direction opposite to that of the cam; this gives the same relative rotation of the follower (but in the opposite direction) as we would have obtained for a particular rotation of the cam. The movement of the trace point corresponding to this rotation is determined from the displacement diagram (or can be calculated if the displacement function is prescribed). Thus, the trace point locations are obtained for an adequate number of inverted follower positions called station points and the smooth curve drawn through these trace points is the pitch curve. Once the trace points are obtained, the follower outlines corresponding to these station points can be drawn, as shown in Figs. 9.4-1a and 9.4-1b, for a translating roller and a flat-face follower mechanism, respectively. The cam profile is given by the envelope of the follower

9.4. CAM SYNTHESIS (GRAPHICAL APPROACH)

profiles as indicated. The following sections present the method for typical problems in graphical cam synthesis. Only roller and flat-face followers are discussed; for a knife-edge follower the pitch curve (with the knife edge as the trace point) itself represents the cam profile. We present cases of F-type cam mechanisms only. The reader is advised to solve a few cases of P-type cam synthesis problems.

9.4.1 Radial Translating Follower

The procedure for laying out cam profiles for mechanisms with a radial translating follower is illustrated with the help of a specific example where the follower is a flat-face one. In cases of roller followers the follower profiles will be circles, with r_r as radius and trace points as the centers.

Problem 9.4-1

A radial translating flat-face follower of a cam mechanism has a lift of 30 mm. The rise takes place with simple harmonic motion for $180°$ of cam rotation. Then there is a dwell for $30°$ and the return motion (also simple harmonic) is for $120°$. The remaining $30°$ of cam rotation is a dwell. The base circle radius of the cam is 30 mm and the cam rotates in the CCW direction. Layout the cam profile and determine the minimum width of the follower face with an allowance of 3 mm at both ends.

Solution

First, the displacement diagram is drawn as shown in Fig. 9.4-2a. Using six divisions for θ_{ri} and θ_{re}, each interval between station points 0 and 6 (during the rise) corresponds to $30°$ of cam rotation and each interval between station points 7 and 13 (during the return) corresponds to $20°$ of cam rotation. Each dwell corresponds to $30°$ of cam rotation. The method of obtaining the displacement diagram has been explained in Fig. 9.2-5.

The cam center O_2 in Fig. 9.4-2b is chosen in a manner such that the projections can be taken directly from the displacement diagram to locate the positions of the trace points on the axis of follower translation corresponding to station points 0 to 13. This implies that the abscissa of the displacement diagram, when extended, touches the base circle of the cam, i.e., O_2 is 30 mm below the abscissa.

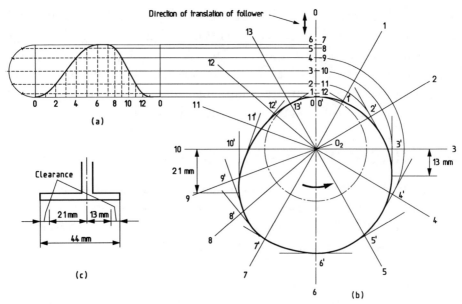

Figure 9.4-2

In the inverted situation, the follower (along with its guide) rotates in the CW direction and the radial lines, marked $1, 2, 3, \ldots 13$, are drawn through O_2, representing the positions of the follower axis (inverted) corresponding to the station points 1 to 13. Station point 0 represents the starting position. Obviously, the angles between consecutive radial lines from 0 to 7 is $30°$ and that from 7 to 13 is $20°$. The angle between lines 13 and 0 is again $30°$.

Thus, corresponding to each station point the distance of the trace point is now known from the cam center O_2, which is transferred to each moving axis of the follower. The inverted trace point locations so obtained are designated $1', 2', 3', 4', \ldots, 13'$. After the trace points are obtained, the follower face (represented by a straight line through the trace point and perpendicular to the inverted follower axis) corresponding to each station point can be drawn. Finally, the curve drawn tangential to all these lines gives the cam profile. To maintain the clarity of the diagrams only fourteen station points have been used, but in practice, to obtain reasonably accurate results, a large scale drawing with many more station points should be prepared.

From Fig. 9.4-2b it is seen that the point of contact at station 0 is

9.4. CAM SYNTHESIS (GRAPHICAL APPROACH)

the trace point (which is the point of intersection of the follower axis and the follower face); for other positions, the point of contact and the trace point are different. The point of contact shifts towards the right of the trace point as the follower moves from 0 to 1, 2, and so on. The point of contact is farthest to the right of the trace point at the station point 3. Afterwards, the point of contact moves to the left, reaching farthest to the left of the trace point at the station point 10. From measurement we obtain the distances of the extreme positions of the contact point from the trace point as 13 mm and 21 mm to the right and to the left, respectively. So the minimum width required for the follower face is (Fig. 9.4-2c):

$$13 \text{ mm} + 21 \text{ mm} + 2 \times 3 \text{ mm} = 40 \text{ mm}.$$

It should be noted that the first step is to obtain the trace point corresponding to each station point. The curve passing through these points is the pitch curve and represents the cam profile in the case of a knife-edge follower. In case of a roller follower, as the trace point is the roller center, circles representing the roller should be drawn at every station point after locating the inverted trace points. The curve tangential to these circles then gives the cam profile.

9.4.2 Offset Translating Follower

The method of construction used for laying out the cam profiles for radial translating followers is modified when the translating follower axis is offset (i.e., does not pass through the cam center). First, an offset circle with a radius equal to the amount of offset is constructed with cam center as the center. The follower axis positions (inverted) should be always tangential to the offset circle. An example using roller follower illustrates the procedure.

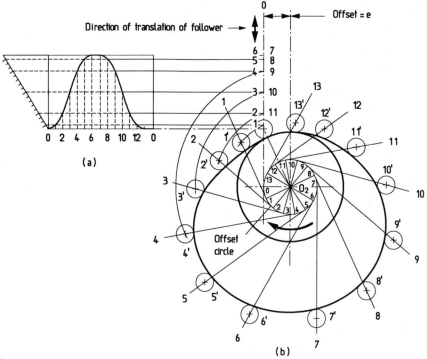

Figure 9.4-3

Problem 9.4-2

A translating roller follower is offset to the left of the cam center by 15 mm. The cam has a base circle radius of 30 mm and the follower has a lift of 40 mm. The cam rotates in the CW direction. The follower has a parabolic motion both during the rise and the return, and during each of these periods, the cam rotates through 150°. The duration of dwell before and after the rise is represented by a 30° rotation of the cam. Assuming the roller radius to be 5 mm, layout the cam profile.

Solution

The displacement diagram is drawn in Fig. 9.4-3a, with six divisions for both rise and return, each interval being 25° of cam rotation. Next, the vertical line, representing the follower axis at the 0th station point (i.e., the datumn position), is drawn. The intersection of this vertical line and the extension of the abcissa of the displacement diagram is the roller center (trace point) at the 0th station point. The

9.4. CAM SYNTHESIS (GRAPHICAL APPROACH)

cam center O_2 should be at a distance of the prime circle radius (30 mm + 5 mm = 35 mm) from the station point 0, and at the same time the follower axis should be at a distance of 15 mm (to the left) from O_2. Thus, O_2 is located (Fig. 9.4-3b). The offset circle is then drawn and divided into the same number of divisions as in the displacement diagram. The magnitudes of the angular intervals should also correspond to those in the displacement diagram. The trace points $1'$, $2'$, $3'$, ... $13'$ are located on these inverted follower positions 1, 2, 3,... 13, keeping their distances from the cam center unchanged. This is explained in Fig. 9.4-3b. Circles (with radius equal to 5 mm) are then drawn, with these trace points as centers. Finally, the cam profile is obtained by drawing a curve tangential to the circles representing the roller as shown in Fig. 9.4-3b.

9.4.3 Oscillating Follower

The procedure is modified when the cam mechanism has an oscillating follower. In such cases the inverted positions of the follower hinge are first identified for the station points. Next, the corresponding positions of the trace point are determined and the follower outline (a circle in the case of roller follower and a straight line in the case of flat-face follower) is drawn for each position of the trace point. The cam profile is given by the envelope to this family of circles or straight lines. The example below illustrates the procedure for a cam mechanism with an oscillating roller follower.

Problem 9.4-3

Figure 9.4-4a shows the displacement diagram of an oscillating roller follower driven by a plate (disc) cam rotating in the CCW direction. In Fig. 9.4-4b the follower, with its hinge at O_4 is shown. The cam center is at O_2. The base circle radius of the cam, the follower length and the roller radius are prescribed. It is desired that the follower should oscillate through an angle ψ_L. The oscillation of the follower takes place with modified uniform motion, during both the rise and return phases, as the cam rotates through $150°$ in each period. The dwell is for $30°$ at the end of the rise and also at the end of the return. Illustrate the procedure for laying out the cam profile.

496   CHAPTER 9. CAM MECHANISMS

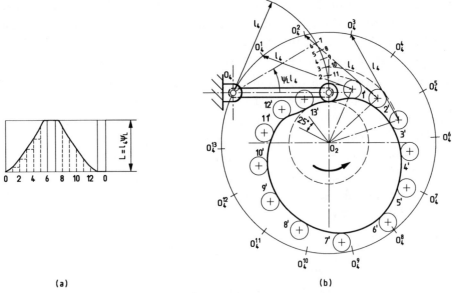

(a)   (b)

Figure 9.4-4

Solution

The lift in the displacement diagram is given by $L = l_4\psi_L$, where l_4 is the follower length.[15] The trace point in this problem is the roller center. The displacement diagram is completed first and both the rise and the return portions are each divided into six equal parts. The locations of the trace point, corresponding to station points 0, 1, 2,... 13, are marked on the arc of oscillation of the trace point and designated 0, 1, 2,... 13. To reduce the error in measuring along the arc, the points should be located successively, by measuring the displacements from 0 to 1, 1 to 2, 2 to 3 and so on. Of course, no problem arises when the displacement diagram is in terms of ψ.

Next, holding the cam fixed, the follower hinge O_4 is rotated in a direction opposite to the rotation of the cam, maintaining a fixed distance from the cam center O_2. Thus, O_4^1, O_4^2, ... O_4^{13} are the inverted positions of the follower hinge corresponding to the station points 1, 2,... 13. The inverted positions of the trace point $1'$, $2'$, $3'$, ... $13'$ can then be located, as the distances of each trace point position from O_2

---

[15]The displacement diagram could be directly θ vs. ψ and the inclination of the follower with its datumn position could be directly used in the construction.

9.5. CAM SYNTHESIS (ANALYTICAL APPROACH)

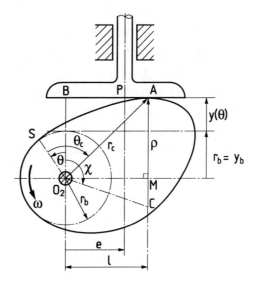

Figure 9.5-1

and O_4 are known. This is explained in Fig. 9.4-4b, for $1'$, $2'$ and $3'$.

Once the trace point positions (inverted) are located, circles with these positions as centers, and radius, equal to roller radius, are drawn. The smooth curve touching these circles (i.e., the envelope) represents the cam profile.

9.5 Cam Synthesis (Analytical Approach)

As already mentioned, the accuracy achieved with graphical approach is low and such procedures are not acceptable in many cases. Furthermore, extensive use of computer, in design has also provided impetus to adopt analytical procedures in designing cams as far as possible. In this section we present a generalized approach, applied to different cases, for deriving cam profile equations in parametric form.

9.5.1 Flat-Face Translating Follower

Figure 9.5-1 shows a cam mechanism with a flat-face translating follower, the axis of which is offset from the cam center O_2. This figure is similar to Fig. 9.3-1. At the beginning of the rise period the follower touches the base circle and the contact point is at S on the cam surface. The indicated position is reached after the cam rotates through an angle θ in the CCW direction. The given quantities are r_b, e and $y(\theta)$, and we want to determine the equation of the cam profile in polar coordinate (r_c, θ_c), where θ_c is measured from the datumn line O_2S. The subscript c indicates that these coordinates are for the points on the cam profile. Other important parameters, viz., ρ and l, have already been determined in Section 9.3 and can be used here.

From Fig. 9.5-1,

$$\theta_c - \theta = \tan^{-1}(l/O_2B).$$

But $O_2B = r_b + y(\theta)$. Hence, using Equation (9.3.7) for l in the above equation, we get

$$\theta_c = \theta + \tan^{-1}[y'(\theta)\{r_b + y(\theta)\}]. \tag{9.5.1a}$$

Again, from Fig. 9.5-1,

$$r_c = [l^2 + O_2B^2]^{1/2}$$

or

$$r_c = [\{y'(\theta)\}^2 + \{r_b + y(\theta)\}^2]^{1/2}. \tag{9.5.1b}$$

Equations (9.5.1a) and (9.5.1b) are the parametric equations of the cam profile. The equation for the radius of curvature of the cam profile is given by Equation (9.3.5).

Problem 9.5-1

The flat-face translating follower of a cam mechanism rises with cycloidal motion when the cam rotates through an angle π, the total lift being 50 mm. The maximum allowed eccentricity of the driving effort from the axis of translation during rise is 20 mm and the radius of curvature of the rise part of the cam profile should not be less than 50 mm at any point, in order to avoid excessive contact stress. Determine the equation for the rise part of the cam profile.

Solution

Before determining the cam profile, certain basic dimensions have to be determined from the constraints and the displacement function.

9.5. CAM SYNTHESIS (ANALYTICAL APPROACH)

The displacement function is

$$\bar{y}(z) = z - \frac{1}{2\pi}\sin 2\pi z$$

where

$$z = \theta/\theta_{ri} \quad \text{with} \quad \theta_{ri} = \pi.$$

Hence, from Equation (9.2.5),

$$y(\theta) = 50\left[\frac{\theta}{\pi} - \frac{1}{2\pi}\sin 2\theta\right] \text{ mm},$$

$$y'(\theta) = \frac{50}{\pi}(1 - \cos 2\theta) \text{ mm},$$

and

$$y''(\theta) = \frac{50}{\pi}(2\sin 2\theta) \text{ mm}, \quad \text{with} \quad 0 \le \theta \le \pi.$$

From the above relations (during rise),

$$[y'(\theta)]_{\max} = 31.85 \text{ mm}.$$

Since $\epsilon_{\max} = 20$ mm, the minimum offset required is determined using Equation (9.3.8a) as follows:

$$e_{\min} = (31.85 - 20) \text{ mm} = 11.85 \text{ mm}.$$

Let us take $e = 12$ mm. Again, from Equation (9.3.6),

$$r_{b_{\min}} = \rho_{\min} - [y(\theta) + y''(\theta)]_{\min}.$$

Now,

$$y(\theta) + y''(\theta) = \frac{50}{\pi}\left(\theta - \frac{1}{2}\sin 2\theta + 2\sin 2\theta\right); \quad 0 \le \theta \le \pi$$

and the minimum value of the above expression during rise is zero. Hence,

$$r_b = r_{b_{\min}} = 50 \text{ mm}.$$

The parametric equation in polar form (with O_2S as the reference) is given by Equations (9.5.1a) and (9.5.1b). Thus,

$$\theta_c = \theta + \tan^{-1}\left[\frac{50}{\pi}(1 - \cos 2\theta)/\left\{50 + 50\left(\frac{\theta}{\pi} - \frac{1}{2\pi}\sin 2\theta\right)\right\}\right]$$

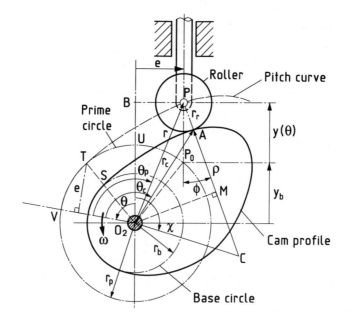

Figure 9.5-2

and

$$r_c = [\{\frac{50}{\pi}(1-\cos 2\theta)\}^2 + \{50 + 50(\frac{\theta}{\pi} - \frac{1}{2\pi}\sin 2\theta)\}^2]^{1/2} \text{ mm}.$$

9.5.2 Translating Roller Follower

Figure 9.5-2 shows a cam mechanism with an offset translating roller follower. The point of intersection of the prime circle and the axis of the follower movement, denoted by P_0, indicates the lowest position of the trace point (i.e., the roller center). The corresponding point on the pitch curve is at T. As explained in the figure, the point corresponding to U at the lowest position of the follower is at V. Thus, the amount of cam rotation (from the start of the rise up to the configuration indicated in this figure) θ is given by $\angle P_0 O_2 T = \angle U O_2 V = \theta$. As already noted in Fig. 9.3-2, $O_2 B = y_b + y(\theta)$, where

9.5. CAM SYNTHESIS (ANALYTICAL APPROACH)

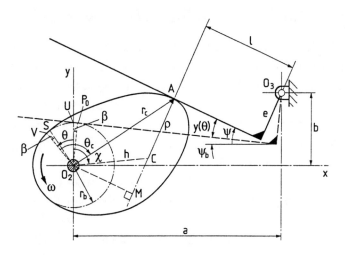

Figure 9.5-3

$y_b = (r_p^2 - e^2)^{1/2}$. The parametric equation of the pitch curve in polar form (with O_2V as the reference line) is as follows:

$$\theta_p = \theta + \tan^{-1}(e/O_2B)$$

$$r = [O_2B^2 + e^2]^{1/2}$$

or

$$\theta_p = \theta + \tan^{-1}[e\{(r_p^2 - e^2)^{1/2} + y(\theta)\}] \tag{9.5.2a}$$

$$r = [\{(r_p^2 - e^2)^{1/2} + y(\theta)\}^2 + e^2]^{1/2} \tag{9.5.2b}$$

We can determine the polar coordinates of the point A on the cam profile as follows:

$$\theta_c = \theta + \tan^{-1}[\{e + r_r \sin\phi\}/\{(r_p^2 - e^2)^{1/2} + y(\theta) - r_r \cos\phi\}] \tag{9.5.3a}$$

$$r_c = [\{(r_p^2 - e^2)^{1/2} + y(\theta) - r_r \cos\phi\}^2 + (e + r_r \sin\phi)^2]^{1/2} \tag{9.5.3b}$$

where ϕ is the pressure angle given by Equation (9.3.13). The radius of curvature of the cam profile is given by Equation (9.3.20).

9.5.3 Oscillating Flat-Face Follower

Figure 9.5-3 shows an oscillating flat-face follower. The distance of the flat face from the follower hinge O_3 is e. The cam center is at O_2, which is also the origin of an x-y coordinate system, as shown in Fig. 9.5-3. The x and y coordinates of O_3 are a and b, respectively. The lowest position of the follower is reached when the follower touches the base circle. At the configuration indicated in the figure, this point of tangency (between the base circle and the follower face) is at P_0. At the start of the rise period, the point of tangency on the cam is at S. As explained in the figure, the point corresponding to U at the start of the rise period is at V. Thus, $\angle P_0 O_2 S = \angle U O_2 V = \theta$, where θ is the cam rotation. The inclination of the follower at its lowest position from the horizontal is ψ_b (measured in the CW direction, as indicated in the figure). When the cam rotates through θ, the follower rises to an angle ψ. If the displacement function is $y(\theta)$, then

$$\psi = \psi_b + y(\theta).$$

When a, b, e and r are given, an expression for ψ_b can be determined in terms of these. Considering the points of contact at A, the magnitude of the normal component (perpendicular to the common tangent at A) of velocity of the point on the follower, $v^n_{A_3}$, must now be the same as that of point A_2 on the cam. Thus,

$$v^n_{A_3} = v^n_{A_2}.$$

But, $v^n_{A_3} = \dot{\psi}.l$ and $v^n_{A_2} = \omega.O_2M = \dot{\theta}.O_2M$. Since $\dot{\psi} = \dot{\theta} \cdot y'(\theta)$, we get

$$l\dot{\theta}.y'(\theta) = \dot{\theta}.O_2M$$

or

$$O_2M = y'(\theta)l. \tag{9.5.4}$$

Again, from Fig. 9.5-3,

$$O_2M = a\cos\psi - b\sin\psi - l. \tag{9.5.5}$$

Using Equations (9.5.4) and (9.5.5),

$$l = (a\cos\psi - b\sin\psi)/[1 + y'(\theta)]. \tag{9.5.6}$$

9.5. CAM SYNTHESIS (ANALYTICAL APPROACH)

The x and y coordinates of point A can be expressed as follows:

$$x_A = a - l\cos\psi - e\sin\psi \tag{9.5.7a}$$

$$y_A = b + l\sin\psi - e\cos\psi \tag{9.5.7b}$$

Thus, in polar form the parametric equation of the cam profile will be

$$\theta_c = \theta + \tan^{-1}(x_A/y_A), \tag{9.5.8a}$$

and

$$r_c = (x_A^2 + y_A^2)^{1/2}. \tag{9.5.8b}$$

To determine the expression for the radius of curvature of the cam profile we replace the cam, for the instant, by a circular cam with its center at C and a radius of curvature ρ (see Section 9.3 for details). Using $O_2C = h$ from Fig. 9.5-3, we can write the following equations:

$$a - e\sin\psi - l\cos\psi - \rho\sin\psi - h\sin\chi = 0 \tag{9.5.9a}$$

$$b - e\cos\psi + l\sin\psi - \rho\cos\psi - h\cos\chi = 0 \tag{9.5.9b}$$

Differentiating Equation (9.5.9b) with respect to θ and treating ρ and h as constants (see Section 9.3),

$$e\sin\psi.\psi' + l\cos\psi.\psi' + l'\sin\psi + \rho\sin\psi.\psi' - h\sin\chi = 0$$

because $d\chi/d\theta = -1$. Hence,

$$h\sin\chi = (e\sin\psi + l\cos\psi + \rho\sin\psi).\psi' + l'\sin\psi. \tag{9.5.10}$$

Differentiating Equation (9.5.6) with respect to θ, we get

$$l' = \begin{matrix} -\{(a\sin\psi + b\cos\psi).\psi'/[1+y'(\theta)]\} \\ -\{(a\cos\psi - b\sin\psi).y''(\theta)/[1+y'(\theta)]^2\} \end{matrix}.$$

or

$$l' = -\frac{(a\sin\psi + b\cos\psi).\psi' + ly''(\theta)}{1 + y'(\theta)}. \tag{9.5.11}$$

Substituting $h\sin\chi$ from Equation (9.5.10), l' from Equation (9.5.11) and l from Equation (9.5.6), in Equation (9.5.9a), an expression for ρ can be determined. After algebraic manipulations we get

$$\rho = a\sin\psi + b\cos\psi + \frac{a\cos\psi - b\sin\psi}{(1+\psi')^3}.\psi''. \tag{9.5.12}$$

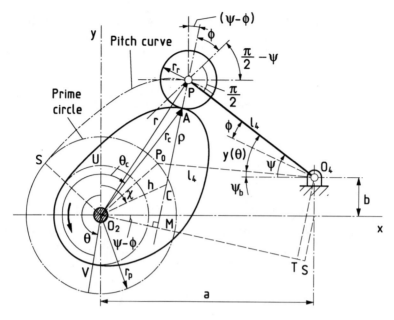

Figure 9.5-4

9.5.4 Oscillating Roller Follower

The procedure for analytical determination of a cam profile in the case of an oscillating roller follower is also along the line discussed so far. In this case the location of the follower hinge O_4 is given in terms of its x and y coordinates a and b, as shown in Fig. 9.5-4. The lowest position of the follower is determined by the point of intersection, P_0, of the circular arc, with O_4 as center and the follower length l_4 (given) as the radius with the prime circle. The instant shown in the figure is attained when the cam has rotated through an angle θ in the CCW direction, where $\angle P_0 O_2 S = \angle U O_2 V = \theta$. The follower angle is given by

$$\psi = \psi_b + y(\theta). \qquad (9.5.13)$$

The x and y coordinates of the trace point P can be written as follows:

$$x_P = a - l_4 \cos\psi \qquad (9.5.14a)$$

$$y_P = b + l_4 \sin\psi \qquad (9.5.14b)$$

9.5. CAM SYNTHESIS (ANALYTICAL APPROACH)

The pressure angle ϕ can be found by following the technique used so far. The components of velocities of a point on the cam at A, and point P along the common normal, are the same. Hence,

$$l_4 \dot{\psi} \cos \phi = O_2 M . \dot{\theta}$$

or

$$l_4 \psi' \cos \phi = O_2 M. \quad (9.5.15)$$

Now from the figure,

$$O_2 M = O_2 S - TS - MT$$

or

$$O_2 M = a \cos(\psi - \phi) - b \sin(\psi - \phi) - l_4 \cos \phi.$$

Using the above relation in Equation (9.5.15),

$$l_4 \psi' \cos \phi = a \cos(\psi - \phi) - b \sin(\psi - \phi) - l_4 \cos \phi.$$

Finally, we get

$$\tan \phi = \frac{l_4(1 + \psi') - (a \cos \psi - b \sin \psi)}{a \sin \psi + b \cos \psi}. \quad (9.5.16)$$

Once ϕ is determined from the above equation, coordinates of the contact point A can be obtained as follows:

$$x_A = x_P - r_r \sin(\psi - \phi) \quad (9.5.17a)$$

$$y_A = y_P - r_r \cos(\psi - \phi) \quad (9.5.17b)$$

In polar form the parametric equation for the cam profile becomes as given below (with $O_2 V$ as the reference line):

$$\theta_c = \theta + \tan^{-1}(x_A/y_A) \quad (9.5.18a)$$

$$r_c = (x_A^2 + y_A^2)^{1/2} \quad (9.5.18b)$$

The radius of curvature of the cam profile is obtained as described below. The basic principle is same as that followed in the previous cases. From Fig. 9.5-4, we get

$$a - l_4 \cos \psi - \rho_P \sin(\psi - \phi) - h \sin \chi = 0 \quad (9.5.19a)$$

and
$$b + l_4 \sin\psi - \rho_P \cos(\psi - \phi) - h\cos\chi = 0 \qquad (9.5.19b)$$
where ρ_P is the length CP (radius of curvature of the pitch curve at P). Differentiating Equation (9.5.19b) with respect to θ, we get (using $d\chi/d\theta = -1$)
$$h\sin\chi = l_4 \cos\psi.\psi' + \rho_p \sin(\psi - \phi)(\psi' - \phi').$$
Substituting $h\sin\chi$ from the above equation into Equation (9.5.19a),
$$a - l_4 \cos\psi - \rho_p \sin(\psi - \phi) - l_4 \cos\psi.\psi' - \rho_p \sin(\psi - \phi)(\psi' - \phi') = 0$$
or
$$\rho_p \sin(\psi - \phi)(1 + \psi' - \phi') = a - l_4 \cos\psi(1 + \psi') = 0$$
or
$$\rho_p = \frac{a - l_4 \cos\psi(1 + \psi')}{\sin(\psi - \phi).(1 + \psi' - \phi')}. \qquad (9.5.20)$$

Substituting ϕ in the above equation from Equation (9.5.16), and ψ from the equation obtained by differentiating both sides of Equation (9.5.16) with respect to θ (and ψ' from the equation obtained by differentiating the prescribed displacement function with respect to θ) in the above equation, ρ_p can be calculated. The radius of curvature of the cam profile at A will be given by
$$\rho = \rho_p - r_r. \qquad (9.5.21)$$

9.6 Spatial Cam Mechanisms

In spatial cam mechanisms the movements of the points are not confined to parallel planes. A generalized treatment of a spatial cam follower mechanism is beyond the scope of the book[16]. However, cylindrical cams, which are commonly used, can be treated without any difficulty. Since cylindrical cams can be developed onto a flat surface, graphical construction of the groove pattern (which is equivalent to a cam profile in the case of disc cams) is possible. This section will be devoted to such cam mechanisms only.

---
[16] For details of spatial cam mechanisms, see Chakraborty, J. and Dhande, S.G.: " Kinematics and Geometry of Planar and Spatial Cam Mechanisms." Wiley Eastern (P) Ltd., New Delhi , 1977.

9.6.1 Basic Features of Cylindrical Cam Mechanisms

Cylindrical cam mechanisms are basically of two types - (i) with translating roller follower and (ii) with oscillating roller follower. Figures 9.6-1a and b show these two types of cam mechanisms. The information about the required follower motion is engraved onto the cylindrical cam surface in the form of a continuous groove of constant depth and width. A roller with diameter equal to the groove width is mounted on either a translating follower or an oscillating follower. In cases of oscillating followers the configuration is designed so that the extreme positions of the trace point (i.e. the roller center) lie on the generating line of the cylinder in the plane containing the follower at all positions. This is indicated in Fig. 9.6-1b. It should be noted that the plane of follower motion touches the cylindrical surface of the cam. The radius of the cylinder, passing through the middle of the groove, is equivalent to base circle radius, r_b (Fig. 9.6-1b). Usually the path of the trace point, in cases of oscillating followers, does not deviate much from the longitudinal direction and it is possible to derive an approximate expression for the minimum required r_b so that the pressure angle does not exceed a prescribed maximum (in cases of translating followers the expression is exact as the follower moves

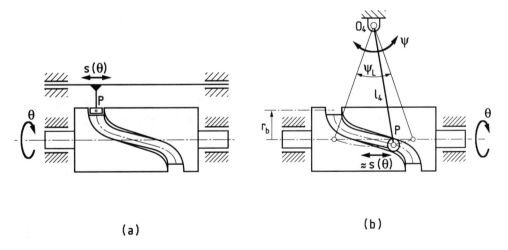

Figure 9.6-1

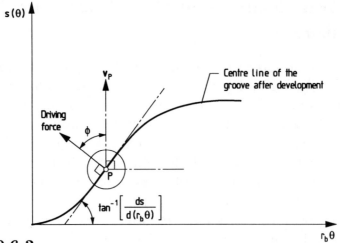

Figure 9.6-2

along the longitudinal direction).

Figure 9.6-2 shows the center line of the groove after the base cylinder (i.e., the cylinder of radius r_b) is developed. The follower movement is $s(\theta)$, along the longitudinal direction (in the case of an oscillating follower it is, of course, approximately equal to ψl_4, l_4 being the follower length as shown). From the figure it is easy to see that

$$\frac{ds}{d(r_b\theta)} = \tan\phi$$

or

$$\frac{1}{r_b}\cdot\frac{ds}{d\theta} \equiv \frac{s'}{r_b} = \tan\phi. \qquad (9.6.1)$$

From Equation (9.6.1) it is seen that ϕ becomes maximum when s' is maximum. Hence, the minimum required r_b can be expressed as follows:

$$r_{b_{min}} = \frac{\mid s' \mid_{max}}{\tan\phi_{max}} \quad \text{(in case of translating follower)} \qquad (9.6.2a)$$

$$r_{b_{min}} \approx \frac{\mid s' \mid_{max}}{\tan\phi_{max}} \quad \text{(in case of oscillating follower)} \qquad (9.6.2b)$$

where ϕ_{max} is the maximum allowable pressure angle.

9.6. SPATIAL CAM MECHANISMS

Problem 9.6-1

In a cylindrical cam mechanism with oscillating roller follower, the follower swings through $30°$ during rise, with simple harmonic motion, when the cam rotates through $90°$. The follower length is 75 mm and the pressure angle during rise should not exceed $50°$. Determine the base cylinder radius.

Solution

As the motion during rise is simple harmonic, the approximately equivalent displacement function can be written as follows:

$$s(\theta) \approx \frac{l_4 \psi_L}{2}\left(1 - \cos\frac{\pi\theta}{\theta_{ri}}\right)$$

Using the given values,

$$s(\theta) \approx 19.6(1 - \cos 2\theta) \text{ mm}.$$

Hence, $s'(\theta) \approx 39.2 \sin 2\theta$ mm. The maximum value of $s'(\theta)$ during rise is, therefore, 19.6 mm when $\theta = \pi/4$. The maximum permissible value of ϕ is $50°$. Hence, $\tan \phi_{max} = 1.192$. Finally, from Equation (9.6.2b),

$$r_{b_{min}} \approx \frac{39.2}{1.192} \text{ mm} = 32.89 \text{ mm}.$$

9.6.2 Graphical Synthesis of Cylindrical Cams

The procedure for laying out the groove center line (which is equivalent to the cam profile in case of cylindrical cams) is quite straightforward and is described below. The basic principle is to first develop the base cylinder surface and then the groove center line is laid out from the prescribed displacement diagram using the principle of inversion.

To begin, the displacement diagram[17] is drawn to a scale with which the groove has to be laid out. The locations of the trace point for various station points are then determined as shown in Fig. 9.6-3b for translating followers. In the cases of translating followers the locus of the trace point is nothing but the displacement diagram itself.

---

[17] $r_b \theta$ vs. $s(\theta)$ in the case of a translating follower, as shown in Fig. 9.6-3a. But for an oscillating follower it is $r_b \theta$ vs. $\psi(\theta)$.

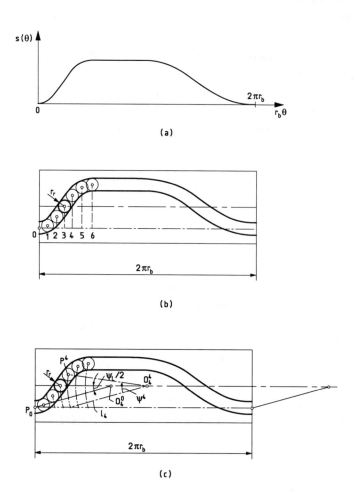

Figure 9.6-3

9.6. SPATIAL CAM MECHANISMS

When an oscillating follower is used, first the locus[18] of the follower hinge O_4 is drawn so that it bisects the follower swing angle ψ_L (Fig. 9.6-3c). Then, the locations of O_4 corresponding to the station points are determined. The positions of a trace point can be found out by rotating the follower (in the direction required) through an angle ψ which corresponds to the displacement for that station point. Fig. 9.6-3c shows this for the station point 4. After the trace points are located, circles with radius r_r are drawn with the inverted positions of the trace point as centers. Two lines touching these circles represent the groove outlines.

9.6.3 Analytical Synthesis of Cylindrical Cams

The analytical approach for determining the equation of the groove center line directly uses the displacement function. In the case of a translating follower the x and y coordinates of a tracepoint location are as follows (Fig. 9.6-4a):

$$x(\theta) = r_b \theta$$

$$y(\theta) = s(\theta)$$

When an oscillating follower is employed, the parametric equation can be derived as follows, using Fig. 9.6-4b. A rotation of θ of the cam corresponds to the location of the follower hinge, which has the following coordinates:

$$x_{O_4} = l_4 \cos(\psi_L/2) + r_b \theta$$

$$y_{O_4} = l_4 \sin(\psi_L/2)$$

Using the above expressions, the x and y coordinates of the corresponding trace point can be written in the following forms:

$$x(\theta) = x_{O_4} - l_4 \cos(\psi - \psi_L/2)$$

$$y(\theta) = y_{O_4} + l_4 \sin(\psi - \psi_L/2)$$

---

[18] Because of inversion the cylinder surface is kept fixed and the follower is moved in the opposite direction.

512 CHAPTER 9. CAM MECHANISMS

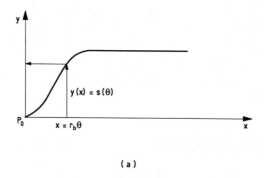

(a)

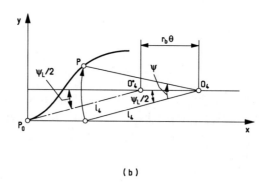

(b)

Figure 9.6-4

9.7. EXERCISE PROBLEMS

or,
$$x(\theta) = l_4 \cos(\psi_L/2) + r_b\theta - l_4 \cos(\psi - \psi_L/2) \quad (9.6.3)$$

$$y(\theta) = l_4 \sin(\psi_L/2) + l_4 \sin(\psi - \psi_L/2)$$

where $\psi(\theta)$ is the prescribed displacement function.

Problem 9.6-2

Find out the equation of the groove center line of the rise part of the cam of Problem 9.6-1. Assume the smallest possible cam.

Solution

Since the motion of the oscillating follower is simple harmonic with a swing of $30°$ (during rise),

$$\psi(\theta) = \frac{\pi}{12}(1 - \cos 2\theta),$$

as the cam rotation during rise is $\pi/2$. The minimumm possible r_b has already been determined to be 32.89 mm. Let us take it as 33 mm. The radius of the cylinder must now be 33 mm $+w/2$, where w is the roller width (or thickness). From the given data, $\psi_L = \pi/6$ and $l_4 = 75$ mm. Using Equation (9.6.3) the parametric equation of the groove center line, when the base cylinder is developed, can be written as follows:

$$x(\theta) = 72.44 + 33\theta - 75 \cos[\frac{\pi}{12} \cos 2\theta] \text{ mm}$$

$$y(\theta) = 19.41 - 75 \sin[\frac{\pi}{12} \cos 2\theta] \text{ mm}$$

9.7 Exercise Problems

9.1 A flat-face follower is offset to the left by 1.2 cm. The base circle radius of the cam is 2.5 cm. The desired displacement of the follower y for any cam rotation θ is listed in Table 9.7-1. Lay out the cam profile and determine the length of the follower face (by trial from the drawing) with 0.3 cm clearance on either side. The cam rotates in the clockwise direction.

Table 9.7-1

Cam rotation θ (degrees)	Follower displacement y(cm)	Cam rotation θ (degrees)	Follower displacement y(cm)
0	0.00	180	3.75
30	0.25	210	3.50
60	0.92	240	2.83
90	1.87	270	1.87
120	2.83	300	0.92
150	3.50	330	0.25

9.2 A roller follower is offset to the right by 1.2 cm. The lift of the follower is 4 cm. The base circle radius of the cam is 2.5 cm and the roller radius is 1 cm. The cam rotates in the counterclockwise direction. Lay out the cam profile if

(a) the rise is for $150°$ of cam rotation, the first $60°$ being at constant acceleration and the rest at constant deceleration,

(b) the dwell is for $30°$,

(c) the return is for $150°$ of cam rotation, the first $90°$ being at constant retardation and the rest at constant acceleration, and

(d) the second dwell is for $30°$. Also find the maximum pressure angle during the rise and return.

9.3 Lay out the cam profile in Fig. 9.7-1 so that the roller follower oscillates through an angle of $30°$ as indicated in the figure. The follower rises for $150°$ of cam rotation in simple harmonic motion. The dwell is for $30°$ of cam rotation, the return for $150°$ of cam rotation with cycloidal motion, and the second dwell for $30°$ of cam rotation.

9.4 The oscillating flat-face follower shown in Fig. 9.7-2 is to rise through an angle of $20°$, with modified uniform motion in $180°$ of cam rotation. This is followed by a dwell for $30°$, and return with modified uniform motion, for $150°$ of cam rotation. Lay

9.7. EXERCISE PROBLEMS

out the cam profile and find the length of the follower face with a 0.3 cm clearance on each side.

9.5 A follower rises through a distance L with cycloidal motion for $0 \leq \theta \leq \theta_{ri}$, where θ is the cam rotation. Determine the value of θ for which the acceleration of the follower is maximum. Find the velocity and jerk for $\theta = \theta_{ri}/2$. The cam rotates at a constant speed ω.

9.6 The boundary conditions of a polynomial-curve cam are

$$y = 0, \quad \dot{y} = 0 \quad \text{at} \quad \theta = 0,$$

$$y = L, \quad \dot{y} = 0 \quad \text{at} \quad \theta = \theta_{ri}.$$

Determine the displacement, velocity, acceleration, and jerk equations.

9.7 A plate cam drives an offset flat-face follower with simple harmonic motion for a rise of 5 cm. The follower moves out and returns in one revolution without any dwell. The base circle radius is 2.5 cm, and the offset is 1 cm. Find the parametric

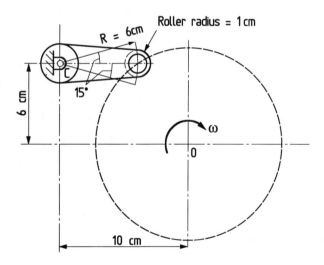

Figure 9.7-1

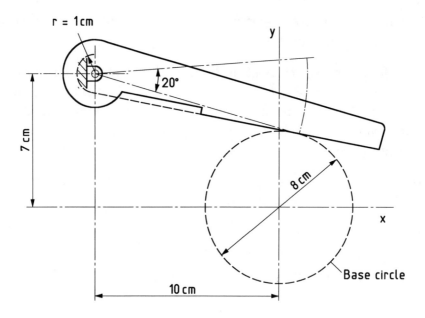

Figure 9.7-2

equation of the cam profile and the theoretical length of the follower face. Draw the cam profile using the parametric equation.

9.8 A flat-face radial follower has a lift of 4 cm. The follower rises the first 1 cm at a constant acceleration for 60° of cam rotation, and then rises 2 cm at a constant velocity for the next 60° of cam rotation, and finally, rises the remaining 1 cm at a constant retardation for another 60° of cam rotation. After a dwell for 45° of cam rotation, the follower returns with simple harmonic motion for the next 90° of cam rotation, followed by a dwell. Determine the minimum base circle radius of the cam in order to ensure positive radius of curvature everywhere on the cam profile and the minimum length of the follower face. Also obtain the equation of the cam profile.

9.9 For the displacement function mentioned in Problem 9.4-1, find out the optimum offset with a roller follower. If the maximum pressure angle is 25°, what is the minimum radius of the prime

9.7. EXERCISE PROBLEMS

circle?

9.10 The oscillating roller follower of a disc cam mechanism rises with cycloidal motion during a $120°$ rotation of the cam. The angle of rise is $30°$. During the next $60°$ rotation of the cam there is a dwell, after which the follower returns with simple harmonic motion as the cam rotates through $150°$. The last $30°$ of cam rotation is again a dwell. If the length of the follower is 150 mm, the roller radius is 10 mm, and the pressure angle is not to exceed $30°$ during the cycle, find out

   (a) the minimum permissible base circle radius and

   (b) the distance of the follower hinge from the cam center. Use Flocke's approximation.

9.11 Determine the parametric equation of the cam profile for the system shown in Fig. 9.7-2 if the oscillating follower has to execute a harmonic motion with a total swing of $45°$. Plot the cam profile. Also, obtain the cam profile by the graphical approach and compare the two cam profiles.

9.12 The displacement y of a translating roller follower driven by a disc cam is $y = 2.5(1 - \cos\theta)$ cm, where θ is the cam rotation. If the maximum permissible pressure angle during rise is $20°$ and the largest possible roller radius is 1 cm, determine the smallest base circle radius of the cam. What is the offset required corresponding to this radius?

Chapter 10

GEARS

Gears and gear trains play a very important role in mechanical engineering. Gear trains are also among mankind's earliest machine elements. Even around 2500 BC, China's Yellow Emperor was said to have used a south-pointing chariot that consisted of a complex differential gear train. However, until the 15th Century A.D., all gears were one-of-a-kind machine element cut crudely by hand and no theory existed so far as the tooth shape was concerned. Subsequently, enormous amount of work has been done towards developing very sophisticated geometrical theories. Thus, the theory of gears has emerged as a separate branch of study. It should be also noted that, perhaps, in this branch the theory and practice are most closely linked.

The most crucial factor in a geared system is the tooth shape. In the past century, considerable amount of debate used to take place about the best possible tooth shape. Gradually, involute has been accepted as the most suitable tooth shape for parallel axis gears. The variety of gears has also increased and many different types of gears exist today. Since discussion on all aspects of gear technology is not possible, this chapter will be devoted primarily to the theory of gearing. Furthermore, emphasis will be given to the two dimensional situation. In the latter part of the chapter, a brief discussion on different types of gears and gear trains will be presented.

520   CHAPTER 10. GEARS

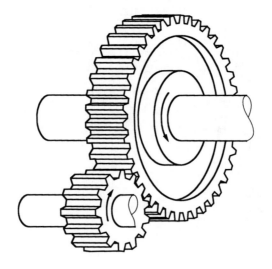

Figure 10.1-1

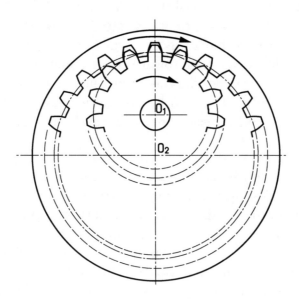

Figure 10.1-2

10.1 Gear Tooth Action and The Law of Gearing

Gears are used for transmitting motion and power from one rotating shaft to another. One major exception is the rack and pinion mechanism. The most common configuration is that in which rotary motion of one shaft is transmitted to another shaft parallel to the first one. The simplest and most commonly used approach is to connect the two shafts with the help of a pair of gears (Fig. 10.1-1). The teeth are straight and parallel to the shaft axes. Such type of straight-tooth gears are called *spur gears*. The wheel with the larger number of teeth is called the *gear* whereas that with the smaller number of teeth is usually referred to as the *pinion*. It is obvious that two shafts connected by a gear and a pinion, as shown in Fig. 10.1-1, rotate in opposite directions. This is the most common configuration in which gears are used, though, employing *internal gear*, the connected shafts can be made to rotate in the same direction (Fig. 10.1-2). When two gears are engaged and rotate together, the teeth of one gear pass in and out of mesh with those of the other gear. This engagement takes place around a point on the line of centers $O_1 O_2$ (Fig. 10.1-3). Thus, if during a given time period t the number of teeth from each gear passing through the engagement zone is equal to N, then the gears 1 and 2 make (N/N_1) and (N/N_2) revolutions, respectively, N_1 and N_2 being the numbers of teeth on these gears. Hence the average angular velocities of these two gears can be expressed as follows:

$$(\omega_1)_{av} = \frac{2\pi N}{N_1 t} \tag{10.1.1a}$$

$$(\omega_2)_{av} = \frac{2\pi N}{N_2 t} \tag{10.1.1b}$$

Either one of ω_1 and ω_2 can be assumed to be positive when the other will be negative. Using Equations (10.1.1a) and (10.1.1b), the relation between the average angular velocities can be expressed as follows:

$$(\omega_1)_{av}/(\omega_2)_{av} = -(N_2/N_1) \tag{10.1.2}$$

The above relation is true for all types of gears irrespective of the shape of the teeth. For a particular pair of a gear and a pinion the ratio of the average angular velocities of the connected shafts is a constant (equal to the ratio of the numbers of teeth with a negative sign). However, the ratio of the instantaneous angular velocities of the shafts is not a constant for all tooth shapes. In general

$$\omega_1/\omega_2 = -(N_2/N_1) + p(\nu t)$$

where $p(\nu t)$ is a periodic function of time, with zero average, and a frequency ν that depends on the speed of a gear and its number of teeth. Therefore, if the driving shaft rotates at a constant speed, the rotation of the driven one will have a fluctuating component. This leads to vibration and fatigue failure. In the past, the speeds at which machineries operated were slow. But with the development of technology, operational speeds of machines are increasing and the above mentioned problem needs to be addressed. To avoid any vibration of the shafts arising out of the geared connection, the ratio of the angular speeds of the two shafts need to be constant.

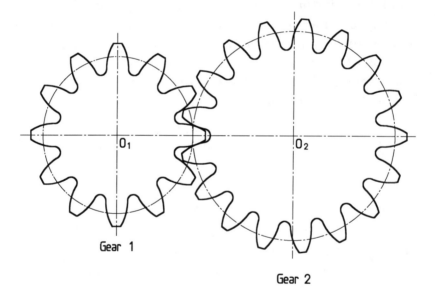

Figure 10.1-3

10.1. GEAR TOOTH ACTION AND THE LAW OF GEARING

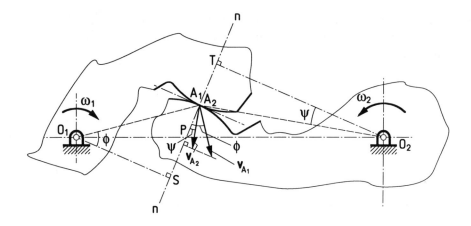

Figure 10.1-4

10.1.1 Fundamental Law of Gearing

As explained above, it has become a necessary condition that the ratio of the angular speeds of the two shafts connected by a pair of gear and pinion remain constant. Thus, Equation (10.1.2) must be replaced by the following equation:

$$\omega_1/\omega_2 = -(N_2/N_1) \qquad (10.1.3)$$

It will now be shown that if the shapes of the contacting teeth satisfy some specific conditions, then Equation (10.1.3) can be achieved. Let us consider the contact between the teeth of gears 1 and 2 at any instant, as indicated in Fig. 10.1-4. Gear 1 is the driving gear and it rotates with an angular velocity ω_1. Thus, the veloity of a particle A_1 on this gear at the point of contact is v_{A_1} (with a magnitude of $\omega_1.O_1A_1$), which is normal to the line O_1A_1. The adjacent particle A_2 on gear 2 moves in such a way that there is no velocity difference between A_1 and A_2 along the direction of common normal nn to the contacting surfaces at the point of contact. This is the condition of contact between two rigid bodies. Hence, the component of v_{A_2} along nn should be same as that of v_{A_1}. Furthermore, the magnitude of v_{A_2} is equal to $\omega_2.O_2A_2$, and v_{A_2} should be normal to the line O_2A_2, where ω_2 is the instantaneous angular velocity of gear 2. Now let us drop two normals O_1S and O_2T from gear centers O_1 and O_2, on nn, as shown in Fig. 10.1-4. The above conditions can be expressed mathematically as follows:

$$v_{A_1} \cos \phi = v_{A_2} \cos \psi \tag{10.1.4a}$$

$$v_{A_1} = \omega_1 . O_1 A_1 \tag{10.1.4b}$$

$$v_{A_2} = \omega_2 . O_2 A_2 \tag{10.1.4c}$$

Using Equations (10.1.4b) and (10.1.4c) in Equation (10.1.4a), we get

$$\omega_1 . O_1 A_1 \cos \phi = \omega_2 . O_2 A_2 \cos \psi.$$

But from the figure, $O_1 A_1 \cos \phi = O_1 S$ and $O_2 A_2 \cos \psi = O_2 T$. Hence, the above equation can be written as (sign of ω is omitted, only the magnitude is considered)

$$\omega_1 . O_1 S = \omega_2 . O_2 T$$

$$\text{or,} \; \omega_1/\omega_2 = O_2 T / O_1 S. \tag{10.1.5}$$

If P is the point of intersection of the common normal nn (to the teeth surfaces at the contact point) and the line of centers $O_1 O_2$, then

$$\Delta O_1 SP \cong \Delta O_2 TP$$

since both of these are right-angled triangles with $\angle O_1 PS = \angle O_2 PT$. Therefore, $O_2 T / O_1 S = O_2 P / O_1 P$. Using this in Equation (10.1.5), we finally get the speed ratio as follows:

$$i = \omega_1/\omega_2 = O_2 P / O_1 P \tag{10.1.6}$$

The above equation implies that if the common normal to the surfaces of the contacting teeth, drawn at the contact point, passes through the same point (which is called *the pitch point*) on the line of centers at all instants, then the ratio of the angular velocities of the driving

10.1. GEAR TOOTH ACTION AND THE LAW OF GEARING

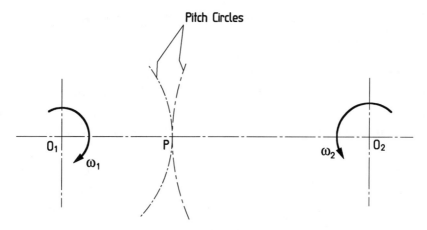

Figure 10.1-5

and the driven gears will be constant.[1] This condition is commonly referred to as the *fundamental law of gearing*. For any shape of the teeth of one gear there exists a unique shape of the teeth of the other gear that satisfies the law of gearing. One profile is called the *conjugate* of the other.

If the tooth profile of one gear is chosen arbitrarily, it is possible to find out the conjugate profile. Since tooth shape depends on the size of the gear, one way to standardize is to specify the tooth profile of a *rack* (which is a segment of a gear with infinite diameter) and then define a system of gears having tooth profiles which are conjugate to the chosen rack. The rack tooth profile is known as the *basic rack* for the system of gears. Various organizations, such as the American Gear Manufacturers Association (AGMA) and the International Organization for Standardization (ISO), use this method. The most commonly used tooth profile is the involute, due to a number of advantages; however, a few other shapes are also used.

---

[1]The reader should note that P is the relative instantaneous center of the two gears, a fact which follows easily from the application of the Kennedy-Aronhold theorem. Hence, Equation (10.1.6) could have been derived easily by using $\boldsymbol{v}_P^{(1)} = \boldsymbol{v}_P^{(2)}$.

10.1.2 Path of Contact, Pitch Circle and Line of Action

The locus of the contact points between a pair of teeth during the rotation of a pair of engaged gears is called the *path of contact*. As a consequence of the law of gearing, the path of contact has to pass through the pitch point. In the special case of involute tooth profile, the path of contact is a straight line.

If two cylinders are conceived with O_1 and O_2 as centers and O_1P and O_2P as the radii (Fig. 10.1-5), then the motions of the two gears 1 and 2 will be identical to the motions of the two cylinders, if one rolls against the other. This is obvious from Equation (10.1.6). These imaginary cylinders are called *pitch cylinders* and the circles representing these cylinders in the view shown in Fig. 10.1-5 are called *pitch circles*. It should be remembered that a pitch circle is not an invariant property of a gear, as it depends on the meshing gear as well. A unique situation arises when the mesh is with its basic rack. The corresponding pitch circle is called the *standard pitch circle*. When two gears mesh properly, without any backlash, the pitch circle coincides with the standard pitch circle.

The common normal drawn to the contacting teeth surfaces at the point of contact also indicates the direction of the normal forces acting on the teeth. In the absence of friction, this force represents the total driving effort. Therefore, this common normal is sometimes referred to as the *line of action*. In case of involute gears, the line of action and the path of contact are the same.

10.1.3 Circular Pitch, Diametral Pitch and Module

The circular pitch (p_r) of a gear at any radius r is defined as the distance between the corresponding points of two adjacent teeth, measured along the circle of radius r (Fig. 10.1-6). Thus,

$$p_r = 2\pi r/N. \tag{10.1.7}$$

Generally the *circular pitch* is measured along the pitch circle of a

10.1. GEAR TOOTH ACTION AND THE LAW OF GEARING

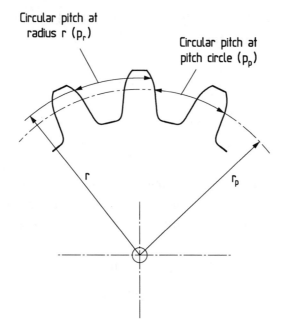

Figure 10.1-6

gear and, is represented by p_p, and

$$p_p = 2\pi r_p/N. \qquad (10.1.8)$$

It is obvious that the pitch of a rack (p) is constant, as indicated in Fig. 10.1-7. If a gear is in mesh with the rack, the circular pitch of the gear, measured along the pitch circle, will be equal to the pitch of the rack. Thus,

$$p_p = p. \qquad (10.1.9)$$

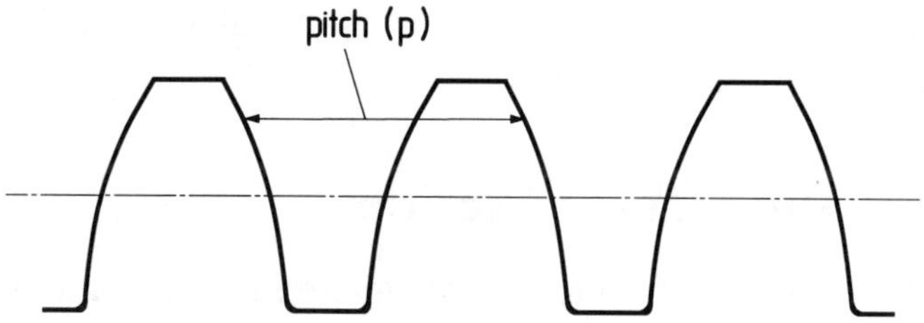

Figure 10.1-7

The *diametral pitch* (P) is the ratio of the number of teeth on a gear to its pitch circle diameter in inches. Hence,

$$P = \frac{N}{2r_p} (r_p \text{ in inches}) \quad (10.1.10)$$

From Equations (10.1.8) and (10.1.10),

$$Pp_p = \pi. \quad (10.1.11)$$

In metric system, the specification is found by the ratio of the pitch diameter in mm to the number of teeth. This ratio is called the *module (m)*. The module is thus expressed as follows:

$$m = \frac{2r_p}{N} (r_p \text{ in mm}) \quad (10.1.12)$$

From Equations (10.1.8) and (10.1.12), we get

$$\frac{p_p}{m} = \pi. \quad (10.1.13)$$

10.2 Involute Spur Gears

In the previous section it was shown that when rotational motion is smoothly transmitted between two parallel shafts through physical contact between two teeth attached to these shafts, the line of action must always pass through a fixed point (known as the pitch point) on the line of centers. Using any assumed form of the teeth on one gear, the shape of the conjugate teeth on the other gear can be determined. Of the many possible conjugate profiles only the involute and cycloid have been standardized. However, except in watches and special cases, the involute profile is used universally for all gears transmitting motion and power.

10.2.1 Involute Teeth Action

Rotary motion of one shaft can be transmitted to a parallel shaft with a fixed transmission ratio by using a pair of pulleys connected by a crossed wire, as shown in Fig. 10.2-1. If the wires are tight enough

10.2. INVOLUTE SPUR GEARS

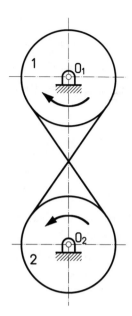

Figure 10.2-1

to prevent slipping, then the angular velocity ratio is equal to the inverse ratio of the pulley diameters. It is obvious that the angular velocity ratio does not depend on the center distance O_1O_2. From the following, we will see that the motion transmitted by teeth with an involute profile is identical to that by the crossed wire.

For convenience, let us imagine that a plate is attached to pulley 1 (as shown in Fig. 10.2-2a) and one side of the wire is removed. Then a marker is attached to the wire at point Q. As pulley 1 rotates, the marker traces a straight line AB in space. However, the marker traces an involute on the plate attached to pulley 1. The same involute could also be generated by the point Q, if the wire was cut at this point and unwrapped from pulley 1 (keeping the wire always in taut condition). Another involute will be traced by Q on a plate attached to pulley 2, as shown in Fig. 10.2-2b. Now, if the plates are cut along the involutes and the wire is removed, the involute plate on one pulley can drive the involute plate on the other, producing the same relative motion between the pulleys as that with the crossed wire (Fig. 10.2-3). From Equation (10.1.6), the angular speed ratio

$$\omega_1/\omega_2 = O_2P/O_1P. \qquad (10.2.1)$$

It should be remembered that the direction of rotation of the two pulleys are opposite and the angular velocity ratio will be a negative quantity. From Fig. 10.2-3,

$$\Delta O_2 BP \cong \Delta O_1 AP$$

$$\text{and} \quad O_2P/O_1P = O_2B/O_1A = r_{b_2}/r_{b_1}$$

where r_{b_1} and r_{b_2} are the radii of the pulleys, which are also the *base circles* for the involutes. Using the above relation in Equation (10.2.1) and noting that O_1P and O_2P are nothing but the pitch circle radii (Fig. 10.2-3), we get the angular speed ratio

$$\omega_1/\omega_2 = r_{b_2}/r_{b_1} = r_{p_2}/r_{p_1}. \qquad (10.2.2)$$

In the present situation the path of contact is indistinguishable from the line of action AB (the direction along which the contact force

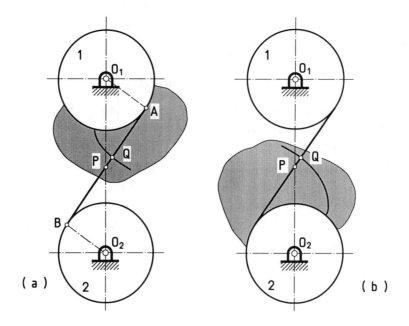

Figure 10.2-2

10.2. INVOLUTE SPUR GEARS

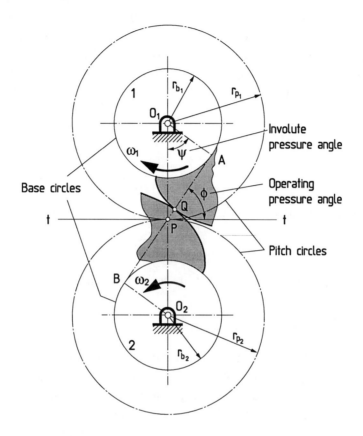

Figure 10.2-3

between the teeth acts in the ideal friction free condition), which is the common tangent to the two base circles. Line tt is the common tangent to the pitch circles at the pitch point P and the inclination of AB with line *tt* is called the *operating pressure angle* (ϕ). The angle between the lines O_1Q and O_1A is called the *involute pressure angle* (ψ). This angle refers to the location of the point on the involute where the contact is taking place at the instant.

When two gears are in mesh, the circular pitch along their pitch circles (i.e., p_{p_1} and p_{p_2}) are equal. Therefore,

$$p_{p_1} = \frac{2\pi r_{p_1}}{N_1} = p_{p_2} = \frac{2\pi r_{p_2}}{N_2} \qquad (10.2.3a)$$

$$\text{or, } r_{p_2}/r_{p_1} = N_2/N_1. \qquad (10.2.3b)$$

Using this in Equation (10.2.2) we get,

$$\omega_1/\omega_2 = N_2/N_1$$

(it will be negative when ω_1 and ω_2 represent angular velocities).

10.2.2 Involute Nomenclature

Already a number of definitions connected with gearing action have been presented in the previous section. A few more quantities, often referred to in the context of involute spur gears, are discussed below. Figure 10.2-4 shows two involute spur gears in mesh. The outer circle up to which the teeth of a gear exist is called the *addendum circle* and the height of the teeth above the pitch circle is called the *addendum* (a). The depth of the teeth-roots below the pitch circle[2] is called the *dedendum* (b) and the corresponding circle is called the *dedendum circle*. The common tangent AB to the base circles is the line of action (and also the path of contact in cases of involute gears). From Fig. 10.2-4 it is obvious that the contact will start at point E, where the addendum circle of the driven gear intersects the path of contact, and it will end at point F, where the addendum circle of the driving

---

[2]Since the involute does not exist inside the base circle, the portion of the tooth profile below the base circle is a straight radial line.

10.2. INVOLUTE SPUR GEARS

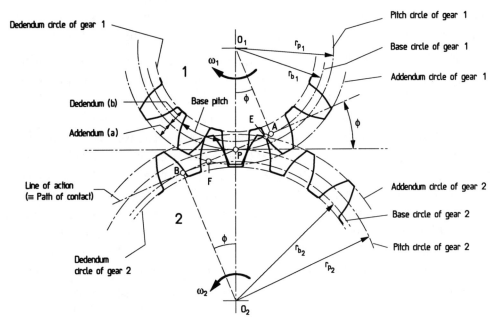

Figure 10.2-4

gear intersects the path of contact. The length of the segment EF is called the *length of action* (sometimes also called the *length of path of contact*). The *base pitch* (p_b) is defined as the distance from a point on one tooth to the corresponding point on an adjacent tooth, measured along the base circle as shown in Fig. 10.2-4. From the figure, the base circle radius of gear 1 is equal to $O_1 A$ whereas the pitch circle radius of the gear is $O_1 P$. Thus,

$$r_{b_1} = r_{p_1} \cos \phi \qquad (10.2.4a)$$

and

$$p_{b_1} = \frac{2\pi r_{b_1}}{N_1} = \frac{2\pi r_{p_1} \cos \phi}{N_1} = p_{p_1} \cos \phi. \qquad (10.2.4b)$$

The base circle radius and the number of teeth are the properties of a single gear and are fixed. On the other hand, the pitch point (and therefore the pitch circle radius) depends on the center distance of a meshing pair of gears. Thus p_p and ϕ are determined by the center distance. However, when a gear is in mesh with its basic rack the operating pressure angle is equal to the pressure angle of the rack.

Figure 10.2-5a shows the positions of a pair of teeth at the beginning of contact and also at the end of contact. Point C is the point on the tooth on gear 1 at the pitch circle radius when the contact is first established. The same point takes the location D when the contact ends. The corresponding points on gear 2 are indicated by G and H, respectively. Arcs CPD and GPH are the arcs on the pitch circles through which the mating teeth move from start to end of the contact. These arcs are known as the *arcs of action*. These arcs are equal, as the relative motion between the pitch cylinders is a pure rolling one. The angles subtended by these arcs at their respective centers are called the *angles of action* (θ_1 and θ_2 in Fig. 10.2-5a). The angles of action are again divided into the *angles of approach* (α) and the *angles of recess* (β). The threshold between the approach and the recess is the instant when the contact is at the pitch point. The ratio of the arc of action to the circular pitch is called the *contact ratio*

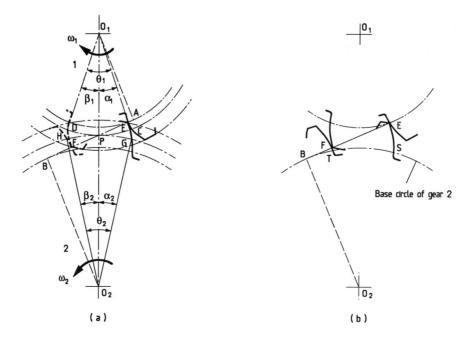

Figure 10.2-5

10.2. INVOLUTE SPUR GEARS

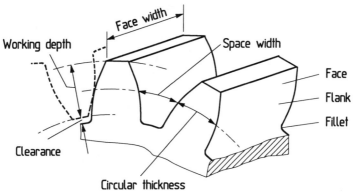

Figure 10.2-6

(m_c). It is a measure of the average number of pairs of teeth that are in contact. In the case of involute gears it is also equal to the ratio of the length of action to the base pitch, as is evident from Fig. 10.2-5b. The length of the path of contact EF is equal to the arc ST of the base circle. If a string wrapped on the base circle along the arc BTS is unwrapped (keeping it always taut), point S goes to E, T to F and S and E (and T and F) are on the same involute.

Figure 10.2-6 shows a few more quantities associated with involute spur gears, which are defined below. Space width is the length of the arc between the closest sides of two adjacent teeth measured along the pitch circle. On the other hand, the length of the arc between two sides of the same tooth measured along the pitch circle is called the *circular thickness* (tooth thickness at pitch circle). Both the space width and the tooth thickness at standard pitch circle are equal to half the circular pitch. The surface of a tooth between the pitch circle and the addendum circle is called the *face* and that between the pitch circle and the root circle (dedendum circle) is called the *flank*. The dedendum of a gear minus the addendum of the mating gear is called the *clearance* and the sum of the addenda of mating gears is called the *working depth*. The space width minus the circular thickness is called the *backslash*.

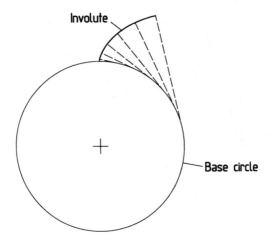

Figure 10.2-7

10.2.3 Involutometry

The involute is defined as the locus of a point on a taut string that is unwrapped from a cylinder. Figure 10.2-7 shows how an involute is generated. The circle that represents the cylinder is called the base circle. It is obvious that the normal to any point on an involute is a tangent to the base circle. In the following, we shall discuss some basic properties of the involute curve, which will be useful later.

Figure 10.2-8 shows an involute generated from a base circle of radius r_b starting at point A. The radius of curvature of the involute at any point I is given by

$$\rho = (r^2 - r_b^2)^{1/2} \qquad (10.2.5)$$

where $r = OI$. The *involute pressure angle* at I is defined as the angle between the driving force at I (which is normal to the involute, i.e., along IB) and the direction in which the point I moves (which is normal to OI, where O is the hinge point about which the tooth rotates). From the figure it can easily be seen that $\angle IOB = \psi$. The angle by which the tangent to the involute rotates while going from A (where it is radial) to I is defined as the *roll angle*, θ. Again from the figure, we find that $\angle AOB = \theta$. Hence, since the length $IB(=\rho)$

10.2. INVOLUTE SPUR GEARS

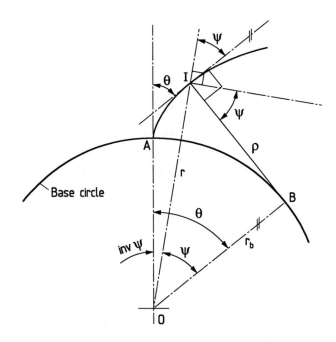

Figure 10.2-8

is equal to the arc AB,

$$\theta = \rho/r_b \qquad (10.2.6)$$

where θ is in radians. Again, from the right-angled triangle IOB,

$$\tan \psi = \rho/r_b.$$

Using Equation (10.2.6) in the above equation, we get

$$\theta = \tan \psi. \qquad (10.2.7).$$

The difference between the roll angle and the involute pressure angle is given by $\theta - \psi = \tan \psi - \psi$. This function is very important in the study of involute gearing and is given the name *involute function*. Thus,

$$\mathrm{inv}\, \psi = \tan \psi - \psi. \qquad (10.2.8)$$

When ψ is given, the value of inv ψ is given by Equation (10.2.8). However, the determination of ψ with inv ψ given, is more complicated. For this reason, inv ψ is tabulated. The value of ψ can be also determined from inv ψ quite accurately with the help of the following two steps:

$$\lambda = (\mathrm{inv}\, \psi)^{2/3} \qquad (10.2.9a)$$

$$\sec \psi = 1 + 1.04004\lambda + 0.32451\lambda^2 - 0.00321\lambda^3$$
$$-0.00894\lambda^4 + 0.00319\lambda^5 - 0.00048\lambda^6 \qquad (10.2.9b)$$

Substituting $r = r_b + \Delta r$ in Equation (10.2.5) yields

$$\rho = (2r_b.\Delta r + \Delta r^2)^{1/2}. \qquad (10.2.10)$$

From the above equation it is clear that ρ increases indefinitely as r_b increases. When the size of the gear (i.e., r_b) increases, it becomes a straight rack in the limit when $r_b \to \infty$. It is seen from Equation (10.2.10) that ρ also tends to ∞ and the tooth profile becomes a

10.2. INVOLUTE SPUR GEARS

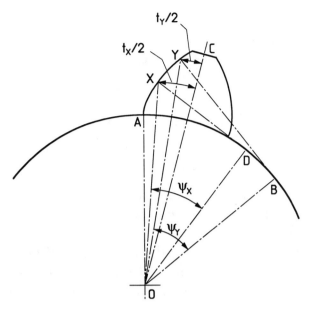

Figure 10.2-9

straight line. Thus, the tooth profile of a basic rack meshing with a pinion with involute teeth will be a straight line.

Figure 10.2-9 shows a complete involute tooth. If the tooth thickness corresponding to a point X is given, then that corresponding to a point Y can be determined, as explained below. From the figure,

$$\angle AOC = \angle AOX = \frac{t_X}{2OX} = \operatorname{inv} \psi_X + \frac{t_X}{2OX}. \tag{10.2.11}$$

Similarly,

$$\angle AOC = \angle AOY + \frac{t_Y}{2OY} = \operatorname{inv} \psi_Y + \frac{t_Y}{2OY}. \tag{10.2.12}$$

From Equations (10.2.11) and (10.2.12),

$$t_Y = 2OY[\frac{t_X}{2OX} + \operatorname{inv} \psi_X - \operatorname{inv} \psi_Y]. \tag{10.2.13}$$

Again, from Fig. 10.2-9,

$$OX \cos \psi_X = OD = r_b$$

and

$$OY \cos \psi_Y = OB = r_b.$$

Hence,

$$\psi_X = \cos^{-1}(r_b/OX) \quad (10.2.14a)$$

and

$$\psi_Y = \cos^{-1}(r_b/OY). \quad (10.2.14b)$$

Thus, with r_b, t_X, OX and OY known, t_Y can be determined from Equations (10.2.13), (10.2.14a) and (10.2.14b).

Problem 10.2-1

The base circle, pitch circle and outside radii of an involute spur gear are 16.45 mm, 17.5 mm and 19 mm, respectively. The tooth thickness along the pitch circle is 2.58 mm. Determine the thickness of the top land.

Solution

Equation (10.2.13) can be used to determine the top land thickness using $OX = 17.5$ mm and $OY = 19$ mm, with $t_X = 2.58$ mm. To obtain inv ψ_X and inv ψ_Y, first ψ_X and ψ_Y are determined using Equations (10.2.14a) and (10.2.14b) as follows:

$$\psi_X = \cos^{-1}(16.45/17.5) = 0.348 \text{ radians}$$

$$\psi_Y = \cos^{-1}(16.45/19) = 0.524 \text{ radians}.$$

From Equations (10.2.8), inv $\psi_X = \tan(0.348) - 0.348 = 0.0148$ and inv $\psi_Y = \tan(0.524) - 0.524 = 0.0539$. Using these in Equation (10.2.13), the thickness of the top land

$$t_Y = 2 \times 19[\frac{2.58}{2 \times 17.5} + 0.0148 - 0.0539] \text{ mm}$$

$$= 1.315 \text{ mm}.$$

Problem 10.2-2

The thickness of an involute tooth at a radius of 48.2 mm is 4.9 mm and the corresponding involute pressure angle is 18°. Calculate the tooth thickness along the base circle.

10.2. INVOLUTE SPUR GEARS

Solution

If the base circle radius is r_b then Equation (10.2.14a) yields

$$18° = \cos^{-1}(r_b/48.2)$$

or, $r_b = 48.2 \cos 18°$ mm $= 45.84$ mm.

At the base circle the involute pressure angle is zero. From Equation (10.2.13), the tooth thickness along the base circle can be written as

$$t = 2 \times 45.84 \left(\frac{4.9}{2 \times 48.2} + \text{inv } 0.314 - \text{inv } 0 \right) \text{ mm}$$

where

$18°(=\psi_X)$ has been replaced by 0.314 radians,

or, $t = 2 \times 45.84 \left(\dfrac{4.9}{2 \times 48.2} + 0.0107 - 0 \right)$ mm

$= 5.64$ mm.

10.2.4 Advantages of Involute Teeth

There are a number of advantages of involute teeth that are responsible for the overwhelming popularity of this form of tooth profile. These are discussed below:

(i) As we already mentioned, in cases of involute gears the line of action is the common tangent to the two base circles. Therefore, the direction of the line of action for involute gears is fixed. This implies that a constant driving effort that acts along this fixed line of action, does not produce any dynamic moment about the gear centers as the moment arm is constant. Such a situation helps in reducing dynamic loading on the shafts.

(ii) Another important advantage of involute gears is that the conjugate action of a pair of gears is not affected if the center-distance of the two gears is changed. Thus, the gears transmit rotation at a fixed angular velocity ratio, as the fundamental law of gearing is always satisfied. Figure 10.2-10 shows a pair of involute teeth

542   CHAPTER 10. GEARS

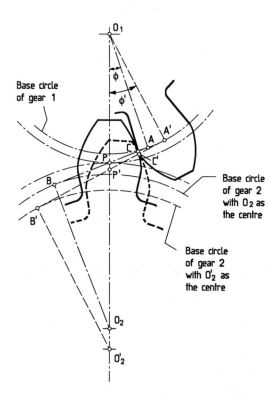

Figure 10.2-10

10.3. CHARACTERISTICS OF INVOLUTE ACTION

(belonging to the pair of gears 1 and 2) in mesh. The gear centers are at O_1 and O_2 and P is the pitch point (the intersection of the common tangent AB to the base circles with the line of centers O_1O_2). The angular velocity ratio is given by O_1P/O_2P. When the center of gear 2 is shifted to O_2', the pitch point shifts to P'. The conjugate action remains valid, as the tooth profiles are involutes with the same base circles. From the similarity of the triangles,

$$O_1P/O_2P = O_1A/O_2B \quad \text{and} \quad O_1P'/O_2P' = O_1A'/O_2B'.$$

Again, $O_1A = O_1A' = r_{b_1}$ and $O_2B = O_2B' = r_{b_2}$.

Hence, $O_1P/O_2P = O_1P'/O_2P'$ and the angular velocity ratio is not changed. However, the increase in the center distance increases the operating pressure angle and the amount of backlash and it decreases the length of path of contact.

(iii) The most crucial point in favour of involute gears arises, however, from manufacturing considerations. Gears are mostly produced by the generating process using a cutter with a conjugate profile. Since the basic rack of an involute gear has straight edges, it is easy to make a shaping tool in the form of a basic rack. Thus, high accuracy of the generated involute profile can be achieved.

10.3 Characteristics of Involute Action

In the previous section, basic features and some important points regarding involute gears were discussed. This section will be devoted to a more quantitative analysis of various aspects relating to involute gears.

10.3.1 Contact Ratio

If the rotational motion needs to be transmitted by a pair of gears in a continuous manner, there must be at least one pair of teeth in contact at all times. In reality, certain amount of overlap is present

544   CHAPTER 10. GEARS

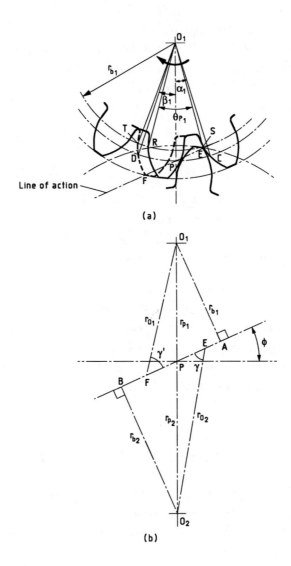

Figure 10.3-1

10.3. CHARACTERISTICS OF INVOLUTE ACTION

between the actions of two consecutive pairs of teeth. In other words, the contact between one pair of teeth continues for sometime after the next pair of teeth comes in contact. To provide a quantitative measure of the amount of overlap, the term contact ratio is introduced. In Section 10.2.2, contact ratio was defined as the ratio of the arc of action to the circular pitch. If α_1 and β_1 are the angles of approach and recess, respectively, and θ_{p_1} is the *angular pitch* $(= 2\pi/N_1)$ of the gear, then the contact ratio is given by

$$m_c = (\alpha_1 + \beta_1)/\theta_{p_1} \tag{10.3.1}$$

Figure 10.2-5a is redrawn in Fig. 10.3-1a, in a modified form. Gear 1 (the driver), represented by the solid line, indicates the position where contact is first made with a tooth of gear 2 at E (the point of intersection of the line of action with the addendum circle of gear 2). When the contact is first made, C is a point on the tooth of gear 1 at the pitch circle radius. The tooth of gear 1, shown by broken line, represents the position where the contact ends at point F (the point of intersection of the line of action with the addendum circle of gear 1). Points S and T are the points on the tooth of gear 1 (on its base circle) at the start and end of contact, respectively. From Fig. 10.3-1a,

$$\alpha_1 + \beta_1 = \angle CO_1D = \angle SO_1T = \frac{\text{arc } ST}{r_{b_1}} \text{ in radians}$$

Since E and F are two points on two involutes with the same base circle, the length EF will be equal to the arc ST. Hence,

$$\alpha_1 + \beta_1 = \frac{EF}{r_{b_1}} \text{ in radians} \tag{10.3.2}$$

Again,

$$\theta_{p_1} = \frac{\text{Base pitch}}{r_{b_1}} = \frac{SR}{r_{b_1}} = \frac{p_{b_1}}{r_{b_1}} \text{ in radians.} \tag{10.3.3}$$

From Equations (10.3.1), (10.3.2) and (10.3.3),

$$m_c = \frac{EF}{p_{b_1}}. \tag{10.3.4}$$

The essentials of Fig. 10.3-1a are shown in Fig. 10.3-1b. The length of the path of contact

$$EF = (EF + AE) + (EF + BF) - (AE + BF + EF)$$
$$= AF + BE - AB.$$

If r_{O_1} and r_{O_2} are the outer circle (same as the addendum circle) radii of gears 1 and 2, respectively, then the above relation yields

$$EF(r_{O_1}^2 - r_{b_1}^2)^{1/2} + (r_{O_2}^2 - r_{b_2}^2)^{1/2} - C\sin\phi \qquad (10.3.5)$$

where $C = r_{p_1} + r_{p_2} = O_1P + O_2P = O_1O_2$ (center distance).
The base pitch of gear 1

$$p_{b_1} = 2\pi r_{b_1}/N_1.$$

From Equations (10.2.3a) and (10.2.4) we get

$$p_{b_1} = p_{p_1}\cos\phi = p_{p_2}\cos\phi = p_{b_2}.$$

Thus, like the circular pitch, the base pitches of two meshing gears are equal and can be represented by p_b. From Equations (10.3.4) and (10.3.5), the contact ratio can be expressed as either

$$m_c = \frac{(r_{O_1}^2 - r_{b_1}^2)^{1/2} + (r_{O_2}^2 - r_{b_2}^2)^{1/2} - C\sin\phi}{2\pi r_{b_1}^2/N_1} \qquad (10.3.6a)$$

or

$$m_c = \frac{(r_{O_1}^2 - r_{b_1}^2)^{1/2} + (r_{O_2}^2 - r_{b_2}^2)^{1/2} - C\sin\phi}{2\pi r_{b_2}^2/N_2}. \qquad (10.3.6b)$$

Normally the contact ratio is not a whole number. If the ratio is 1.4 it means that there are, alternately, one pair and two pairs of teeth in contact and the time average of it is 1.4. Though theoretically the minimum acceptable value of m_c is unity, in practice 1.4 is recommended as the minimum acceptable value. From Figs. 10.2-5a and 10.3-1a it is seen that by increasing the addendum, the length of the path of contact EF can be increased, which will lead to a larger contact ratio, resulting in smoother operation. However, the amount by

10.3. CHARACTERISTICS OF INVOLUTE ACTION

which the addendum can be increased without causing other problems is limited, as will be seen later.

Figure 10.3-2 shows the positions of the tooth on gear 1 when the contact is first established at E (in solid) and when the contact is at the pitch point P (in dotted). During this period the gear rotates through the angle of approach α_1. From the figure, arc SQ is equal to the length EP for involute gears and it is obvious that

$$\alpha_1 = \text{arc } SQ/r_{b_1} \text{ (in radians).}$$

Hence,

$$\alpha_1 = EP/r_{b_1}.$$

Now, from Fig. 10.3-1b, using the property of $\triangle AO_2P$, we get

$$EP = r_{O_2}\cos(\gamma + \phi)/\cos\phi$$

where

$$\gamma = \sin^{-1}(r_{p_2}\cos\phi/r_{O_2}).$$

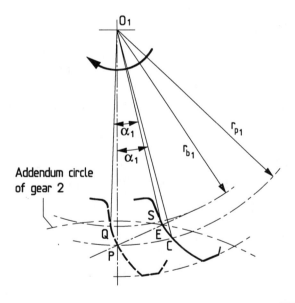

Figure 10.3-2

Hence,
$$\alpha_1 = r_{O_2}\cos(\gamma + \phi)/(r_{b_1}\cos\phi). \qquad (10.3.7a)$$

Similarly,
$$\beta_1 = PF/r_{b_1}$$
$$= r_{O_1}\cos(\gamma' + \phi)/(r_{b_1}\cos\phi) \qquad (10.3.7b)$$

where
$$\gamma' = \sin^{-1}(r_{p_1}\cos\phi/r_{O_1}).$$

The expressions for α_2 and β_2 are obtained by interchanging the subscripts 1 and 2 in Equations (10.3.7a) and (10.3.7b), respectively.

Problem 10.3-1

The pitch circle radii of two involute spur gears in mesh are 51.2 mm and 63.9 mm. The outer circle radii are 57.2 mm and 69.9 mm, respectively, and the operating pressure angle is 20°. Determine (a) the length of the path of contact, (b) the contact ratio and (c) the angles of approach and recess for the gear and pinion. The number of teeth in the pinion is 24.

Solution

Let us consider the pinion to be gear 1. Then, from the given data $r_{p_1} = 51.2$ mm, $r_{O_1} = 57.2$ mm, $r_{p_2} = 63.9$ mm, $r_{O_2} = 69.9$ mm and $\phi = 20°$. The center distance $C = r_{p_1} + r_{p_2} = 115.1$ mm. From Equation (10.2.4), $r_{b_i} = r_{p_i}\cos 20°$; $i = 1,2$. Hence, $r_{b_1} = 51.2\cos 20°$ mm $= 48.11$ mm and $r_{b_2} = 63.9\cos 20°$ mm $= 60.05$ mm.

(a) Using Equation (10.3.5), the length of the path of contact
$$\begin{aligned} EF &= [(57.2^2 - 48.11^2)^{1/2} + (69.9^2 - 60.05^2)^{1/2} \\ &\quad - 115.1\sin 20°]\text{ mm} \\ &= (30.94 + 35.78 - 39.37)\text{ mm} \\ &= 27.35\text{ mm}. \end{aligned}$$

(b) Since the number of teeth in the pinion is 24,
$$p_b = 2\pi \times 48.11/24\text{ mm} = 12.6\text{ mm}.$$

From Equation (10.3.6a),
$$m_c = EF/p_b = 27.35/12.6 = 2.17.$$

10.3. CHARACTERISTICS OF INVOLUTE ACTION

(c) From Equation (10.3.7a), for the pinion

$$\alpha_1 = 69.9\cos(\gamma + 20°)/(48.11 \times \cos 20°) \text{ radians}$$
$$\text{with } \gamma = \sin^{-1}(63.9\cos 20°/69.9) = 59.21°$$
$$\text{or, } \alpha_1 = 0.289 \text{ radians}$$
$$= 16.56°.$$

Again from Equation (10.3.7b),

$$\beta_1 = 57.2\cos(\gamma' + 20°)/(48.11\cos 20°) \text{ radians}$$
$$\text{with } \gamma' = \sin^{-1}(51.2\cos 20°/57.2) = 57.26°$$
$$\text{or, } \beta_1 = 0.279 \text{ radians}$$
$$= 15.99°.$$

For the gear, the angle of approach

$$\alpha_2 = r_{O_1}\cos(\gamma + \phi)/(r_{b_2}\cos\phi) \text{ radians}$$
$$\text{with } \gamma = \sin^{-1}(r_{p_1}\cos\phi/r_{O_1}).$$
$$\text{Using the data } \alpha_2 = 12.81°.$$

Similarly,

$$\beta_2 = r_{O_2}\cos(\gamma' + \phi)/(r_{b_2}\cos\phi)$$
$$\text{with } \gamma' = \sin^{-1}(r_{p_2}\cos\phi/r_{O_2}). \text{ Substitution of values yields}$$
$$\beta_2 = 13.29°.$$

The length of the path of contact and the contact ratio for the case of a pinion in mesh with a rack can be determined in a similar way. Figure 10.3-3 shows a pinion of base circle radius r_{b_1} and outer circle radius r_{O_1} in mesh with a rack of addendum a_r. For the rack, the pitch circle takes the form of a straight line, which is called the pitch line, and touches the pitch circle of the pinion at the pitch point P. The line of action is tangent to the base circle of the pinion and passes through the pitch point. Since the line of action is perpendicular to the contacting teeth surfaces at the contact point, the rack tooth profile is normal to the line of action. The addendum circle for the rack is a straight line parallel to its pitch line at a distance a_r (Fig. 10.3-3). In this case the length of the path of contact is EF, as shown in Fig. 10.3-3, which can be determined in the following way:

$$EF = EP + PF \qquad (10.3.8)$$

where, from Fig. 10.3-3,

$$EP = a_r / \sin\phi. \qquad (10.3.9)$$

Next, consider the right-angled triangle O_1FA. From this triangle

$$AF = (O_1F^2 - O_1A^2)^{1/2} = (r_{O_1}^2 - r_{b_1}^2)^{1/2}.$$

Again considering the right-angled triangle O_1PA, we can write

$$AP = O_1A \tan\phi = r_{b_1} \tan\phi.$$

Hence,

$$PF = AF - AP = (r_{O_1}^2 - r_{b_1}^2)^{1/2} - r_{b_1} \tan\phi. \qquad (10.3.10)$$

From Equations (10.3.8), (10.3.9) and (10.3.10), the length of contact

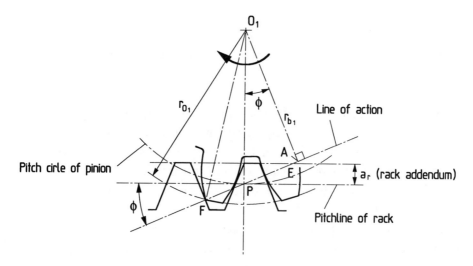

Figure 10.3-3

10.3. CHARACTERISTICS OF INVOLUTE ACTION

$$EF = \frac{a_r}{\sin \phi} + (r_{O_1}^2 - r_{b_1}^2)^{1/2} - r_{b_1} \tan \phi. \qquad (10.3.11)$$

Dividing this by the base pitch p_b, the contact ratio is obtained as follows:

$$m_c = \frac{1}{p_b}[(r_{O_1}^2 - r_{b_1}^2)^{1/2} + \frac{a_r}{\sin \phi} - r_{b_1} \tan \phi] \qquad (10.3.12)$$

Problem 10.3-2

A rack is in mesh with a pinion of 52 mm pitch circle radius. The operating pressure angle and the addendum for both the rack and pinion are 20° and 5 mm, respectively. Determine the length of the path of contact and the contact ratio if the module is 4.

Solution

From the data, $r_{p_1} = 52$ mm, $r_{O_1} = (52+5)$ mm $= 57$ mm, $\phi = 20°$ and $a_r = 5$ mm. Since the module is 4, from Equation (10.1.13) the circular pitch

$$p_p = \pi \times 4 \text{ mm} = 12.57 \text{ mm}.$$

With a 20° operating pressure angle, the base pitch

$$p_b = p_p \cos 20° = 5.9 \text{ mm}.$$

The base circle radius can also be determined, as

$$r_{b_1} = r_{p_1} \cos 20° = 48.86 \text{ mm}.$$

The length of the path of contact according to Equation (10.3.11)

$$\begin{aligned} EF &= [\frac{5}{\sin 20°} + (57^2 - 48.86^2)^{1/2} - 48.86 \tan 20°] \text{ mm} \\ &= 26.19 \text{ mm}. \end{aligned}$$

The contact ratio is

$$m_c = EF/p_b = 26.19/12.57 = 2.08.$$

Problem 10.3-3

Two identical involute spur gears are in mesh. The module is 4 and the gears have 24 teeth each. If the operating pressure angle is 20°, determine the minimum value of addendum needed to ensure continuous transmission of motion.

Solution

For continuous transmission of motion, the minimum permissible value of the contact ratio is 1. If a is the minimum addendum, then

$$r_{O_1} = r_{p_1} + a; \quad r_{O_2} = r_{p_2} + a.$$

Hence from the data,

$$r_{p_1} = r_{p_2} = 48 \text{ mm}$$

and with $\phi = 20°$,

$$r_{b_1} = r_{b_2} = 48 \cos 20° \text{ mm} = 45.11 \text{ mm}.$$

The center distance

$$C = r_{p_1} + r_{p_2} = 96 \text{ mm}.$$

Using any of the two equations (10.3.6a and b) and taking $m_c = 1$, we get the following relation:

$$1 = \frac{2 \times \{(48 + a)^2 - 45.11^2)\}^{1/2} - 96 \sin 20°}{2\pi \times 45.11/24}$$

Rearranging the above equation,

$$a^2 + 96a - 229.18 = 0$$

which yields

$$a = 2.33 \text{ mm}.$$

The other root being negative is not acceptable.

10.3. CHARACTERISTICS OF INVOLUTE ACTION

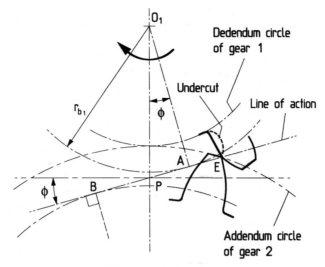

Figure 10.3-4

10.3.2 Interference and Undercutting

In the previous section it was mentioned that though an increase in the addendum increases the contact ratio, there are limitations to the extent to which the addendum can be increased. In this section the effects of excessive addendum will be presented.

In the cases of involute gears, the conjugate action is ensured when the contact is between two involute surfaces. At the same time it should be remembered that involute exists only above the base circle. Since generally the root circle (or the dedendum circle) is smaller than the base circle, the portion of a tooth below the base circle does not have an involute profile. Most often it is radial and joins the root circle smoothly with the help of a circular corner. Since this portion of the tooth does not have an involute profile, it is not intended to come in contact with the teeth of any meshing gear. If a contact does take place, the tip of the contacting teeth will mate with the noninvolute portion, resulting in *interference*. The tip will then scoop out some material from the teeth below the base circle, which is called *undercutting*. Such a situation is shown in Fig. 10.3-4. The intersection (point E) of the addendum circle of the driven gear (gear 2) and the line of action is beyond the point A, where the line of action touches the base circle of the driver. Therefore, during the period when the contact point is on EA, there is interference and undercutting. Similarly, if the addendum circle of the driver is large enough to place its point of intersection (F) with the line of action beyond B, the tip of the tooth of gear 1 will interfere with the noninvolute portion of the tooth on gear 2.

In the case of a pinion meshing with a rack, the interference will occur when the addendum of the rack is excessively large. Such a situation is shown in Fig. 10.3-5. On the other hand, no interference can occur due to excessive addendum of the pinion because the point of tangency (B) of the line of action with the base circle of the rack lies to the left at infinity. However, there is also a constraint on the addendum of any gear. This arises because at a particular height above the base circle and for a given module, the thickness of the tooth becomes zero (in other words, with a particular addendum the tooth becomes pointed).

The phenomenon of undercutting can be better understood from a different point of view. Figure 10.3-6 shows a rack in mesh with a pinion. The relative motion between the rack and the pinion is such that the pitch line of the rack can be considered to be rolling over the

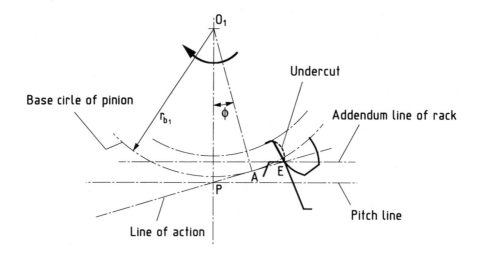

Figure 10.3-5

10.3. CHARACTERISTICS OF INVOLUTE ACTION

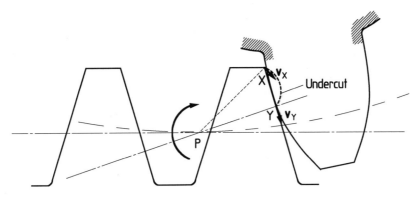

Figure 10.3-6

pitch circle of the pinion. Thus the point X on the tip of the rack has a velocity v_X (perpendicular to the line PX) with respect to the pinion as shown. The digging effect of the tip into the tooth of the pinion is quite obvious. All points on the rack tooth up to the point Y (on the line of action) have similar digging effects. Undercutting is generally undesirable, as it weakens the tooth. Apart from this, the determination of contact ratio is also difficult when undercutting has occurred.

The interference is avoided when the contact between the teeth of the two gears in mesh remains confined to the segment AB, where A and B are the points where the line of action touches the base circles. For this reason, these points are also called the *interference points*. To determine the maximum addenda for a pair of gears tha operate without interference, we consider the situation in which the addendum circles pass through the interference points, as shown in Fig. 10.3-7. Under this condition, the length of the path of contact is AB and is given by

$$AB = (r_{p_1} + r_{p_2}) \sin \phi = C \sin \phi. \qquad (10.3.13)$$

From the figure, the maximum addendum circle (outer circle) radii can be determined as follows:

$$r_{O_1}^{max} = (r_{b_1}^2 + AB^2)^{1/2} = (r_{b_1}^2 + C^2 \sin^2 \phi)^{1/2} \qquad (10.3.14a)$$

$$r_{O_2}^{max} = (r_{b_2}^2 + AB^2)^{1/2} = (r_{b_2}^2 + C^2 \sin^2 \phi)^{1/2} \qquad (10.3.14b)$$

Problem 10.3-4

Two involute spur gears of module 3 and with 18 and 24 teeth, operate at a pressure angle of 20°. Find the maximum addendum for the gears so that no interference occurs.

Solution

First, let us determine the pitch circle diameters of the gears. From Equation (10.1.12),

$$r_{p_1} = \frac{3 \times 18}{2} \text{ mm} = 27 \text{ mm}$$

$$\text{and} \quad r_{p_2} = \frac{3 \times 24}{2} \text{ mm} = 36 \text{ mm}.$$

Since $r_b = r_p \cos \phi$ we get using $\phi = 20°$

$$r_{b_1} = 27 \cos 20° \text{ mm} = 25.37 \text{ mm}$$

$$\text{and} \quad r_{b_2} = 36 \cos 20° \text{ mm} = 33.83 \text{ mm}.$$

With $C = r_{p_1} + r_{p_2} = (27 + 36)$ mm $= 63$ mm, Equations (10.3.14a) and (10.3.14b) yield

$$r_{O_1}^{max} = (25.37^2 + 63^2 \sin^2 20°)^{1/2} \text{ mm} = 33.29 \text{ mm}$$

$$\text{and } r_{O_2}^{max} = (33.83^2 + 63^2 \sin^2 20°)^{1/2} \text{ mm} = 40.11 \text{ mm}.$$

Hence, the maximum permissible addenda are

$$a_1^{max} = (r_{O_1}^{max} - r_{p_1}) = (33.29 - 27) \text{ mm} = 6.29 \text{ mm}$$

$$\text{and} \quad a_2^{max} = (r_{O_2}^{max} - r_{p_1}) = (40.11 - 36) \text{ mm} = 4.11 \text{ mm}.$$

It can be easily shown that the tendency of interference at the base of the teeth of the driving gear, for a given addendum of the driven gear, increases as the size of the driven gear increases. Therefore, if there is no interference when a pinion is in mesh with a rack, there will be no interference with any gear with the same addendum, since the rack represents the largest possible gear (with $r_p = \infty$).

10.3.3 Standardization of Involute Gears

The importance of a proper selection of addendum has been made amply clear in the previous sections. This, along with the need to ensure interchangeability, has led to the standardization of gears and a brief description is presented below.

In the past an operating pressure angle of $14\frac{1}{2}°$ had been used. This is because of the fact that most gears were cast and $\sin 14\frac{1}{2}° \approx \frac{1}{4}$, which was very convenient for pattern layout. More recently, the tendency has been to choose an operating pressure angle of $20°$, as fewer teeth can be cut without undercutting. Because of the tendency towards higher pressure angle, angles of $22.5°$ and $25°$ are also sometimes used, especially in aerospace gearing. The addendum and dedendum of gears are also standardized by relating them to the module. The following table gives some information:

Table 10.3-1: Standard Metric Gears

Parameter	British	German
Addendum (a)	$1.000\ m$	$1.000\ m$
Dedendum (b)	$1.250\ m$	1.157 or $1.167\ m$
Operating Pressure Angle (ϕ)	$20°$	$20°$

As we already mentioned, meshing gears must have same circular pitch. This, along with Equation (10.1.13), shows that meshing gears must be of same module (and, therefore, same addendum and dedendum too). The term *standard gear* is used to classify those gears that are cut by the *standard cutters*. Such standard gears are *interchangeable*. However, nonstandard gears are also used, especially in automobile and aerospace industries. The major disadvantage of an involute profile is the possibility of interference, between the tip of a gear tooth and the flank of the meshing gear tooth. Nonstandard system of gearing has been developed to avoid interference and consequent undercutting. Two of these nonstandard systems are popular since standard cutters can be used. In one, known as *extended center distance system*, the cutter is withdrawn by a certain amount

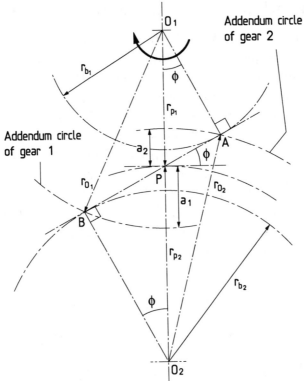

Figure 10.3-7

from the blank so that the addendum of the cutter passes through the interference points of the gear (the smaller one, i.e., the pinion) being cut. In case of such gears the center distance cannot be calculated from the module and number of teeth and is therefore considered non-standard. The pressure angle is also increased by this method. In the other system, the cutter is advanced into the blank of the larger gear by the same amount by which it will be withdrawn from the blank of the smaller gear (pinion). As a result, the addendum of the pinion increases and that of the gear decreases by an equal amount. The center distance and the operating pressure angle remain the same as those in the case of standard gears. This system is known as the *long and short addendum system*.

10.3. CHARACTERISTICS OF INVOLUTE ACTION

10.3.4 Minimum Number of Teeth

An examination of Fig. 10.3-7 shows that for a pair of standard involute gears, the tip of the gear interferes with the flank of the pinion tooth first. This is because the addenda of a meshing pair of standard gear are equal ($a_1 = a_2$). Interference starts just when the addendum circle of the larger gear (gear 2 in Fig. 10.3-7) passes through the point A. The number of teeth on the pinion (for a prescribed transmission ratio) for this situation (which is the minimum number of teeth without interference) can be determined as follows.

If the addendum is equal to $a(= a_1 = a_2)$, then

$$a = O_2A - O_2P. \tag{10.3.15}$$

Considering the ΔO_2AP with ϕ as the operating pressure angle,

$$\begin{aligned} O_2A^2 &= O_2P^2 + PA^2 - 2O_2P.PA\cos(\pi/2 + \phi) \\ &= O_2P^2 + PA^2 + 2O_2P.PA\sin\phi. \end{aligned}$$

Now, from ΔO_1AP we can write $PA = O_1P\sin\phi$. Hence, the above relation can be written as

$$\begin{aligned} O_2A^2 &= O_2P^2 + O_1P^2\sin^2\phi + 2O_2P.O_1P\sin^2\phi \\ &= O_2P^2[1 + (O_1P/O_2P)(O_1P/O_2P + 2)\sin^2\phi]. \end{aligned}$$

Using Equation (10.3.15) with the above relation, we get

$$a = O_2P[1 + (O_1P/O_2P)(O_1P/O_2P + 2)\sin^2\phi]^{1/2} - O_2P$$

$$= r_{p_2}[\{1 + (1/i)(1/i + 2)\sin^2\phi\}^{1/2} - 1]. \tag{10.3.16}$$

Normally the addendum is expressed as a fraction f of the module, as indicated in Table 10.3-1. Therefore,

$$a = fm = f.(2r_{p_2}/N_2). \tag{10.3.17}$$

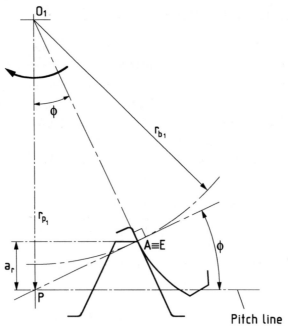

Figure 10.3-8

Using Equation (10.3.17) in Equation (10.3.16) we get

$$2f = N_2[\{1 + \mu(\mu + 2)\sin^2\phi\}^{1/2} - 1]$$

where $\mu = 1/i = 1/\text{transmission ratio} = \omega_2/\omega_1 = N_1/N_2$.

Rewriting the above equation and replacing N_2 by N_1/μ, we finally arrive at the following expression for the minimum number of teeth on the pinion to avoid interference:

$$N_1^{min} = \frac{2f\mu}{[1 + \mu(\mu + 2)\sin^2\phi]^{1/2} - 1} \qquad (10.3.18)$$

For a pinion meshing with a rack, the situation where the interference is just starting is shown in Fig. 10.3-8. From the figure,

$$a_r = AP\sin\phi = O_1P\sin^2\phi.$$

With $a_r = fm = f.2O_1P/N_1$, we get the minimum number of teeth on the pinion without causing interference as

10.3. CHARACTERISTICS OF INVOLUTE ACTION

$$N_1^{min} = 2f/\sin^2\phi. \tag{10.3.19}$$

Problem 10.3-5

A pair of standard spur gears have 12 and 24 teeth with a module of 12. Find out whether there will be any interference or not. The pressure angle is equal to 20°.

Solution

From Table 10.3-1 we find that $f = 1$. Using the data,

$$\mu = N_1/N_2 = 12/24 = 0.5.$$

Substituting these and $\phi = 20°$ in Equation (10.3.18), the minimum number of teeth on the pinion, for the given transmission ratio, is determined as follows:

$$N_1^{min} = \frac{2 \times 0.5}{[1 + 0.5(0.5 + 2)\sin^2 20°]^{1/2} - 1}$$
$$= 14.16$$

Since the number of teeth on the pinion is less than 14.16, interference will occur.

Problem 10.3-6

Find out the minimum number of teeth on a standard involute gear with operating pressure angle 20°, which will mesh with any standard gear of the same module without interference.

Solution

It has been already shown that if a standard gear is in mesh with its standard rack without interference, there will be no interference when it meshes with any other standard gear. Hence, from Equation (10.3.19), the minimum number of teeth of a standard gear meshing with a rack without interference is determined to be

$$N_1^{min} = 2/\sin^2 20° = 17.1$$

because $f = 1$ for a standard gear. Thus, the required number of teeth is equal to 18.

Problem 10.3-7

While transmitting rotational motion from one shaft to another with a pair of standard spur gears, a pinion with 16 teeth is mounted on the driving shaft. If the operating pressure angle is 20°, obtain the maximum speed reduction without having any interference.

Solution

The interference starts when the number of teeth on the pinion satisfies the relation (10.3.18)

$$N_1 = \frac{2f\mu}{[1 + \mu(\mu + 2)\sin^2 \phi]^{1/2} - 1}.$$

From the data $N_1 = 16$, $f = 1$ and $\phi = 20°$. Using these and rearranging the above equation,

$$1 + \mu(\mu + 2)\sin^2 20° = \left(\frac{2\mu}{16} + 1\right)^2$$

or, $0.101\mu^2 - 0.016\mu = 0$ or, $\mu = 0.158$. Hence, the maximum reduction possible without interference is $1/\mu = 6.31$.

10.3.5 Backlash

Backlash can be qualitatively defined as the amount of play between a mating pair of gear teeth in assembled condition. Because of manufacturing errors the teeth profiles do not conform exactly to the theoretical involute shapes; the gear dimensions and the center distance are also never exactly equal to the specified values. Besides these, any change in temperature also results in variations in the gear dimensions. Because of all these, the tooth thickness of each gear is so chosen that contact takes place on one face only. Otherwise, there is a possibility of jamming, overheating, damage and objectionable noise. The small gap between the noncontacting surfaces of the teeth of the gears is the backlash. If the directions of rotation of the gears do not change, backlash does not create any problem. However, excessive backlash may lead to not only inaccuracy in the system, but also noise and impact loading. Choice of proper amount of backlash is, therefore, extremely important.

10.3. CHARACTERISTICS OF INVOLUTE ACTION

The amount of backlash can be specified in two different ways as follows:

Circular Backlash: The *circular backlash* B of a pair of gears is defined as the difference between the space width of one gear and the circular thickness of the other. This can also be shown to be equal to the possible rocking movement of a point on the pitch circle of a gear when the other one is held fixed. If t_{p_1} is the circular thickness of gear 1 and w_{p_1} is its space width, then according to the definitions

$$t_{p_1} + w_{p_1} = p_p. \qquad (10.3.20)$$

Hence, the difference between the space width of gear 1 and the circular thickness of gear 2 (which is equal to the circular backlash, B, and is also equal to the difference between the space width of gear 2 and the circular thickness of gear 1) becomes

$$B = w_{p_1} - t_{p_2} = p_p - t_{p_1} - t_{p_2}. \qquad (10.3.21)$$

The rocking angles of the gears because of this backlash are as follows:

$$\Delta\beta_1 = B/r_{p_1}; \quad \Delta\beta_2 = B/r_{p_2} \qquad (10.3.22)$$

Normal Backlash: Backlash can be quantified as the gap between teeth along the common normal. This is called *normal backlash*. Figure 10.3-9 shows a pair of gears in which the contact point lies on the line of action EF. The other interior common tangent to the base circles is $E'F'$. Since the noncontacting faces of the teeth are also involutes with the same base circles, $E'F'$ must be normal to the noncontacting faces at their closest points and B_n is the gap measured along $E'F'$, as indicated in Fig. 10.3-9. From this figure the amount of rocking angular motion $\Delta\beta_2$ of gear 2, because of the normal backlash B_n, can be written as

$$\Delta\beta_2 = \frac{B_n}{r_{b_2}}. \qquad (10.3.23)$$

From Equations (10.3.22) and (10.3.23) a relation between B and B_n can be found as follows:

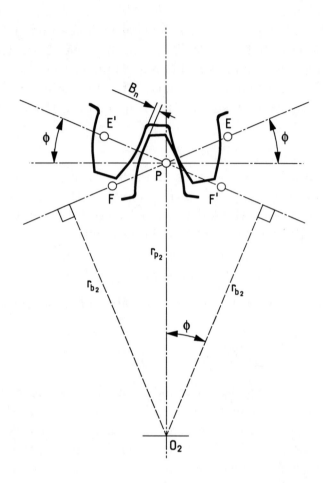

Figure 10.3-9

10.3. CHARACTERISTICS OF INVOLUTE ACTION

$$B_n = \Delta\beta_2 r_{b_2} = Br_{b_2}/r_{p_2} = B\cos\phi. \qquad (10.3.24)$$

When the profiles of the teeth of a meshing pair of standard gears are considered to be ideal, a backlash can occur due to excess center distance and the amount of backlash can be calculated as discussed below.

When two involute standard gears, cut by standard cutters, mesh without any backlash the pitch circles are referred to as standard (cutting) pitch circles,[3] as shown in Fig. 10.3-10a. The corresponding center distance and operating pressure angle are C and ϕ, respectively. When the centers are moved apart from their standard positions by ΔC, the center distance is $C'(=C+\Delta C)$. The line of action now intersects the line of centers at a point P' (different from P) and the operating pressure angle increases to ϕ', as shown in Fig. 10.3-10b. Since the pitch point is shifted to P', the *operating pitch circles* will have radii, r'_{p_1} and r'_{p_2}, but the transmission ratio remains unaffected.

---

[3]This is also the case where a standard gear is in mesh with its basic rack, as already mentioned in Section 10.1.2.

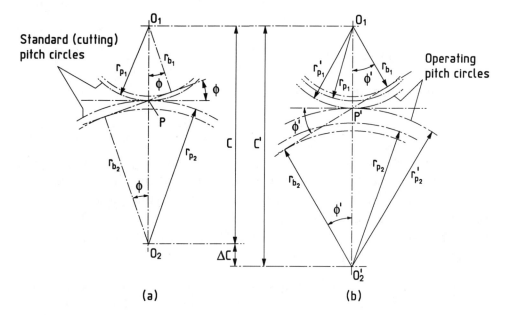

Figure 10.3-10

Thus,
$$i = \frac{\omega_1}{\omega_2} = \frac{N_2}{N_1} = \frac{r'_{p_2}}{r'_{p_1}} \qquad (10.3.25)$$

Furthermore, the new center distance
$$C' = r'_{p_1} + r'_{p_2}. \qquad (10.3.26)$$

From Equations (10.3.25) and (10.3.26),
$$r'_{p_1} = \left(\frac{N_1}{N_1 + N_2}\right)C' \qquad (10.3.27a)$$

$$\text{and } r'_{p_2} = \left(\frac{N_2}{N_1 + N_2}\right)C' \qquad (10.3.27b)$$

Again from Fig. 10.3-10a and b,
$$C' = r'_{p_1} + r'_{p_2} = \frac{r_{b_1} + r_{b_2}}{\cos \phi'}$$

$$\text{or, } C' = \frac{(r_{p_1} + r_{p_2})\cos \phi}{\cos \phi'} = C \cdot \frac{\cos \phi}{\cos \phi'} \qquad (10.3.28)$$

$$\text{and } \Delta C = C' - C = C\left(\frac{\cos \phi}{\cos \phi'} - 1\right). \qquad (10.3.29)$$

Now, from Equation (10.3.1) we can write
$$B = p'_p - t'_1 - t'_2 \qquad (10.3.30)$$

where p'_p is the *operating circular pitch* given by $2\pi r'_{p_1}/N_1$ (or $2\pi r'_{p_2}/N_2$), t'_1 is the tooth thickness of gear 1 on its operating pitch circle and t'_2 is the tooth thickness of gear 2 on its operating pitch circle. If t_1 and t_2 are the thicknesses of the teeth on gears 1 and 2 at their respective standard pitch circles, then Equation (10.2.13) can be used to determine t'_1 and t'_2 as follows[4]:

---

[4]The involute angle ψ becomes equal to pressure angle ϕ when the point on the involute coincides with the pitch point. Since in both the conditions (with center distance C and $C + \Delta C$) the tooth thickness is measured at the respective pitch points, ψ_X and ψ_Y in Equation (10.2.13) can be replaced by ϕ and ϕ', respectively.

10.3. CHARACTERISTICS OF INVOLUTE ACTION

$$t'_1 = 2r'_{p_1}(t_1/2r_{p_1} + \text{inv } \phi - \text{inv } \phi') \qquad (10.3.31a)$$

and

$$t'_2 = 2r'_{p_2}(t_2/2r_{p_2} + \text{inv } \phi - \text{inv } \phi') \qquad (10.3.31b)$$

Substituting these in Equation (10.3.30) and rearranging, we get

$$B = 2C'(\text{inv } \phi' - \text{inv}\phi). \qquad (10.3.32)$$

Problem 10.3-8

A standard gear with 24 teeth is in perfect mesh with another gear with 60 teeth. The pressure angle and the module are 20° and 4, respectively. If the center distance is increased by 0.5 mm, calculate the circular and normal backlash.

Solution

The center distance C, when the gears mesh without any backlash, is found as

$$C = r_{p_1} + r_{p_2}.$$

Since the number of teeth of the gears and the module are known as $N_1 = 24$ and $N_2 = 60$, respectively, using Equation (10.1.12),

$$C' = \frac{m}{2}(N_1 + N_2) = 2(24 + 60)\text{mm} = 168\text{mm}.$$

Hence, $C' = C + \Delta C = 168.5$ mm. As from the data, $\phi = 20°$, using Equation (10.3.28),

$$\cos \phi' = \frac{C}{C'} \cos \phi = \frac{168}{168.5} \cos 20° = 0.936904215.$$

Thus, $\phi' = 20.462$. From the definition given in Equation (10.2.8),

$$\text{inv } \phi \quad = \quad \tan \phi - \phi \text{ (in radians)} = 0.0149$$
$$\text{and} \quad \text{inv } \phi' \quad = \quad \tan \phi' - \phi' \text{ (in radians)} = 0.016.$$

Therefore, finally using Equation (10.3.32), we get the circular backlash

$$B = 2 \times 168.5(0.016 - 0.0149) \text{ mm} = 0.371 \text{ mm}.$$

The normal backlash B_n can be determined using Equation Equation (10.3.24) as follows:[5]

$$B_n = B \cos \phi' = 0.371 \cos 20.462 \text{ mm} = 0.348 \text{ mm}$$

10.3.6 Sliding Phenomenon of Gear Teeth

When motion is transmitted with the help of a pair of spur gears, the relative motion between the contacting teeth is a combination of rolling and sliding, except when the contact point coincides with the pitch point. When the contact point coincides with the pitch point, the relative motion is a pure rolling one. As the contact point moves away from the pitch point, a certain amount of relative sliding also takes place, that increases with the distance from the pitch point. Since the phenomenon of wear depends on the amount of sliding it is often desirable to find out the magnitude of the sliding velocity. Apart from purely kinematic reasons, information about the sliding phenomenon is important since in cases of very high speed gears, the product of Hertz stress and the maximum sliding velocity has been found to be a very useful design criterion.

Figure 10.3-11 shows a pair of teeth in contact at a point Q, between the starting point E and the pitch point P. If Q_1 and Q_2 are two adjacent particles at Q on gear 1 and 2, respectively, then the sliding velocity between the two teeth at the instant is given by the difference between the velocities of point Q_1 and Q_2. The sliding velocity is in a direction along the common tangent at the point of contact. Therefore, in the case of involute teeth, the difference between the velocities of Q_1 and Q_2 must be always in a direction normal to the line of action.

From Fig. 10.3-11 the transverse components of v_{Q_1} and v_{Q_2}, perpendicular to the line of action, can be expressed as follows:

---

[5]It should be noted that for the condition in which backlash is present, the operating pressure angle is equal to ϕ'.

10.3. CHARACTERISTICS OF INVOLUTE ACTION

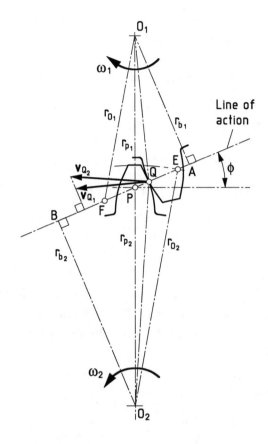

Figure 10.3-11

$$v^t_{Q_1} = AQ.\omega_1$$
$$\text{and} \quad v^t_{Q_2} = BQ.\omega_2$$

Hence, the relative sliding velocity at Q,

$$v^s_Q = v^t_{Q_2} - v^t_{Q_1} = BQ.\omega_2 - AQ.\omega_1.$$

Writing BQ as $(BP+PQ)$ and AQ as $(AP-PQ)$, the expression for the relative sliding velocity becomes

$$v^s_Q = BP.\omega_2 - AP.\omega_1 + PQ(\omega_2 - \omega_1).$$

But as $\omega_1/\omega_2 = O_2P/O_1P = BP/AP$, the above relation can be written as

$$v^s_Q = PQ.(\omega_2 - \omega_1). \tag{10.3.33}$$

Problem 10.3-9

Two standard involute spur gears of module 4 with 24 and 36 teeth are in mesh. The pressure angle is 20° and the pinion is rotating at 1400 rpm. Determine the magnitude of the maximum relative sliding velocity.

Solution

From the data,

$$\omega_1 = \frac{2\pi \times 1400}{60} \text{ rad/s} = 146.61 \text{ rad/s}.$$

Since $\omega_1/\omega_2 = N_2/N_1$,

$$\omega_2 = \omega_1 \frac{N_1}{N_2} = 146.61 \times \frac{24}{36} \text{ rad/s} = 97.74 \text{ rad/s}.$$

Next we have to determine the maximum distance of the point of contact from the pitch point. Referring to Fig. 10.2-5, it is obvious that the maximum distance from P will be either PE or PF (whichever is larger). Considering ΔO_2PE in Fig. 10.3-11,

$$r^2_{O_2} = r^2_{p_2} + PE^2 + 2r_{p_2}.PE.\sin\phi. \tag{a}$$

10.3. CHARACTERISTICS OF INVOLUTE ACTION

From the data,

$$r_{p_2} = \frac{mN_2}{2} = \frac{4 \times 36}{2} \text{ mm} = 72 \text{ mm}$$

and $\quad r_{O_2} = r_{p_2} + m = 76$ mm.

Hence, Equation (a) can be rewritten as follows:

$$PE^2 + 49.25PE - 592 = 0 \quad (PE \text{ in mm})$$

So, $PE = 9.99$ mm.
Similarly, considering $\Delta O_1 FP$,

$$r_{O_1}^2 = r_{p_1}^2 + PF^2 + 2r_{p_1} PF \sin\phi. \tag{b}$$

Now, $r_{p_1} = r_{p_2}(N_1/N_2) = 72 \times \frac{24}{36}$ mm $= 48$ mm.
So, $r_{O_1} = r_{p_1} + m = (48 + 4)$ mm $= 52$ mm.
Again rewriting Equation (b) as

$$PF^2 + 32.83PF - 400 = 0,$$

we get

$$PF = 9.46 \text{ mm}.$$

Hence, from Equation (10.3.33), the magnitude of maximum sliding velocity is equal to

$$\begin{aligned} 9.99 \quad &\times \quad (146.61 - 97.74) \text{ mm/s} \\ &= \quad 488.2 \text{ mm/s} \\ &\approx \quad 0.49 \text{ m/s}. \end{aligned}$$

This occurs when the contact is at point E.

10.4 Cycloidal (Trochoidal) Gears

Due to a number of advantages, discussed in Section 10.2.4, the involute is, except for a few special applications, the universal tooth form for all parallel axis gears. However, the supremacy of the involute profile was often challenged in the past. In fact in the fifteenth century it was discovered that the fundamental law of gearing is satisfied by tooth profiles given by the *cycloidal family* (or in other words *trochoidal*) of curves. UnTil the last century, the use of cycloidal gears was not infrequent. The gears with trochoidal tooth profile have a few special advantages and are still used in some special situations, as will be discussed later.

10.4.1 Trochoidal Curves

Figure 10.4-1 shows how the trochoidal curves (cycloidal family of curves) are generated. When a circle (usually called the *generating or rolling circle*) rolls on a straight line, as indicated in Fig. 10.4-1a, every point on the generating circle generates a *cycloid*. If an xy coordinate system is chosen as shown, then the point A of the circle that coincides with the origin O describes a cycloid whose equation

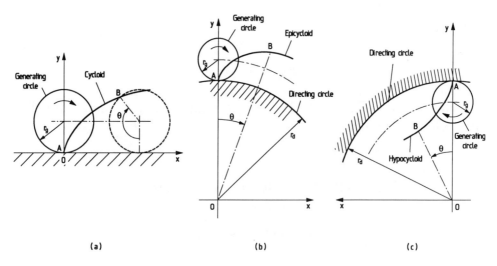

Figure 10.4-1

10.4. CYCLOIDAL (TROCHOIDAL) GEARS

can be written in the following parametric form:

$$x = r_g(\theta - \sin\theta) \tag{10.4.1a}$$

$$y = r_g(1 - \cos\theta) \tag{10.4.1b}$$

where r_g is the radius of the generating circle and θ is the angle of rotation of the generating circle in radians.

When a generating circle rolls on the outside of another circle, called the *directing circle*, each point on the generating circle describes an *epicycloid*, as shown in Fig. 10.4-1b. Taking an xy coordinate system with its origin at the center of the directing circle a point on the generating circle, which coincides with the point of intersection of the y axis and the directing circle, generates an epicycloid whose equation in parametric form is as follows:

$$x = (r_g + r_d)\sin\theta - r_g \sin[(1 + r_d/r_g)\theta] \tag{10.4.2a}$$

$$y = (r_g + r_d)\cos\theta - r_g \cos[(1 + r_d/r_g)\theta] \tag{10.4.2b}$$

where r_g and r_d are the radii of the generating and the directing circles, respectively, and θ is the angle indicated in Fig. 10.4-1b.

The points on a generating circle describe *hypocycloids* when the circle rolls inside a directing circle as shown in Fig. 10.4-1c. It is obvious that for generating a hypocycloid, the radius of the generating circle has to be less than that of the directing circle (i.e., $r_g < r_d$). The xy coordinate is chosen as shown. The point on the generating circle that coincides with the point of intersection of the y axis and the directing circle describes a hypocycloid with the following equation in parametric form:

$$x = (r_d - r_g)\sin\theta - r_g \sin[(r_d/r_g - 1)\theta] \tag{10.4.3a}$$

$$y = (r_d - r_g)\cos\theta + r_g \cos[(r_d/r_g - 1)\theta] \tag{10.4.3b}$$

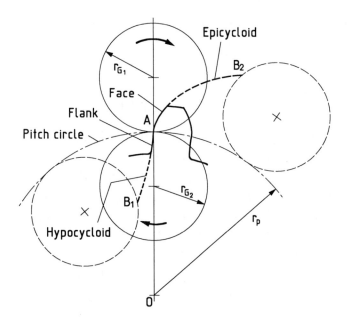

Figure 10.4-2

10.4.2 Cycloidal (Trochoidal) Teeth

The formation of a tooth of a cycloidal gear is indicated in Fig. 10.4-2. The gear has the pitch circle of radius r_p, with O as the center. The portion of the tooth profile above the pitch circle is called the *face* and is a part of the epicycloid generated by a generating circle with radius r_{G_1}. The portion of the profile below the pitch circle, called the *flank*, is represented by a part of the hypocycloid generated by a generating circle with radius r_{G_2}, as shown in Fig. 10.4-2. It is obvious that the epicycloid and the hypocycloid have to meet at a point A on the pitch circle. Thus, unlike the involute profile, a cycloidal profile involves a change in curvature. The forms of the face and the flank of a cycloidal tooth depend on the ratios r_{G_1}/r_p and r_{G_2}/r_p, respectively. To minimize the contact stress it is desirable for the profile to possess as little curvature as possible. Hence, the tooth form should be constructed with the largest possible generating circle for a given size of the pitch circle. Normally, the generating circle

10.4. CYCLOIDAL (TROCHOIDAL) GEARS

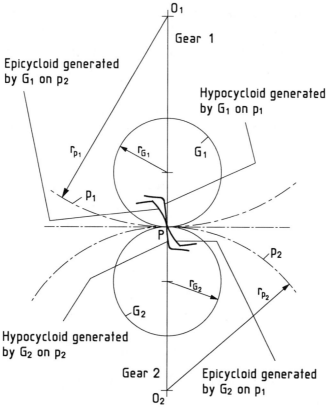

Figure 10.4-3

radius used is

$$r_G = 0.33 \text{ to } 0.4 r_p. \tag{10.4.4}$$

It will be shown later that to satisfy the fundamental law of gearing, the contacting portions of the teeth profiles of two gears in mesh must be generated by the same generating circles. This is clearly indicated in Fig. 10.4-3. The flank of a tooth of gear 1 is a hypocycloid generated by rolling the generating circle G_1 on the inside of the pitch circle p_1, and the corresponding face of the tooth of gear 2 is the epicycloid generated by rolling the generating circle G_2 on the outside of the pitch circle p_2. Similarly, the face of the tooth of gear 1 and the flank of the tooth of gear 2 are generated by the generating circle G_2. It is obvious that in the case of a rack in mesh with a cycloidal gear, both the face and the flank are cycloids obtained by rolling the respective generating circles on the pitch line. Therefore, unlike involute profiles, the cycloidal rack teeth profile is not straight.

Though there is no need for the generating circles to be of the same size so far as a particular pair of gear is concerned, in a system of interchangeable cycloidal gears, one common size for the generating circles should be chosen. As the size of the generating circle is limited by the size of the pitch circle of the gear, the common size for the generating circle should correspond to the allowable size for the smallest possible gear of the system. In practice, if N_{min} is the number of teeth on the smallest gear, the radius of the generating circle is usually taken as half of the pitch circle radius of the smallest gear. Thus,

$$r_G = \frac{r_{p_{min}}}{2} = \frac{mN_{min}}{4}. \qquad (10.4.5)$$

In many systems of cycloidal gears N = 11. Hence,

$$r_G = 2.75m \text{ mm}. \qquad (10.4.6)$$

From Fig. 10.4-2 it is obvious that the tooth of a cycloidal gear spreads towards its root.

10.4.3 Cycloidal Tooth Action

Figure 10.4-4 represents two gears in mesh. The pitch circles touch each other at the pitch point P. The hypocycloid P-m has been generated by rolling the generating circle G_1 inside the pitch circle of gear 1, and the flank of a tooth of gear 1 is formed by a portion of P-m suitably located. Rolling G_1 on the outside of the pitch circle of gear 2, the epicycloid P-s is obtained and a part of it forms the face of a tooth on gear 2. The generating circle G_2 generates the hypocycloid P-n when rolled inside the pitch circle of gear 2 and it forms the flank of the tooth on gear 2. The epicycloid P-r is obtained by rolling G_2 on the outside of the pitch circle of gear 1 and a portion of it forms the face of the tooth on gear 1.

If gear 1 rotates in the CW direction and drives gear 2, as indicated in Fig. 10.4-4, the contact begins at A, where G_1 intersects the

10.4. CYCLOIDAL (TROCHOIDAL) GEARS

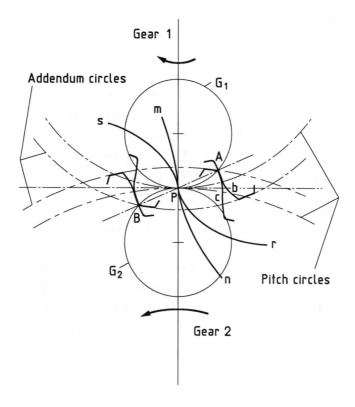

Figure 10.4-4

addendum circle of gear 2. This is because the face of the tooth of gear 2 and the contacting flank of the tooth on gear 1 are both generated by circle G_1 and hence, the contact point of these two profiles must be always on the generating circle G_1. As the two gears move, the contact point will proceed towards P along the arc AP. At the same time the end of the flank b and the end of the face c also approach P simultaneously. Beyond the point P, the contact takes place between the face of the tooth on gear 1 and the flank of the tooth on gear 2. The contact point proceeds along the arc of G_2 and the contact ends at B, where G_2 intersects the addendum circle of gear 1.

To demonstrate how the fundamental law of gearing is satisfied, let us consider Fig. 10.4-5. The generating circle G_1 rolls along the pitch circles to generate the contacting teeth profile. At the same time the pitch circles also roll over each other at P. Hence, the points on the pitch circles and G_1 at P have the same velocity (v_P), as shown, and the mutual contact point of the circles remains fixed at P. So far as the motion of G_1 with respect to either of the pitch circles is concerned, the relative instantaneous center will be at P. Hence the trajectory, a point Q on G_1 that generates on either gear 1 or 2, must be normal to the line PQ. So the normal on the common tangent to the teeth profiles at the contact point passes through P, which is

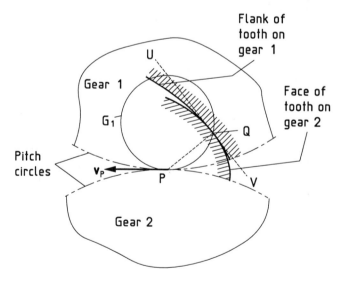

Figure 10.4-5

10.4. CYCLOIDAL (TROCHOIDAL) GEARS

a fixed point on the line of centers. Thus, the fundamental law of gearing is satisfied and the teeth are conjugate.

Unlike in the case of involute teeth, the operating pressure angle (the angle made by the instantaneous line of action QP with the common tangent to the pitch circles at P) varies. It is maximum when the contact is at either A or B and it is zero when the contact is at the pitch point P (Fig. 10.4-4).

10.4.4 Advantages and Disadvantages of Cycloidal Teeth

When compared with the gears with involute teeth, the cycloidal teeth have three distinct advantages. Since there is no interference and undercutting problem, the number of teeth on a cycloidal gear can be as low as 6. In some special extreme situations the number of teeth can be even 3 or 4! This makes cycloidal gears very suitable for gearing with large transmission ratios and it also makes cycloidal gears very popular for clock mechanisms. Cycloidal teeth are stronger than standard involute teeth because the cycloidal teeth have spreading flanks and the involute teeth have radial flanks. The cycloidal teeth have less sliding action. This results in less wear and more life as compared to involute teeth.

However, there are many disadvantages of the cycloidal teeth. Manufacture of accurate cycloidal gears is difficult, since the teeth of a cycloidal rack cutter do not have straight edges. The conjugate action in cases of cycloidal gears is possible with only one theoretically correct center distance. Any error in the center distance destroys the conjugate action. This is so because the teeth profiles are generated using the pitch circles and any change in the center distance alters the sizes of the pitch circles. In cycloidal gears the path of contact is not a straight line and the operating pressure angle continuously varies. This introduces dynamic loading on the bearings, which results in undesirable vibration in the system.

Problem 10.4-1

The driving pinion of a pair of cycloidal gears in mesh has 12 teeth, whereas that on the driven gear has 48 teeth, the module being

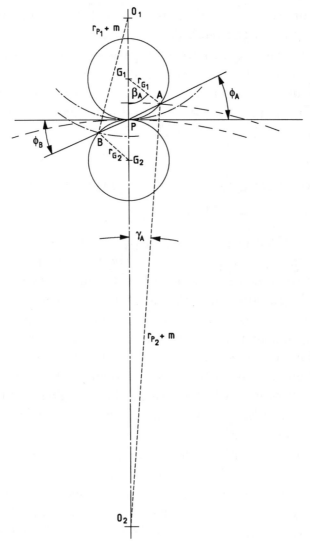

Figure 10.4-6

4. The generating circles are of same size and its radius is 0.4 times the radius of the pinion. Both the pinion and the gear have standard addendum (equal to module). Determine the maximum operating pressure angle.

Solution

The radii of the pitch circles can be determined using Equation

10.4. CYCLOIDAL (TROCHOIDAL) GEARS

(10.1.12), as follows:

$$r_{p_1} = mN_1/2 = 4 \times 12/2 = 24 \text{ mm}$$
$$r_{p_2} = mN_2/2 = 4 \times 48/2 = 96 \text{ mm}$$

The radius of the generating circles

$$r_{G_1} = r_{G_2} = 0.4 \times 24 \text{ mm} = 9.6 \text{ mm}.$$

Figure 10.4-6 shows two meshing gears with pitch circle radii r_{p_1} and r_{p_2}. The addendum circles are of radii $(r_{p_1} + m)$ and $(r_{p_2} + m)$. The beginning and the end of contact are at A and B, respectively.

Now, $O_2G_1 = O_2P + PG_1 = r_{p_2} + r_{G_1}$, $O_2A = r_{p_2} + m$ and $G_1A = r_{G_1}$.

Using the property of $\Delta O_2 G_1 A$

$$O_2A^2 + O_2G_1^2 - 2O_2A.O_2G_1.\cos\gamma_A = AG_1^2$$
and $$G_1O_2^2 + G_1A^2 - 2G_1O_2.G_1A\cos\beta_A = O_2A^2.$$

Substituting the numerical values we get

$$\gamma_A = 4.36° \text{ and } \beta_A = 52.17°.$$

Next, consider ΔPG_1A and we get

$$PA = 2r_{G_1}\sin(\beta_A/2).$$

Substituting the numerical values in the above expression,

$$PA = 8.44 \text{ mm}.$$

Again considering $\Delta O_2 PA$, we can write

$$\frac{O_2A}{\sin(90° + \phi_A)} = \frac{PA}{\sin\gamma_A}$$

or, $\sin(90° + \phi_A) = (O_2A/PA)\sin\gamma_A$.

Substituting the values, we obtain

$$\sin(90° + \phi_A) = (100/8.44)\sin 4.36° = 0.900746$$

or, $\phi_A = 25.74°$.

Using a similar procedure, we find

$$\phi_B = 23.73°.$$

Hence, the maximum operating pressure angle is at the beginning of the contact and is equal to $25.74°$.

10.5 Determination of Conjugate Profile

According to the fundamental law of gearing, the common normal to the mating surfaces of two bodies hinged to the frame at any point of contact must always pass through a fixed point on the line joining the two hinge centers. This action is often termed as conjugate action and two profiles satisfying this condition are called conjugate profiles. It has been already shown that both involutes and trochoids are curves that fulfill the requirements of conjugate action. However, in general, if any profile is prescribed, its conjugate profile can be determined. In this section, both the graphical and analytical procedures for the determination of the conjugate of a prescribed profile are presented.

10.5.1 Graphical Approach

The graphical approach has the advantage that the mathematical equation of the prescribed profile (which may be difficult to obtain in many cases) is not required and very complex shapes can be treated by this procedure. Figure 10.5-1 shows two bodies, 1 and 2, hinged to the frame at O_1 and O_2, respectively. If S_1T_1 is the profile of body 1, it is desired to find out the profile of body 2 so that the common normal to the contacting surfaces at the point of contact always passes through P.

10.5. DETERMINATION OF CONJUGATE PROFILE

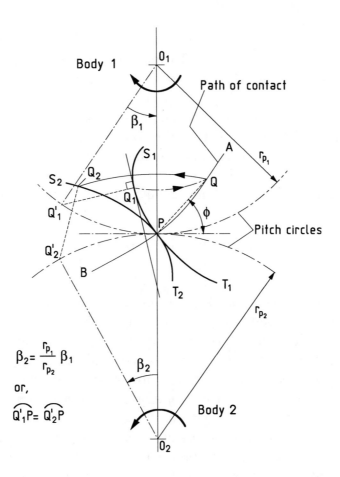

Figure 10.5-1

Let us take any point Q_1 on the prescribed profile S_1T_1 and determine the corresponding point of the conjugate profile. A tangent is first drawn to the prescribed profile at Q_1, and a line perpendicular to the tangent is drawn through Q_1 that intersects the circular arc[6] with O_1 as the center and $O_1P(=r_{p_1})$ as the radius at Q'_1. When body 1 with the profile S_1T_1 is rotated back so that Q'_1 comes to point P, Q_1 takes the position Q. To find out Q, first a circular arc is drawn through Q_1 with O_1 as the center and point Q is marked on this arc so that $PQ = Q'_1Q_1$. The point Q indicates the location of the contact point in space when point Q_1 on the profile S_1T_1 touches its conjugate. To determine the corresponding point Q_2 on the conjugate profile we should note that the relative motion between the pitch circles is one of pure rolling. So the point on the pitch circle of body 2, Q'_2, corresponding to Q'_1 can be obtained using the condition

$$\text{arc } PQ'_1 = \text{arc} PQ'_2$$
$$\text{or, } \beta_1 r_{p_1} = \beta_2 r_{p_1}$$

as indicated in Fig. 10.5-1. The point Q_2 on the conjugate profile, which corresponds to Q_1, also occupies the location Q when Q_1 touches Q_2. To find out the position of Q_2, first we draw a circular arc through Q with O_2 as the center and locate Q_2 on this arc so that $Q'_2Q_2 = PQ$. If the procedure is repeated for a large number of points on the prescribed profile S_1T_1 we get enough numbers of points to draw the conjugate profile S_2T_2, as shown in the figure. During the process all locations of Q can be used to draw the path AB along which the contact proceeds. The instantaneous values of the operating pressure angle ϕ can be determined as shown in Fig. 10.5-1.

10.5.2 Analytical Approach

Like all graphical procedures, graphical construction of conjugate profile also lacks accuracy. Furthermore, in cases where the process needs to be repeated many times, the graphical procedure becomes tedious.

---

[6]This may be treated as the pitch circle of body 1.

10.5. DETERMINATION OF CONJUGATE PROFILE

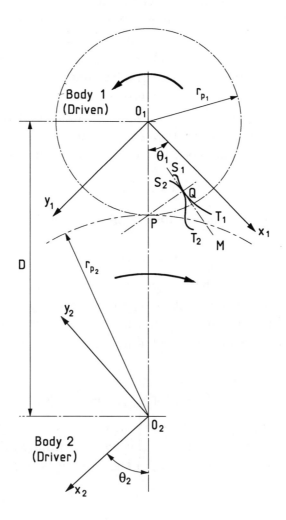

Figure 10.5-2

The analytical approach can result in much higher accuracy. As computers have become very easily available, the profiles for which closed form analytical expressions are not available can be also tackled using numerical methods. Computer graphics and computerized drafting eliminate the need to draw the developed profile.

Figure 10.5-2 shows two bodies (could be gears also) in contact and hinged at O_1 and O_2. The pitch point P, as noted earlier, is also the relative instantaneous center of rotation of the two bodies. The coordinate systems $x_1 y_1$ and $x_2 y_2$ are fixed to the bodies 1 and 2, respectively, their origins being at O_1 and O_2 as indicated. The profile of body 1 (of the teeth in case of gears)[7] in the $x_1 y_1$ system is

$$y_1 = f_1(x_1) \qquad (10.5.1)$$

It is desired to obtain the conjugate profile[8] in the $x_2 y_2$ system

$$y_2 = f_2(x_2). \qquad (10.5.2)$$

Since the conjugate action implies pure rolling of the imagined pitch circles on each other, $r_{p_1}\theta_1 = r_{p_2}\theta_2$, where θ_1 and θ_2 are the angles made by the line of centers with the x-axes of the two coordinate systems and r_{p_1} and r_{p_2} are the pitch circle radii, as indicated in the figure. Hence,

$$\theta_1 + \theta_2 = (1 + r_{p_1}/r_{p_2})\theta_1 = \lambda \theta_1 \qquad (10.5.3)$$

where $\lambda = 1 + r_{p_1}/r_{p_2}$. The center distance is $C(= r_{p_1} + r_{p_2})$. The inclination between the x-axes of the two coordinate systems is $(\theta_1 + \theta_2)$ or $\lambda \theta_1$. The coordinates of O_2 in the $x_1 y_1$ system can be expressed as follows:

$$x_1^{O_2} = C \cos \theta_1 \qquad (10.5.4a)$$

$$y_1^{O_2} = C \sin \theta_1 \qquad (10.5.4b)$$

---

[7] In cases of gears it is convenient to choose the x-axis so as to pass through the center line of the prescribed tooth profile.

[8] For conjugate action it is unimportant which one is driving and which one is driven.

10.5. DETERMINATION OF CONJUGATE PROFILE

From Fig. 10.5-3, the equations for transformation between the coordinate systems will become

$$\begin{aligned} x_2 &= x_1 \cos(\theta_1 + \theta_2) + y_1 \sin(\theta_1 + \theta_2) - C \cos \theta_2 \\ &= x_1 \cos \lambda\theta_1 + y_1 \sin \lambda\theta_1 - C \cos(\lambda\theta_1 - \theta_1) \\ &= (x_1 - C \cos \theta_1) \cos \lambda\theta_1 + (y_1 - C \sin \theta_1) \sin \lambda\theta_1. \end{aligned}$$
(10.5.5a)

Similarly,

$$y_2 = -(x_1 - C \cos \theta_1) \sin \lambda\theta_1 + (y_1 - C \sin \theta_1) \cos \lambda\theta_1. \qquad (10.5.5b)$$

Now for conjugate action it is necessary for the common normal to the profiles at Q to pass through the pitch point P. From Equation (10.5.1), the slope of the common tangent QM (in the $x_1 y_1$ system),

$$\frac{dy_1}{dx_1} = f'_1(x).$$

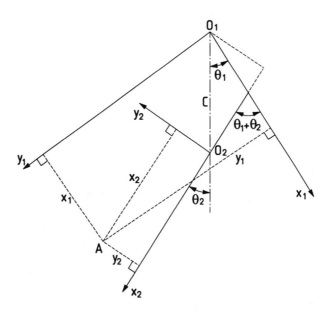

Figure 10.5-3

Hence, the slope of the common normal QP is $[-1/f_1'(x_1)]$. The coordinates of P (in the $x_1 y_1$ system) are

$$x_1^P = r_{p_1} \cos \theta_1$$

and

$$y_1^P = r_{p_1} \sin \theta_1.$$

Thus, the equation of the line QP (in the $x_1 y_1$ system) can be written as follows:

$$y_1 - r_{p_1} \sin \theta_1 = -\frac{1}{f_1'(x_1)}(x_1 - r_{p_1} \cos \theta_1). \qquad (10.5.6)$$

The coordinates of the point of contact Q can be determined by the simultaneous solution of Equations (10.5.1) and (10.5.6). Once x_1^Q and y_1^Q are obtained they can be transformed to the coordinates of the corresponding point of the conjugate profile (in the $x_2 y_2$ system) by using the transformation equations (10.5.5a) and (10.5.5b). The equation of the conjugate tooth profile is obtained in parametric form

$$x_2 = x_2(\theta_1)$$

and

$$y_2 = y_2(\theta_1). \qquad (10.5.7)$$

Sometimes the prescribed profile is also given in parametric form.

Problem 10.5-1

The teeth of a 12-tooth pinion have a trapezoidal profile with an included angle of $30°$, as shown in Fig. 10.5-4. The pitch circle radius of the pinion is 250 mm. Determine the tooth profile of a mating gear that has 24 teeth. The addendum and dedendum of the given tooth are 12.5 mm and 16 mm, respectively.

Solution

The coordinate axis x_1 is chosen to bisect the tooth on gear 1 and the extensions of the sides of the trapezoidal profiles meet at point S on the x_1 axis. From the given data,

10.5. DETERMINATION OF CONJUGATE PROFILE

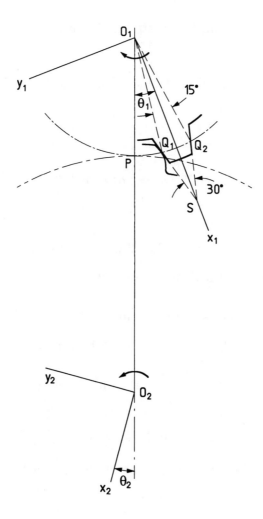

Figure 10.5-4

$$\angle Q_1 O_1 Q_2 = \frac{360°}{2 \times 12} = 15°.$$

The coordinates of S can be written as follows:

$$x_1^S = O_1 S = O_1 Q_1 \cos 7.5° + Q_1 S \cos 15°$$
$$y_1^S = 0$$

where

$$Q_1 S \sin 15° = O_1 Q_1 \sin 7.5°$$

and

$$O_1 Q_1 = r_{p_1} = 250/2 \text{ mm} = 125 \text{ mm}.$$

Substituting the numerical values we get

$$x_1^S = 185 \text{ mm} \quad \text{and} \quad y_1^S = 0.$$

Again, we have $C = r_{p_1} + r_{p_2} = (125 + \frac{1}{2} m . N_2)$ mm

$$= (125 + \frac{1}{2} \frac{250}{12} . 24) \text{ mm} = 375 \text{ mm},$$

$$\theta_2 = \theta_1 \frac{12}{24} = \theta_1/2 \quad \text{and} \quad \lambda = (1 + r_{p_1}/r_{p_2}) = 1.5.$$

The equation for the contacting face of the prescribed profile can be written as follows:

$$y_1 - y_1^S = -\tan 15°(x_1 - x_1^S)$$

$$\text{or, } y_1 = 49.6 - 0.268 x_1 = f_1(x_1) \tag{a}$$

Hence, $-f_1'(x_1) = 0.268$. The equation for the common normal, given by Equation (10.5.6), can be now written in the form

$$y_1 - 125 \sin \theta_1 = 3.73(x_1 - 125 \cos \theta_1). \tag{b}$$

10.6. INTERNAL GEARS

The simultaneous solution of Equations (a) and (b) yields the coordinates of the point of contact Q as follows:

$$x_1 = 12.4 + 116.5 \cos\theta_1 - 31.25 \sin\theta_1 \text{ mm}$$
$$y_1 = 46.28 - 31.3 \cos\theta_1 + 8.37 \sin\theta_1 \text{ mm}$$

Using the transformation equations (10.5.5a) and (10.5.5b) we get the equation of the conjugate profile in the following parametric form:

$$\begin{aligned} x_2 &= (12.4 - 258.5\cos\theta_1 - 31.25\sin\theta_1)\cos(1.5\theta_1) \\ &+ (46.28 - 31.3\cos\theta_1 - 366.63\sin\theta_1)\sin(1.5\theta_1) \text{ mm} \\ y_2 &= -(12.4 - 258.5\cos\theta_1 - 31.25\sin\theta_1)\sin(1.5\theta_1) \\ &+ (46.28 - 31.3\cos\theta_1 - 366.63\sin\theta_1)\cos(1.5\theta_1) \text{ mm}. \end{aligned}$$

10.6 Internal Gears

Figure 10.1-2 shows that an arrangement is possible in which the pitch circle of the gear surrounds that of the pinion. In the larger gear the teeth are cut on the inside of a ring and such a gear is called an *internal gear* (or *annular*, or *ring gear*). Normally, the smaller pinion is the driver. The tooth space of an internal gear resembles the tooth of a similar external gear and vice versa. Though on principle, internal gears can have any conjugate profile geometry, only involute gears will be considered in this section. Before discussions on the kinematic and geometric aspects are undertaken, it is desirable to present the advantages and disadvantages of internal gears.

10.6.1 Advantages and Disadvantages

The most important point in favour of internal gears is the relatively shorter center distance and the resulting compactness of the gear pair. The tooth forms of internal gears are stronger as the teeth width increases towards the root. Since the surface of an internal gear tooth is concave, the level of contact stress is much lower compared to that in cases of external gears. Besides this, the magnitude of the relative sliding velocity between the surfaces of a contacting pair of teeth is also smaller since both the gear and the pinion rotate in the same direction. Therefore, the wearing action of the teeth is less severe. The driving action is smoother in cases of internal gears because the contact ratio is greater for the same transmission ratio. Over and above these advantgages, the internal gear acts as a natural guard, which may be important in some situations.

There are a few disadvantages of internal gears. The ratio of the pitch circle diameters of the gear and the pinion must not be less than about 1.5. Otherwise a secondary interference between the teeth of the two gears takes place while the tooth of the pinion enters and comes out of the tooth space of the internal gear, as will be shown later. The other disadvantages arise out of the mounting difficulties when internal gears are used. Internal gears are also difficult to produce, since hobbing is not possible. The processes available for manufacturing internal gears are limited; these are casting, shaping, milling with formed cutters and, for accurate production, generation using a pinion cutter. Broaching can also be used for producing internal gears with small diameters.

10.6.2 Internal Gear Tooth Shape

In the case of external gears the tooth profile can be defined as the one that has conjugate action with the basic rack. Unfortunately, an internal gear cannot mesh with a rack and a different approach is necessary to define the tooth shape. Only those internal gears that mesh with pinions having involute teeth will be considered.

Let us now determine the required tooth shape of an internal gear that meshes with the involute teeth, satisfying the law of gearing. Figure 10.6-1 shows an involute pinion with its center at O_1 and in mesh with an internal gear with its center at O_2. Their pitch circles (with radii r_{p_1} and r_{p_2}, respectively) touch each other at the pitch point P and roll against each other. To satisfy the law of gearing the common normal to the touching profiles of the teeth in contact at point Q must always pass through P. The speed ratio is dependent on the ratio of the number of teeth and the ratio of the pitch circle

10.6. INTERNAL GEARS

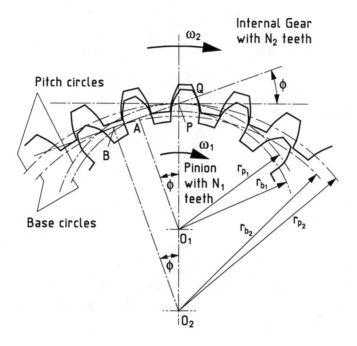

Figure 10.6-1

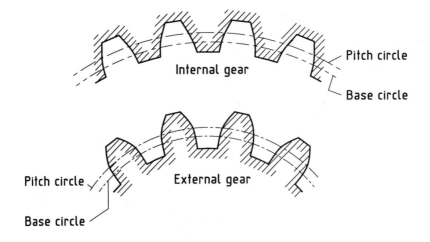

Figure 10.6-2

radii, as follows:

$$\omega_1/\omega_2 = N_2/N_1 = r_{p_2}/r_{p_1}. \qquad (10.6.1)$$

The only difference with the case of a pair of external gears is that the directions of rotation of the gears are the same. From Fig. 10.6-1 we get the center distance

$$C = r_{p_2} - r_{p_1}. \qquad (10.6.2)$$

The base circle radius of the involute pinion is r_{b_1}, as indicated in Fig. 10.6-1. To determine the position where a tooth of the involute pinion touches that of the internal gear for the position shown, we draw a tangent to the base circle of the pinion through P. This line intersects the profile of a tooth of the pinion at Q and, since the tooth profile of the pinion is an involute, the line PQ must be normal to the tangent to the profile at Q. Since the common normal must be perpendicular to both the surfaces at the point of contact, PQ must be also normal to the surface of the contacting tooth of the internal gear and the point of contact must be Q. Further, since the line of action PQ touches the base circle radius of the pinion, this line is fixed and the point of tangency A is also fixed. Let us drop a normal

10.6. INTERNAL GEARS

to the extension of QA from O_2 and let it intersect the line of action at B. Now, if we draw a circle with O_2 as the center and O_2B as the radius, the line of action touches it at B. Since the line BQ is also normal to the tooth profile of the internal gear at all instants, the profile of the internal gear tooth is also an involute, with $O_2B(=r_{b_2})$ as the radius of the base circle with its center at O_2. However, the teeth of the internal gear lie outside the profile whereas those of an external gear lie inside the profile, as indicated in Fig. 10.6-2. From Fig. 10.6-1 we can easily derive the following relationships:

$$r_{b_1}/r_{b_2} = r_{p_1}/r_{p_2} = N_1/N_2 \qquad (10.6.3a)$$

$$r_b = r_p \cos\phi \qquad (10.6.3b)$$

where ϕ is the operating pressure angle. Base pitches of both gears are defined in the standard manner, as given below:

$$p_b = 2\pi r_b/N \qquad (10.6.4)$$

Using this with Equation (10.6.3a) we get

$$p_{b_1}/p_{b_2} = (r_{b_1}/N_1)/(r_{b_2}/N_2) = 1.$$

Hence, the base pitches and, using Equation (10.6.3b), the circular pitches of the two gears must be the same. In this case also, the line of action is the common tangent to the two base circles and hence, fixed.

Though the internal gear cannot mesh with a real rack it is possible to conceive an imaginary rack in mesh with the internal gear. Thus, the tooth shape of an internal gear can be standardized in the same manner in which the teeth of external involute gears are specified.

Problem 10.6-1

The base circle radii of an internal gear and an involute pinion in mesh are 175 mm and 90 mm, respectively. The center distance is 90.45 mm. Determine the operating pressure angle.

Solution

From Equation (10.6.3b)

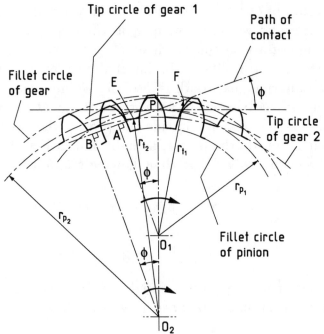

Figure 10.6-3

$$r_{p_1} = r_{b_1}/\cos\phi \text{ and } r_{p_2} = r_{p_2}/\cos\phi$$

and from Equation (10.6.2)

$$C = r_{p_2} - r_{p_1} = (r_{b_2} - r_{b_1})/\cos\phi.$$

Hence,

$$\cos\phi = (r_{b_2} - r_{b_1})/C.$$

Using the given data in the above equation,

$$\phi = \cos^{-1}(0.9396) = 20°.$$

10.6.3 Contact Ratio, Interference, Secondary Interference and Undercutting

Contact ratio of an external gear pair is defined as the rotation of either gear during one meshing cycle (between one pair of teeth), divided by the angular pitch of the same gear. This has been shown to be the same as the ratio of the length of the contact path and the base pitch. For internal gears also the contact ratio can be defined in a similar manner. Figure 10.6-3 shows the details of the contact zone. When gear 1 (the pinion) is the driver, the contact begins at E where the tip circle of the internal gear intersects the path of contact (which is the common tangent to the base circles of the two gears). The point of contact proceeds along the path of contact (also the line of action), and the contact between a particular pair of teeth ends at F, where the tip circle[9] of gear 1 intersects the path of contact. The length of the path of contact

$$EF = EP + PF$$
$$= (BP - BE) + (AF - AP).$$

Now, from Fig. 10.6-3

$$BP = O_2B \tan\phi = r_{b_2} \tan\phi,$$
$$AP = O_1A \tan\phi = r_{b_1} \tan\phi,$$
$$BE^2 = O_2E^2 - O_2B^2 = r_{t_2}^2 - r_{b_2}^2$$
$$\text{and} \quad AF^2 = O_1F^2 - O_1A^2 = r_{t_1}^2 - r_{b_1}^2.$$

Thus, the length of the path of contact

$$EF = (r_{b_2} - r_{b_1})\tan\phi + \sqrt{(r_{t_1}^2 - r_{b_1}^2)} - \sqrt{(r_{t_2}^2 - r_{b_2}^2)}.$$

Finally, the contact ratio

---

[9] For an extrnal gear the tip circle is nothing but the addendum circle and the tip circle radius is nothing but the outside radius of the gear.

$$m_c = [(r_{b_2} - r_{b_1})\tan\phi + \sqrt{(r_{t_1}^2 - r_{b_1}^2)} - \sqrt{(r_{t_2}^2 - r_{b_2}^2)}]/p_b. \qquad (10.6.5)$$

where p_b is the base pitch.

Problem 10.6-2

An involute gear with 40 teeth and module 4 meshes with an internal gear, resulting in an operating pressure angle of 20° and a speed ratio 2. Determine the contact ratio. The difference between the tip circle radii and the corresponding pitch circle radii is equal to the module.

Solution

From the given data the following values can be calculated:

$$N_1/N_2 = 2, \quad \text{or,} \quad N_2 = 2 \times 40 = 80$$

Since $m = 4$, $r_{p_1} = (4 \times 40/2)$ mm $= 80$ mm and $r_{p_2} = 160$ mm. Using Equation (10.6.3b), $r_{b_1} = 80\cos 20°$ mm $= 75.18$ mm and $r_{b_2} = 150.35$ mm. From the given information, $r_{t_1} = r_{p_1} + 4$ mm $= 84$ mm and $r_{t_2} = r_{p_2} - 4$ mm $= 156$ mm. The base pitch $p_b = p_{b_1} = (2\pi \times 75.18/40)$ mm $= 11.81$ mm. Substituting the numerical values in Equation (10.6.5),

$$\begin{aligned}
m_c &= [(150.35 - 75.18)\tan 20° + \sqrt{(84^2 - 75.18^2)} \\
&\quad - \sqrt{(156^2 - 150.35^2)}]/11.81 \\
&= 1.97.
\end{aligned}$$

In Fig. 10.6-3 points A and B, where the line of action touches the base circles of the gears in mesh, represent the interference points (as in the case of external gears). The conjugate action is lost if the path of contact extends beyond these points. To avoid interference, i.e., contact where the profile is noninvolute, the point E must be above A. This condition is satisfied if BE is larger than BA, i.e.,

$$\sqrt{(r_{t_2}^2 - r_{b_2}^2)} > (r_{b_2} - r_{b_1})\tan\phi. \qquad (10.6.6)$$

Unlike the case of external gears, there is no possibility of any interference beyond F because both of the interference points lie in

10.6. INTERNAL GEARS

the same side of E. Theoretically it is possible to make the tip circle radius of the pinion indefinitely larger without causing interference. In practice, however, the tooth thickness of the pinion reduces to zero at a certain height above the base circle and a practical limit is imposed. To enable a contact to take place during the whole period when the contact moves from E to F, the fillet circle of the pinion must be below the point E and that of the internal gear must be above F, as shown in Fig. 10.6-3. Usually a margin of $0.025m$ between the tip circle and the corresponding fillet circle is provided. Thus, if r_{f_1} and r_{f_2} are the fillet circle radii of the pinion and the gear, respectively,

$$\begin{aligned} r_{f_1} &= O_1E - 0.025m \\ &= (O_1A^2 + AE^2)^{1/2} - 0.025m \\ &= [r_{b_1}^2 + (BE - BA)^2]^{1/2} - 0.025m. \end{aligned}$$

Finally,

$$r_{f_1} = [r_{b_1}^2 + \{\sqrt{(r_{t_2}^2 - r_{b_2}^2)} - (r_{b_2} - r_{b_1})\tan\phi\}^2]^{1/2} - 0.025m \quad (10.6.7a)$$

and

$$r_{f_2} = [r_{b_2}^2 + \{\sqrt{(r_{t_1}^2 - r_{b_1}^2)} - (r_{b_2} - r_{b_1})\tan\phi\}^2]^{1/2} - 0.025m \quad (10.6.7b)$$

To avoid interference, conditions given by Equations (10.6.6) and (10.6.7) must be satisfied.

Problem 10.6-3

In the pair of gears described in Problem 10.6-2, the pinion has an outside diameter of 170 mm. Determine the limiting fillet circle radii of the gears and the minimum tip circle radius of the internal gear so that no interference takes place.

Solution

In the limiting case, Equation (10.6.6) becomes

$$\sqrt{(r_{t_2}^2 - r_{b_2}^2)} = (r_{b_2} - r_{b_1})\tan\phi.$$

Substituting the values of r_{b_1}, r_{b_2} and ϕ we get $r_{t_2} = 152.82$ mm. From the data, $r_{t_1} = (170/2)$ mm $= 85$ mm. In the limiting case of Equation (10.6.6), as mentioned above, Equation (10.6.7a) takes the form

$$r_{f_1} = r_{b_1} - 0.025m.$$

Hence, $r_{f_1} = (75.18 - 0.025 \times 4)$ mm $= 75.08$ mm. Substituting the numerical values in Equation (10.6.7b),

$$\begin{aligned} r_{f_2} &= [150.35^2 + \{\sqrt{(85^2 - 75.18^2)} + 27.36\}^2]^{1/2} + 0.1 \text{ mm} \\ &= 164.71 \text{ mm}. \end{aligned}$$

Satisfying Equations (10.6.6) and (10.6.7) does not ensure free running of an internal gear system, as there is a second type of interference that can occur in such a system. This is called secondary interference (or tip interference or fouling). This type of interference imposes a lower limit to the difference in the number of teeth of the meshing pair of gears. Consider a point on the contacting pinion-tooth profile at the tip circle. Its trajectory with respect to the internal gear (because of their relative movement, which is a pure rolling of the pinion pitch circle on the gear pitch circle) is a hypocycloid, as indicated by the broken line in Fig. 10.6-4. Since the teeth surfaces of the internal gear are concave, it can so happen that the path of F_1 can pass through the tip of the internal gear (Fig. 10.6-5a). To investigate whether secondary interference takes place or not, let us find

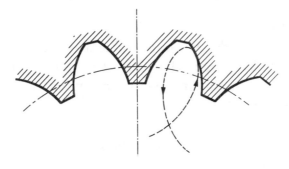

Figure 10.6-4

10.6. INTERNAL GEARS

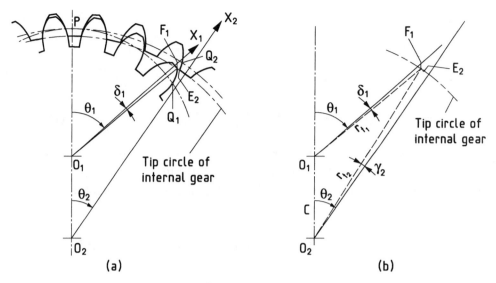

Figure 10.6-5

out the position, relative to the internal gear, of the point where the trajectory of F_1 crosses the tip circle of the internal gear, and compare it with E_2, a point on the gear tooth profile at the tip circle of the gear. Figure 10.6.5a shows the situation in which point F_1 is on the tip circle of the internal gear. At this instant the angular positions of the teeth of the pinion and the internal gear from the line of centers are θ_1 and θ_2, respectively, as indicated. The points Q_1 and Q_2 of the pinion and the gear teeth on the respective pitch circles, coincide at P when the gears are rotated back to the datumn (or reference) starting position. Hence, the amounts of rotations of the pinion and the gear needed to bring them to the positions shown are represented by the arcs PQ_1 and PQ_2, respectively, which must be equal. If t_1 and t_2 are the tooth thicknesses at the pitch circles of the pinion and the gear[10], respectively, from Fig. 10.6-5a,

$$\text{arc } PQ_1 = r_{p_1}\theta_1 + t_1/2$$
$$\text{and} \quad \text{arc } PQ_2 = r_{p_2}\theta_2 - t_2/2.$$

---

[10]In the ideal situation $t_1 = t_2$ when the gears mesh perfectly without backlash and the pitch circles coincide with the corresponding standard pitch circles.

Hence,

$$r_{p_2}\theta_2 - t_2/2 = r_{p_1}\theta_1 + t_1/2$$

or, $\quad \theta_2 = [r_{p_1}\theta_1 + (t_1+t_2)/2]/r_{p_2} = (r_{p_1}\theta_1 + t)/r_{p_2}$ \hfill (10.6.8)

where $t = t_1 = t_2 =$ tooth thickness.

Next, consider the $\Delta O_1 O_2 F_1$ as shown in Fig. 10.6-5b. From the properties of a triangle we can obtain the following relations:

$$O_2 F_1^2 = r_{t_2}^2 = O_1 F_1^2 + O_1 O_2^2 + 2 O_1 F_1 . O_1 O_2 \cos(\theta_1 + \delta_1)$$

or, $\quad \cos(\theta_1 + \delta_1) = (r_{t_2}^2 - r_{t_1}^2 - C^2)/2 r_{t_1} C.$ \hfill (10.6.9)

Again we can write

$$O_1 F_1 / \sin(\theta_2 - \gamma_2) = O_2 F_1 / \sin(\theta_1 + \delta_1)$$

or, $\quad \sin(\theta_1 + \delta_1)/r_{t_2} = \sin(\theta_2 - \gamma_2)/r_{t_1}.$ \hfill (10.6.10)

Now for the pinion, δ_1 is half the angle subtended by the tip of the tooth at O_1. Using Equations (10.2.13) and (10.2.14) we can express this as follows (when the gears mesh properly without any backlash):

$$\delta_1 = \frac{t}{2r_{p_1}} + \text{inv } \phi - \text{inv}[\cos^{-1}(r_{b_1}/r_{t_1})] \tag{10.6.11}$$

For internal gears the tooth thickness is equal to the space width of the profile, generated as mentioned earlier. It can be easily shown, following a little modification of the method used in Section 10.2.3, that $\delta_2 (= \angle X_2 O_2 E_2$, not shown in the figure) can be expressed as follows:

$$\delta_2 = \frac{t}{2r_{p_2}} - \text{inv } \phi + \text{inv}[\cos^{-1}(r_{b_2}/r_{t_2})] \tag{10.6.12}$$

From Equation (10.6.9),

10.6. INTERNAL GEARS

$$\theta_1 = \cos[(r_{t_2}^2 - r_{t_1}^2 - C^2)/2r_{t_1}C] - \delta_1. \qquad (10.6.13)$$

The value of γ_2 is found from Equation (10.6.10) as

$$\gamma_2 = \theta_2 - \sin^{-1}[r_{t_1}\sin(\theta_1 + \delta_1)/r_{t_2}]. \qquad (10.6.14)$$

The distance of F_1 from E_2 at the closest approach

$$\epsilon = r_{t_2}(\gamma_2 - \delta_2),$$

which must be more than $0.05m$ in order to provide adequate clearance. The condition to avoid secondary interference is thus

$$r_{t_2}(\gamma_2 - \delta_2) \geq 0.05m. \qquad (10.6.15)$$

This condition is generally satisfied when the difference between the numbers of teeth on the gear and the pinion is at least 8. However, it is always advisable to check for secondary interference, following the procedure described above, for each individual case in which the difference between the number of teeth is suspiciously low.

Problem 10.6-4

An involute gear with 72 teeth and module 4 meshes with an internal gear with 80 teeth. The pressure angle is 20° and the difference between the tip circle radius and the corresponding pitch circle radius is equal to the module. Investigate whether secondary interference occurs or not.

Solution

To check condition (10.6.15) we have to determine r_{t_2}, γ_2 and δ_2. From the data the pitch circle radii, the tip circle radii and the base circle radii result as follows:

$$\begin{aligned}
r_{p_1} &= N_1 m/2 = (72 \times 4/2) \text{ mm} = 144 \text{ mm} \\
r_{p_2} &= N_2 m/2 = (80 \times 4/2) \text{ mm} = 160 \text{ mm} \\
r_{t_1} &= r_{p_1} + m = 148 \text{ mm} \\
r_{t_2} &= r_{p_2} - m = 156 \text{ mm} \\
r_{b_1} &= r_{p_1} \cos\phi = 144 \cos 20° \text{ mm} = 135.32 \text{ mm} \\
r_{b_2} &= r_{p_2} \cos\phi = 160 \cos 20° \text{ mm} = 150.35 \text{ mm}
\end{aligned}$$

The tooth thicknesses at the pitch circle for both the gears are equal to half the circular pitch. Thus,

$$t = \pi m = 12.57 \text{ mm}.$$

From Equation (10.6.11),

$$\begin{aligned}\delta_1 &= 12.57/(2 \times 144) + \text{inv } 20° - \text{inv }[\cos^{-1}(135.32/148)] \text{ radians} \\ &= (0.0436 + 0.0149 - 0.0259) \text{ radians} \\ &= 0.0326 \text{ radians}.\end{aligned}$$

From Equation (10.6.12),

$$\begin{aligned}\delta_2 &= 12.57/(2 \times 160) - \text{inv } 20° + \text{inv }[\cos^{-1}(150.35/156)] \text{ radians} \\ &= (0.0393 - 0.0149 + 0.0067) \text{ radians} \\ &= 0.0311 \text{ radians}.\end{aligned}$$

Equation (10.6.13) yields θ_1 as

$$\theta_1 = \cos^{-1}[(156^2 - 148^2 - 16^2)/(2 \times 148 \times 16)] - 0.0326 \text{ radians}$$

where

$$C = r_{p_2} - r_{p_1} = (160 - 144) \text{ mm} = 16 \text{ mm}$$

or, $\theta_1 = 1.0608$ radians. The rotation of the internal gear is given by Equation (10.6.8) as

$$\begin{aligned}\theta_2 &= (144 \times 1.0608 + 12.57)/160 \text{ radians} \\ &= 1.0333 \text{ radians}.\end{aligned}$$

Once θ_2 is known, γ_2 can be determined using Equation (10.6.14) as

$$\begin{aligned}\gamma_2 &= 1.0333 - \sin^{-1}[148\sin\{(1.0608 + 0.0326) \times 180/\pi\}/156] \text{ radians} \\ &= 0.0312 \text{ radians}.\end{aligned}$$

Now,

$$r_{t_2}(\gamma_2 - \delta_2) = 156 \times (0.0312 - 0.0311) \text{ mm} = 0.0156 \text{ mm},$$

which is less than 0.05 times the module (= 0.2 mm). Thus, though there is no secondary interference, the clearance is less than desirable.

In the case of internal gears there is no danger of conventional undercutting at the tooth fillets, however large the cutter addendum may be. Of course, there is a possibility of undercutting at the tip of the teeth of an involute gear. The following condition must be satisfied if the conventional undercutting at the tooth tip is to be avoided:

$$r_{t_2}^2 \geq r_{b_2}^2 + [(r_{b_2} - r_{b_c})\tan\phi + \sqrt{(r_{f_c}^2 - r_{b_c}^2)}]^2 \qquad (10.6.16)$$

where r_{b_c} is the base circle radius and r_{f_c} is the fillet circle radius of the cutter. The derivation of condition (10.6.16) is left as an exercise.

Tip interference also may result in some undercutting of the teeth of internal gears. The conditions for undercutting can be derived in a manner similar to that for tip interference, discussed earlier in this seciton. A minimum clearance of $0.02m$ is prescribed in place of $0.05m$ to ensure no undercutting.

10.7 Noncircular Gears

If the distance of the pitch point from the centers of a pair of gears in mesh changes during the rotation of the gears, the angular velocity ratio also varies accordingly. This happens when the pitch curves are not circles and the gears are noncircular. There are situations in which a continuously variable angular velocity ratio is required. In many such applications noncircular gears are adopted. Sometimes only an overall change in the angular speed characteristics is specified, whereas there are cases where the angular speed characteristics are expected to vary according to specified functions. The pitch curves of noncircular gears can be closed (as in case of elliptic gears) or open. The open type of gears can be rotated only for a portion of a revolution. Usually one gear is rotated at a constant speed and the meshing gear rotates according to a specified function.

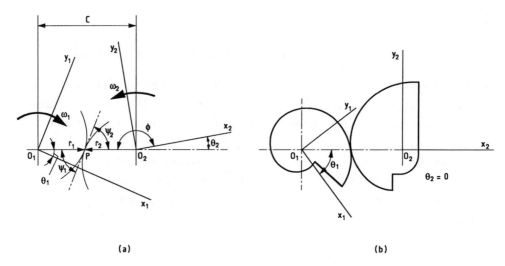

Figure 10.7-1

Kinematic synthesis of a pair of noncircular gears involves two steps. In the first step the geometry of the pitch curves are determined so that the required characteristics of the angular speed ratio are achieved. The second step involves the determination of the tooth profiles.

10.7.1 Determination of Pitch Curves

The problem of the determination of the pitch curves, for which the rotation of one shaft is coordinated with that of the other in a prescribed manner, will be illustrated with the help of an example. The following three conditions are used for the purpose.

(i) The pitch point lies on the line of centers.

(ii) The center distance is constant.

(iii) The pitch curves touch each other on the line of centers and have the same slope at the point of contact.

10.7. NONCIRCULAR GEARS

Problem 10.7-1
Determine the equations of the pitch curves of two noncircular gears such that the rotation of one is proportional to the natural logarithm of that of the other.

Solution
Referring to Fig. 10.7-1a, let O_1 and O_2 be the gear centers and $x_1 y_1$ and $x_2 y_2$ coordinate systems be rigidly attached to gears 1 and 2, respectively. The radii of the pitch curves at the pitch point are r_1 and r_2, as indicated. Thus,

$$r_1 + r_2 = C \text{ (constant) and } dr_1 + dr_2 = 0. \qquad (a)$$

From the prescribed condition,

$$\theta_2 = K \ln \theta_1 \text{ or, } \theta_1 = e^{\theta_2/K}, \qquad (b)$$

where θ_1 and θ_2 are the rotations of the x_1 and x_2 axes from the line of centers and K is a constant of proportionality. If ϕ is the angle made by the radius r_2 with x_2 axis

$$\theta_2 = \pi - \phi = K \ln\theta_1 \text{ or, } \theta_1 = \exp[(\pi - \phi)/K]$$

$$\text{So, } d\phi = -(K/\theta_1)d\theta_1$$

$$\text{or, } d\theta_1 = -d\phi[\exp\{(\pi - \phi)/K\}/K]. \qquad (c)$$

Since the slopes of the pitch curves at the pitch point P are the same, $\psi_1 = \psi_2$ or, $\tan\psi_1 = \tan\psi_2$. Since ψ represents the angle of the tangent to the pitch curve at P with the radius vector, $\tan\psi = r(d\theta/dr)$. Hence,

$$r_1(d\theta_1/dr_1) = r_2(d\theta_2/dr_2). \qquad (d)$$

Using Equations (a) and (c) in (d),

$$r_1(d\theta_1/dr_1) = \left(\frac{C - r_1}{-dr_1}\right)\left(-\frac{K}{\theta_1}d\theta_1\right)$$

$$\text{or, } r_1 = (C - r_1)(K/\theta_1)$$

$$\text{or,} \quad r_1 = C/(1+\theta_1/K). \tag{e}$$

Substituting r_1, $d\theta_1$ and dr_1 in terms of r_2, $d\phi$ and dr_2 in Equation (d) as

$$r_1 = C - r_2, \; dr_1 = -dr_2 \text{ and } d\theta_1 = -d\phi[\exp\{(\pi-\phi)/K\}/K],$$

we get

$$\frac{C-r_2}{-dr_2}[-d\phi \exp\{(\pi-\phi)/K\}/K] = r_2\frac{d\phi}{dr_2}$$

$$\text{or,} \exp\{(\pi-\phi)/K\} = Kr_2/(C-r_2)$$

$$\text{or,} \; r_2 = C \exp\{(\pi-\phi)/K\}/[K + \exp\{(\pi-\phi)/K\}]. \tag{f}$$

Equations (e) and (f) define the two pitch curves that generate the desired motion. The pitch curve profiles are shown in Fig. 10.7-1b. The reader is advised to check that the pitch curves roll over each other without slipping.

Problem 10.7-2

Referring to Fig. 10.7-1a, let the pitch curve of gear 1 be a logarithmic spiral, $r_1 = b \exp(a\theta_1)$. Find the matching pitch curve of gear 2 and the relation between the rotations of the two gears.

Solution

Since $r_1 = b \exp(a\theta_1)$, we get

$$dr_1 = ba \exp(a\theta_1)d\theta_1$$

and

$$\tan\psi_1 = r_1(d\theta_1/dr_1) = 1/a.$$

Furthermore, $r_1 + r_2 = C$ and $\psi_1 = \psi_2$, i.e., $r_1(d\theta_1/dr_1) = r_2(d\theta_2/dr_2) = 1/a$. Thus,

$$dr_2/r_2 = ad\phi$$

10.7. NONCIRCULAR GEARS

$$\text{or, } \ln r_2 = a\phi + C$$

$$\text{or, } r_2 = b_1 \exp(a\phi)$$

$$\text{where } b_1 = \exp(C_1).$$

When $\theta_1 = 0$, $\theta_2 = 0$ (or, $\phi = \pi$) and $r_1 = b$ (or, $r_2 = C - b$). So $C - b = b_1 \exp(a\pi)$ and $b_1 = (C - b)\exp(-a\pi)$. Finally,

$$r_2 = (C - b)\exp\{a(\phi - \pi)\} = (C - b)\exp(-a\theta_2).$$

This represents the matching pitch curve and is seen to be another logarithmic spiral.

The relation between the rotations of the gears can be expressed by the following ratio:

$$(d\theta_2/d\theta_1) = r_1/r_2 = b\exp(a\theta_1)/[(C - b)\exp(-a\theta_2)]$$

$$\text{or, } \exp(-a\theta_2)d\theta_2 = [b/(C - b)]\exp(a\theta_1)d\theta_1$$

Integrating both sides we finally get

$$-\exp(a\theta_2) = [b/(C - b)]\exp(a\theta_1) + C$$

where C_2 is a constant. This is the relation between the rotations of the gears.

10.7.2 Tooth Profile of Noncircular Gears

Tooth profiles of noncircular gears can be determined using the crosswire and pulley analogy used in Section 10.2.1. To startwith, the shapes of the pulleys resulting in identical angular velocity ratio is obtained by rolling the pitch curves over each other. Due to the loss of symmetry, two sets of pulleys need to be considered for the two sides of the teeth. In this section the procedure is illustrated using the pitch curves obtained in Problem 10.7-1.

In Section 10.2.1 it was seen that the wire represents the line of action and is tangential to the pulleys. It passes through the pitch point and the angle it makes with the common tangent to the pitch circles at the point is called the operating pressure angle. Figure 10.7-1b shows the pitch curves, which are redrawn in Fig. 10.7-2. The line TT representing the wire makes an angle ϕ with the normal to the line of centers. As the pitch curves roll on each other the relative locations of the wire in the x_1y_1 and x_2y_2 coordinate systems are represented by two families of straight lines. The envelopes to these two families give the shapes of the pulleys because the wire is always tangential to the pulleys. From Problem 10.7-1, the polar equation of the pitch curve of gear 1 in the x_1y_1 system is given by

$$r_1 = C/(1 + \theta_1/K). \tag{10.7.1}$$

The distance of the wire TT from O_1 is $r_1 \cos \phi$ and the wire makes an angle $(\theta_1 + \pi/2 - \phi)$ with the x_1-axis. The equation of line TT in the x_1y_1 system is given by

$$x_1 \cos(\theta_1-\phi)+y_1 \sin(\theta_1-\phi) = r_1 \cos \phi = CK \cos \phi/(K+\theta_1). \tag{10.7.2}$$

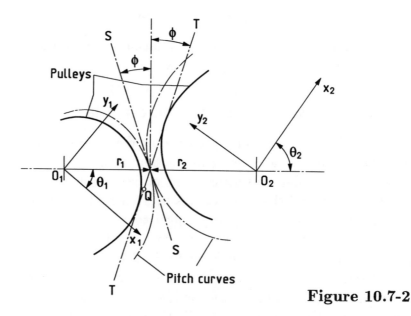

Figure 10.7-2

10.7. NONCIRCULAR GEARS

The above equation represents (in parametric form) the whole family of straight lines in the x_1y_1 system with θ_1 as the parameter. We get the envelope (i.e., the pulley on shaft 1) as the simultaneous solution of Equation (10.7.2) and the following equation obtained by differentiating Equation (10.7.2) with respect to θ_1:

$$-x_1 \sin(\theta_1 - \phi) + y_1 \cos(\theta_1 - \phi) = -CK \cos\phi/(K+\theta_1)^2 \quad (10.7.3)$$

The parametric equation of the pulley is

$$x_1 = [CK \cos\phi/(K+\theta_1)^2][(\theta_1+K)\cos(\theta_1-\phi) + \sin(\theta_1-\phi)] \quad (10.7.4a)$$

and

$$y_1 = [CK \cos\phi/(K+\theta_1)^2][(\theta_1+K)\sin(\theta_1-\phi) - \cos(\theta_1-\phi)]. \quad (10.7.4b)$$

Now let Q be a point on the wire TT, which is used to generate the tooth profile of gear 1. Let Q leave the pulley surface at $\theta_1 = \theta_{1_o}$. Its coordinates at that instant (x_{1_o}, y_{1_o}) can then be obtained from Equation (10.7.4) by substituting $\theta_1 = \theta_{1_o}$. The location of Q for any value of θ_1 can be obtained by considering the portion of the wire, not in contact with the corresponding pulley surface, as a straight line. The distance of Q from the point of tangency (contact with the pulley surface) should be equal to the arc length of the pulley between angles θ_{1_o} and θ_1. Thus, the coordinates of Q can be expressed as follows:

$$x_1 = x_{1_o} - \sin(\theta_1 - \phi) \int_{\theta_1}^{\theta_{1_o}} ds \quad (10.7.5a)$$

$$y_1 = y_{1_o} + \cos(\theta_1 - \phi) \int_{\theta_1}^{\theta_{1_o}} ds \quad (10.7.5b)$$

where $ds^2 = dx_1^2 + dy_1^2$. Using Equations (10.7.4a) and (10.7.4b) for dx_1 and dy_1, respectively, we get

$$ds = [CK \cos\phi/(\theta_1 + K)^3][2 + (\theta_1 + K)^2]d\theta_1.$$

Hence,

$$\int_{\theta_1}^{\theta_{1_o}} ds = CK\cos\phi[1/(\theta_1+K)^2 - 1/(\theta_{1_o}+K)^2$$

$$+ \ln\{(\theta_{1_o}+K)/(\theta_1+K)\}]. \qquad (10.7.6)$$

Substituting the above in Equations (10.7.5a) and (10.7.5b), the parametric equation of the top face of the profile of the tooth on gear 1 is obtained.

Since Q also traces the profile (in the x_2y_2 system) of the tooth on gear 2, the bottom face of the tooth on gear 2 can be obtained by transforming the position coordinates of Q to the x_2y_2 system. The bottom face of the profile of the tooth on gear 1 can be obtained in a similar manner with the help of a wire SS, as shown in Fig. 10.7-2. The parametric equation of the corresponding pulley on shaft 1 can be shown to be as follows:

$$x_1 = [CK\cos\phi/(\theta_1+K)^2][(\theta_1+K)\cos(\theta_1+\phi) + \sin(\theta_1+\phi)] \quad (10.7.7a)$$

$$y_1 = [CK\cos\phi/(\theta_1+K)^2][(\theta_1+K)\sin(\theta_1+\phi) + \cos(\theta_1+\phi)] \quad (10.7.7b)$$

10.8 Helical Gears

When a pair of spur gears are in mesh, for each tooth pair in contact the length of the contact line is equal to the width of the gear face (say B). Therefore, the total length of the contact line is either B or $2B$, depending on whether one or two pairs of teeth are in contact. Since the total length of contact abruptly changes from B to $2B$ and $2B$ to B during operation, running of the system is not smooth. This problem is reduced if *helical gears* are used. Helical gears can be used to transmit motion between parallel shafts (as shown in Fig. 10.8-1a) like spur gears and such gears are called *parallel helical gears*. Helical gears can be also used to transmit motion between shafts whose axes are neither parallel nor intersecting (as shown in Fig. 10.8-1b) and such gears are called *crossed helical gears*.

10.8. HELICAL GEARS

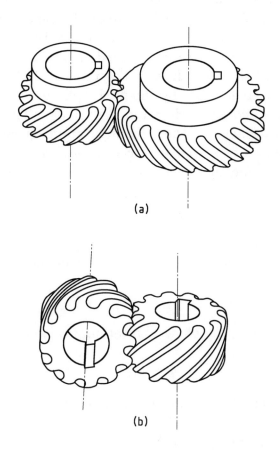

Figure 10.8-1

10.8.1 Geometry of Helical Gears

When a plane rolls on a base cylinder, a line lying on the plane and parallel to the cylinder axis generates the surface of a spur gear tooth (Fig. 10.8-2a). In the case of helical gears the generating line is not parallel to the axis of the cylinder, but to a helix, as shown in Fig. 10.8-2b. The surface so geneated is called an *involute helicoid*. The intersection of a helical tooth with the pitch cylinder is called the *pitch helix*. The angle made by the pitch helix with the gear axis is called the *helix angle* ψ, as shown in Fig. 10.8-3. The *lead angle* is the complementary angle to the helix angle and is given by $(90° - \psi)$. Figure 10.8-3 shows two circular pitches p_p and p_{p_n}, measured in the plane of rotation and in the normal plane, respectively. From the figure,

$$p_{p_n} = p_p \cos \psi. \tag{10.8.1}$$

Correspondingly, the *normal module*, m_n, is defined as

$$p_{p_n} = \pi m_n \tag{10.8.2a}$$

and the module along the plane of rotation, m, is defined as

$$p_p = \pi m \tag{10.8.2b}$$

From Equations (10.8.1), (10.8.2a) and (10.8.2b) it is obvious that

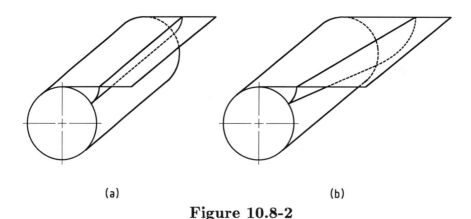

Figure 10.8-2

10.8. HELICAL GEARS

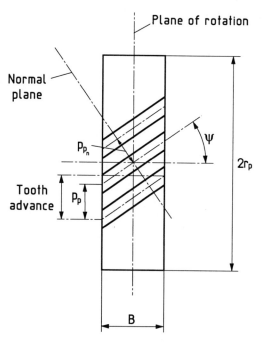

Figure 10.8-3

$$m_n = m \cos \psi. \qquad (10.8.3)$$

The *tooth advance* (or, *face advance*) can be defined as the distance along the pitch circle through which a helical tooth moves from the initial position (where the contact begins at one side) to the final position (where the contact ends at the other side). This is shown in Fig. 10.8-3. When the tooth advance is more than the circular pitch, overlapping action is achieved, which results in a smoother operation. Usually, the tooth advance is at least 15% more than the circular pitch. To obtain this, the following condition must be satisfied:

$$B \geq \frac{1.15 p_p}{\tan \psi} \qquad (10.8.4)$$

The direction in which the helical gear teeth slope is called the *hand* of the gear. When the teeth go towards the left while going from the bottom to the top face with the gear axis vertical, the gear

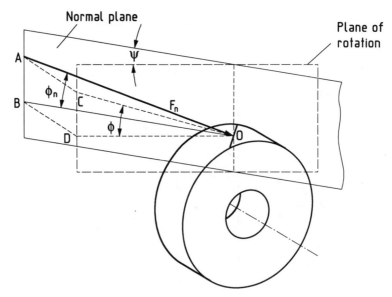

Figure 10.8-4

is called *left handed* (otherwise the gear is called *right handed*). The helical gear shown in Fig. 10.8-3 is a left handed gear.

In the case of helical gears the pitch circle radius (radius of the pitch cylinder) and the number of teeth are related, as in the case of a spur gear where the spur gear profile is obtained as a cross section of the helical gear in the plane of rotation. Thus,

$$r_p = mN/2. \qquad (10.8.5)$$

Like other parameters it is also necessary to distinguish between the pressure angle[11], ϕ, in the plane of rotation and the *normal pressure angle*, ϕ_n, in the normal plane. The relation between the two pressure angles can be obtained by considering Fig. 10.8-4. The total normal force acting on a tooth is $F_n(= AO)$ and ϕ_n is its inclination with line OB, which is in the normal plane and tangential to the pitch cylinder at O. The angle ϕ is the inclination of the component of F_n in the plane of rotation with the line OD, which is a tangent to the pitch cylinder at O in the plane of rotation.

---

[11] Also called the transverse pressure angle.

10.8. HELICAL GEARS

Now from Fig. 10.8-4, $OD = OB\cos\psi$, $\tan\phi_n = (AB/OB)$ and $\tan\phi = (CD/OD) = (AB/OD)$ as $CD = AB$. Hence,

$$\tan\phi_n/\tan\phi = OD/OB = \cos\psi$$

or, $\tan\phi_n = \tan\phi\cos\psi$. (10.8.6)

For a standard tooth profile, ϕ_n is generally 20°.

As in the case of spur gears thad dendum for helical gears is also equal to the normal module m_n. Therefore, the tip circle radius (or addendum circle radius or outer radius) of a helical gear of module m (in the plane of rotation) with N number of teeth is expressed as

$$r_o = Nm/2 + m_n.$$

The dedendum for helical gears is equal to 1.25 m_n. The base circle radius $r_b = r_p\cos\phi$.

10.8.2 Equivalent Spur Gear and Virtual Number of Teeth

It is possible to study many characteristics of a helical gear by replacing it with an *equivalent spur gear* obtained from a cross section of the helical gear along the normal plane. This provides an approximate but simplified calculation procedure as the relevant spur gear formula can be made of use.

Figure 10.8-5 shows a sectional view with the section XX along the normal plane. The section of the pitch cylinder of the helical gear is represented by the ellipse of minor axis $2r_p$ and major axis $2r_p/\cos\psi$, r_p being the pitch circle radius of the helical gear. On the periphery of this ellipse the cross sectional profiles of the teeth vary. The profile at the top of the minor axis can be approximately taken to be the profile of a spur gear of pitch circle radius $r_{p_{eq}}$, that is equal to the radius of curvature of the ellipse at P (Fig. 10.8-5). The radius of curvature of the ellipse at P is given by $r_p/\cos^2\psi$. Hence,

$$r_{p_{eq}} = r_p/\cos^2\psi. \qquad (10.8.7)$$

A spur gear having a pitch circle radius $r_{p_{eq}}$ is called the *equivalent spur gear*. The number of teeth of such a spur gear is called the *virtual number of teeth* N_v, that is given by

$$N_v = 2\pi r_{p_{eq}}/p_{p_n}. \qquad (10.8.8)$$

Using Equations (10.8.2a), (10.8.3) and (10.8.7) in the above equation, we get

$$N_v = (2\pi r_p/\cos^2\psi)/(\pi m \cos\psi) = (2r_p/m)/\cos^3\psi$$
$$= N/\cos^3\psi. \qquad (10.8.9)$$

It should be remembered that the above treatment is approximate but sufficiently accurate for all practical requirements if $\psi \leq 20°$. When the helix angle is larger (as is generally the case with crossed helical gears), the following expression for N_v is used:

$$N_v = N[(\tan\phi - \phi)/(\tan\phi_n - \phi_n)] = N(\text{inv }\phi/\text{inv}\phi_n) \qquad (10.8.10)$$

where ϕ and ϕ_n are expressed in radians.

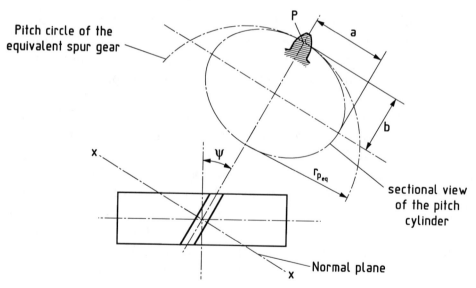

Figure 10.8-5

10.8. HELICAL GEARS

For the selection of the proper cutter number while cutting helical gears the value of N_v should be used in place of N.

10.8.3 Parallel Helical Gears

For proper meshing of a pair of helical gears with their axes parallel, certain conditions must be satisfied. Let us consider the cross section of a pair of parallel helical gears along the plane of rotation (transverse cross section). The profiles will be identical with the cross sections of a corresponding pair of spur gears (with the same pitch circle radii) in mesh. From Section 10.2.1 we know that the circular pitches p_{p_1} and p_{p_2} have to be equal for proper meshing in the case of spur gears. Therefore, the condition for correct meshing in the case of parallel helical gears is that the transverse circular pitches of the two gears are equal. Thus,

$$p_{p_1} = p_{p_2}. \qquad (10.8.11)$$

The pitch cylinder radii and the operating pressure angle ϕ (in the plane of rotation, i.e., the transverse plane) are given by the following relations:

$$r_{p_1} = \frac{N_1 C}{N_1 + N_2} \qquad (10.8.12a)$$

$$r_{p_2} = \frac{N_2 C}{N_1 + N_2} \qquad (10.8.12b)$$

$$\cos \phi = (r_{b_1} + r_{b_2})/C \qquad (10.8.13)$$

where C is the center distance, r_{b_1} and r_{b_2} are the base circle radii. From Equation (10.2.4a) we can now write

$$\cos \phi = r_{b_1}/r_{p_1} = r_{b_2}/r_{p_2}.$$

It is also quite obvious that the helix angles of the gears in proper mesh must be equal and the gears should be of opposite hand. From Equations (10.8.1), (10.8.11) and the condition of equality of the helix angles, we can also write

$$p_{p_{n_1}} = p_{p_{n_2}}. \qquad (10.8.14)$$

The contact ratio, m_c, of a pair of spur gears in mesh was defined by Equation (10.3.1) as the ratio of the angle of rotation of one of the gears (during which any particular pair of teeth remains in contact) and the angular pitch of the gear. The contact ratio of a pair of parallel helical gears in mesh can be also defined in a similar way. However, an extra amount of contact ratio results due to the tooth advance in the case of helical gears. The contact period for one pair of teeth lasts from the initial contact at one extreme transverse plane of the gears (one face) to the final contact at the other extreme transverse plane (i.e., the other face). Thus, the total angle of rotation during which the contact between any pair of teeth lasts, θ_c, can be expressed as

$$\theta_c = \theta_P + \theta_F \qquad (10.8.15)$$

where θ_P is the rotation of the gear during which the contact between any pair of teeth is maintained in the plane of rotation (the transverse plane). This is the same as $(\alpha_1 + \beta_1)$ given by Equation (10.3.2) for a spur gear. θ_F is the helical rotation between the two faces of the meshing gears. The expression for θ_F can be easily derived as

$$\theta_F = B \tan \psi / r_p. \qquad (10.8.16)$$

The *total contact ratio*, m_c, is now obtained by dividing θ_c by the angular pitch as

$$m_c = (\theta_P + \theta_F)/\theta_p = m_P + m_F \qquad (10.8.17)$$

where $m_P (= \theta_P/\theta_p)$ is the *profile contact ratio* and $m_F (= \theta_F/\theta_c)$ is the *face contact ratio*. Further, m_P can be expressed directly from Equation (10.3.6a) and m_F is obtained by dividing the R.H.S. of Equation (10.8.16) by θ_p. Taking any of the two gears (say 1) for reference,

$$m_c = \frac{(r_{O_1}^2 - r_{b_1}^2)^{1/2} + (r_{O_2}^2 - r_{b_2}^2)^{1/2} - C \sin \phi}{2\pi r_{b_1}/N_1} + \frac{B \tan \psi}{p_{p_1}}.$$

10.8. HELICAL GEARS

(10.8.18)

In the case of a pair of spur gears the recommended minimum value of m_c is 1.4. In the case of helical gears the total contact ratio is the sum of the profile and face contact ratios. The profile contact ratio can be smaller and even a value of 1 is considered sufficient. However, to achieve the full benefit of the helical action a value of m_c around 2 is often recommended.

Problem 10.8-1

A pair of parallel helical gears with 20 and 35 teeth are in mesh. The gears are 35 mm wide and the helix angle is 25°. The normal operating pressure angle is 20°. The normal module is 4 mm. Determine the contact ratio.

Solution

From the data, $N_1 = 20$ and $N_2 = 35$. With $m_n = 4$ mm, $m = (4/\cos 25°)$ mm $= 4.41$ mm. Hence,

$$r_{p_1} = mN_1/2 = 44.1 \text{ mm and } r_{p_2} = mN_2/2 = 77.18 \text{ mm}.$$

For standard gears the addendum is equal to m_n. Hence, $r_{O_1} = (44.1 + 4)$ mm $= 48.1$ mm and $r_{O_2} = (77.18 + 4)$ mm $= 81.18$ mm.

The normal pressure angle ϕ_n is given as 20°. Hence, the pressure angle in the plane of rotation ϕ can be obtained using Equation (10.8.6) as

$$\tan \phi = \tan 20°/\cos 25° = 0.4016.$$

Hence, $\phi = 21.88°$. Once ϕ is known the base circle radii can be determined as follows:

$$r_{b_1} = r_{p_1} \cos \phi = (44.1 \cos 21.88°) \text{ mm} = 40.92 \text{ mm}$$
$$r_{b_2} = r_{p_2} \cos \phi = (77.18 \cos 21.88°) \text{ mm} = 71.62 \text{ mm}$$

The transverse circular pitch of gear 1 is determined using Equation (10.8.2b) as

$$p_{p_1} = \pi m = 4.41\pi \text{ mm} = 13.85 \text{ mm}.$$

Considering any transverse cross section, the center distance is the same as the sum of the pitch circle radii of the two gears. Thus, $C = r_{p_1} + r_{p_2} = 121.28$ mm. Substituting all these values in Equation (10.8.18), the contact ratio results as

$$m_c = \frac{25.28 + 38.22 - 45.2}{12.86} + 1.18 = 1.42 + 1.18 = 2.6.$$

The method for checking interference in a pair of helical gears is similar to that adopted for a pair of spur gears in mesh. Thus, the transverse cross sectional profiles are to be checked. The phenomenon of undercutting in a helical gear is the same as that in spur gears. The transverse section of the helical gear must be studied to ensure there is no undercutting. The backlash in a pair of helical gears can be defined in three different ways. The backlash along the plane of rotation is called the *circular backlash*. Besides this, the backlash can be defined in the normal plane and also along the common normal. Further discussion on backlash is avoided and a standard reference can be consulted. One important disadvantage of helical gears is the side thrust that is generated. Herringbone gears eliminate this problem.

10.8.4 Crossed Helical Gears

Until assembly, there is no difference between a crossed helical gear and a parallel helical gear. When a pair of helical gears is used to transmit motion between nonparallel and nonintersecting shafts the gears are referred to as crossed helical gears. The contact between each pair of meshing teeth takes place at a point[12] and any significant transmission of power is not advisable. Only in instruments and in other situations where the power involved is less, use of crossed helical gears is recommended.

For crossed helical gears to mesh properly only one condition must be satisfied, which is that the normal circular pitch p_{p_n} (or normal module m_n) for the two gears must be the same. The hands of the

---

[12]Of course with time the teeth surfaces wear out and the contact extends over a small area. The contact stress is much higher compared to cases where initial contact is along a line.

10.8. HELICAL GEARS

gears need not be opposite nor is it necessary for the helix angles to be equal. The circular pitches of the gears in their respective planes of rotations are neither necessarily nor usually equal. Figure 10.8-6 shows two pairs of crossed helical gears in mesh. In Fig. 10.8-6a the two gears are of the same hand (the line ST represents a pair of teeth in mesh at an instant) and

$$\Sigma = \psi_2 + \psi_1 \tag{10.8.19a}$$

where Σ represents the angle between the shaft axes and ψ_1 and ψ_2 are the helix angles. When the hands of the helical gears in mesh are opposite, as shown in Fig. 10.8-6b, the angle between the shaft axes is given by

$$\Sigma = \psi_2 - \psi_1. \tag{10.8.19b}$$

For a pair of crossed helical gears to mesh correctly the only requirement is that they must have equal normal pitch or equal normal module. Thus, for two gears 1 and 2 in mesh,

$$p_{p_{n_1}} = p_{p_{n_2}} \tag{10.8.20}$$

$$\text{or,}\ m_{n_1} = m_{n_2}. \tag{10.8.21}$$

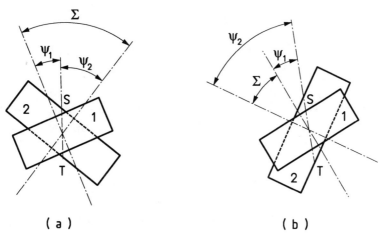

Figure 10.8-6

However, unlike parallel helical gears, the crossed helical gears need not have equal helix angles and the transverse modules m (in the respective plane of rotation) of the two gears are not equal. Therefore, the normal module m_n is used for specifying the size. The pitch circle radius is given by Equation (10.8.5) as

$$r_p = m_n N/2\cos\psi. \qquad (10.8.22)$$

The outside radius is obtained by adding m_n to r_p. Thus,

$$r_o = m_n N/2\cos\psi + m_n. \qquad (10.8.23)$$

It is obvious that the ratio of the pitch circle diameters of crossed helical gears is not in proportion to their teeth ratio. The angular velocity ratio can be expressed as follows:

$$i = \omega_1/\omega_2 = N_2/N_1 = r_{p_2}\cos\psi_2/r_{p_1}\cos\psi_1 \qquad (10.8.24)$$

The distance between the shaft axes

$$C = r_{p_1} + r_{p_2}. \qquad (10.8.25)$$

Using Equation (10.8.22) in (10.8.25), we get

$$\begin{aligned} C &= (m_n/2)(N_1/\cos\psi_1 + N_2/\cos\psi_2) \\ &= (m_n N_1/2)(1/\cos\psi_1 + N_2/N_1\cos\psi_2) \\ &= (m_n N_1/2)(1/\cos\psi_1 + i/\cos\psi_2). \end{aligned}$$

A very frequent situation is that in which the shafts are at right angles, i.e., $\Sigma = 90°$. Assuming the gears to be of the same hand, Equation (10.8.19a) yields

$$\psi_2 = 90° - \psi_1.$$

Hence, the expression for the center distance becomes

$$C = (m_n N_1/2)(\sec\psi_1 + i\operatorname{cosec}\psi_1). \qquad (10.8.26)$$

10.8. HELICAL GEARS

To achieve maximum compactness we must select the helix angle in such a way that C becomes minimum. Differentiating C with respect to ψ_1 and equating it to zero we get the following equation for optimum helix angle ψ_{1_o}:

$$\sec \psi_{1_o} \tan \psi_{1_o} - i \csc \psi_{1_o} \cot \psi_{1_o} = 0$$

$$\text{or,} \quad \sin^3 \psi_{1_o} / \cos^3 \psi_{1_o} = \tan^3 \psi_{1_o} = i$$

$$\text{or,} \quad \tan \psi_{1_o} = i^{1/3}. \tag{10.8.27}$$

The minimum center distance becomes

$$C_{min} = (m_n N_1/2)(\sec\psi_{1_o} + i \csc\psi_{1_o})$$

$$= (m_n N_1/2) \sec^3 \psi_{1_o}$$

$$= (m_n N_1/2)(1 + i^{2/3})^{3/2}. \tag{10.8.28}$$

Problem 10.8-2

In a pair of crossed helical gears of the same hand the shaft angle is 90°. The speed ratio is 3, the normal module is 2.5 mm and the number of teeth on the gear with the larger number of teeth[13] is 75. Determine the helix angles of the gears for maximum compactness and the corresponding center distance.

Solution

If the first gear is the smaller gear, then $N_2 = 75$ and $N_1 = 75/3 = 25$. The angular velocity ratio $i = \omega_1/\omega_2 = N_2/N_1 = 3$.

To achieve maximum compactness the center distance should be minimum and let the corresponding optimum helix angle of gear 1 be ψ_{1_o}. Then, from Equation (10.8.27),

$$\psi_{1_o} = \tan^{-1}(3^{1/3}) = 55.26°.$$

---

[13]Since the pitch circle diameters are not in proportion to the number of teeth it may so happen that the gear with the larger number of teeth may be smaller in diameter. Therefore, the larger gear need not have the larger number of teeth.

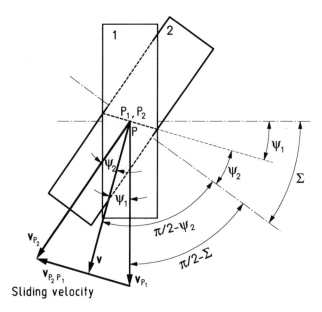

Figure 10.8-7

Hence, $\psi_{2_o} = 90° - 55.26° = 34.74°$, as the gears are of the same hand. From Equation (10.8.28) the minimum center distance is determined as follows:

$$\begin{aligned} C_{min} &= (2.5 \times 25/2)(1 + 3^{2/3})^{3/2} \text{ mm} \\ &= 168.92 \text{ mm} \end{aligned}$$

The pitch cylinders and the teeth of a pair of crossed helical gears in mesh have point contact and there is sliding along the direction of relative motion. Figure 10.8-7 shows such a pair. The particles on the two gears at the contact point (at the instant under consideration the contact is at the pitch point P) are P_1 and P_2. The velocities of these points are v_{P_1} and v_{P_2} as shown, which are such that the velocity difference between the contacting points P_1 and P_2 is along the common tangent to the teeth surfaces at the contact point and in the plane containing v_{P_1} and v_{P_2}. The components of v_{P_1} and v_{P_2} along the common normal at P must be equal ($= v$), as shown in Fig.

10.9. BEVEL GEARS

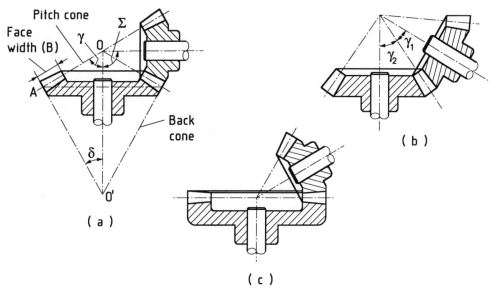

Figure 10.9-1

10.8-7. From the figure the magnitude of the relative velocity, i.e., the sliding velocity when the contact is at the pitch point P, can be expressed as

$$v_{P_2 P_1} = v_{P_1} \sin \psi_1 + v_{P_2} \sin \psi_2. \tag{10.8.29}$$

The sliding velocity in the case of crossed helical gears is large and is a special disadvantage.

Determination of contact ratio and the condition for avoiding interference are somewhat more involved and the interested reader can consult more specialized books for this purpose.

10.9 Bevel Gears

Pairs of bevel gears are used to connect pairs of shafts with intersecting axes, as indicated in Fig. 10.9-1a. The *shaft angle*, Σ, is defined as the angle between the shaft center lines in the plane containing the engaged teeth. The shaft angle is mostly 90°, though it can be more than or less than 90°, as indicated in Figs. 10.9-1b and c, respectively.

In cases of spur and helical gears the pitch surface is a cylinder. In the case of a bevel gear the pitch surface is a cone called the *pitch cone* (Fig.10.9-2a). The relative motion is equivalent to the rolling of the pitch cones against each other. The development of the involute tooth surface of a bevel gear from a *base cone* is shown in Fig. 10.9-2b. A generating point A is always at a fixed distance from the apex of the cone and, hence, moves on a spherical surface. The curve generated is called a *spherical involute*. When two bevel gears meet, their respective pitch cones touch along a line and every point on a bevel gear remains at a fixed distance from the common tip of the pitch cones, termed as the *apex* (Fig. 10.9-2a). Thus, all points have spheric motion with the apex as the center. The semi apex angle is called the *pitch-cone angle*, γ, and can be determined when the angular velocity ratio and the shaft angle are known. The *pitch diameter* of a bevel gear is the diameter of the pitch cone at the large end. The module is defined in the same manner as was in the case of spur gears. Hence, if $2r_p$ is the pitch diameter and N is the number of teeth, the module m is given by

$$m = 2r_p/N. \qquad (10.9.1)$$

Similarly, if p_p is the circular pitch, then

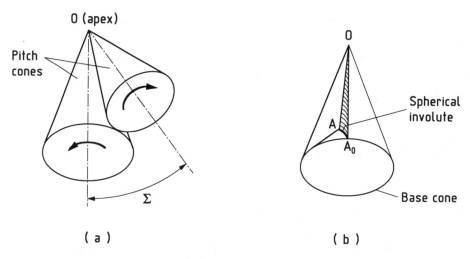

Figure 10.9-2

10.9. BEVEL GEARS

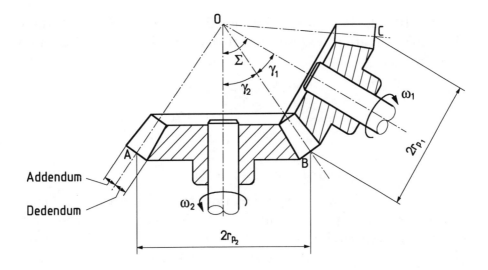

Figure 10.9-3

$$p_p = 2\pi r_p/N. \tag{10.9.2}$$

When i is the angular velocity ratio, then

$$i = \omega_1/\omega_2 = N_2/N_1 = r_{p_2}/r_{p_1}. \tag{10.9.3}$$

Figure 10.9-3 shows a pair of bevel gears with pitch radius r_{p_1} and r_{p_2} in mesh with a shaft angle of Σ. The *length of the pitch cone* (also called *cone distance*) is equal to $OB(= OA = OC)$. Since $\Sigma = \gamma_1 + \gamma_2$, where γ_1 and γ_2 are the pitch-cone angles of the gears,

$$\begin{aligned}\sin \gamma_2 &= \sin(\Sigma - \gamma_1) \\ &= \sin \Sigma \cos \gamma_1 - \cos \Sigma \sin \gamma_1.\end{aligned}$$

Dividing both sides of the above equation by $\sin \Sigma \sin \gamma_1$ and rearranging we get

$$\frac{1}{\sin \Sigma}\left(\frac{\sin \gamma_2}{\sin \gamma_1} + \cos \Sigma\right) = \frac{1}{\tan \gamma_1}.$$

From Fig. 10.9-3, $\sin \gamma_1 = r_{p_1}/OB$ and $\sin \gamma_2 = r_{p_2}/OB$. Using these in the above equation and substituting r_{p_2}/r_{p_1} by i from Equation (10.9.3), we get

$$\frac{1}{\sin \Sigma}(i + \cos \Sigma) = \frac{1}{\tan \gamma_1}.$$

Finally,

$$\tan \gamma_1 = \sin \Sigma/(i + \cos \Sigma). \qquad (10.9.4)$$

Figure 10.9-1 shows a few other important features. The cone generated by a line, perpendicular to the line generating the pitch cone (at the pitch circle), is called the *back cone*. The *back cone angle*, δ, and the pitch cone angles are complementary. The back cone angle is also known as the *face angle*. The length of a tooth measured along the pitch cone generator is called the *face width*, B. Generally, B lies between 8m to 10m. The concept of virtual number of teeth is quite useful, as in the case of helical gears. The *virtual number of teeth*, N_v, (also called the *formative number of teeth*) for a bevel gear is defined as the number of teeth a spur gear (with the radius equal to the length of the back cone $O'A$, as shown in Fig. 10.9-1) would have. Since $O'A = r_p/\sin \delta = r_p/\cos \gamma$,

$$N_v = 2\pi . O'A/p_p = 2\pi r_p/p_p \cos \gamma.$$

Using Equation (10.9.2) in the above equation,

$$N_v = N/\cos \gamma. \qquad (10.9.5)$$

It is obvious that the virtual number of teeth is generally not a whole number.

A *crown gear* is a bevel gear in which the pitch cone angle is equal to 90° and bears the same relation to a bevel gear as a rack does to a spur gear. Therefore, the bevel gear system has been developed so that the teeth generated are conjugate to the teeth of a crown gear having flat sides. Figure 10.9-4 shows a crown gear; the sides of the teeth lie in planes passing through the center of the sphere (the apex). The path of contact is in the form of a figure 8 and only a portion of the path of contact is used, depending on the height of the teeth.

10.9. BEVEL GEARS

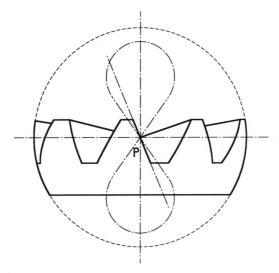

Figure 10.9-4

The addendum a and the dedendum b are measured perpendicular to the pitch cone at the outside of the gear, as indicated in Fig. 10.9-3. For standard bevel gears

$$a = m \text{ mm}$$

$$\text{and } b = 1.2m \text{ mm.} \tag{10.9.6}$$

The minimum number of teeth to avoid undercutting is determined in a way similar to that used in the cases of spur or helical gears. Of course the virtual number of teeth instead of actual number of teeth is used in the equation to determine N_{min}. In most practical situations with 20° pressure angle $N_{vmin} = 14$.
Hence,

$$N_{min} = 14 \cos \gamma. \tag{10.9.7}$$

Problem 10.9-1

A pair of shafts with interesecting axes are connected by a pair of bevel gears. The speed ratio and the shaft angle are 2.5 and 75°, respectively. If the gears are to be standard bevel gears with 20° pressure angle, determine the smallest possible pair of gears with a module of 5.

Solution

From the given data, $i = 2.5$ and $\Sigma = 75°$. From Equation (10.9.4)

$$\tan \gamma_1 = \sin 75°/(2.5 + \cos 75°) = 0.35.$$

Hence, $\gamma_1 = 19.3°$. Since $\gamma_1 + \gamma_2 = \Sigma = 75°$, $\gamma_2 = 75° - 19.3° = 55.7°$. The pinion being the smaller gear, it has to satisfy the smallest number of teeth condition:

$$N_1 = 14 \cos \gamma_1 = 14 \cos 19.3° = 13.2 = 14.$$

Since $N_2/N_1 = 2.5$ we get $N_2 = 14 \times 2.5 = 35$.

The pitch circle diameters are

$$\begin{aligned} 2r_{p_1} &= 14 \times 5 \text{ mm} = 70 \text{ mm} \\ \text{and} \quad 2r_{p_2} &= 35 \times 5 \text{ mm} = 175 \text{ mm}. \end{aligned}$$

Bevel gears can also be provided with spiral teeth. The tooth elements of *spiral bevel gears* (or *curved tooth bevel gear*) ideally should be spirals but in practice they are segments of circular arcs because of the ease of manufacture. The meshing action is smoother when spiral bevel gears are used instead of straight tooth bevel gears.

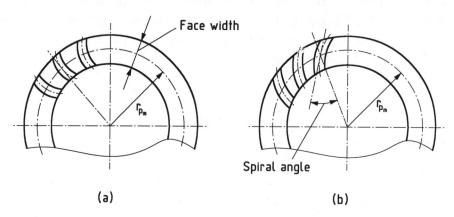

Figure 10.9-5

10.10. WORM GEARS

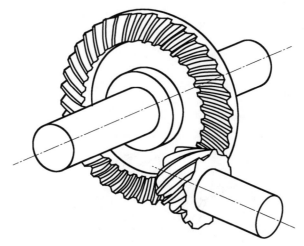

Figure 10.9-6

In a special case of curved tooth bevel gears the tooth curve is tangential to the radius at the midwidth location as shown in Fig. 10.9-5a. Such gears are called *Zerol bevel gears*. The general case of a spiral bevel gear is shown in Fig. 10.9-5b. The *spiral angle* is defined as the angle between the tangent to the tooth curve at the midwidth location and the corresponding radial line as shown. Thus Zerol bevel gear is a spiral gear with zero spiral angle. The *hypoid gears* resemble spiral bevel gears as is evident from Fig. 10.9-6. However, the axis of the pinion is offset and, so, does not interesect that of the gear. To maintain line contact in this condition the pitch surface of a hypoid gear is a hyperboloid of revolution instead of a cone as in the case of spiral bevel gears. Hypoid gears operate more quietly and can be used for higher reduction ratio. Transmission ratios of 60 and higher are not uncommon using hypoid gears.

10.10 Worm Gears

When the helix angle of a helical gear is too large and a tooth makes one or more complete revolutions on the pitch cylinder, it takes the form of a *worm*. It resembles a screw and its teeth are called *threads*. The other gear in mesh is much larger in diameter and is called a *worm*

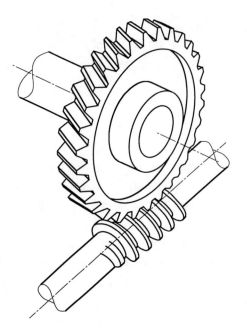

Figure 10.10-1

10.10. WORM GEARS

gear. The axes of the worm and the worm gear in mesh are at right angles, as indicated in Fig 10.10-1. The worm gear is generally the driven member of the pair and is made to envelop the worm partially. Thus, the load bearing capacity (so the power transmission capacity) is increased. Figure 10.10-2a shows the scheme of single-enveloping worm gear in which the wheel (i.e., the worm gear) partially envelops the worm. In the case of a *double-enveloping* worm gear, the worm is also curved longitudinally to fit the wheel curvature, as indicated in Fig. 10.10-2b. The worm can also be *single-thread* or *multiple-thread*. If L is the *lead* of the worm and p is the *axial pitch*, then

$$L = np \tag{10.10.1}$$

where n is the number of threads. The slope of the threads, at the pitch diameter, λ, is called the *lead angle*. If d_{p_w} is the pitch diameter of the worm, then

$$\tan \lambda = L/(\pi d_{p_w}). \tag{10.10.2}$$

The distance between the axes of the worm and the wheel (gear)

$$C = (d_{p_w} + d_{p_g})/2 \tag{10.10.3}$$

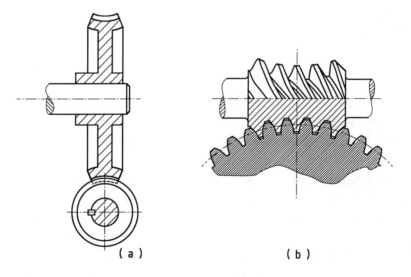

Figure 10.10-2

where d_{p_g} is the pitch diameter of the worm gear. The *circular pitch* (in the plane of rotation) of the worm gear and the axial pitch of the worm must be equal and can be expressed in the same manner used in the case of spur gears. Thus,

$$p = \pi d_{p_g}/N_g \qquad (10.10.4)$$

where N_g is the number of teeth on the worm gear.

For proper meshing of the worm and the gear, the following conditions must be satisfied:

(i) The axes of the worm and the gear must be at right angles.

(ii) The lead angle of the worm must be equal to the helix angle of the worm gear.

(iii) The axial pitch of the worm is equal to the circular pitch of the worm gear in its plane of rotation.

For each rotation of the worm, the worm gear circumference advances by n number of teeth. Thus, the worm gear rotates through (n/N_g) rotation. Hence, the speed ratio of the worm and the gear can be expressed as

$$\omega_w/\omega_g = N_g/n. \qquad (10.10.5)$$

A worm-gear set is usually not reversible and the input motion needs to be given to the worm. This is very useful in many situations where self locking is necessary to prevent bodies from coming down under gravity after vertical positioning has taken place. Therefore, in hoists and cranes worm gears are used frequently. Equation (10.10.5) shows that by making N_g sufficiently large and using a single-thread worm, a large speed reduction is possible in one stage. Such speed reduction units are quite popular. Since the relative sliding motion between the threads of a worm and the teeth of a meshing worm gear is substantial, energy loss through friction is greater compared to other types of gearing.

10.11. GEAR TRAINS

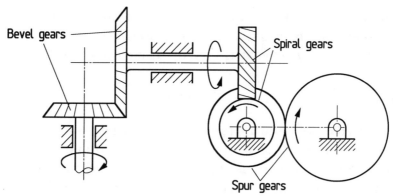

Figure 10.11-1

10.11 Gear Trains

A combination of gears of similar or different types used for transmitting motion from one shaft to another is known as a *gear train*. Figure 10.11-1 shows a gear train consisting of bevel, spiral and spur gears. When the axes of rotation of all the gears in a train are fixed, with respect to the frame, the train is called an *ordinary train*. The train shown in Fig. 10.11-1 is an ordinary train. When at least one of the gear axes rotates relative to the frame in addition to the gear's own rotation about its own axis, the train is called a *planetary gear train*, or an *epicyclic gear train*. The term "epicyclic" comes from the fact that points on gears with moving axes of rotation describe epicyclic paths. The ordinary gear trains are further classified into two types: (i) *simple gear train* and (ii) *compound gear train*. The most important kinematic analysis of the gear trains involves the determination of the motion of all members when that of the input member is prescribed. Often it is necessary to find out the ratio of angular velocities of the *input* (the first) *gear* and the *output* (the last) *gear of a train*. The input gear and the output gear of a train are also referred to as the *driving* and the *driven* gears, respectively. The speed ratio[14]

$$R = \omega_{\text{driver}}/\omega_{\text{driven}}$$

---

[14]The symbol i was earlier used to indicate the speed ratio of two mating gears.

is called the *train ratio*. The reciprocal of this quantity is often referred to as the *train value*. In the following sections the simple, compound and epicyclic gear trains are discussed separately.

10.11.1 Simple Gear Trains

When each shaft of a gear train carries one gear only the system is called a simple gear train. Figures 10.11-2a, b and c show a few examples of simple gear trains. In cases of simple gear trains the train ratio and the train value depend only on the numbers of teeth on the driving and the driven gear. If the numbers of teeth on gears 1, 2, 3 and 4 of the train shown in Fig. 10.11-2b are N_1, N_2, N_3 and N_4, respectively, then

$$\omega_1/\omega_2 = -N_2/N_1,$$
$$\omega_2/\omega_3 = -N_3/N_2$$
$$\text{and} \quad \omega_3/\omega_4 = -N_4/N_3.$$

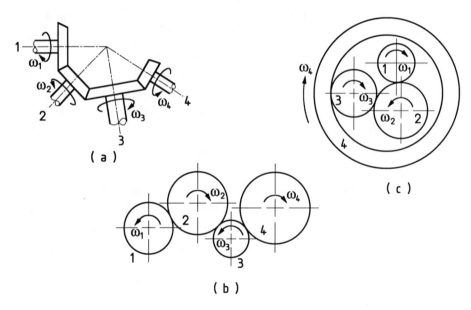

Figure 10.11-2

10.11. GEAR TRAINS

Multiplying all the terms on both sides of the three equations, we get the train ratio as

$$R = \omega_1/\omega_2 = -N_4/N_1.$$

In general, the train ratio of a simple gear train without any internal gear can be expressed as

$$R = \omega_{\text{driver}}/\omega_{\text{driven}} = (-1)^{n+1}(N_{\text{driven}}/N_{\text{driver}}) \qquad (10.11.1a)$$

where n is the number of gears between the driving and the driven shafts.

When a simple gear train consists of both external and internal gears (as shown in Fig. 10.11-2c),

$$R = \omega_{\text{driver}}/\omega_{\text{driven}} = (-1)^{n+1}(N_{\text{driven}}/N_{\text{driver}}) \qquad (10.11.1b)$$

where n is the number of external gears between the driving and the driven shafts.

It is obvious from Equation (10.11.1) that the range of train ratio in cases of simple gear trains is rather limited because of the limitations on the number of teeth of a gear in a train. A much wider range of the train ratio can be achieved in cases of compound gear trains, as will be shown in the following section.

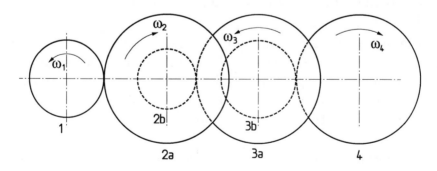

Figure 10.11-3

10.11.2 Compound Gear Trains

In cases of compound gear trains at least one shaft contains more than one active gear. The gears mounted on one shaft rotate at the same angular velocity. In such cases the numbers of teeth of the intermediate gears influence the train ratio. Figure 10.11-3 shows a compound gear train consisting of 6 gears mounted on four shafts; gears 1 and 4 may be considered to be the driving and the driven gears, respectively. Gears 1 and 4 are mounted on shafts 1 and 4, respectively, while the intermediate shafts 2 and 3 carry gears 2a and 2b and 3a and 3b. If N_i is the number of teeth of gear i and ω_j is the angular velocity of shaft j, then

$$\omega_1/\omega_2 = -N_{2a}/N_1,$$
$$\omega_2/\omega_3 = -N_{3a}/N_{2b}$$
$$\text{and} \quad \omega_3/\omega_4 = -N_4/N_{3b}.$$

Multiplying both sides of the above equation,

$$R = \omega_1/\omega_4 = -(N_4/N_{3b})(N_{3a}/N_{2b})(N_{2a}/N_1). \qquad (10.11.2)$$

In general,

$$R = (-1)^n \prod_{i=1}^{n} R_i \qquad (10.11.3)$$

where R_i is the ratio of each stage and n is the number of stages.

Problem 10.11-1

If the maximum and minimum permissible numbers of teeth in a gear train are 80 and 20, respectively, determine the maximum possible train ratio of a 3 stage compound gear train with two intermediate shafts. Compare this with the maximum possible train ratio of a simple gear train with two intermediate shafts.

Solution

To achieve maximum train ratio we should have maximum possible speed ratio of each stage, as evident from Equation (10.11.3). In cases of two intermediate shafts, from Equation (10.11.3),

10.11. GEAR TRAINS

$$R = (-1)^3 R_1.R_2.R_3.$$

The maximum possible value of the speed ratio of any stage is

$$N_{max}/N_{min} = 80/20 = 4.$$

Hence,

$$R_{max} = -4 \times 4 \times 4 = -64.$$

In the case of a simple gear train,

$$R_{max} = -N_{4max}/N_{1min} = -80/20 = -4.$$

Compound gear trains are used to make gear boxes from which multiple output speeds can be achieved using a constant input speed and various combinations of mating gears. Figures 10.11-4a and b show a 3 speed and a 9 speed gearbox, respectively, using sliding composite gears. In the case of the single stage gearbox shown in Fig. 10.11-4a, three possible output speeds,

$$\omega_o = -\omega_i(N_a/N_d), \ -\omega_i(N_b/N_e) \text{ and } -\omega_i(N_c/N_f)$$

are possible by shifting the composite gear to the appropriate locations. In the case of a two stage gear, box shown in Fig. 10.11-4b, nine output speeds are possible for the nine combinations of the positions of the two sliding composite gears. It should be noted that since the module used for all gears in a train is generally same, the following condition must be satisfied in cases of trains using sliding composite gear. Sums of the numbers of teeth of the driving and the corresponding driven gears must be the same. Hence, for the system shown in Fig. 10.11-4a, $N_a + N_d = N_b + N_e = N_c + N_f$.

10.11.3 Epicyclic Gear Trains

In cases of simple and compound gear trains it is not difficult to visualize the motions of the gears and the determination of the speed ratio is easy. However, in cases of epicyclic gear trains it is often difficult

642   CHAPTER 10. GEARS

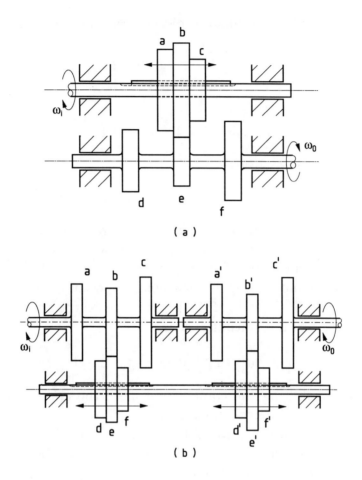

Figure 10.11-4

10.11. GEAR TRAINS

to visualize the motions of the gears. That is why a systematic procedure, without any requirement to visualize the gear movement, should be developed for analyzing epicyclic gear trains. Such a method is being presented below.

The basic principle of these methods is that the relative motion between a pair of gears is always the same whether the axes of rotation are fixed or moving. The most commonly used method of solving epicyclic gear train problems is the *tabular method*. In this method the system is rotated enbloc as a rigid body, thereby having no relative motion between the different members comprising the system. Next, the system is considered as an ordinary gear train by supressing the motion responsible for making it epicyclic. The resultant motion is determined by adding the two motions. To illustrate the principle let us analyse the motion of a simple epicyclic (planetary) gear train shown in Fig. 10.11-5. The system consists of a central gear 1 (called the *sun*) and another gear 2 in mesh with 1 (called the *planet*). Gear 2 is carried by the arm OA hinged at O on the shaft of gear 1, as shown. Since there are three moving bodies connected by three lower pairs (one hinge between the arm and gear 2 at A, one hinge between the frame and the shaft of gear 1 at O and the third hinge between the arm and the shaft of gear 1 at O) and one higher pair (between

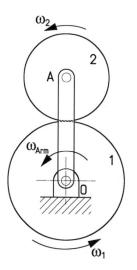

Figure 10.11-5

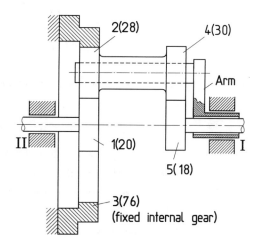

Figure 10.11-6

the two gears), the system possesses two degrees-of-freedom. Hence, the motions of two bodies must be prescribed in order to have a definite motion of the system. If gear 1 and the arm rotate with angular velocities ω_1 and ω_{Arm}, respectively, let us determine the angular velocity of gear 2 if the number of teeth on gears 1 and 2 are N_1 and N_2. Table 10.11.1 shows the two steps in which

Table 10.11.1

	Condition of motion	Angular velocities (CCW positive)		
		ω_1	Arm	ω_2
1.	All members locked and rotated at speed x	x	x	x
2.	Arm locked and gear 1 rotated at speed y	y	0	$-y \cdot \frac{N_1}{N_2}$
	Resultant	$x+y$	x	$x - y \cdot \frac{N_1}{N_2}$

the system is provided motion. The resultant shows that if gear 1 is rotated at speed $x + y$ and the arm is rotated at speed x, gear 2 rotates at speed $x - y.(N_1/N_2)$. From the prescribed conditions we know

$$x + y = \omega_1$$

and

$$x = \omega_{Arm}.$$

Hence, $y = \omega_1 - \omega_{Arm}$. Substituting this in the expression of ω_2 we get

$$\omega_2 = \omega_{Arm} - (\omega_1 - \omega_{Arm})(N_1/N_2).$$

A few problems on epicyclic gear trains have been solved by using the above mentioned tabular method in the following examples. CCW rotation has been assumed to be positive.

Problem 10.11-2

Figure 10.11-6 shows an epicyclic gear train in which the numbers of teeth of various gears are indicated within parenthesis. If shaft I rotates at 60 rad/s (in the CW direction when viewed from the right

10.11. GEAR TRAINS

end), determine the speed and the direction of rotation of shaft II. Gear 3 is a fixed internal gear with 76 teeth in mesh with gear 2.

Solution

Let us view the system from the right end. First, the system is rotated enbloc at x rad/s in the CCW direction so that every constituting element rotates at the same speed. Next, the arm is kept fixed and shaft I is rotated at y rad/s in the CCW direction. The resultant motion is obtained from the table shown below. Two different conditions of motion are considered:

1. All members locked and the system rotated enblock at x rad/s in the CCW direction.

2. Arm fixed and shaft I rotated at y rad/s in the CCW direction.

Condition of motion	Angular velocities (CCW positive) rad/s				
	Shaft I and Gear 5	Gear 4 and Gear 2	Arm	Gear 3	Gear 1 and Shaft II
1.	x	x	x	x	x
2.	y	$-y\frac{18}{30}$	0	$-y\cdot\frac{18}{30}\cdot\frac{28}{76}$	$y\cdot\frac{18}{30}\cdot\frac{28}{20}$
Resultant	$(x+y)$	$(x-y\cdot\frac{18}{30})$	x	$(x-y\cdot\frac{18}{30}\cdot\frac{28}{76})$	$(x+y\cdot\frac{18}{30}\cdot\frac{28}{20})$

From the given input conditions we now have

$$x + y = -60$$

and

$$x - y\cdot\frac{18}{30}\cdot\frac{28}{76} = 0.$$

Solving the above equations we get $x = -10.862$ rad/s and $y = -49.138$ rad/s. The motion of shaft II is given by

$$(x - y\cdot\frac{18}{30}\cdot\frac{28}{20})$$

which becomes equal to -52.138 rad/s after substituting the values of x and y. Hence, shaft II rotates at 52.138 rad/s in the CW direction (when viewed from the right end).

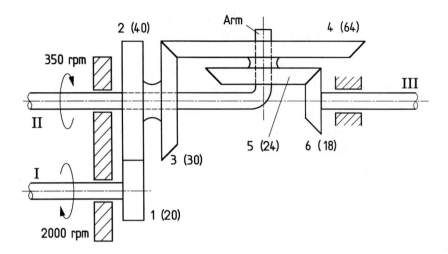

Figure 10.11-7

Problem 10.11-3

Figure 10.11-7 shows an epicyclic gear train in which shafts I and II rotate at 2000 and 350 rpm, respectively, in the directions indicated. Determine the direction and speed of rotation of shaft III.

Solution

Let us view from the right end and take the CCW direction as positive. Shaft I and gear 1 are not a part of the epicyclic gear train. The input condition can be modified as gear 2, rotating at 1000 rpm in the CW direction (when viewed from the right end) and shaft I, along with gear 1 are removed. Next, the following table is prepared adopting the usual procedure.

10.11. GEAR TRAINS

Condition of motion	Rotational speed r.p.m.		(CCW positive)
	Shaft II and Arm	Gear 2 and Gear 3	Gear 6 and Shaft III
1. All elements locked and rotated at x rpm	x	x	x
2. Arm locked and gear 2 rotated at y rpm	y	0	$-y \cdot \frac{30}{64} \cdot \frac{24}{18}$
Resultant	x	$x+y$	$-y \cdot \frac{30}{64} \cdot \frac{24}{18}$

From the prescribed input condition we get

$$x = -350 \text{ rpm}$$

and

$$x + y = -1000 \text{ rpm}.$$

Solving, we get $y = -650$ rpm and substituting the values of x and y we get the rotational speed of shaft III as 56.25 rpm. Since it is positive, the direction of rotation is CCW when viewed from the right end.

Problem 10.11-4

In the gear train shown in Fig. 10.11-8, the gear C is fixed, gear B is connected to the input shaft I and gear F is connected to the output shaft II. The arm A, carrying the compound gears D and E, turns freely on the output shaft. If the input shaft rotates at 1000 rpm in the direction shown (CCW when viewed from the right) determine the speed and the direction of rotation of the output shaft. $N_B = 20$, $N_c = 80$, $N_D = 60$, $N_E = 30$ and $N_F = 32$.

Solution

The table below shows the details of motion of each member for the different situations mentioned in the first column.

Condition of motion	Rotational speed (CCW positive) r.p.m.			
	Shaft I and Gear B	Gear C	Arm A	Gear F and Shaft II
1. All elements locked and rotated at x rpm	x	x	x	x
2. Arm A locked and Shaft I rotated at y rpm	y	$-\frac{20}{80}y$	0	$-\frac{20}{60}\cdot\frac{30}{32}y$
Resultant	$x+y$	$x-\frac{1}{4}y$	x	$x-\frac{5}{16}y$

Now, from the prescribed condition,

$$x + y = 1000 \text{ rpm}$$

and

$$x - \frac{1}{4}y = 0.$$

Solving, we get $x = 200$ rpm and $y = 800$ rpm. Hence, the motion of the output shaft II is given by $x - \frac{5}{16}y = -50$ rpm (i.e., in the CW direction when viewed from the right).

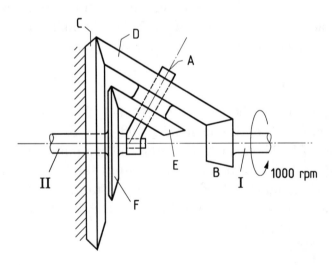

Figure 10.11-8

The fact that an epicyclic gear train possesses two degrees-of-freedom has application in the rear-axle differential of an automobile or any differential gear system where the output must be a resultant from two input motions.

10.12 Exercise Problems

10.1 A spur gear has a module of 5 mm. Determine the circular pitch. If the gear contains 32 teeth, determine the pitch circle diameter of the gear.

10.2 A pair of meshing spur gears has 25 and 45 teeth with a module of 4 mm. The pinion rotates at 1500 rpm. Determine (a) the pitch circle diameters, (b) the center distance and (c) the speed of the gear.

10.3 A pair of meshing standard spur gears has 25 and 40 teeth, with a module of 5 mm and an operating pressure angle of $20°$. Determine (a) the outside diameters of the gear and the pinion, (b) the center distance, (c) the base circle radii, (d) the contact ratio and (e) the angles of approach and recess for the gear and the pinion.

10.4 Determine the thicknesses of the teeth at the base circle and at the tip, in the Exercise Problem 10.3.

10.5 The tooth of an involute standard spur gear has a thickness of 6.65 mm at a radius of 102 mm. If the pressure angle is $20°$, at what radius will the involutes representing the two sides of the tooth intersect each other?

10.6 Two equal spur gears with a module of 3 mm mesh with a pressure angle of $20°$. Estimate (a) the length of action and (b) the contact ratio if the gears have 50 teeth each.

10.7 Determine the number of teeth of the smallest pinion that will mesh with a standard rack with $14\frac{1}{2}°$ pressure angle.

10.8 Two equal gears with module 5 mm are in mesh and have 35 teeth each. If the operating pressure angle is 20°, determine the minimum value of the addendum to ensure continuous transmission of motion.

10.9 Determine the *approximate* number of teeth in a 20° involute spur gear so that the base circle diameter is equal to the dedendum circle diameter.

10.10 Speed is intended to be reduced from 1800 rpm in a single stage transmission using two standard involute spur gears with 20° operating pressure angle. If the pinion has 18 teeth, determine the minimum possible speed of the driven shaft without causing any interference.

10.11 A 20° pinion with 20 teeth and a module of 4 mm drives a gear with 45 teeth. If the center distance of the gears is 132.5 mm, estimate the operating pressure angle.

10.12 A 3mm-module, 20° pinion with 20 teeth, meshes with a rack without backlash. If the gear center is shifted by 0.75 mm away from the rack estimate, the amount of backlash.

10.13 Calculate the circular and normal backlash in the case of Exercise Problem 10.11.

10.14 A 3mm-module, 20° pinion with 24 teeth, drives a gear with 72 teeth without backlash. If the center distance is increased by 1%, determine the resulting circular and normal backlash.

10.15 Two standard 4mm-module, 20° involute gears with 18 and 24 teeth are in perfect mesh. If the pinion speed is 1800 rpm determine the magnitude of the maximum relative sliding velocity.

10.16 A pair of 4mm-module cycloidal gears in mesh reduce the speed by a factor of 3.5. The pinion has 14 teeth and the generating circles are of the same size and its radius is equal to half the pitch circle radius of the pinion. The pinion and the gear have

10.12. EXERCISE PROBLEMS

standard addendum equal to the module. Determine the maximum value of the operating pressure angle and the situation in which it occurs.

10.17 The tooth of a pinion is of the shape of a pin, with circular cross section of radius h, with its center on the pitch circle of radius r_{p_1}. Show that the equation of the mating tooth profile is given by

$$x_2 = r_{p_1} \cos(C\theta_1/r_{p_2}) + h\sin(\frac{1}{2} + r_{p_1}/r_{p_2})\theta_1$$
$$-C\cos(\theta_1.r_{p_1}/r_{p_2})$$

and

$$y_2 = -r_{p_1} \sin(C\theta_1/r_{p_2}) + h\cos(\frac{1}{2} + r_{p_1}/r_{p_2})\theta_1$$
$$+C\cos(\theta_1.r_{p_1}/r_{p_2})$$

where r_{p_2} is the pitch circle radius of the gear, C is the center distance and θ_1 is the angle made by the line of centers with the x_1-axis (passing through the center of the pinion and the center of the circular tooth).

10.18 The teeth of a 18-tooth pinion have rectangular profile. The pitch circle radius of the pinion is 300 mm. The addendum and the dedendum of the given tooth are 16 mm and 20 mm, respectively. Determine the tooth profile of a mating gear with 30 teeth.

10.19 The base circle radii of an internal gear and involute pinion in mesh are 200 mm and 125 mm, respectively. The operating pressure angle is 20°. Determine the pitch circle radii of the gear and the pinion and the center distance.

10.20 Determine the equations of the pitch curves of two gears so that the relation between the rotations of the pinion and the gear, θ_1 and θ_2, is given by $\theta_2 = \tan\theta_1$.

10.21 The pitch curve of a gear is given by the polar equation of an ellipse, with one focus as the origin

$$r = a(1-e^2)/(1-e\cos\theta)$$

where a is the semi major axis and e is the eccentricity. If the distance between the centers is $2a$, show that the pitch curve of the mating gear is a similar ellipse. Determine the relation between the rotations of the two gears.

10.22 A pair of parallel helical gears with 25 and 40 teeth are in mesh. The gears are 40 mm wide and the helix angle is 25°. The normal operating pressure angle and the normal module are 20° and 4 mm, respectively. Determine the contact ratio.

10.23 In a system involving a pair of crossed helical gears with opposite hands the shaft angle is 120°. The speed ratio is 2.5 and the normal module is 3 mm. If the number of teeth on the gear with larger number of teeth is 75, determine the helix angles of the gears for maximum compactness. Also, determine the corresponding center distance.

10.24 A 4.23 mm-module straight bevel pinion of 21 teeth is in mesh with a gear with 27 teeth. The shaft angle is 90°. Determine the pitch angles, the addendum and dedendum and the face width of each gear.

10.25 A pair of shafts at right angles are connected by a pair of standard straight bevel gears so that the speed ratio is 3. If the operating pressure angle and the module are 20° and 4, respectively, estimate the smallest possible pair of gears.

10.26 A worm and worm gear connects two nonintersecting shafts at a right angle. Derive the expressions for the diameters of the worm and the worm gear in terms of the center distance, speed ratio of the shafts and the lead angle of the worm.

10.27 Figure 10.12-1 shows the schematic view of a spur gear differential in which gears 1 and 2 (connected to the shafts I and II, respectively) are identical. Gears 3 and 4 are also identical and are carried by the frame, that is free to rotate about the shafts I and II, as indicated (shafts I and II are obviously coaxial). The gears make contact in the sequence 1-3-4-2. Show that the

10.12. EXERCISE PROBLEMS

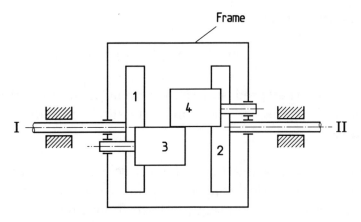

Figure 10.12-1

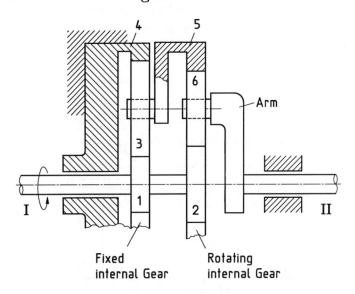

Figure 10.12-2

speed of rotation of the frame is equal to the arithmetic mean of the speeds of shafts I and II.

10.28 In the epicyclic gear train shown in Fig. 10.12-2, gears 1 and 2 are rigidly mounted on shaft I. The internal gear 4 is fixed and gear 3 rotates freely on a pin carried by the rotating internal gear 5. Gear 6 meshes with gears 2 and 5 and is carried by an arm rigidly mounted on shaft II. The numbers of teeth on the gears are as follows:

$$N_1 = 20, N_2 = 24, N_4 = 80, N_5 = 80$$

Determine the speed and the direction of rotation of shaft II if shaft I rotates at 2000 rpm in the direction indicated in the figure.

10.29 A clever system was patented (shown schematically in Fig. 10.12-3) in which electrical connection can be made to a device placed on a continuously rotating platform from a power supply fixed to the ground without causing continuous twisting of the cable. To achieve this it is necessary that the carrier rotates at half the speed at which the platform rotates. Determine the necessary condition to be satisfied by the numbers of teeth on the various gears.

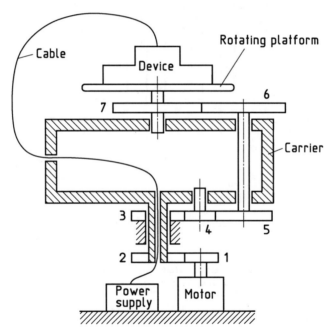

(After D. A. Adams, U.S. Patent No. 3,586,413)

Figure 10.12-3

10.12. EXERCISE PROBLEMS

10.30 The gear B of the planetary gear train shown in Fig. 10.12-4 is fixed. The arm A carries a planet gear C. The arm A and the gears D and E are free to turn on the shaft. The numbers of teeth on the gears are as follows:

$$N_B = 50, N_C = 20, N_D = 49 \text{ and } N_E = 51$$

The gears B, D and E are cut from gear blanks of similar diameter. If the arm A is given one turn in the CCW direction, when viewed from the right, determine the directions and the number of turns the gears D and E undergo.

10.31 In the gear train shown in Fig. 10.12-5 the sun gear S rotates at 500 rpm and the planet carrier (i.e., the arm A) rotates at 100 rpm in the same direction. Determine the speed and the direction of rotation of the planet gear P if the module of the gears is 3 and the diameter of the fixed internal gear F is to be as close to 250 mm as possible.

10.32 In the gear train shown in Fig. 10.12-6, the shaft I rotates at 500 rpm in the direction indicated. Determine N_A/N_C if the shaft II rotates at 5 rpm in the direction indicated. What will

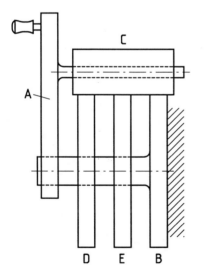
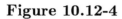

Figure 10.12-4

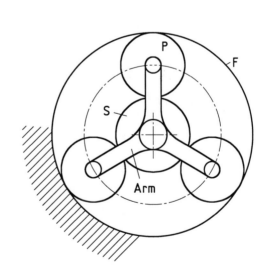

Figure 10.12-5

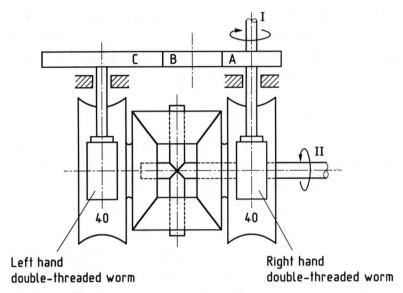

Figure 10.12-6

be the value of N_A/N_C if the direction of rotation of the shaft II is reversed? The numbers of teeth on the gears are indicated in the figure.

BIBLIOGRAPHY

1. Artobolevsky, I.I., Theory of Mechanisms (in Russian), Nauka, Moscow, 1965.

2. Artobolevsky, I.I., Mechanisms in Modern Engineering Design, Vols. I-V, (in English), Mir Publishers, Moscow, 1977.

3. Beggs, J.S., Mechanism, McGraw-Hill, New York, 1955.

4. Beggs, J.S., Advanced Mechanism, Macmillan, London, 1966.

5. Beyer, R., Technische Raumkinematic (in German), Springer-Verlag, Berlin, 1963.

6. Buckingham, E., Analytical Mechanics of Gears, Dover, New York, 1949.

7. Colbourne, J.R., The Geometry of Involute Gears, Springer-Verlag, New York, 1987.

8. Dijksman, E.A., Motion Geometry of Mechanisms, Cambridge University Press, London, 1976.

9. Dittrich, G. and Braune, R., Getriebetechnik in Beisplielen (in German), R.Oldenbourg Verlag, Munich, 1978.

10. Dizioglu, B., Getriebelehre, Vols. 1-4 (in German), Friedr. Vieweg & Sohn, Braunschweig, 1966.

11. Dudley, D.W., Gear Handbook, McGraw-Hill, New York, 1962.

12. Erdman, A.G. and Sandor, G.N., Mechanism Design: Analysis and Synthesis, Vol. 1, Prentice Hall, Englewood Cliffs, New Jersey, 1984.

13. Ghosh, A. and Mallik, A.K., Theory of Mechanisms and Machines, Affiliated East-West Press (P) Ltd., New Delhi, 1988.

14. Hain, K., Applied Kinematics, McGraw-Hill, New York, 1967.

15. Hall, A.S., Kinematics and Linkage Design, Prentice-Hall, Englewood Cliffs, New Jersey, 1961.

16. Hartenberg, R.S. and Denavit, J., Kinematic Synthesis of Linkages, McGraw-Hill, New York, 1964.

17. Hirschhorn, J., Kinematics and Dynamics of Plane Mechanisms, McGraw-Hill, New York, 1962.

18. Hrones, J. A. and Nelson, G.L., Analysis of the Four-Bar Linkage, MIT Press, Cambridge, Massachusetts, 1951.

19. Hunt, K.H., Kinematic Geometry of Mechanisms, Oxford University Press, Oxford, 1978.

20. Jensen, P.W., Cam Design and Manufacture, The Industrial Press, New York, 1965.

21. Lichtenheldt, W. and Luck, K., Konstruktionslehre der Getriebe (in German), Akademie-Verlag, Berlin, 1979.

22. Litvin, F.L., Theory of Toothed Gearing (in Russian), Science Publishers, Moscow, 1968.

23. Luck, K. and Modler, K.H., Konstruktionslehre der Getriebe (in German), Akademie-Verlag, Berlin, 1990.

24. Martin, G.H., Kinematics and Dynamics of Machines, McGraw-Hill, New York, 1982.

25. Maitra, G.M., Handbook of Gear Design, Tata McGraw-Hill Publishing Co. Ltd., New Delhi, 1985.

26. Paul, B., Kinematics and Dynamics of Planar Machinery, Prentice-Hall, Englewood Cliffs, New Jersey, 1979.

27. Rosenauer, N. and Willis, A.H., Kinematics of Mechanisms, Dover, New York, 1967.

28. Rothbart, H.A., Cams-Design, Dynamics and Accuracy, Wiley, New York, 1956.

29. Sandor, G.N., and Erdman, A.G., Advanced Mechanism Design: Analysis and Synthesis, Vol. 2, Prentice-Hall, Englewood Cliffs, New Jersey, 1984.

30. Shigley, J.E. and Uicker (Jr.), J.J., Theory of Machines and Mechanisms, McGraw-Hill, New York, 1980.

31. Suh, C.H. and Radcliffe, C.W., Kinematics and Mechanism Design, Wiley, New York, 1978.

32. Tao, D.C., Applied Linkage Synthesis, Addison-Wesley, Reading, Massachusetts, 1964.

33. Tao, D.C., Fundamentals of Applied Kinematics, Addison-Wesley, Reading, Massachusetts, 1967.

34. Volmer, J., Getriebetechnik (Lehrbuch) (in German), VEB Verlag Technik, Berlin, 1979.

35. Volmer, J., Getriebetechnik, Aufgabensammlung, (in German), VEB Verlag Technik, Berlin, 1979.

36. Volmer, J., Getriebetechnik: Kurvengetriebe (in German), VEB Verlag Technik, Berlin, 1989.

INDEX

A-matrix, 388
Acceleration analysis, 167
    auxiliary point method, 168, 191
    Euler-Savary equation method, 210
    Goodman's indirect method, 168, 198
    graphical method, 182
    method of acceleration difference, 185
    method of normal component, 168, 188
    of spatial linkage, 410
Acceleration image, 45
Accuracy (precision) point, 232, 233
Addendum, 532
Addendum circle, 532
Advanced cam curves, 463
Alt's construction, 262, 282
Angles of action, 534
Angles of approach, 534
Angles of recess, 534
Angular pitch, 545
Arcs of action, 534
Aronhold-Kennedy theorem, 53
Auxiliary line, 180
Auxiliary point, 180, 191
Axode, 62
Backlash, 535, 562
    circular, 563, 622
    normal, 563
Ball's point, 87
    curve, 88
Base pitch, 533
Bennett's linkage, 377, 391
Bevel gear, 627
    apex, 628
    back cone, 630
        angle, 630
    base cone, 628
    crown gear, 630
    face angle, 630
    face width, 630
    hypoid, 633
    pitch cone, 628
        angle, 628
        length of (cone distance), 629
        pitch diameter, 628
    shaft angle, 627
    spherical involute, 628
    spiral, 632
    spiral angle, 633
    virtual number of teeth, 630
    Zerol, 633
Binary link, 11
Bloch's method, 284, 294
Bobilliear construction, 106
Branch and order defect, 306
Burmester point, 103, 239
Burmester theory, application of, 280

Cams:
    base circle, 450
    constant breadth, 448
    disc/plate/radial, 446
    dual/conjugate, 448
    features and nomenclature, 450
    pitch curve, 450
    prime circle, 451
    wedge, 446
Cam mechanisms:
    cylindrical, 507
    F-, 451
    follower:
        curved-face, 444
        flat-face, 444
        knife-edge, 444
        offset, 451
    oscillating, 446
    roller, 444
    trace point, 450
    translating, 446
        offset, 446
        radial, 446
    P-, 452
    plane, 443, 446
    pressure angle, 453, 470
    space, 443, 506
    types of, 443, 446
Cam profile synthesis:
    analytical approach, 497
    cylindrical cams, 506, 507
    graphical approach, 490
Cardan circles, 67, 106
Cardan motion, 67, 73
Cayley's construction, 341
Center point, 95
Center-point curve, 95
Centro, 49
Centrode:
    fixed and moving, 58, 62, 330
    relative, 62
    tangent and normal, 68
Chain (kinematic):
    closed, 11
    compound, 11
    simple, 11
Chebyshev's polynomial, 234
Chebyshev's straight-line mechanism, 358
Circle diagram, 57
Circle point, 95
Circle-point curve, 90, 95
Circular pitch, 526
    operating, 566
Cognate linkages, 338
    extension of, 346
Coincident points, 46
Collineation axis, 107
Compatibility condition, 293, 295
Complex mechanism:
    high degree of complexity, 184
    low degree of complexity, 184
Contact ratio, 534, 543, 597
Coriolis acceleration, 48
Coupler, 24
Coupler cognate, 350
Coupler curve, 30, 327
    circle of singular foci, 329
    crunode, 329
    cusp, 329
    symmetry, 329
Crank, 24
Crank rocker, 24, 125
Crossed-slider trammel, 120
Crunode, 89, 329
Cubic of stationary curvature, 82
Cycloid, 572
Cycloidal (trochoidal) curves, 572

INDEX   663

directing circle, 573
epicycloid, 573
generating (rolling) circle, 572
hypocycloid, 573
Cycloidal (trochoidal) gears, 572
Cycloidal (trochoidal) teeth, 574
    action, 576
    face, 574
    flank, 574
Cylindric pair, 7
Davis automobile steering gear, 26
Dead-center problems, 232, 258
Dedendum, 532
Dedendum circle, 532
Degrees-of-freedom, 5, 111-115, 117, 120, 121, 376
Deltoid (kite) chain, 128
Diametral pitch, 526
Dimensional synthesis, 231
    analytical method, 283
    classification of, 231
    graphical method, 237
    optimization method, 298
Displacement analysis, 135
    analytical method, 143
    graphical method, 135
    of spatial linkage, 402
Displacement diagram, 454
Displacement functions, 456
Double-crank, 24, 125
Double point, 89
Double- rocker, 24, 125
Dual angle, 434
Dual numbers, 432
Dwell, 456
Dwell mechanism, 320, 327
Elements, 9
Elliptic trammel, 25
Epicyclic gear train, 637, 641
    planet, 643
    sun, 643
Equivalent linkage, 26, 185
Equivalent spur gear, 618
Euler-Savary equation, 73, 185, 210
    first form of, 74
    second form of, 76
Evans' straight-line mechanism, 358
Extended center distance system, 557
Fixed hinges, 24
Floating link, 184
Flocke's method, 480
Follower motion, 446, 454
    cycloidal, 462
    parabolic, 460
    polynomial functions, 463
    simple harmonic, 459
    uniform, 459
Force-closed, 10
Form-closed, 10
Four-bar linkage, 24
Four-link planar mechanisms, 22
Frame, 12
    line, 24
Freudenstein's equation, 288
Freudenstein's method, 287, 295
Freudenstein's theorem, 107
Function generation, 232, 243
Fundamental law of gearing, 523
Galloway mechanism, 128
Gear:
    addendum, 532
    addendum circle, 532
    angular pitch, 545
    base pitch, 533
    bevel, 627
        spiral, 632
        Zerol, 633

circular pitch, 526
   operating, 566
circular thickness, 535
clearance, 535
conjugate profile, 525
   determination of, 582
crown, 630
cycloidal, 572
dedendum, 532
dedendum circle, 532
diametral pitch, 526
face, 535
flank, 535
helical, 612
   crossed, 622
   parallel, 619
hypoid, 633
interchangeable, 557
internal, 521, 591
involute profile, 519, 526, 528
involute spur, 528
noncircular, 605
nonstandard, 557
module, 526
space width, 535
spur, 521, 528
standard, 557
standardization, 557
working depth, 535
worm, 633
Gear tooth action, 521
   angles of action, 534
   angles of approach, 534
   angles of recess, 534
   arcs of action, 534
   backlash, 535, 562
   contact ratio, 534, 543, 597
   interference and undercutting, 553, 597
   interference points, 555
   length of action, 533
   length of path of contact, 533
   line of action, 526
   path of contact, 526
   pitch circle, 526
      operating, 565
      standard, 526, 562
   pitch cylinder, 526
   pitch point, 524
   pressure angle:
      involute, 532
      operating, 532
Gear trains, 637
   epicyclic (planetary), 637, 641
   ordinary, 637
      simple, 638
      compound, 640
   train ratio, 638
   train value, 638
Geared five-link mechanism, 349
Geared-linkage, 293
Gearing, law of, 521
Geometric inversion, 146
Goodman's indirect method, 168, 198
Grashof chain, 123
   inversions of, 125
Grashof criterion, 122
   extension of, 128
Grashof linkage, 123
Ground pivots, 24
Grubler criterion, 111
Hartmann's construction, 211
Hart's straight-line mechanism, 373
Helical gears, 612
   contact ratio, 620
   face, 620
   profile, 620

INDEX

total, 620
crossed, 612, 622
equivalent spur gear, 617
geometry of, 614
   hand, 615
   helix angle, 614
   involute helicoid, 614
   lead angle, 614
   module along plane of rotation, 614
   normal module, 614
   normal pressure angle, 616
   pitch helix, 614
   tooth advance, 615
   transverse pressure angle, 616
   virtual number of teeth, 617
Herringbone gear, 622
Higher pair, 10
Homogeneous transformation matrix, 386
Hooke's joint, 383
Hypoid gear, 633
IC velocity, 67
Image pole, 91
Inflection circle, 70, 211
Inflection pole, 73
Instantaneous center
   of acceleration, 52
   of velocity, 49
   relative, 51
Interference and undercutting, 553, 597
   minimum number of teeth, 559
   secondary, 597, 600, 603
Interference points, 555
Internal gears, 521, 591
   contact ratio, 597
   interference and undercutting, 597

secondary interference, 597
tooth shape, 592
Inverse, 368
Inversors, 368
Involute:
   action, 543
   base circle, 530
   function, 538
   nomenclature, 532
   roll angle, 536
   spur gear, 528
   spherical, 628
   teeth action, 528
Involute helicoid, 614
Involutometry, 536
Kempe-Burmester focal mechanism, 121
Kinematic analysis, 5
   analytical method, 215
Kinematic chain, 11
Kinematic diagram, 12, 13
Kinematic inversion, 21
Kinematic pair, 5
Kinematics, 2
Kinematic synthesis, 5
Kinetics, 2
Kutzbach equation, 111, 376
Lead, 635
Lead angle, 635
Lift, 456
Limit position analysis, 419, 424
Link, 11
   binary, 11
   quaternary, 11
   singular, 11
   ternary, 11
Linkage, 12
Link parameters, 233

Long and short addendum system, 558
Lower pair, 10
Machine, 1
Matrix method, 386
Mechanical error, 162, 308
    deterministic approach, 309
    stochastic approach, 311
Mechanism, 1
    constrained, 12
    planar, 16
    spatial, 16, 375
    spherical, 16, 375
Mobility, 111
Mobility criteria, 111
    failure of, 119
    spatial linkages, 419
Module, 526
Motion transfer point, 180, 184
Motion generation, 231, 237
Non-Grashof chain, 126
Nonlinearity coefficient, 296
Noncircular gears, 605
    pitch curves, 606
    tooth profile, 609
Nonstandard gears, 557
    extended center distance system, 557
    long and short addendum system, 558
Number synthesis, 117
Offset slider-crank, 25
Oldham's coupling, 22, 25
Open chain, 11
Opposite poles, 93
Opposite-pole quadrilateral, 90, 93
Optimization method, 232, 298
Osculating circle, 82
Overclosed linkage 120, 354

Parallel motion generator, 350
Parallelogram chain, 127
Path curvature, 70
Path generation, 232, 241
Peucellier mechanism, 371
Pinion, 521
Pitch circle, 526
    operating, 565
    standard, 526, 565
Pitch cylinder, 526
Pitch point, 524
Plagiograph mechanism, 151
Planar pair, 9
Point-position reduction, 275
Pole, 60, 90
    image, 91
    of velocity diagram, 173
    polygon, 61
        fixed, 61
        moving, 62
    relative, 247
    triangle, 90
Precision (accuracy) point, 232, 233
Precision point approach, 232
Pressure angle, 453, 470
    involute, 532
    normal, 616
    operating, 532
    transverse, 616
Principle of transference, 432
Prismatic pair, 6
Quaternary link, 11
Quick-return mechanism, 135
    for slotting machine, 162
Quick-return ratio, 135
    of 4R linkage, 156
Rack, 525
    basic, 525
Rack and pinion, 521

INDEX 667

Redundant degree-of-freedom, 114
Relative pole, 247
Revolute pair, 6
Return, 456
Rise, 456
Roberts-Chebyshev theorem, 338
Roberts straight-line mechanism, 358
Scale factor, 233
Scotch-yoke mechanism, 22, 25
Screw pair, 6
Secondary interference, 597
Singular link, 11
Slider-crank mechanism, 2, 25
    offset, 25
Spatial cam mechanism, 506
Spatial linkages, 375
    degrees-of-freedom, 376
    displacement equation, 378
    kinematic synthesis, 411
    matrix method, 386
    mobility analysis, 419
    type identification, 419
    velocity and acceleration analysis, 410
Spherical linkages, 375, 381, 425
Spheric pair, 8
Spur gears, (see gears), 521
Standardization of gears, 557
    standard cutters, 557
    standard gear, 557
Stephenson's chain, 118
Straight-line linkages (approximate), 358
    Chebyshev's, 358
    Evans', 358
    Roberts', 358
    Watt's, 358
Straight-line linkages (exact), 367
    Hart's, 373

Peucellier, 371
Stretch-rotation operation, 356
Structural error, 232
Structure diagram,
Symbolic notation, 13
Synthesis of cam profile:
    analytical approach, 497
    cylindrical cams, 506, 507, 509
    graphical approach, 490
Synthesis of linkages:
    analytical methods (four positions), 293
        Bloch's method, 294
        Freudenstein's method, 295
    analytical methods (three positions), 283
        Bloch's method, 284
        Freudenstein's method, 287
    classification of problems, 231
    dead-center problems, 232, 258
    dimensional, 231
    exact and approximate, 232
    graphical methods (four positions), 275
        Burmester theory, 280
        point position reduction, 275
    graphical methods (three positions), 237
        function generation, 243
        motion generation, 237
        path generation, 241
    number, 117
    optimization method, 232, 298
    spatial mechanism, 411
    type, 29
Ternary link, 11
Thales circle, 333
Transmission angle, 159, 264
Trochoidal curves, 572

Type synthesis, 29
Uncertainty (folding) configuration, 127
Undercutting, 553, 597
Undulation point, 87
Velocity analysis, 167
    auxiliary-point method, 168, 179
    graphical method, 168
    method of instantaneous center, 168, 169
    method of velocity difference, 173
    of spatial linkage, 410
Velocity diagram, 173
    pole of, 173
Velocity image, 45
Virtual number of teeth, 617
Volmer's nomogram, 264
Watt's chain, 118
Watt's straight-line mechanism, 358
Worm gear, 633
    axial pitch, 635
    circular pitch, 636
    double-enveloping, 635
    lead, 635
    lead angle, 635
    multiple-thread, 635
    pitch diameter, 636
    single-enveloping, 635
    single-thread, 635
    threads, 633
    worm, 633
Zerol bevel gear, 633